METHODS IN MOLECULAR BIOLOGY

For further volumes:
http://www.springer.com/series/7651

For over 35 years, biological scientists have come to rely on the research protocols and methodologies in the critically acclaimed *Methods in Molecular Biology* series. The series was the first to introduce the step-by-step protocols approach that has become the standard in all biomedical protocol publishing. Each protocol is provided in readily-reproducible step-by-step fashion, opening with an introductory overview, a list of the materials and reagents needed to complete the experiment, and followed by a detailed procedure that is supported with a helpful notes section offering tips and tricks of the trade as well as troubleshooting advice. These hallmark features were introduced by series editor Dr. John Walker and constitute the key ingredient in each and every volume of the *Methods in Molecular Biology* series. Tested and trusted, comprehensive and reliable, all protocols from the series are indexed in PubMed.

Plant Pathology

Methods and Protocols

Edited by

Nicola Luchi

Institute for Sustainable Plant Protection, National Research Council (IPSP-CNR), Sesto Fiorentino, Florence, Italy

Humana Press

Editor
Nicola Luchi
Institute for Sustainable Plant
Protection
National Research Council (IPSP-CNR)
Sesto Fiorentino, Florence, Italy

ISSN 1064-3745 ISSN 1940-6029 (electronic)
Methods in Molecular Biology
ISBN 978-1-0716-2516-3 ISBN 978-1-0716-2517-0 (eBook)
https://doi.org/10.1007/978-1-0716-2517-0

This Humana imprint is published by the registered company Springer Science+Business Media, LLC, part of Springer Nature.
The registered company address is: 1 New York Plaza, New York, NY 10004, U.S.A.

Dedication

To my wife Laura for her love, patience, and encouragement

To my little daughters, Serena and Agnese

To my parents and to my brother who always supported me

Preface

Plant pathology is an applied science that studies diseases caused by infectious microorganisms in relation to environmental and host conditions. Plant diseases are responsible of considerable economic and environmental losses in agriculture and forestry. In the last years, the impact of climate change and globalization led plant pathogens to became more harmful, posing a serious threat to food security and the environment. In this scenario, it is crucial to study the biology of plant-pathogen interaction and how this is influenced by host physiology, which is strictly related to the impact of human activities on the environment. In this changing world, the study of plant disease and the development of innovative and sustainable methods for its detection and management represent an important challenge to prevent the risk of new possible disease outbreaks in new environments, to improve plant productivity, and to protect ecosystems' biodiversity.

In recent years, the knowledge on plant diseases become increased even for the use of advanced molecular techniques that provided new insights in plant pathogen detection and plant-pathogen interaction, providing a significant support for diseases control strategies.

The aim is to provide a volume that covers the latest developments in different areas of plant pathology. The contents encompass the expertise of a broad range of researchers with long expertise in plant pathology. The volume is divided into seven parts. The three chapters in Part I provide traditional methods for isolation and identification of invasive pathogens and root disease. In Part II are included three chapters on novel and rapid DNA extraction protocols from different samples (plant tissue, soil, insects). In Part III, the ten chapters are focused on molecular detection protocols (based on PCR, real-time PCR, high-resolution melting analysis, digital PCR) for identification and quantification plant pathogens, including fungal and bacterial invasive species. The three chapters of Part IV describe the application of metabarcoding in plant pathology, while in Part V, three chapters cover studies on plant-pathogen interactions. Part VI concentrates on population genomics of plant pathogens. The final part of the volume (Part VII) is devoted to biocontrol of plant pathogens.

I hope this volume will be useful for the community of plant pathologists, providing innovative methods and approaches for the detection, identification, and control of plant diseases.

Firenze, Italy *Nicola Luchi*

Acknowledgments

The editor gratefully acknowledges all contributing authors for sharing their expertise and making this volume possible, and expresses his thanks to the series editor, Prof. John Walker, whose help and guidance has been fundamental in the preparation of this book.

Acknowledgement

Contents

Contributors

SRĐAN GORAN AĆIMOVIĆ • *CenterVirginia Polytechnic Institute and State University, School of Plant and Environmental Sciences, Alson H. Smith Jr. Agricultural Research and Extension Center, Winchester, VA, USA*
ERIK ANDREASSON • *Department of Plant Protection Biology, Swedish University of Agricultural Sciences, Alnarp, Sweden*
ANTONIO ARONADIO • *Laboratory of Phytopathological Diagnostics and Molecular Biology, Plant Protection Service of Tuscany, Pistoia, Italy*
MATTHEW P. AYRES • *Department of Biological Sciences, Dartmouth College, Hanover, NH, USA*
RAFFAELLA BALESTRINI • *National Research Council, Institute for Sustainable Plant Protection (CNR-IPSP), Torino, Italy*
LINDA BARTOLINI • *Laboratory of Phytopathological Diagnostics and Molecular Biology, Plant Protection Service of Tuscany, Pistoia, Italy*
GUILLAUME J. BILODEAU • *Canadian Food Inspection Agency, Nepean, ON, Canada*
PIERLUIGI BONELLO • *Department of Plant Pathology, The Ohio State University, Columbus, OH, USA*
SARA CAMPIGLI • *Department of Agriculture, Food, Environment and Forestry (DAGRI), Plant Pathology and Entomology Section, University of Florence, Florence, Italy*
GABRIELE CENCETTI • *National Research Council, Institute of Bioscience and Bioresources, (CNR-IBBR), Sesto Fiorentino, Italy*
ANNE CHANDELIER • *Walloon Agricultural Research Centre, Department Life Sciences, Crops and Forests Health Unit, Gembloux, Belgium*
AAKASH CHAWADE • *Department of Plant Breeding, Horticum, Swedish University of Agricultural Sciences, Alnarp, Sweden*
CAROLINA CORNEJO • *Swiss Federal Research Institute WSL, Birmensdorf, Switzerland*
DANIELE DA LIO • *Department of Agriculture, Food and Environment (DAFE), University of Pisa, Pisa, Italy*
JULIO JAVIER DIEZ • *Sustainable Forest Management Research Institute, University of Valladolid-INIA, Palencia, Spain*
HATICE TUĞBA DOĞMUŞ LEHTIJÄRVI • *Faculty of Forestry, Isparta University of Applied Sciences, Isparta, Turkey*
MUKESH DUBEY • *Department of Forest Mycology and Plant Pathology, Uppsala Biocenter, Swedish University of Agricultural Sciences, Uppsala, Sweden*
TIM J. DUMONCEAUX • *Agriculture and Agri-Food Canada Saskatoon Research and Development Centre, Saskatoon, SK, Canada; Department of Veterinary Microbiology, University of Saskatchewan, Saskatoon, SK, Canada*
MILOŇ DVOŘÁK • *Department of Forest Protection and Wildlife Management, Faculty of Forestry and Wood Technology, Mendel University in Brno, Brno, Czech Republic*
CARRIE J. FEARER • *Department of Plant Pathology, The Ohio State University, Columbus, OH, USA*
PAULO N. FIRMINO • *Forest Research Centre (CEF), School of Agriculture, University of Lisbon, Lisbon, Portugal*

GIOVANA PRADO FORTUNA MACAN • *Department of Plant Breeding, Swedish University of Agricultural Sciences, Alnarp, Sweden*
CRISTINA FRANCIA • *Laboratory of Phytopathological Diagnostics and Molecular Biology, Plant Protection Service of Tuscany, Pistoia, Italy*
GIORGIO GAMBINO • *National Research Council, Institute for Sustainable Plant Protection (CNR-IPSP), Torino, Italy*
KATARINA GAŠIĆ • *Cornell University, Plant Pathology and Plant-Microbe Biology Section, School of Integrative Plant Sciences, Hudson Valley Research Laboratory, Highland, NY, USA; Institute for Plant Protection and Environment, Department of Plant Diseases, Laboratory for Phytopathology, Belgrade, Serbia*
DARIO GAUDIOSO • *Department of Agriculture, Food, Environment and Forestry (DAGRI), Molecular Plant Pathology Lab, University of Florence, Florence, Italy*
JENNIFER HODGETTS • *Elsoms Seeds Ltd., Spalding, Lincolnshire, UK*
MARILIA HORTA JUNG • *Phytophthora Research Centre, Department of Forest Protection and Wildlife Management, Faculty of Forestry and Wood Technology, Mendel University in Brno, Brno, Czech Republic*
ANDREA IENCO • *National Research Council, Institute of Chemistry of OrganoMetallic Compounds, (CNR-ICCOM), Sesto Fiorentino, Italy*
RENAUD IOOS • *ANSES Plant Health Laboratory, Unit of Mycology, Malzéville, France*
THOMAS JUNG • *Phytophthora Research Centre, Department of Forest Protection and Wildlife Management, Faculty of Forestry and Wood Technology, Mendel University in Brno, Brno, Czech Republic*
PRUTHVI B. KALYANDURG • *Department of Plant Breeding, Swedish University of Agricultural Sciences, Alnarp, Sweden; Department of Plant Breeding, Horticum, Swedish University of Agricultural Sciences, Alnarp, Sweden*
SAMMAR KHALIL • *Department of Biosystems and Technology, Swedish University of Agricultural Sciences, Alnarp, Sweden*
QUIRIN KUPPER • *Swiss Federal Research Institute WSL, Birmensdorf, Switzerland*
ASKO LEHTIJÄRVI • *Isparta University of Applied Sciences, Sütçüler Prof. Dr. Hasan Gürbüz Vocational School, Isparta, Turkey*
QINSONG LIU • *Key Laboratory of Southwest China Wildlife Resources Conservation (Ministry of Education), College of Life Science, China West Normal University, Nanchong, China*
NICOLA LUCHI • *Institute for Sustainable Plant Protection, National Research Council (IPSP-CNR), Sesto Fiorentino, Florence, Italy*
ANTONINO MALACRINÒ • *Institute for Evolution and Biodiversity, Westfälische Wilhelms-Universität Münster, Münster, Germany*
GUIDO MARCHI • *Department of Agriculture, Food, Environment and Forestry (DAGRI), Plant Pathology and Entomology Section, University of Florence, Florence, Italy*
FELICIA MENICUCCI • *National Research Council, Institute of Chemistry of OrganoMetallic Compounds, (CNR-ICCOM), Sesto Fiorentino, Italy*
MARCO MICHELOZZI • *National Research Council, Institute of Bioscience and Bioresources, (CNR-IBBR), Sesto Fiorentino, Italy*
DUCCIO MIGLIORINI • *National Research Council, Institute for Sustainable Plant Protection, (CNR-IPSP), Sesto Fiorentino, Florence, Italy*
CHRYSTEL OLIVIER • *Agriculture and Agri-Food Canada Saskatoon Research and Development Centre, Saskatoon, SK, Canada*
FUNDA OSKAY • *Faculty of Forestry, Çankırı Karatekin University, Çankırı, Turkey*

EMILY OSTERMANN • *Department of Chemical and Biomolecular Engineering, North Carolina State University, Raleigh, NC, USA*
ELEONORA PALAGANO • *National Research Council, Institute of Bioscience and Bioresources, (CNR-IBBR), Sesto Fiorentino, Italy*
NIDHI PAREEK • *Department of Microbiology, School of Life Sciences, Central University of Rajasthan, Bandarsindri, Kishangarh, Ajmer, Rajasthan, India*
CHIARA PASTACALDI • *Department of Agriculture, Food, Environment and Forestry (DAGRI), Molecular Plant Pathology Lab, University of Florence, Florence, Italy*
RAJESH PAUL • *Department of Chemical and Biomolecular Engineering, North Carolina State University, Raleigh, NC, USA*
EDEL PÉREZ-LÓPEZ • *Université Laval, Faculté des Sciences de l'Agriculture et de l'Alimentation, Département de Phytologie, Québec City, QC, Canada*
ANA PÉREZ-SIERRA • *Forest Research, Alice Holt Lodge, Farnham, Surrey, UK*
IRENE PERRONE • *National Research Council, Institute for Sustainable Plant Protection (CNR-IPSP), Torino, Italy*
CARLA S. PIMENTEL • *Forest Research Centre (CEF), School of Agriculture, University of Lisbon, Lisbon, Portugal*
EDOARDO PIOMBO • *Department of Forest Mycology and Plant Pathology, Uppsala Biocenter, Swedish University of Agricultural Sciences, Uppsala, Sweden*
REINALDO PIRE • *Universidad Centroccidental Lisandro Alvarado, Post-Grado de Agronomía, Barquisimeto, Venezuela*
SIMONE PROSPERO • *Swiss Federal Research Institute WSL, Birmensdorf, Switzerland*
KAROLINA PUSZ-BOCHENSKA • *Department of Biology, University of Saskatchewan, Saskatoon, SK, Canada; Agriculture and Agri-Food Canada Saskatoon Research and Development Centre, Saskatoon, SK, Canada*
AIDA RAIO • *National Research Council, Institute for Sustainable Plant Protection, (CNR-IPSP), Sesto Fiorentino, Florence, Italy*
DOMENICO RIZZO • *Laboratory of Phytopathological Diagnostics and Molecular Biology, Plant Protection Service of Tuscany, Pistoia, Italy*
CRISTINA ROSA • *Department of Plant Pathology and Environmental Microbiology, The Pennsylvania State University, State College, PA, USA*
ELISABETTA ROSSI • *Department of Agriculture, Food and Environment (DAFE), University of Pisa, Pisa, Italy*
RICARDO DELGADO SANTANDER • *Cornell University, Plant Pathology and Plant-Microbe Biology Section, School of Integrative Plant Sciences, Hudson Valley Research Laboratory, Highland, NY, USA*
ALBERTO SANTINI • *National Research Council, Institute for Sustainable Plant Protection, (CNR-IPSP), Sesto Fiorentino, Florence, Italy*
FABIANO SILLO • *National Research Council, Institute for Sustainable Plant Protection (CNR-IPSP), Torino, Italy*
DAMIANO STEFANUCCI • *Department of Agriculture, Food, Environment and Forestry (DAGRI), Molecular Plant Pathology Lab, University of Florence, Florence, Italy*
POORVA SUNDARARAJAN • *Department of Plant Breeding, Horticum, Swedish University of Agricultural Sciences, Alnarp, Sweden*
STEFANIA TEGLI • *Department of Agriculture, Food, Environment and Forestry (DAGRI), Molecular Plant Pathology Lab, University of Florence, Florence, Italy*
ÉMILIE D. TREMBLAY • *Agriculture and Agri-Food Canada, Ottawa, ON, Canada*

ANNA MARIA VETTRAINO • *Department for Innovation in Biological, Agro-food and Forest Systems (DIBAF), University of Tuscia, Viterbo, Italy*
RAMESH R. VETUKURI • *Department of Plant Breeding, Swedish University of Agricultural Sciences, Alnarp, Sweden; Department of Plant Breeding, Horticum, Swedish University of Agricultural Sciences, Alnarp, Sweden*
QINGSHAN WEI • *Department of Chemical and Biomolecular Engineering, North Carolina State University, Raleigh, NC, USA; Emerging Plant Disease and Global Food Security Cluster, North Carolina State University, Raleigh, NC, USA*
STEPHEN C. WHISSON • *Cell and Molecular Sciences, The James Hutton Institute, Invergowrie, Dundee, UK*
TYLER J. WIST • *Agriculture and Agri-Food Canada Saskatoon Research and Development Centre, Saskatoon, SK, Canada*
MUHAMMAD AWAIS ZAHID • *Department of Plant Protection Biology, Swedish University of Agricultural Sciences, Alnarp, Sweden*
CRISTINA ZAMORA-BALLESTEROS • *Sustainable Forest Management Research Institute, University of Valladolid-INIA, Palencia, Spain*

Part I

Plant Disease Diagnosis: Traditional Methods

Chapter 1

Obtaining and Maintaining Cultures of Pinewood Nematodes *Bursaphelenchus xylophilus* from Wild Dauers

Carla S. Pimentel, Paulo N. Firmino, and Matthew P. Ayres

Abstract

The establishment of laboratory isolates of the pinewood nematode *Bursaphelenchus xylophilus*, the causal agent of the pine wilt disease, has been crucial to research on this important forest pathogen. Here we describe a simple, low-cost, and easy way to obtain samples of wild populations of *B. xylophilus* by culturing dauers extracted directly from the insect vector.

Key words *Botrytis cinerea*, Forest pathogen, Malt extract agar, Pine forests, Pinus, Wood fiber

1 Introduction

The pinewood nematode *Bursaphelenchus xylophilus* (Steiner & Buhrer) Nickle, native to North America, is the causal agent of the pine wilt disease and considered one of the most important invasive pests in Eurasian pine forests. The pine wilt disease is vector-borne, and the pinewood nematode has evolved a phoretic association with pine sawyer beetles of the genus *Monochamus* (Coleoptera, Cerambycidae). *Bursaphelenchus xylophilus* travel within the trachea of *Monochamus* adults as dormant dauers (fourth-stage dispersal juveniles (JIV)), which are their vehicle to exit depleted pine tree hosts and to colonize other pines that are susceptible [1]. The establishment of laboratory isolates of *B. xylophilus* has been crucial to research on this important forest pathogen, including its virulence [2, 3] and host resistance [4, 5]. However, growth conditions and in vitro serial passage of laboratory isolates can be selective for its characteristics, and thus the choice of a growth medium is crucial [6–9]. It is an important consideration that laboratory isolates should retain the main characteristics of the wild populations and the adequacy of laboratory-adapted reference isolates for the study of "real-world" pathogenesis [7, 10, 11].

Nicola Luchi (ed.), *Plant Pathology: Methods and Protocols*, Methods in Molecular Biology, vol. 2536,
https://doi.org/10.1007/978-1-0716-2517-0_1,

There has not been much consideration of potential adaptive selection/evolution of *B. xylophilus* lab isolations to chosen media, and techniques for maintaining lab cultures of this pathogen are not standardized. OEPP/EPPO [12] advises using mycelium of *Botrytis cinerea* Pers. (1794), as food to obtain adults from fourth-stage dispersal juveniles. After isolation, laboratory cultures are commonly grown on this fungus which is itself growing on malt extract agar (MA) or potato dextrose agar (PDA) [2, 13]. However, *B. xylophilus* lab cultures are often being maintained in other media, such as *B. cinerea* grown on barley grains [14, 15].

When trying to culture *B. xylophilus* from the wild dauers extracted from insect vectors trapped in a forest setting, we found that the standard agar-based media used for maintaining lab populations of the pathogen led to very poor results, with most attempted cultures failing. Adding pine groundwood to these media dramatically improved the survival and growth of wild populations of *B. xylophilus* in the laboratory. This methodology worked similarly well for nematodes originating from eastern North America, within the pathogen native range, and in Portugal, where it is invasive [16]. Here we describe a methodology to easily obtain a potentially representative sample from wild populations of *B. xylophilus*. The system involves culturing dauers extracted directly from the insect vector. Dauers phoretic on single *Monochamus* represent a nematode population from a single pine host. Such cultures, when originating from a reasonable size sample of pine sawyers trapped during a few weeks in one location, will be representative of *B. xylophilus* infesting a wider area of forest, due to the mobility of this large beetle [17]. Materials used are easily available, and the methods are easy to apply in any geographical area. The methodology allows the collection and culturing of a sample of the *B. xylophilus* population infesting an affected forest in just a few weeks.

2 Materials

2.1 Obtaining B. xylophilus Dauers from Wild Populations

2.1.1 Trapping Monochamus spp.

1. Lindgren Multi-Funnel Traps with 12 black funnels [18].
2. Liquid Fluon®.
3. Dry 5-L containers.
4. Bark beetle pheromones (e.g., ipsenol).
5. Volatiles released from host trees (e.g., α-pinene).
6. Monochamol can also be included.

2.1.2 Extracting Nematode Dauers from Monochamus spp.

1. Kimwipes.
2. 60-degree angle filtering funnels with narrow stems, 50-mm top diameter.

3. Clear PVC tubing adaptable to the funnels' stem diameter.
4. Day's pinchcock tubing clamps.
5. Funnel supports.
6. Small conical tubes (15–50 mL).
7. Conical bottom tubes rack.
8. Pasteur Pipettes and pipette rubber pump.
9. Milli-Q™ water.
10. Ethanol 70%.

2.2 Establishing and Maintaining Lab Populations from the Wild Dauers

1. Fresh pine wood shavings.
2. Malt extract and agar.
3. Milli-Q™ water.
4. Disposable sterilized 60-mm Ø Petri dishes.
5. Glassware: 1–2-L beakers and Erlenmeyer flasks, glass rods.
6. Lab spoons/spatula.
7. Parafilm® M sealing film.
8. Cultures of *B. cinerea*.
9. Work should be performed in a biosafety level one microbiology laboratory, equipped with laminar flux hoods, autoclaves, and growth chambers. A dissecting microscope should be available.

3 Methods

3.1 Obtaining B. xylophilus Dauers from Wild Populations

3.1.1 Trapping Monochamus spp.

1. The bottom funnel of the Multi-Funnel Traps should be brushed with a thin layer of Liquid Fluon® and attached with a dry 5-L container (*see* **Note 1**). The beetle pheromones and pine trees volatiles should be attached to a funnel in the middle of the trap (*see* **Notes 2** and **3**).
2. Traps should be placed in forest spots with recently dead pines from storms, drought, bark beetles, and/or pine wilt disease (*see* **Note 4**). Traps should be deployed close to the main stem, at a height between 1 and 2 meters. Rope may be used to hang them on branches. Traps should be deployed sufficiently early in the season so that the first flying pine sawyers are trapped, since these earlier flying beetles are the ones with the maximum number of phoretic dauers. Phoresy is neglectable after a few weeks, so there is little use on continuing trapping afterward. The pine sawyers' flight period tends to start earlier and get longer at lower latitudes, but the phoretic patterns are quite consistent across different geographical areas [19, 20].

3. Traps should be checked every 2–3 days to obtain live insects, and on each sampling occasion, fresh pine shoots should be placed inside the traps' cups to provide substrate and food for insects until the next trap check. Extending the checking cycles beyond this period may also accumulate unwanted insect species, including predators, which degrade the intended data collection.
4. Baits should be replaced whenever required according to suppliers' instructions.

3.1.2 Extracting Nematodes from Monochamus *spp.*

1. Funnels should be assembled in the supports, with a piece of tubing attached to the stem and closed with a Day's pinchcock clamp.
2. Each single beetle should be dissected and body parts wrapped in a Kimwipes tissue and placed in an individual funnel filled with Milli-Q™ water so that it just covers the wrapped beetle, for a maximum of 24 h. For dissection, we cut each beetle in four to five pieces with a sharp scissor.
3. About 1 mL of water should be collected in a conical tube, by slightly opening the clamp (*see* **Note 5**).
4. The conical tube should be left sitting for a few hours (ideally 24 h) so that the content in suspension sediments to the bottom (*see* **Note 6**).
5. The sedimented nematodes can be collected with a Pasteur Pipette. The dauers carried by each *Monochamus* can then be assessed and counted using a dissecting microscope (*see* **Note 7**).
6. Dauers can be maintained for weeks at 2 °C, before inoculation.

3.2 Establishing and Maintaining Lab Populations from the Wild Dauers

3.2.1 Preparation of the Media

1. 100–150 g of dry pinewood (*see* **Note 8**), ground to 4–8-mm Ø particle size—a coffee grinder can be used for this purpose (*see* **Note 9**).
2. 25-g malt extract and 25-g agar should be added to 1-L Milli-Q™ water and mixed in an Erlenmeyer flask (*see* **Note 10**).
3. Groundwood should be placed in a separated beaker flask (*see* **Note 11**), moistened with Milli-Q™ water and first autoclaved for 30 min (*see* **Note 12**). After this first round of sterilization, the Erlenmeyer flask containing the malt extract agar mix should also be placed in the autoclave, and sterilization should go for an extra 15 min.
4. The malt extract agar mix should be poured over the beaker containing the ground pinewood. The wood fiber should be imbibed in the malt extract agar mix with the help of a glass rod.

Fig. 1 Pinewood/malt agar media inoculated with *B. cinerea* for culturing pinewood nematodes

5. The malt extract agar wood fiber mix should be distributed into the Petri dishes with the help of a spoon, ensuring to cover the whole area (Fig. 1).
6. The media should be inoculated with fully developed *B. cinerea* mycelium and placed on a growth chamber at 25 °C in the dark, until the fungal mat covers about half of the surface of the media, which should take a bit over 1 week (*see* **Note 13**).

3.2.2 Obtaining and Maintaining Lab Cultures of B. xylophilus

1. Dauers should be inoculated in the fungal mat media. The higher the number of dauers inoculated, the higher the probability of obtaining lab cultures, but this is not straightforward. Often inoculation of less than ten dauers results to thriving lab cultures, while the opposite is also true.
2. After inoculation plates should be sealed with Parafilm, to prevent them from drying or getting contaminated or the nematodes from escaping.
3. Plates should be incubated in growth chambers in the dark at 25–30 °C, which is optimal for *B. xylophilus* colonies' growth [21], up to a maximum of 3 weeks, or until all the fungal mat is consumed.
4. Nematodes should be extracted from the media with the same methodology as described in Subheading 3.1.2. However, the media wrapped in a Kimwipes tissue should be left sitting on

the Milli-Q™ water-filled funnels for a longer time—up to 48 h—since nematodes take a longer time to exit this substract.

5. The presence of nematodes should be checked as described previously, and the species of the adult nematodes should be confirmed based on male spicule, female vulva flap, and tail terminus [22].
6. The same fungal media can be used to maintain the obtained lab cultures of *B. xylophilus* [16].

4 Notes

1. Generally these traps are commercially available with a smaller container more adapted to bark beetles but can be easily attached to modified dry 5-L containers. Spraying the inner surface of the bottom funnel of the trap with Fluon® leaves a slippery surface, which together with the larger containers allows more efficient trapping of large insects [19, 20, 23, 24].
2. Adult longhorn beetles are attracted to stressed trees or recently cut logs for mating and oviposition [25]. They can be baited with a combination of bark beetle pheromones and the volatiles released from host trees (monoterpenes and ethanol) [24, 26] which are commercially available.
3. A male-produced aggregation pheromone—monochamol (2-undecyloxy-1-ethanol) [27]—was isolated and has repeatedly shown to increase the efficiency of trapping for several *Monochamus* species when combined with plant volatiles and bark beetle pheromones [28].
4. *Monochamus* spp. adults emerge from and search for damaged, stressed, or recently dead pine trees. According to our experience, healthy forests lead to significant captures of pine sawyer beetles, no matter how many traps are deployed.
5. This is an adaptation of the Baermann funnel technique [12], the nematode dauers tending to quickly exit the beetle, falling to the bottom of the funnel stem, close to the clamp.
6. A conical tubing is the most suited because the nematodes tend to sediment in a concentrated spot at the bottom after a few hours and can be easily collected with a Pasteur Pipette.
7. *B. xylophilus* dauers are highly aggregated in their insect vector, with most of the trapped beetles having non-phoretic nematode while a few having hundreds to thousands. These pine sawyer beetles with high phoresy are concentrated at the beginning of the flight season (*see* Subheading 3.1.1, **step 1**, and refs. 19 and 20). This should be considered when counting the nematodes. Thus, to check if each sample (and so the extracted

Monochamus) has nematode dauers, take a few millimeters of the bottom of the conical tube after sedimentation with a Pasteur Pipette, and check if there are dauers with a dissecting microscope. If the dauers are zero to few, the number can be registered right away. If there are hundreds to thousands of dauers, then the solution can be remixed and a subsample of a known volume counted, being the total number of nematodes on the pine sawyer beetle extrapolated from the total volume of the solution.

8. Wood shavings from any local pines can be used. The method was tested with two very different pine species, the European *Pinus pinaster* and North American *P. strobus*, with equally good results [16].
9. The beneficial effect of adding groundwood to growth media is probably related to the texture changes of the media. There were strong effects from granularity of the wood fiber, with coarser fiber being better for nematode colonies' growth and survival [16].
10. Dry pinewood does not have much nutrition; thus, we need to add nutrients (malt extract) and an agent to increase cohesivity of the media (agar). Malt extract proved to lead to better results when culturing *B. xylophilus*, recently obtained from the wild, than the other most common nutrient source used in standard media—potato dextrose extract [16].
11. Dry wood has different densities, making a different relation mass/volume, and that should be taken into consideration when selecting the portion of wood fiber to be used in the media. For example, *P. pinaster* wood is about 1/3 more dense than *P. strobus* wood, and thus, 150 g of its wood will occupy the same volume as 100 g of *P. strobus*. The consideration should be that the wood fiber will absorb and be imbibed in 1 L of malt extract agar solution [16].
12. Wood is a good thermal insulator to heat, and thus to facilitate conductivity and achieve thorough sterilization, it should be moistened and autoclaved for a longer time than the malt extract agar mix.
13. *B. xylophilus* has a mycophagous phase when it is quite polyphagous, but *B. cinerea* stand out as a fungal species that allows high levels of colony survival and fast population growth [16, 29].

Acknowledgments

This work was supported by Fundação para a Ciência e a Tecnologia (FCT), Portugal, with the project PTDC/AGR-CFL/

098869/2008 and funding to Carla Pimentel through the grants SFRH/BPD/46995/2008 and SFRH/BPD/80867/2011 and the contract DL 57/2016/CP1382/CT0009. This research was funded by the Forest Research Centre, a research unit funded by FCT (UIDB/00239/2020). Extra funding was provided by Fundação Luso-Americana para o Desenvolvimento (FLAD) in support to the FCT project PTDC/AGR-CFL/098869/2008.

References

1. Giblin-Davis RM, Davies KA, Morris K et al (2003) Evolution of parasitism in insect-transmitted plant nematodes. J Nematol 35: 133–141
2. Ma HB, Lu Q, Liang J et al (2011) Functional analysis of the cellulose gene of the pine wood nematode, *Bursaphelenchus xylophilus*, using RNA interference. Genet Mol Res 10903: 1931–1941
3. Mori Y, Miyahara F, Tsutsumi Y et al (2008) Relationship between resistance to pine wilt disease and the migration and proliferation of pine wood nematodes. Eur J Plant Pathol 1224:529–538
4. Kuroda K (2004) Inhibiting factors of symptom development in several Japanese red pine (*Pinus densiflora*) families selected as resistant to pine wilt. J For Res 9:217–224
5. Pimentel CS, Firmino PN, Calvão T et al (2017) Pinewood nematode population growth in relation to pine phloem chemical composition. Plant Pathol 665:856–864
6. Bizzarri MF, Bishop AH, Dinsdale A et al (2008) Changes in the properties of *Bacillus thuringiensis* after prolonged culture in a rich medium. J Appl Microbiol 104:60–69
7. Fux CA, Shirtliff M, Stoodley P et al (2005) Can laboratory reference strains mirror "real-world" pathogenesis? Trends Microbiol 132: 58–63
8. Kikuchi T, Akker SE, Jones JT (2017) Genome evolution of plant-parasitic nematodes. Annu Rev Phytopathol 55:333–354
9. Woods CM, Woodward S, Redfern DB (2005) In vitro interactions in artificial and wood-based media between fungi colonizing stumps of Sitka spruce. For Pathol 35:213–229
10. Kirkland PD, Hawkes RA (2004) A comparison of laboratory and "wild" strains of bluetongue virus – is there any difference and does it matter? Vet Ital 40:448–455
11. Krokene P, Solheim H (2001) Loss of pathogenicity in the blue-stain fungus *Ceratocystis polonica*. Plant Pathol 50:497–502
12. OEPP/EPPO (2009) PM 7/4(2): *Bursaphelenchus xylophilus*. EPPO Bull 39(3):344–353
13. Sriwati R, Takemoto S, Futai K (2007) Cohabitation of the pine wood nematode, *Bursaphelenchus xylophilus*, and fungal species in pine trees inoculated with *B. xylophilus*. Nematology 9:77–86
14. Aikawa T, Kikuchi T (2007) Estimation of virulence of *Bursaphelenchus xylophilus* (Nematoda: Aphelenchoididae) based on its reproductive ability. Nematology 9:371–377
15. Espada M, Silva AC, van den Akker SE et al (2016) Identification and characterization of parasitism genes from the pinewood nematode *Bursaphelenchus xylophilus* reveals a multilayered detoxification strategy. Mol Plant Pathol 17:286–295
16. Pimentel CS, Firmino PN, Ayres MP (2020) Comparison of methods to obtain and maintain cultures of the pinewood nematode, *Bursaphelenchus xylophilus*. J For Res 25:101–107
17. David G, Giffard B, Piou D et al (2014) Dispersal capacity of *Monochamus galloprovincialis*, the European vector of the pine wood nematode, on flight mills. J Appl Entomol 138:566–576
18. Lindgren BS (1983) A multiple funnel trap for scolytid beetles (Coleoptera). Can Entomol 115:299–302
19. Firmino PN, Calvão T, Ayres MP et al (2017) *Monochamus galloprovincialis* and *Bursaphelenchus xylophilus* life history in an area severely affected by pine wilt disease: implications for forest management. For Ecol Manag 389: 105–115
20. Pimentel CS, Ayres MP, Vallery E et al (2014) Geographical variation in seasonality and life history of pine sawyer beetles *Monochamus* spp: its relationship with phoresy by the pinewood nematode *Bursaphelenchus xylophilus*. Agric For Entomol 16:196–206
21. Pimentel CS, Ayres MP (2018) Latitudinal patterns in temperature-dependent growth rates of a forest pathogen. J Therm Biol 72:39–43

22. Nickle WR, Golden AM, Mamiya Y et al (1981) On the taxonomy and morphology of the pine wood nematode, *Bursaphelenchus xylophilus* (Steiner & Buhrer 1934) Nickle 1970. J Nematol 13:385–392
23. Graham EE, Mitchell RF, Reagel PF et al (2010) Treating panel traps with a fluoropolymer enhances their efficiency in capturing cerambycid beetles. J Econ Entomol 103: 641–647
24. Morewood WD, Hein KE, Katinic PJ et al (2002) An improved trap for large woodboring insects, with special reference to *Monochamus scutellatus* (Coleoptera: Cerambycidae). Can J For Res 32:519–525
25. Hughes AL, Hughes MK (1982) Male size, mating success, and breeding habitat partitioning in the whitespotted sawyer *Monochamus scutellatus* (Say) (Coleoptera: Cerambycidae). Oecologia 55:258–263
26. Miller DR, Allison JD, Crowe CM et al (2016) Pine sawyers (Coleoptera: Cerambycidae) attracted to α-Pinene, Monochamol, and Ipsenol in North America. J Econ Entomol 109: 1205–1214
27. Pajares JA, Alvarez G, Ibeas F et al (2010) Identification and field activity of a male-produced aggregation pheromone in the pine sawyer beetle, *Monochamus galloprovincialis*. J Chem Ecol 36:570–583
28. Macias-Samano JE, Wakarchuk D, Millar JG et al (2012) 2-Undecyloxy-1-ethanol in combination with other semiochemicals attracts three *Monochamus* species (Coleoptera: Cerambycidae) in British Columbia, Canada. Can Entomol 144:821–825
29. Pimentel CS, Firmino PN, Ayres MP (2021) Interactions between pinewood nematodes and the fungal community of pine trees. Fungal Ecol 51:101046

Chapter 2

Survey and Monitoring of *Phytophthora* Species in Natural Ecosystems: Methods for Sampling, Isolation, Purification, Storage, and Pathogenicity Tests

Ana Pérez-Sierra, Marilia Horta Jung, and Thomas Jung

Abstract

Phytophthora species can be found in multiple substrates. Due to dormancy of resting structures and presence of faster-growing antagonists, direct isolation of *Phytophthora* can be difficult to achieve, and indirect baiting methods often reach higher isolation frequencies. In this chapter, different methodologies are described for sampling and for the successful isolation of *Phytophthora* species from natural ecosystems. Sampling methods for soil, roots, bark cankers, and waterbodies are described. Agar recipes and guidance on the selection of suitable tissue to perform direct isolations are provided. A range of different baiting techniques are described for isolation of *Phytophthora* from different substrates. Purification methods to obtain clean and non-mixed cultures and conservation methods for pure cultures are shown. Two pathogenicity test methods for soil infestation and for under-bark inoculation, respectively, are described in detail.

Key words Baiting, Roots, Canker, Soil, Water, Forest decline, Agar media, Soil infestation, Inoculations, Koch's postulates

1 Introduction

The oomycete genus *Phytophthora* arguably comprises some of the most damaging plant pathogens of horticultural, agricultural, forestry, and natural ecosystems. Many species are soilborne infecting roots, the root collar, and sometimes stem bark of plants and can survive for long periods of time in the soil, whereas other species are airborne and affect mainly the aerial parts of plants. *Phytophthora* species are mainly primary pathogens [1] causing declines and diebacks of a wide range of forest and natural ecosystems [2]. The detection and isolation of *Phytophthora* pathogens from these environments requires experience and technical expertise. The first step to successfully isolate *Phytophthora* is to collect the right type of sample for the isolation tests. In this chapter we provide guidance for sampling from natural ecosystems, including soil and root

Nicola Luchi (ed.), *Plant Pathology: Methods and Protocols*, Methods in Molecular Biology, vol. 2536,
https://doi.org/10.1007/978-1-0716-2517-0_2,

sampling, sampling of and isolation from bark cankers, soil baiting using young leaves or green apples as baits, root isolations, isolations from waterbodies using a filtering technique, naturally fallen floating leaves, and in situ baiting. Selective agar media like PARPNH-agar are used for isolation from different plant tissues, and V8 juice agar, carrot agar, or oatmeal agar is used for subculturing, purification, and storage of pure cultures. As it is common to have bacterial contamination or mixed cultures with other oomycetes or fungi during isolation, different methods such as the pancake method and the use of green apples or leaf baits are proposed. In most studies, once the *Phytophthora* species responsible for the decline has been identified, Koch's postulates need to be completed to confirm pathogenicity. Two different pathogenicity test methods for soil infestation and under-bark inoculation are described.

2 Materials

It is recommended to make a list of the tools that are needed when sampling in the field as you might be in remote locations and forgetting some of the tools could jeopardize the success of your sampling trip. It is important to record the GPS (Global Positioning System) coordinates of the sampling locations as you might need to return if more samples are required or *Phytophthora* species of interest are detected. Therefore, a GPS device is highly recommended.

2.1 Sampling

2.1.1 Soil

1. Shovel, spade, and hoe to remove organic layer and dig and excavate the soil.
2. Plastic zip bags to place and transport the collected soil.
3. Permanent marker pen to label the samples.
4. Disinfectant for tools to avoid cross contamination.

2.1.2 Roots

1. Shovel, spade, and hoe to remove organic layer and dig for roots.
2. Pruning shears, hand pruners, or secateurs to cut smaller roots.
3. Hand saw to remove main roots or thicker roots.
4. Plastic sampling bags to place and transport the collected roots.
5. Permanent marker pen to label the samples.
6. Disinfectant for tools to avoid cross contamination.

2.1.3 Bleeding Bark Cankers

1. Chisel, hatchet, or knife to remove outer bark or to extract bark panels.
2. Hammer or mallet to extract bark panels.
3. Screw cap bottles with distilled water to place and transport the bark samples.
4. Permanent marker pen to label the samples.
5. Disinfectant for tools to avoid cross contamination.

2.1.4 Naturally Fallen Leaves in Waterbodies or on Forest Ground

1. Litter picker to collect floating or submerged leaves or leaves from the ground.
2. Plastic zip bags with moistened paper tissue to place and transport the collected leaves.
3. Permanent marker pen to label the samples.

2.1.5 Water

1. Plastic containers (5 L).
2. Funnel.
3. 100-μm pore size mesh to remove any debris.
4. Knapsack sprayer.
5. Filter holders.
6. Filters.
7. Tweezers.
8. 15-mL polypropylene tubes.
9. Cool box.
10. Permanent marker to label the samples.
11. Disinfectant for tools to avoid cross contamination.
12. Waste bag.

2.2 Isolation

2.2.1 Soil Baiting with Leaves

1. Plastic containers or tubs (ca 18 × 25–27 × 13–14 cm).
2. Distilled water.
3. Fly mesh.
4. Sieve.
5. Tissue paper.
6. Young leaves from susceptible tree species to be used as baits (i.e., *Acacia* spp., *Quercus* spp., *Fagus* spp., *Castanopsis* spp., *Ceratonia siliqua*, *Chamaecyparis* spp., *Cinnamomum* spp., *Drimys winteri*, *Lithocarpus* spp., *Litsea* spp., *Machilus* spp., *Neolitsea* spp., *Nothofagus* spp., *Prunus avium*, *Rhododendron* spp.) [3–11].

2.2.2 Soil Baiting with Fruits

1. Green apples (Granny Smith or Golden Delicious).
2. Cork borer.
3. Spatula.
4. Distilled water.
5. Adhesive tape.

2.2.3 In Situ Baiting from Waterbodies

1. Fly mesh.
2. Styrofoam board (ca 25–30 mm thick).
3. Scissors.
4. Cutting knife.
5. Stapler and staples.
6. Big zip bags.
7. Paper tissue.
8. Sprayer with distilled water.
9. Plastic cord.

2.2.4 Agar Media for Phytophthora Isolations

1. Selective PARPNH-agar [12]: V8 agar (V8A)—16 g agar, 2 g $CaCO_3$, 100 mL V8 juice (Campbell's or any proprietary vegetable juice), and 900 mL distilled water. Cornmeal agar (CMA) can be used instead of V8A: 17 g of cornmeal agar and 1 L distilled water.
2. Add to the agar media 10-μg/mL pimaricin, 200-μg/mL ampicillin, 10-μg/mL rifampicin, 25-μg/mL pentachloronitrobenzene (PCNB), 50-μg/mL nystatin, and 50-μg/mL hymexazol. The nystatin suspension and the hymexazol are dissolved in 100-mL sterile distilled water with a temperature <45 °C and stirred on a magnetic stirrer until they are dissolved. The other antibiotics (ampicillin, PCNB (very toxic! handle carefully!), rifampicin, and pimaricin) are dissolved in 5–10 mL of 80% ethanol and stirred on a magnetic stirrer until they are dissolved. All antibiotics are added to the V8 agar when it has cooled down to 45 °C (measure with a thermometer that was sterilized with ethanol!). Then the vial with the agar is shaken to mix the antibiotics uniformly. During pouring of the PARPNH-agar in the Petri dishes, shake the vial regularly to ensure uniform distribution of the antibiotics.

2.3 Purification of Contaminated Cultures

2.3.1 Pancake Method

1. V8A.
2. Carrot agar (CA): 200 g grated carrots, 15 g of agar, 1000 mL distilled water, cold tap water, and a muslin cloth [13].
3. PARPNH-agar.
4. A teaspoon or a scalpel.

2.3.2 Green Apples

1. Green apples (Granny Smith or Golden Delicious).
2. Knife.
3. Adhesive tape.

2.3.3 Purification of Mixed Cultures

1. Sterile 90 mm Petri dishes.
2. Distilled water.
3. Young leaves from susceptible tree species to be used as baits (e.g., *Fagus*, *Castanopsis*, *Lithocarpus*, *Quercus* spp.).

2.4 Storage of Phytophthora

1. V8A medium: *see* Subheading 2.2.4.
2. CA medium: *see* Subheading 2.3.1.
3. Oatmeal agar (OA) medium: 72.5 g oatmeal agar, 1 L distilled water [1].
4. Rye agar (RA) medium: 100 g of rye seeds boiled for 30 min in 0.5 L of distilled water, then squeezed through three layers of cheese cloth. The filtrate is adjusted with distilled water to 1 L; 20 g of agar and 5 g of glucose are added, followed by autoclaving (modified after [1]).
5. Liquid storage in sterile soil extract: 50–100 g forest soil; 900 mL distilled water. Stirrer.
6. Liquid storage in sterile distilled water.
7. Petri dishes and tubes: 35 or 50 mm Petri dishes; 20 or 50 mL screw cap tubes for agar slants; 2–3 mL screw cap tubes.

2.5 Pathogenicity Tests

2.5.1 Soil Infestation Trial

1. 500-mL (1000 mL) wide-mouthed glass bottles with lid or autoclave sacs.
2. 250 mL (500 mL) of fine or medium vermiculite.
3. 20 mL (40 mL) of whole oat grains or millet seeds.
4. 175 mL (350 mL) of vegetable juice broth (200 mL juice, 800 mL distilled water amended with 3 g $CaCO_3$).
5. Actively growing V8A cultures of the respective *Phytophthora* isolates to be tested.
6. Sterilized scalpel and glass stick.
7. Tape to seal the glass lids.
8. Black plastic bags to place the glass bottles with the inoculum.
9. Fine pore sieve to rinse the inoculum before use.
10. Seeds of the respective tree species to be tested.
11. Per host–*Phytophthora* combination: 1 planting tray with 12 separate compartments and 1 plastic box big enough to enclose the tray or 1 inner plastic box for growing the 10–12 seedlings (35 × 25 × 17 cm to 50 × 35 × 27 cm, depending on the size of seedlings), an outer same-sized or slightly bigger plastic box and an electric drill for drilling holes at the bottom of the inner box.

12. Per host–*Phytophthora* combination: one extra plastic box for draining the flooding water after the flooding periods.
13. Sterile autoclaved soil mixture: 1:1:1, v/v/v, of peat, vermiculite, and sand.
14. For rebaiting of the inoculated *Phytophthora* isolates, young leaves from susceptible tree species to be used as baits (i.e., *Quercus* spp., *Fagus* spp., *Castanopsis* spp., *Cinnamomum* spp., *Machilus* spp.).
15. Selective PARPNH-agar for reisolations from baiting leaves, roots, and necrotic bark.

2.5.2 Under-Bark Inoculation Trial

1. Cork borer (0.5 cm diameter), knife, or razor blade.
2. Sterile distilled water.
3. Cotton wool.
4. Parafilm.
5. Aluminum foil paper.
6. Selective PARPNH-agar for reisolations.

3 Methods

Make sure that equipment, tools, and footwear are disinfected between sampling sites. Avoid working under unsuitable climatic conditions, and follow the appropriate risk assessment for each particular sampling site.

3.1 Sampling

3.1.1 Soil

1. Soil samples are collected from around the bases of trees showing early symptoms of decline (or any tree in case of *Phytophthora* diversity studies) (*see* **Note 1**). Soil samples from ca six trees per sampling site are usually sufficient.
2. Remove the leaf litter layer (but not the A horizon) (Fig. 1a, e, g–h).
3. Using the appropriate tools for the respective soil type, take soil samples at 3–4 points around the tree in approximately 0.3–1 m distance from the trunk at 10–30 cm depth (Fig. 1a–i), and bulk all together to one sample of ca 1–3 L per surveyed tree [3, 7]. Make sure to remove stones if present. Do not remove fine roots and small woody roots from the sampled tree as they can also be used for isolations.
4. Label the bag with date, reference number, and host, and record the GPS coordinates.
5. Disinfect tools before sampling the next tree.
6. Store the soil sample at 10–20 °C until it can be tested (it can be kept for up to 3–4 weeks).
7. For long-term storage, keep soils at 4 °C (15 °C with tropical soils) (*see* **Note 2**).

Fig. 1 Sampling of soil in forest stands for *Phytophthora* isolations. (**a**) Removing of the organic litter layer with a shovel. (**b**, **c**) Sampling from a rocky soil with a hoe. (**d**) Sampling of soil with a hoe in a gravelly ephemeral flush. (**e**) Sampling of soil in a tropical rainforest with a sharp-edge hoe to cut through the surface root layer. (**f**) Placing soil with a small shovel into a zip bag after loosening the soil with a small hoe. (**g**, **h**) Sampling a soil–root monolith with a spade in a deep loamy soil. (**i**) Sampling soil on a steep rocky slope with a small shovel

3.1.2 Roots

1. Since *Phytophthora*-infected fine roots and small woody roots become quickly detached, they are sampled together with the surrounding rhizosphere soil as described in Subheading 3.1.1 [4].
2. Thicker roots with necrotic bark lesions are cut with pruning shears, hand pruners, secateurs, or a hand saw and placed into the soil bag or wide-mouthed screw cap bottles with distilled water.
3. Label the bag/bottle with date, reference number, and host, and record the GPS coordinates.
4. Disinfect tools before sampling the next tree.

3.1.3 Bleeding Bark Cankers

1. Prior to collecting the sample, take pictures of the symptoms observed, and label the sampling bottle with date, reference number, and host, and record the GPS coordinates.
2. Identify the area of bark for sampling. This can be done by locating the edge of the lesion by removing the outer bark with a knife or chisel. The edge of the lesion can be identified by the change of color of the inner bark. Living tissue is whitish or greenish, whereas infected tissue has an orange-brown to dark-brown discoloration (Fig. 2a). Usually, the lesion front in the inner bark and cambium area is 10–15 cm ahead of the tarry or rusty exudate spots visible on the outside of the bark (Fig. 2a–c) [2, 10, 14].
3. Using a mallet/hammer and a chisel (blade 3–4 cm wide) with thicker bark or a hatchet or a knife with thinner bark, cut and remove several bark panels of approximately 10–15 × 4 cm each down to the surface of the underlying wood (Fig. 2b–e). Usually, this is most easily achieved by first cutting the outline of the panel to be removed, inserting the chisel or hatchet on one of the narrow edges of the inscribed area, and then gently levering up the overlying bark (with the aid of the mallet if necessary) until the panel can be lifted away. Hold the bark sample during the cutting with the chisel with two fingers so that it does not break into pieces (Fig. 2c, e) (*see* **Note 3**). An ideal sample panel consists of approximately 70–80% lesion material and 20–30% healthy bark.
4. Place the bark panels in a wide-mouthed screw cap bottle with distilled water (Fig. 2f) [10, 14].
5. Disinfect any tools used before sampling the next tree.
6. In the lab, incubate the bottles at 18–20 °C, and change the water 3–4 times per day over a period of 2–3 days to remove polyphenols and reduce bacterial growth. Sterilize the discarded water with bleach or by autoclaving for 20 min at 120 °C.

Fig. 2 Sampling and isolation techniques from bleeding bark cankers. (**a**) Tongue-shaped front (white arrow) of an orange-brown inner bark lesion on an *Alnus glutinosa* tree ca 10 cm above the highest (most recent) tarry exudate spot (red arrow) on the bark surface. (**b**, **c**) Sampling a bark panel with a chisel and a hammer starting ca 10 cm above the highest tarry exudate spot (red arrows); note that the fingers support the bark panel (white arrow) during cutting to avoid it from breaking into pieces. (**d**) Sampling a bark panel with a chisel and a hammer from a bark canker in a slightly higher position. (**e**) Supporting the bark panel during cutting with two fingers (white arrow) to avoid breaking. (**f**) Sampled bark panels in wide-mouthed screw cap bottles with distilled water. (**g**) Flame-shaped orange-brown bark lesion typical of an active *Phytophthora* infection. (**h**) Cutting a bark lesion with a sharp single-edge razor blade into small segments. (**i**) Plating up to 10–15 lesion

3.1.4 Water [15–17]

1. Collect water from a waterbody (river, stream, lake, sea) with a plastic container/bucket. In total 10 L of water per site are needed for the sampling.
2. The water is filtered in situ.
3. Pour the water through a 100 μm pore size mesh (to remove any debris) into the clean knapsack sprayer (Fig. 3a).
4. Pump the water through a 47 mm polypropylene Swinnex filter holder (Fig. 3b) with a 5 μm pore size autoclaved filter (cellulose acetate and cellulose nitrate mixture) (Fig. 3c) (*see* **Note 4**). Use three filters per 10 L of water. Make sure that you remove the filter and change it for a new autoclaved one when you notice that the flow of water decreases and the pumping of the water gets harder which indicates that the filter is already saturated and needs to be changed.
5. Each filter is carefully removed with a pair of sterilized tweezers and placed in a sterilized 15 mL polypropylene tube. The tubes are kept in a cool box during field sampling and stored in the fridge (5–8 °C) or the freezer (−20 °C) before use in isolation tests (*see* Subheading 3.2.6) or metagenomic approaches (not part of this chapter), respectively. This is necessary to make clear that samples stored in the freezer can only be used for metagenomic approaches!
6. Disinfect the sprayer and rinse with water before sampling the next site.
7. Remove used filter holder and place it in a waste bag to be disinfected and autoclaved in the laboratory.

3.1.5 Naturally Fallen Leaves

1. Naturally fallen leaves on the forest ground or in waterbodies (Fig. 4a–c) act as natural baiting leaves [7, 9, 18].
2. Using a litter picker, collect naturally fallen leaves with necrotic lesions from waterbodies or from the forest ground (mainly in humid forests and mangroves), and place them in plastic zip bags (Fig. 4d, e). Cover them with moistened paper tissues to keep them moist during the transport. In total ca 50–100 necrotic leaves per site are needed.

3.1.6 Transport of Samples

1. During warm or sunny periods, samples should be transported from the field to the lab in cool bags and/or air-conditioned

Fig. 2 (continued) segments onto selective PARPNH-agar. (**j**) *Phytophthora* colony growing after 3–4 days from a plated bark lesion segment. (**k, l**) Baiting from a bleeding bark canker. (**k**) Shredded bark canker segments are spread in a plastic tub, fixed with fly mesh and sterile stones to the bottom, and flooded with distilled water. (**l**) Young *Fagus sylvatica* leaves after 4 days floating over bark lesion segments; note the typical brownish lesions on the leaves (arrows) and the orange discoloration of the water from the polyphenols leaching from the bark

Fig. 3 Filtering technique for isolating *Phytophthora* from water samples. (**a**) Placing a 100 μm pore size mesh at the opening of a clean knapsack sprayer before filling it with water to remove debris. (**b**) 47 mm polypropylene Swinnex filter holder. (**c**) 5 μm pore size cellulose acetate/cellulose nitrate filter inside the filter holder showing brownish discoloration after filtering of river water. (**d**) Placing one half of the used filter below a flap cut tangentially from a green apple. (**e**) Multiple brown lesions (arrows) developing on the apple flap which is wrapped with adhesive tape. (**f**) Multiple brown lesions (arrows) inside the apple flap. (**g**) Cross section through a green apple showing a firm brown fruit rot (arrow) developing on both sides of the filter. (**h**) Typical *Phytophthora* colonies growing from small pieces of the fruit rot plated onto selective PARPNH-agar. (**i**) Segments of a filter plated directly onto PARPNH-agar

Fig. 4 Sampling and isolation techniques from naturally fallen leaves. (**a**, **b**) Naturally fallen leaves in forest streams. (**c**) Naturally fallen leaves in a mangrove. (**d**, **e**) Collecting naturally fallen leaves from a stream and a mangrove with a litter picker and placing them into plastic zip bags. (**f–h**) Cutting necrotic leaf lesions with razor blades or scalpels into small segments and plating them onto selective PARPNH-agar

vehicles. Avoid direct contact between samples and cooling elements. Tropical samples should not get colder than 15 °C.

3.2 Isolation

After finishing an isolation test, all soil, plant, and water material has to be sterilized before discarding by autoclaving them at 120 °C for at least 20 min (plant tissues and water samples) or 60 min (soil samples). Water samples can also be sterilized with bleach.

3.2.1 Leaf Baiting from Soil [3, 9–11]

1. Mix the soil sample in a plastic sac thoroughly before baiting (Fig. 5a).
2. Place 200–500 mL of the mixed soil sample in a plastic container/tub, and incubate it at 18–20 °C (temperate and Mediterranean samples) or 20–24 °C (subtropical or tropical samples) in an air-conditioned room at natural daylight (Fig. 5e) or in a walk-in incubator with 12-h light regime (Fig. 5f).
3. In case the soil sample is dry, premoisten it with distilled water 24–48 h before baiting.
4. Flood the soil sample with distilled water. Add the water slowly from one corner of the container to avoid disturbance of the soil (Fig. 5b). Add water so that the waterline is ca 3–4 cm above the soil surface.
5. Remove larger floatıng debris with tweezers or a small sieve and smaller floating particles with tissue paper (Fig. 5c, d). This has to be done very slowly and without touching the soil or roots to avoid causing disturbance of the soil.
6. Wait at least 2 h (or overnight) for the soil particles to settle and other debris to float up to the surface. Then remove the floating particles again slowly with tissue paper (*see* **Note 5**).
7. Select young healthy leaves (3–8 days old, hydromorphic, and soft) of susceptible tree species *Quercus robur* or another susceptible species (*see* list in Subheading 2.2.1) (*see* **Note 6**).
8. Carefully float the leaves on the water surface making sure that the whole surface of the leaf is in touch with the surface of the water. Cover the whole surface with leaves (Fig. 5e, g–i). Do not submerge the leaves since submerged leaves will usually not get infected by *Phytophthora*.
9. If *Phytophthora* is present, necrotic brown or black spots/lesions will appear on the floating leaves after 2–5 days (Fig. 5g–i). With some *Phytophthora* species, this might take up to 7 days. Since *Pythium/Phytopythium* spp. will also colonize the leaves, examine the leaves (bottom side up) under the light microscope at ×80 magnification to clarify whether sporangia formed on the lesions are either *Phytophthora*-like or *Pythium-/Phytopythium*-like (*see* **Notes 7** and **8**).
10. Isolation from necrotic leaf spots/lesions (*see* Subheading 3.2.12).

Fig. 5 Baiting technique of *Phytophthora* from soil samples. (**a**) Thorough mixing of soil samples in plastic sacs. (**b**) Slowly pouring distilled water on a soil sample to avoid disturbance of the soil. (**c**, **d**) Cleaning the water surface from dust and litter particles with several layers of tissue paper. (**e**) Selecting suitable soft baiting leaves and floating them on the water surface. (**f**) Incubation of baiting tubs in an air-conditioned walk-in incubator. (**g**) Young soft leaves of *Quercus robur*, *Nothofagus obliqua*, *Drimys winteri*, and *Chamaecyparis lawsoniana* covering the whole surface of a baiting tub. (**h**) Same baiting tub as in (**g**) after 3 days incubation at 20 °C with most leaves showing necrotic lesions caused by *Phytophthora cinnamomi*. (**i**) Baiting tub with young leaves of *Fagus sylvatica* after 4 days incubation at 20 °C showing multiple necrotic lesions caused by *Phytophthora cactorum* and *Phytophthora plurivora*

3.2.2 Baiting from Soil with Green Apples [3, 19]

1. Select green apples, rinse them with water, and disinfest their surface with 70% methylated spirit (denatured ethanol) before use. Use one apple per soil sample to be tested.
2. With the help of a cork borer, remove two to four cores of the apple (in opposite sides of the apple) (Fig. 6a, b). The cores should be about 1.5–2 cm in diameter and 1.5–2 cm in depth. Discard the cores.
3. Mix thoroughly the soil to be tested. Make sure that stones or debris are removed.
4. With the help of a spatula (Fig. 6a), fill the holes in the apple with the soil to be tested (Fig. 6c).
5. Saturate the soil with distilled water (Fig. 6d).
6. Cover the holes with adhesive tape (Fig. 6e).
7. Label the apple with the sample code, initials, and date.
8. Incubate the apple at 18–20 °C for 4–7 days. If *Phytophthora* is present, a brown, hard fruit rot will develop (Fig. 6e, f).
9. Isolation from the fruit rot lesions (*see* Subheading 3.2.13).

3.2.3 Direct Isolation from Roots Using PARPNH-Agar [3, 4]

1. After thorough mixing of the bulked 1–3 L soil sample per tree (Fig. 5a), take a subsample of 200–400 mL, and flood it in a ca 18 × 12 cm tub with distilled water, and stir it with a spoon or glass stick so that detached fine roots float on the water surface (Fig. 7a) [4] (*see* **Note 9**).
2. Incubate the plastic tub for 48 h at 20 °C to stimulate germination of resting structures in the roots.
3. Remove with a pair of tweezers ca 50 fine long and short roots (<2 mm diam, 3–20 mm length) from the water surface, and also cut sections with fine scissors from submersed small woody roots (2–5 mm diam) and lateral fine roots.
4. Wash the roots thoroughly, surface disinfest them for 10 s in 70% ethanol, and blot them dry.
5. Plate the washed and disinfested fine and small woody roots onto PARPNH-agar (Fig. 7b).
6. With necrotic bark lesions on thicker roots, disinfest the root surface with 70% methylated spirit (denatured ethanol) and cut with a single-edge razor blade or a scalpel thin segments of ca 3–8 × 2–3 mm from the growing edge of the lesion (Fig. 7c), blot them quickly on paper tissue, and plate them onto PARPNH-agar.
7. Incubate the plates at 20 °C in the dark.
8. Examine the plated fine roots, root segments, and lesions after 24, 48, and 72 h under the stereomicroscope at ×20 for developing colonies of *Phytophthora* (Fig. 7d–f), subculture them with a fine spatula-tip needle (Fig. 9a) onto V8A or CA, and incubate them at 20 °C in the dark.

Fig. 6 Baiting technique of *Phytophthora* from soil and fine roots using green apples. (**a**) Materials needed for this isolation technique: distilled water, green apple, spatula, cork borer, pipette, and adhesive tape. (**b**) Removing with the cork borer a core from the apple. (**c**) Filling the hole with the soil sample. (**d**) Moistening the soil with distilled water until saturation using a pipette. (**e**) Fruit rot (arrow) developing after 4–7 days around the soil-filled hole which was covered with an adhesive tape. (**f**) Cross section through the apple showing the firm brown fruit rot (arrows) developing around the soil-filled hole. (**g**) Placing fine roots underneath a flap cut tangentially into a green apple

Fig. 7 Isolation of *Phytophthora* from roots using direct plating and leaf baiting techniques. (**a**) Detached fine roots and debris floating on a flooded soil sample after stirring. (**b**) Plating of detached fine roots after washing, brief drying, and surface disinfesting onto selective PARPNH-agar. (**c**) Active tongue-shaped front of a girdling lesion on a small woody root after removing the outer bark with a sharp sterile scalpel. (**d**, **e**) Typical *Phytophthora cinnamomi* colonies growing from fine root segments and root lesions plated onto PARPNH-agar; note that fungal growth remains limited to the root surface (arrow) and does not progress into the selective agar. (**f**) Typical coralloid, irregular hyphae of *P. cinnamomi* growing from a plated fine root; scale bar = 20 μm. (**g**) Young soft leaves of *Quercus robur*, *Q. suber*, and *Pittosporum undulatum* floating over a flooded woody root of *Banksia grandis*; note that several leaves show brownish lesions (arrows) caused by *P. cinnamomi*. (**h**) Baiting from fine roots using young leaves of *Q. suber* as baits

3.2.4 Leaf Baiting from Roots [4]

1. Collect and prepare fine and small woody roots as described in Subheading 3.2.3 (**steps 1–4**).
2. After washing and surface disinfesting, flood fine and small woody roots in a 90 mm Petri dish or a small tub and thicker roots in bigger tubs or trays with distilled water, and cover the whole water surface with young leaves of susceptible tree species (*see* Subheadings 2.2.1 and 3.2.1) (Fig. 7g, h).
3. If *Phytophthora* is present, necrotic brown or black spots/lesions will appear on the floating leaves after 3–7 days (Fig. 7g). With some *Phytophthora* species, this might take up to 10 days. Since *Pythium/Phytopythium* spp. will also colonize the leaves, examine the leaves (bottom side up) under the light microscope at ×80 magnification to clarify whether sporangia formed on the lesions are either *Phytophthora*-like or *Pythium-/Phytopythium*-like (*see* **Note 7**).
4. Isolation from necrotic leaf spots/lesions (*see* Subheading 3.2.12).

3.2.5 Isolation from Roots with Green Apples [3, 20]

1. Collect and prepare fine and small woody roots as described in Subheading 3.2.3 (**steps 1–4**).
2. Select green apples, rinse them with water, and disinfest their surface with 70% methylated spirit (denatured ethanol) before use. Use one apple per tree to be tested.
3. Cut with a knife a shallow flap from the opposite sides of the apple (Fig. 6g).
4. Place the washed and disinfested fine roots underneath the apple flap (Fig. 6g).
5. Seal the flap with adhesive tape.
6. Label the apple with sample code, initials, and date.
7. Incubate the apple for 3–10 days at 20 °C.
8. Cut the apple to expose developing fruit rot lesions.
9. Isolation from the fruit rot lesions (*see* Subheading 3.2.13).

3.2.6 Isolation from Filters with Green Apples (Modified After [3, 15, 20])

1. Select green apples, rinse them with water, and disinfest their surface with 70% methylated spirit (denatured ethanol) before use. Use one apple per filter.
2. Cut with a knife two shallow flaps from the opposite sides of the apple (Fig. 3d).
3. Cut the filter into two halves and place them underneath the apple flaps (Fig. 3d).
4. Seal the flap with adhesive tape (Fig. 3e).

5. Label the apple with the sample code, initials, and date.
6. Incubate the apple for 3–5 days at 20 °C.
7. Remove the flaps and cut the apple to expose developing fruit rot lesions (Fig. 3f, g).
8. Isolation from the fruit rot lesions (*see* Subheading 3.2.13).

3.2.7 Direct Isolation from Filters Using PARPNH-Agar

1. Cut the filter into small segments and plate them onto PARPNH-agar (Fig. 3i) (*see* **Note 10**).
2. Incubate the plates at 20 °C in the dark.
3. Examine the plated filters after 24, 48, and 72 h under the stereomicroscope at ×20 for developing colonies of *Phytophthora*, subculture them with a fine spatula-tip needle (Fig. 9a) onto V8A or CA, and incubate them at 20 °C in the dark.

3.2.8 Direct Isolation from Bleeding Bark Cankers Using PARPNH-Agar [10, 14]

1. Remove the bark samples from the wide-mouthed screw cap bottles, rinse them in a sieve with distilled water, and blot them dry with tissue paper.
2. Turn the bark sample around, and expose on the cambium/xylem side the margins of the inner bark lesion using a sterilized sturdy single-edge razor blade or scalpel (*see* **Note 11**). Active bleeding cankers have an orange-brown, flame-shaped appearance (Fig. 2g) and a nonacidic and sometimes fruit-like smell.
3. Cut 2–3 mm thin layers of ca 3 cm length from the lesion margin into the lesion, cut them with a single-edge razor blade or a scalpel into small segments (3–8 × 3–8 × 2–3 mm) (Fig. 2h), and plate them onto PARPNH-agar (10–15 pieces placed equidistant onto the surface of a 90 mm PARPNH-agar plate) (Fig. 2i). The remaining parts of the bark lesion can be used for baiting (*see* Subheadings 3.2.9 and 3.2.10).
4. Incubate the plates at 20 °C in the dark.
5. Plates are examined daily under the stereomicroscope at ×20 magnification for outgrowing *Phytophthora* hyphae (Fig. 2j) which are transferred to V8A or CA and incubated at 20 °C in the dark.
6. The resulting colonies are examined daily to ensure that no *Pythium*/*Phytopythium*, fungi, or bacteria are contaminating the pure *Phytophthora* cultures. If contamination is observed, the cultures need to be cleaned (*see* Subheading 3.3).

3.2.9 Leaf Baiting from Bleeding Bark Cankers [10, 14, 21]

1. Cut all parts of the bark lesion remaining after the direct plating or the whole bark lesion in case no direct isolation is performed with a strong and sharp knife into smaller pieces.

2. Flood these bark pieces in a tub or box with distilled water so that the water surface is standing ca 2–3 cm above the bark pieces (Fig. 2k).
3. Fix the bark pieces with a fly mesh and sterilized small stones to the bottom of the tub (Fig. 2k).
4. Cover the whole water surface with young leaves of susceptible tree species (*see* Subheadings 2.2.1 and 3.2.1) (Fig. 2l).
5. Change the water daily or every second day to remove excess polyphenols leaching from the bark samples, and place the baiting leaves back to the water surface.
6. If *Phytophthora* is present, necrotic brown or black spots/ lesions will appear on the floating leaves after 3–7 days (Fig. 2l). With some *Phytophthora* species or inactive lesions, this might take up to 10 days (*see* **Notes** 7 and **12**).
7. Isolation from necrotic leaf spots/lesions (*see* Subheading 3.2.12).

3.2.10 Baiting from Bleeding Bark Cankers with Green Apples (Modified After [3])

1. Select green apples, rinse them with water, and disinfest their surface with 70% methylated spirit (denatured ethanol) before use. Use one apple per bleeding canker to be tested.
2. Cut with a knife a shallow flap from the opposite sides of the apple (Fig. 6g).
3. Place the washed bark lesion pieces underneath the apple flap.
4. Seal the flap with adhesive tape.
5. Label the apple with sample code, initials, and date.
6. Incubate the apple for 3–7 days at 20 °C.
7. Cut the apple to expose developing fruit rot lesions.
8. Isolation from the fruit rot lesions (*see* Subheading 3.2.13).

3.2.11 In Situ Baiting from Waterbodies [6–9, 22]

1. Cut fly mesh into 60 × 30-cm rectangles (Fig. 8a).
2. Fold the fly mesh, and use a stapler to construct a 25 × 30-cm baiting raft with lateral compartments for Styrofoam floaters (Fig. 8b).
3. Cut a Styrofoam board (ca 25–30 mm thick) into 25 cm long stripes, and put two stripes into the lateral compartments, and seal all but one narrow end with staples (Fig. 8b).
4. Place 10–20 non-wounded young leaves of susceptible tree species (*see* list in Subheading 2.2.1) into the baiting raft (Fig. 8c), and seal the edge with staples.
5. Moisten the leaves with distilled water, cover the baiting raft with moistened tissue paper, and place it into a plastic zip bag (Fig. 8d, e).

Fig. 8 In situ baiting techniques from waterbodies. (**a**) Cutting fly mesh into 60 × 30 cm rectangles. (**b**) Constructing with a stapler a 25 × 30 cm baiting raft with two lateral compartments for Styrofoam floaters cut from Styrofoam boards. (**c**) Baiting raft filled with young leaves from susceptible tree species as baits. (**d**) Moistening filled baiting rafts with distilled water. (**e**) Baiting raft with moistened paper tissue inside a plastic bag for the transport to the baiting site. (**f**) Floating the baiting raft in a small forest stream and fixing it with a plastic cord to a large branch. (**g**) A baiting raft in a marine lagoon fixed to a long iron stick (ca 1.8 m) pushed 0.5–1 m into the ground. (**h**) Baiting rafts transported back to the lab in plastic bags with moistened tissue paper. (**i**) Baiting leaves from a baiting raft showing numerous necrotic lesions. (**j**) Cutting necrotic leaf lesions with razor blades or scalpels into small segments and plating them onto selective PARPNH-agar

6. Transport the baiting rafts to the waterbodies in an air-conditioned car and/or in a cool bag. Avoid direct contact between the baiting rafts and cooling elements.
7. For placing the baiting raft into a waterbody, select a location where it cannot easily be spotted by people, e.g., in remote parts of the ecosystems, under overhanging branches or under bridges, etc. (*see* **Note 13**).
8. At the baiting site, take the baiting rafts out of the plastic bag, and put it to float at a place where water flow is calm so that zoospores can easily adhere to the baiting leaves, and fix the raft with a plastic cord to rocks, stems, or branches at the riverbank (Fig. 8f). Take care that the raft cannot be washed ashore during a water flush after a heavy rain.
9. In the tidal zone of rivers, in mangroves, and in lagoons, the cord should be fixed to long iron sticks (ca 1.8 m) pushed 0.5–1 m into the ground (Fig. 8g) at a place where the raft does not fall dry during low tide.
10. Collect the raft after 2–3 days (subtropical and tropical climate), 3–4 days (temperate and Mediterranean climate), or 4–7 days (subboreal to boreal climate).
11. Place the baiting raft in a big zip bag with moistened tissue paper (Fig. 8h), and transport it to the lab in an air-conditioned car and/or in a cool bag. Avoid direct contact between the baiting rafts and cooling elements.
12. Isolation from necrotic leaf spots/lesions (*see* Subheading 3.2.12).

3.2.12 Isolation from Baiting Leaves and Naturally Fallen Leaves Collected from Waterbodies or the Forest Ground [3, 9, 23]

1. Select leaves with brown to black necrotic spots/lesions (Figs. 2l, 5h–i, 7g, and 8i), and confirm the presence of characteristic *Phytophthora* sporangia under the light microscope at ×80 magnification to avoid plating of lesions infected by *Pythium*/*Phytopythium* spp. (*see* **Note 14**).
2. Wash the leaves briefly in distilled water and blot them dry on paper tissue.
3. Cut the necrotic spots/lesions with a razor blade or scalpel into small pieces (2 × 2–5 mm), dip them on tissue paper to remove excess water, and plate them onto PARPNH-agar (Figs. 4g, h and 8j). You can plate 10–15 pieces per 90 mm Petri dish. If leaves are completely brown, cut two transects and plate ca five pieces from each transect. Avoid plating lesions which develop primarily along leaf veins because these are usually caused by *Pythium*/*Phytopythium* spp.
4. Incubate the plates at 20 °C in the dark.
5. Plates are examined daily for outgrowing *Phytophthora* hyphae either under the stereomicroscope at ×20 magnification or in

Fig. 9 Subculturing *Phytophthora* from selective PARPNH-agar. (**a**) Needle with a thin spatula-like tip. (**b–d**) Subculturing of *Phytophthora* with a spatula-tip needle by holding the Petri dishes against a light. (**d**) To avoid shaking of the needle tip, the hand has to be stabilized by pressing the little finger against the edge of the Petri dish. (**e**) Subculturing of *Phytophthora* with a spatula-tip needle under the stereomicroscope at ×20 magnification. (**f**) Cutting a small agar piece with a spatula-tip needle from the edge of a *Phytophthora* colony growing from a plated leaf segment

the absence of a stereomicroscope by holding it against a light source (Fig. 9b–e). *Phytophthora* spp. usually grow from the plated leaf pieces after 24 to max. 48 h.

6. Subculture *Phytophthora* hyphae with a fine spatula-tip needle to V8A or CA (Fig. 9a–f) (*see* **Note 15**). If hyphae from both *Phytophthora* and faster-growing *Pythium*/*Phytopythium* spp.

are present, check under the light microscope at ×80 magnification for areas where only *Phytophthora* hyphae are present, and subculture them under the stereomicroscope onto V8A or CA (*see* **Note 16**).

7. The resulting *Phytophthora* colonies are examined daily to ensure that no *Pythium/Phytopythium*, fungi, or bacteria are contaminating the pure *Phytophthora* cultures. If contamination is observed the cultures need to be cleaned (*see* Subheading 3.3).

3.2.13 Isolation from Apple Baits [3, 19]

1. When a fruit rot is observed, cut the fruits with a sharp knife or scalpel either vertically or transversally to expose the colonized inner flesh (Figs. 3g and 6f). A brown firm fruit rot indicates a *Phytophthora* infection, while brighter soft lesions are usually caused by *Pythium/Phytopythium* spp.
2. Remove with a sterile scalpel pieces (3–5 mm) from the edge of the brown lesion, and plate up to seven pieces per 90 mm PARPNH-agar plate (Fig. 3h).
3. Incubate the plates at 20 °C in the dark.
4. Plates are examined daily under the stereomicroscope or in the absence of a stereomicroscope by holding it against a light source for outgrowing *Phytophthora* hyphae (Fig. 3h) which are immediately transferred to V8A or CA using a fine spatula-tip needle (Fig. 9a).
5. The resulting colonies are examined daily to ensure that no *Pythium* nor bacteria are contaminating the pure *Phytophthora* cultures. If contamination is observed, the cultures need to be cleaned (*see* Subheading 3.3).

3.3 Cleaning of Contaminated Phytophthora Cultures

3.3.1 Cleaning of Bacteria-Contaminated Phytophthora Cultures Using the Pancake Method (Modified After [20, 24])

1. Cut two 5–8 × 8–10-mm pieces from the margins of a contaminated culture.
2. Sterilize and cool down the roundish opposite end of a spoon or a scalpel handle, and move it along the edge of a V8A, CA, or PARPNH-agar plate, and lift the agar up (Fig. 10a, b).
3. Place the cut agar pieces under the lifted agar (Fig. 10b), and remove the spoon/scalpel handle slowly.
4. Incubate the plates upright at 20 °C in the dark until first hyphae appear at the surface after ca 24–48 h (Fig. 10c, d).
5. Cut with a fine spatula-tip needle very shallow pieces of agar from the surface (Fig. 10e–h). First perform the four shallow vertical cuts. Then cut horizontally a very thin piece by slowly moving the spatula end of the needle back and forth. Then lift the piece and transfer it to V8A or CA (*see* **Note 17**).

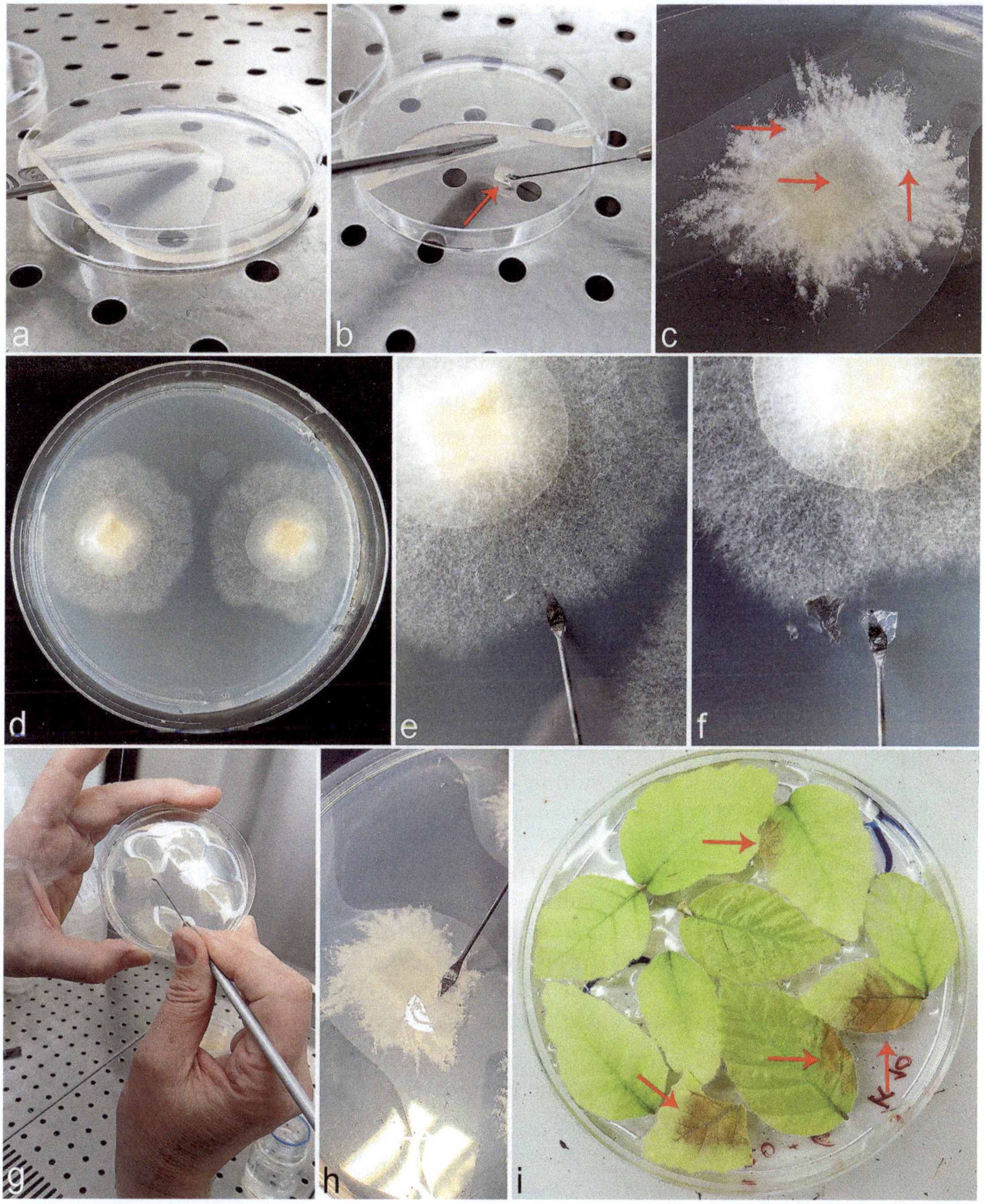

Fig. 10 Cleaning of contaminated *Phytophthora* cultures using the pancake method (**a–h**) or a leaf baiting technique (**i**). (**a**) Lifting the agar of a clean Petri dish with the sterilized and cooled-down, roundish end of a scalpel handle in a laminar flow. (**b**) Placing an agar square, cut from a bacteria-contaminated *Phytophthora* culture, underneath the lifted clean agar. (**c**) *Phytophthora* colony growing underneath the clean agar and first clean hyphae arriving at the upper surface of the agar (arrows). (**d**) Clean *Phytophthora* colonies on the upper surface of the agar developed from agar squares of a bacteria-contaminated culture placed underneath the agar. (**e–h**) Cutting with a fine spatula-tip needle very shallow pieces of agar from the clean colony on the surface of the agar. (**i**) Young soft leaves of *Fagus sylvatica* floating in a 90 mm Petri dish above submerged agar squares cut from a *Phytophthora plurivora* colony contaminated with a *Pythium* species. Note the necrotic lesions caused by zoospores of *P. plurivora*

3.3.2 Cleaning of Bacteria-Contaminated Phytophthora Cultures Using Green Apples (Modified After [3, 19])

1. Cut with a knife a shallow flap from the opposite sides of a green apple (Fig. 6g).
2. Cut squares of the contaminated culture (10 × 10–15 mm), open the apple flap, and place the agar square facing outward. The agar piece is sandwiched under the apple flap.
3. Seal the flap with adhesive tape. This method is very similar to the isolation of *Phytophthora* from fine roots or bleeding cankers using green apples (*see* Subheadings 3.2.5 and 3.2.10; Fig. 6g).
4. Label the apple with the sample code, initials, and date.
5. Incubate the apple for 3–4 days at 20 °C.
6. Cut the apple to expose the flesh, and isolate as described under Subheading 3.2.13 from the edge of the observed fruit rot lesions.

3.3.3 Purification of Phytophthora Cultures Mixed with Fungal Contaminants Using Selective PARPNH-Agar

1. Since the growth of fungi is inhibited or at least significantly reduced on PARPNH-agar, subculture agar squares from the mixed culture onto PARPNH-agar, and incubate the Petri dishes at 20 °C in the dark.
2. After 24–48 h subculture developing *Phytophthora* hyphae under the stereomicroscope onto V8A or CA using a fine spatula-tip needle.
3. The resulting colonies are examined daily to ensure that the *Phytophthora* cultures are pure.

3.3.4 Rebaiting from Phytophthora Cultures Mixed with Oomycete or Fungal Contaminants [3]

1. Two agar squares (ca 15 × 15 mm) from the central part of the mixed culture (in the outer parts may only be the faster-growing *Pythium/Phytopythium* or fungal species) are placed into 90 mm Petri dishes and flooded with distilled water to 1–2 cm depth (otherwise long *Pythium/Phytopythium* hyphae may directly reach the baits) (*see* **Note 18**).
2. After ca 6 h remove the agar squares with a pair of tweezers, exchange the distilled water, and submerge the agar squares again.
3. Float young unwounded leaflets of susceptible Fagaceae (*Fagus*, *Castanopsis*, *Lithocarpus*, *Quercus* spp.) trees above the agar pieces (Fig. 10i), and incubate the Petri dishes at 20 °C at natural daylight.
4. Perform isolations from lesions developing usually within 2–4 days on the baiting leaves as described under Subheading 3.2.12.

3.4 Storage of Phytophthora

3.4.1 In Dry Agar Media (Method A)

1. Isolates are subcultured onto V8A, CA, OA, or RA, in Petri dishes or on agar slants, sealed with Parafilm, and incubated at 20 °C for 3 weeks in the dark so that resting structures can mature.
2. Then the agar slants, the Petri dishes, or agar discs cut from the Petri dishes and placed into 2-mL screw cap tubes (ten discs per tube) are double-bagged and stored at 15–25 °C (tropical species) or 5–14 °C (species from other climatic regions).

3.4.2 In Dry Agar Media (Method B)

1. OA agar slants are prepared in 20–50 mL screw cap tubes.
2. One plug (5 mm diameter) of the *Phytophthora* isolate to be stored is placed on the agar slant, and the cap is slightly screwed.
3. The agar slant tube is incubated at 20 °C in the dark until the surface of the agar slant is fully colonized.
4. Screw the cap tightly, seal it with Parafilm, and incubate at 12–14 °C.

3.4.3 In Agar Media Under Liquid Solution

1. Isolates are subcultured onto V8A, CA, OA, or RA, in Petri dishes, sealed with Parafilm, and incubated at 20 °C for 3 weeks in the dark so that resting structures can mature.
2. Then agar discs are cut from the Petri dishes and placed into 2 mL screw cap tubes (ten discs per tube), and the tube is filled with either autoclaved soil filtrate (50–100 g forest soil flooded with 1 L forest soil, stirred, incubated for 24–48 h, and then filtered through a filter paper; [3]) or autoclaved distilled water after cooling down.
3. The tubes are double-bagged and stored at 15–25 °C (tropical species) or at 5–14 °C (species from other climatic regions).

3.5 Pathogenicity Tests

3.5.1 Soil Infestation Trial: Production of Plants [3, 11, 25, 26]

1. Mix peat, medium-sized vermiculite and sand (1:1:1, v/v/v), place it in autoclave sacs, and autoclave it for 2 h at 120 °C.
2. Fill the autoclaved, cooled-down soil substrate into either big plastic boxes with holes drilled into the bottom for drainage (Fig. 11a) or plant trays with 12 compartments, and insert tubes or flasks (6–8 per plastic box, depending on the size of the box; 1 per tray compartment) as placeholders for the inoculum to be added later (Fig. 11b–d) (*see* **Note 19**). Place the plastic box/tray into an even-sized or bigger plastic box without holes in the bottom (Fig. 11a, d, i–j).
3. Sow the seeds of the plant species to be tested (12–15 per plastic box or 1 per tray compartment (Fig. 11c, d)) and incubate the boxes at ca 18–22 °C (temperate plant species) or 22–25 °C (tropical plant species) in an air-conditioned greenhouse or in an incubator with a 12-h regime of plant grow lights for the whole period of the trial (Fig. 11h).

Fig. 11 Soil infestation trial. (**a**) Plastic boxes for soil infestation trial; inner box with holes drilled into the bottom for drainage of flooding water and even-sized outer box allowing flooding of the inner box. (**b**) Inner plastic box filled with autoclaved peat–vermiculite–sand mixture around tubes (arrow) inserted as placeholders for adding later the *Phytophthora* inoculum. (**c**) Inner plastic box with *Quercus suber* test plants grown from seeds sown between the placeholder tubes. (**d**) Twelve-compartment planting tray inside of a plastic box, with one test plant and one placeholder flask (arrow) per compartment. (**e**) Wide-mouthed glass bottle with mixture of fine and medium vermiculite. (**f**) Whitish *Phytophthora* mycelium after 4 weeks growth at 20 °C in the dark in a vermiculite–millet seed–vegetable juice broth inoculum mixed in a wide-mouthed glass bottle sealed with Parafilm and adhesive tape. (**g**, **h**) Whitish *Phytophthora* mycelium after 4 weeks growth at 20 °C

3.5.2 Soil Infestation Trial: Production of Inoculum [3, 11, 25, 26]

1. Thoroughly mix 250 mL (500 mL) of fine or medium vermiculite (Fig. 11e) with 20 mL (40 mL) of whole oat grains or millet seeds in a 500 mL (1000 mL) wide-mouthed glass bottle with lid or in an autoclave sac; moisten the mixture with 175 mL (350 mL) of vegetable juice broth (200 mL juice, 800 mL distilled water amended with 3 g $CaCO_3$).
2. Close the bottle with a lid and autoclave tape or the autoclave sac by knotting the open ends and autoclave it.
3. After cooling down in the laminar flow overnight, inoculate the mixture with 15–25 agar discs cut from an actively growing V8A culture of the respective *Phytophthora* isolate.
4. Mix the agar discs thoroughly with the substrate mixture using a sterilized glass stick; close the bottle with the lid, Parafilm, and tape (Fig. 11f) or the autoclave sac by knotting the ends; and place the bottle/autoclave sac in a plastic bag, seal it, and incubate it for 4–5 weeks at 20 °C in the dark. A well-developed inoculum is characterized by the presence of whitish aerial mycelium (Fig. 11f–h).

3.5.3 Soil Infestation Trial: Infestation Technique and Flooding Regimes [3, 11, 25, 26]

1. When the plants are at least 3 months old, remove the placeholders from the soil substrate in the plastic box or tray, and fill the holes with the 4–5 week old inoculum (ca 40 cm^3 per plant). Before use, the inoculum has to be rinsed in a sieve with distilled water to remove excess nutrients and thoroughly mixed with a sterile glass stick. Controls receive only rinsed noninfested vermiculite/oat or millet seed/vegetable juice broth mixture at the same rate.
2. Fill the plastic box slowly with distilled water until the water is standing ca 1 cm above the soil (Fig. 12a).
3. After 72 h drain the flooding water by transferring the inner plastic box or the tray into a third plastic box (Fig. 11j). Afterward transfer them back to their outer plastic box.
4. Sterilize after each flooding cycle the drained flooding water with bleach (Fig. 12b, c) or by autoclaving.
5. The 72-h flooding periods are repeated every 3 weeks for the whole duration of the trial.

Fig. 11 (continued) in the dark in a vermiculite–millet seed–vegetable juice broth inoculum mixed in a sealed autoclave sac. (**i**) Soil infestation test with numerous *Phytophthora*–host tree combinations in an air-conditioned room at 20 °C with a 12-h regime of artificial plant grow lights; note that the inner boxes are placed into even-sized outer boxes. (**j**) The three plastic boxes of a single host–*Phytophthora* combination: outer box to allow the flooding of the inner box containing the plants and the third bigger box for draining the flooding water (arrow) after the 72-h flooding period

Fig. 12 Soil infestation trial. (**a**) Rebaiting of the inoculated *Phytophthora* species during a flooding period using young *Quercus suber* leaves as baits. (**b**, **c**) Sterilizing the drained flooding water with bleach. (**d**) Young plants of *Castanea sativa* showing wilting, mortality, and chlorosis at the end of a 5-month soil infestation trial. (**e**) At the end of a 5-month soil infestation trial, the root systems of young *Quercus robur* plants are intertwined and need to be carefully separated and cleaned from soil particles. (**f**) Careful washing of a root system in a tray with distilled water to remove peat and vermiculite particles. (**g**) Root system of a test plant spread out on a white carton for estimating the proportion of necrotic and healthy roots; an extensive lesion has girdled the stem and the tap root so that the whole root system has to be calculated as dead. (**h**) Girdling lesion of a tap root; the arrow shows the border between the healthy brown and the dead black bark tissue. (**i**) Active tongue-

6. At the last flooding cycle, reisolations from the soil should be performed by floating young leaflets of susceptible tree species (e.g., *Fagus sylvatica*, *Quercus robur*, *or Q. suber*) as baits on the surface of the flooding water (Fig. 12a) in order to test whether the respective *Phytophthora* species was still active.

3.5.4 Soil Infestation Trial: Assessment [3, 11, 25, 26]

1. During the trial aboveground symptoms (chlorosis, wilting, collar rot, mortality; Fig. 12d) are continuously monitored and recorded. In case that collar rot symptoms develop, reisolations from necrotic bark are immediately performed (Fig. 12j) before the plant dies and dries out to confirm the inoculated *Phytophthora* species as causal agent.
2. At the end of the trial after usually 3–6 months (with older plants or less aggressive *Phytophthora* species, trials can run up to 12 months), the plants are carefully removed from the boxes/trays (Fig. 12e). Root systems are gently shaken to remove loose peat, vermiculite, and sand particles and then thoroughly but gently washed in a tray with distilled water (Fig. 12f) (*see* **Note 20**).
3. For each seedling the proportion of healthy and dead roots is estimated visually after spreading the roots uniformly on white trays or carton (Fig. 12g) (*see* **Note 21**). The whole root system below a girdling root lesion (Fig. 12h, i) or collar lesion (Fig. 12j) has to be rated as dead regardless of the condition of these roots.
4. The root systems can also be scanned and analyzed with a root measurement software (Fig. 12k) (*see* **Note 22**).
5. Record specific symptoms like root and collar rot lesions and tap root dieback.
6. Perform reisolations from necrotic roots and collar rot lesions (Fig. 12h–j) as described under Subheading 3.2.2 to confirm the inoculated *Phytophthora* species as cause of the disease symptoms, thus fulfilling Koch's postulates.
7. Dry the root systems and shoots for 72 h at 65 °C, and measure the dry weight of small woody roots (diam 2–10 mm), fine roots (diam <2 mm), and shoots.
8. Sterilize all soil substrates and plant parts by autoclaving for 2 h at 120 °C.

Fig. 12 (continued) shaped front of a lesion on a tap root after removing the outer bark with a scalpel. (**j**) Extensive bark lesion which progressed from the collar up the stem, with tongue-shaped brownish lesion front in the inner bark (arrow) and dark discoloration of the necrotic outer bark. (**k**) Scanned roots of one tree seedling from a soil infestation test for biometric root analysis

3.5.5 Under-Bark Inoculation Trial [3, 27]

1. Select the area of the stem (or the log of a mature tree), branch, or shoot where the inoculation is going to take place, and disinfect it with 70% methylated spirit (denatured ethanol).
2. With a disinfected cork borer (5–6 mm), remove a bark disc aseptically (Fig. 13a).
3. With a cork borer (5–6 mm), cut a disc of the V8A from the fresh culture of the *Phytophthora* species to be inoculated. Controls are inoculated with sterile V8A.
4. Place the agar disc facing down onto the place where the bark has been removed (Fig. 13b). The disc should fit nicely into the place and can be covered by the bark disc although this is optional.
5. Alternatively, the bark can be opened by performing with a sterile scalpel a longitudinal cut into which the agar disc is placed (Fig. 13f).
6. Moisten sterile cotton wool with sterile distilled water and cover or wrap it around the inoculation point (Fig. 13c).
7. Wrap the cotton wool with Parafilm (Fig. 13d).
8. Wrap the Parafilm with aluminum foil (Fig. 13e).
9. Incubate in an air-conditioned greenhouse or room (plants, logs) (Fig. 13g, h) or an incubator (shoots, branches) at the conditions required (22–25 °C for subtropical/tropical *Phytophthora* species; 18–20 °C for species from other climatic regions).
10. Assess lesions (Fig. 13i) after removal of the outer bark (Fig. 13j–l) by measuring the length of the lesion or by calculating the area of the lesion (lesion lengths are assessed after 10–14 days for twig inoculations or after 1–3 months for log inoculations).
11. Reisolations are made randomly by plating necrotic tissue (Fig. 13j–l) onto selective PARPNH-agar (*see* Subheading 3.2.3, **step 1**) to confirm that the *Phytophthora* species inoculated is the actual cause of the lesions observed, thus fulfilling Koch's postulates.
12. Sterilize all plant parts by autoclaving for 2 h at 120 °C.

4 Notes

1. Avoid collecting samples from dying or dead trees where high inoculum levels of antagonistic secondary pathogens and saprophytes can make the detection of the primary *Phytophthora* species very difficult.
2. Long-term storage at low temperatures can cause dormancy, and it can take too long to break the dormancy during the short period of a baiting test. Therefore, it is recommended to perform baiting tests from soil samples within 4 weeks of sampling.

Fig. 13 Under-bark inoculation trial. (**a**) Removal of a bark disc (arrow) with a sterile cork borer. (**b**) Placing an even-sized agar disc, cut with a cork borer from a growing *Phytophthora* culture, onto the exposed cambium/xylem area. (**c**) Covering the inoculation point with moistened sterilized cotton. (**d**) Wrapping the cotton with Parafilm. (**e**) Wrapping the Parafilm with aluminum foil. (**f**) Inserting an agar square, cut from a growing *Phytophthora* culture, into a bark slit cut with a sterile sharp scalpel. (**g**) Under-bark inoculation trial with numerous *Phytophthora–Alnus glutinosa* combinations in an air-conditioned greenhouse with 20 °C and natural daylight. (**h**) Same under-bark inoculation trial as in (**g**) after 14 days; note that many plants with girdling bark lesions show wilting and mortality. (**i**) *Alnus glutinosa* stems showing black girdling bark lesion 14 days after under-bark inoculation with *Phytophthora* × *cambivora*. (**j**) *Alnus glutinosa* stems with extensive

However, in soil samples from nurseries, *Phytophthora* species are often suppressed by fungicides and fungi-static chemicals. Therefore, storage of nursery soil samples for 3–6 months can facilitate the isolation of *Phytophthora* since the retention time of most chemicals is usually shorter.

3. Avoid removing the outer bark from the sample panel because this will initiate oxidation of the phloem and potentially compromise isolation of *Phytophthora*. This is especially important if the bark samples are transported between moist paper tissue in sample bags instead of water bottles.
4. Autoclaved filter holders and filters should be prepared in advance in the laboratory according to the number of sampling sites.
5. The water must be as clear as possible to reduce direct contact of antagonistic fungi, *Pythium/Phytopythium*, and bacteria with the baiting leaves. This might not be perfectly achievable with some soil types, in particular organic and silty soils. With organic soils cover the whole soil surface with a fly mesh, and pin it to the ground with small sterile stones.
6. The long-term experience of the authors with baiting tests in a wide range of ecosystems in different countries and continents showed that best isolation results are achieved with young leaves of *Fagus sylvatica*, *F. crenata*, *Quercus glauca*, *Q. ilex*, *Q. robur*, *Q. suber*, and *Q. variabilis*. The use of *Rhododendron* leaves for soil baiting tests is widespread but has the disadvantages that necrotic spots are not clearly visible, sporangia cannot be observed under the microscope, and only a limited range of *Phytophthora* species can be isolated.
7. Necrotic lesions that develop on the baiting leaves primarily along the leaf veins are usually indicating infections by *Pythium/Phytopythium* spp.
8. If only *Pythium/Phytopythium* species can be isolated, discard the baiting water, and dry the soil sample at 20 °C before incubating it for 1–2 months at 20 °C. Then moisten the soil with distilled water for 48 h and repeat the baiting test. The principle is that the dry conditions often eliminate *Pythium/Phytopythium* species or at least inactivate them considerably stronger than *Phytophthora* species so that the latter are having an advantage in a second baiting test.

Fig. 13 (continued) girdling bark lesions 14 days after inoculation with *Phytophthora* × *alni*; reisolations should be performed at the orange-brown lesion fronts (arrows); note that several stems formed adventitious shoots below the lesions. (**k, l**) Removal of the outer bark with a sharp sterile scalpel shows brown necrotic inner bark lesions on *A. glutinosa* stems 14 days after inoculation; reisolations should be performed at the orange-brown lesion fronts (arrows)

9. *Phytophthora*-infected fine roots become very quickly detached. Therefore, isolation frequencies of *Phytophthora* are much higher from detached fine roots in the soil than from fine roots which are still attached to the root system.
10. Due to abundant inoculum of *Pythium/Phytopythium* species in waterbodies and, hence, on the filters, *Phytophthora* hyphae growing from the plated filter segments will most likely be mixed with faster-growing *Pythium/Phytopythium* hyphae making it difficult to obtain pure *Phytophthora* cultures. Therefore, the use of green apples for isolations of *Phytophthora* from water filters is preferable.
11. When using a razor blade or scalpel, always cut away from your body.
12. From older, dark-brown inactive lesions, baiting is usually giving significantly higher isolation frequencies than direct plating.
13. Due to curiosity people tend to remove the baiting rafts or damage them.
14. Plating from necrotic spots/lesions can also be done without prior examination of the sporangia formed on the lesions. However, the isolation frequency of *Phytophthora* spp. might be lower due to *Pythium/Phytopythium* infections.
15. When subculturing *Phytophthora* from PARPNH-agar by holding the Petri dish against a light source, the hand has to be stabilized by pressing the little finger against the edge of the Petri dish in order to avoid shaking of the needle tip.
16. *Pythium/Phytopythium* spp. are often growing submerged in the agar below *Phytophthora* colonies. Therefore, focus with the stereo- and light microscopes deeper into the agar to ensure that only *Phytophthora* hyphae are present. In case that a very slow-growing *Phytophthora* is mixed with *Pythium/Phytopythium* and the *Phytophthora* colonies are too small for subculturing, remove with a fine spatula-tip needle all *Pythium/Phytopythium* hyphae, and check the plates after 6–12 h again to subculture *Phytophthora*. If this is not successful, cut out the mixed colony from the PARPNH-agar, and use it for baiting (*see* Subheading 3.3.4).
17. Bacteria cannot actively grow through the agar, and follow the hyphae passively through the agar to the surface. Therefore, the first hyphae appearing at the surface are clean. Do not cut with the needle through the agar to the bottom to avoid contact with the bacteria.
18. The principle is that *Phytophthora* species are usually faster in producing sporangia and infecting baiting leaves than *Pythium/Phytopythium* species. Furthermore, every zoospore causes a distinct necrotic lesion. Plating all necrotic lesions

onto selective PARPNH-agar increases the probability to plate some lesions that were only caused by the *Phytophthora* isolate and, hence, obtain a pure *Phytophthora* culture.

19. Growing all plants of a host–isolate combination in a large single box does not fulfill the requirement of true replicates like separate compartments in a tray. However, we prefer this trial design with pseudo-replicates because it allows uniform inoculum distribution and undisturbed root development which facilitates the root assessment at the end of the trial.
20. Most likely roots will be entangled with other roots and vermiculite or peat from the substrate. Avoid breaking and detaching of roots during washing. Submersing the root systems overnight in a bucket with distilled water before washing to soak and loosen soil particles is recommended.
21. Etching the tray or carton with 2 × 2-cm rectangles facilitates the estimation of proportions of healthy and dead roots.
22. A biometric root analysis of scanned pictures cannot reliably differentiate between necrotic dead roots and roots which are healthy but dark discolored due to the surrounding organic substrate. Moreover, since root systems react to infections by producing replacement roots, in trials which run shorter than the time needed for dead roots to become detached, the total root length and number of root tips can be higher in an infected plant than in a healthy control tree. Therefore, accurate assessments of the proportion of healthy and dead roots are crucial.

Acknowledgments

This work was supported by the European Regional Development Fund financing the Project *Phytophthora* Research Centre Reg. No. CZ.02.1.01/0.0/0.0/15_003/0000453 and the Tree Health & Biosecurity Research Programme of the Science and Innovation Strategy for Forestry in Great Britain.

References

1. Erwin DC, Ribeiro OK (eds) (1996) Phytophthora diseases worldwide. APS Press, St. Paul
2. Jung T, Pérez-Sierra A, Durán A et al (2018a) Canker and decline diseases caused by soil- and airborne *Phytophthora* species in natural and semi–natural ecosystems. Persoonia 40: 182–220
3. Jung T, Blaschke H, Neumann P (1996) Isolation, identification and pathogenicity of *Phytophthora* species from declining oak stands. Eur J For Pathol 26:253–272
4. Jung T, Colquhoun IJ, Hardy GESJ (2013) New insights into the survival strategy of the invasive soilborne pathogen *Phytophthora cinnamomi* in different natural ecosystems in Western Australia. For Pathol 43:266–288
5. Jung T, Orlikowski L, Henricot B et al (2016) Widespread *Phytophthora* infestations in European nurseries put forest, semi-natural

and horticultural ecosystems at high risk of Phytophthora diseases. For Pathol 46: 134–163

6. Jung T, Chang TT, Bakonyi J et al (2017a) Diversity of *Phytophthora* species in natural ecosystems of Taiwan and association with disease symptoms. Plant Pathol 66:194–211
7. Jung T, Durán A, Sanfuentes von Stowasser E et al (2018b) Diversity of *Phytophthora* species in Valdivian rainforests and association with severe dieback symptoms. For Pathol 48: e12443
8. Jung T, La Spada F, Pane A et al (2019) Diversity and distribution of *Phytophthora* species in protected natural areas in Sicily. Forests 10:259
9. Jung T, Scanu B, Brasier CM et al (2020) A survey in natural forest ecosystems of Vietnam reveals high diversity of both new and described *Phytophthora* taxa including *P. ramorum*. Forests 11:93
10. Jung T (2009) Beech decline in Central Europe driven by the interaction between *Phytophthora* infections and climatic extremes. For Pathol 39:73–94
11. Pérez-Sierra A, López-García C, León M et al (2013) Previously unrecorded low temperature *Phytophthora* species associated with *Quercus* decline in a Mediterranean forest in Eastern Spain. For Pathol 43:331–339
12. Jung T, Hansen EM, Winton L et al (2002) Three new species of *Phytophthora* from European oak forests. Mycol Res 106:397–411
13. Brasier CM (1967) Physiology of reproduction in *Phytophthora*. PhD thesis, University of Hull, UK, 220 pp
14. Jung T, Blaschke M (2004) Phytophthora root and collar rot of alders in Bavaria: distribution, modes of spread and possible management strategies. Plant Pathol 53:197–208
15. Hong C, Richardson PA et al (2002) Comparison of membrane filters as a tool for isolating pythiaceous species from irrigation water. Phytopathology 92:610–616
16. Scibetta S, Schena L, Chimento A et al (2012) A molecular method to assess *Phytophthora* diversity in environmental samples. J Microbiol Methods 88:356–368
17. Català S, Pérez-Sierra A, Abad-Campos P (2015) The use of genus-specific amplicon pyrosequencing to assess *Phytophthora* species diversity using eDNA from soil and water in northern Spain. PLoS One 10:1–14
18. Jung T, Horta Jung M, Webber JF et al (2021) The destructive tree pathogen *Phytophthora ramorum* originates from the Laurosilva forests of East Asia. J Fungi (Basel) 7:226
19. Campbell WA (1949) A method of isolation *Phytophthora cinnamomi* directly from soil. Plant Dis Rep 33:134–135
20. Ribeiro OK (ed) (1978) A source book of the genus *Phytophthora*. Cramer, Vaduz
21. Corcobado T, Cech TL, Brandstetter M et al (2020) Decline of European beech in Austria: involvement of *Phytophthora* spp. and contributing biotic and abiotic factors. Forests 11: 895
22. Hüberli D, Hardy GESJ, White D et al (2013) Fishing for *Phytophthora* from Western Australia's waterways: a distribution and diversity survey. Australas Plant Pathol 42:251–260
23. Jung T, Blaschke H, Osswald W (2000) Involvement of soilborne *Phytophthora* species in Central European oak decline and the effect of site factors on the disease. Plant Pathol 49: 706–718
24. Sleeth B (1945) Agar medium and technique for isolating *Pythium* free of bacteria. Phytopathology 35:1030–1031
25. Jung T, Horta Jung M, Scanu B et al (2017) Six new *Phytophthora* species from ITS Clade 7a including two sexually functional heterothallic hybrid species detected in natural ecosystems in Taiwan. Persoonia 38:100–135
26. Jönsson U, Jung T, Rosengren U et al (2003) Pathogenicity of Swedish isolates of *Phytophthora quercina* to *Quercus robur* in two different soils. New Phytol 158:355–364
27. Pérez-Sierra A, León M, Alvarez LA et al (2010) Outbreak of a new *Phytophthora* sp. associated with severe decline of almond trees in eastern Spain. Plant Dis 94:534–541

Chapter 3

Field and Laboratory Procedures for *Fusarium circinatum* Identification and Diagnosis

Cristina Zamora-Ballesteros, Reinaldo Pire, and Julio Javier Diez

Abstract

Fusarium circinatum is a serious invasive pathogen affecting conifers and causes the disease commonly known as pine pitch canker. Due to the outbreak in European countries, regulations stipulate that Member States must conduct annual official surveys for the fungus on their territory and report the results to the European Commission. Here, we describe the field and laboratory protocols used for the identification and diagnostic of the pathogen.

Key words Agarose gel electrophoresis, Identification, Internal transcribes spacer of the rRNA, PCR amplification, Pine pitch canker, Real-time PCR, Sampling, Sanger sequencing

1 Introduction

Fusarium circinatum Nirenberg & O'Donnell is a quarantine organism included in the EPPO (European and Mediterranean Plant Protection Organization) A2 list and regulated in the European Union (Commission Decision 2007/433/EC). Over 60 *Pinus* species are susceptible to the pathogen, which induces resinous bleeding cankers on trunks and branches of trees, a disease commonly referred to as pine pitch canker (PPC) [1, 2]. These damages lead to trunk deformations (Fig. 1) reducing drastically the value of timber that becomes useless for wood industry. On seedlings, *F. circinatum* causes pre- and post-damping off, which results in significant economic losses in forest nurseries, where mortality rates can reach 90% [3, 4]. The pathogen is present in many countries of North and South Hemisphere, where causes serious problems in nurseries, pine plantations, and natural forests [5]. Since no effective field intervention measures for the PPC control or eradication are available, it is now extremely important to prevent its introduction into countries hitherto free of the pathogen. Hence, continuous surveillance of reproductive material,

Nicola Luchi (ed.), *Plant Pathology: Methods and Protocols*, Methods in Molecular Biology, vol. 2536,
https://doi.org/10.1007/978-1-0716-2517-0_3,

Fig. 1 (**a**) Dieback by *Fusarium circinatum* in a *Pinus radiata* tree. (**b**) Pitch canker on *P. radiata* tree. (**c**) Cutting a log with a chainsaw showing dark areas due to the *F. circinatum* growth. (**d**) Trunk deformation due to *F. circinatum* infection. (**e**) Micelia of *F. circinatum* showing typical hyphae and Polyphialide. (**f**) Polyphialide of *F. circinatum* containing non-septate obovoid microconidia

nurseries and plantations prompt control measures, and realistic contingency plans are of primary importance [6]. Visual inspections are not sufficient for a preliminary diagnosis due to the endophytic behavior of the pathogen under certain conditions [7]. In addition, some conifer pathogens cause symptoms very similar to those caused by *F. circinatum* in seedlings (e.g., *Phytophthora* spp., *Pythium* spp.) and in mature trees (e.g., *Diplodia sapinea*, *Caliciopsis pinea*) [8]. Therefore, sampling for an early detection of *F. circinatum* is essential to prevent its local establishment, spread, and movement into disease-free areas [9]. Morphological and molecular identification approaches must also consider the similar features shared by *F. circinatum* with close species belonging to its same American clade, the *Fusarium fujikuroi* species complex (FFSC) [10].

Sampling methods for *F. circinatum* detection and identification strategies are summarized in these protocols. Sampling protocols for seeds, seedlings, twigs, branches, shoots, soil samples, spore traps, and insects from different studies are detailed. Likewise, methodology for morphological and molecular identification is herein presented. These include conventional PCR using a multiplex approach, as well as the description of some real-time PCR protocols with a target-specific region located in the intergenic spacer region.

2 Materials

2.1 Soil or Roots Collection

1. Sharpshooter shovel or auger.
2. Bucket.
3. Gloves.
4. Plastic sealed bags.
5. Paper bags.
6. Permanent marker.

2.2 Collection of Branches, Needles, Cones, Canker, or Seeds

1. Gloves.
2. Telescopic pruner.
3. Razor blade.
4. Plastic sealed bags.
5. Paper bags.
6. Permanent marker.

2.3 Spore Trapping

1. Rotating arm spore trap (e.g., ROTTRAP 120 Miloň Dvořák, Boršov nad Vltavou, Czech Republic) with a 12 V/19 Ah battery.
2. Wooden rods (pieces of matches).
3. Double-sided non-woven tape.
4. Disposable polypropylene 2-mL microtubes.

2.4 Insect Vector's Trapping

1. Baiting logs.
2. Slit traps, cross traps, sliding, or multiple funnel traps.
3. The substances employed as baits can be either (E)-pityol, ethanol and α-pinene, ipsdienol pheromone, or Galloprotec Pack.

2.5 Isolation and Identification of F. circinatum

1. 90-mm Petri plates.
2. Sterile lab paper.
3. Sterilized tweezers.
4. Sterilized loop.
5. Disposable polypropylene 1.5-mL microcentrifuge tubes.
6. 1% (v/v) Tween-80.
7. Parafilm or any other transparent seal.
8. Microscopy slides and cover slips.
9. Ultrasonic bath (e.g., Ultrasons J. P. Selecta).
10. A basic light microscope with 10×, 20×, and 40× objective.
11. Different media must be prepared for isolation and identification of *F. circinatum*. All media are autoclaved at 121 °C for 40 min, cooled, and poured into 90-mm Petri plates, unless otherwise stated. Antibiotic solutions should be added once the media has reached 45 °C by passing the solution through a 0.22 -μm filter for sterilization. The following media were prepared:
    - Potato dextrose agar (PDA): Dissolve 20 g dextrose, 20 g agar in 1 L broth from 250 g white potatoes mixed with distilled water (a commercial preparation of PDA can also be used).
    - Potato dextrose agar supplemented with streptomycin (PDAS): PDA ingredients supplemented with 0.5 g streptomycin in 1 L.
    - Potato dextrose agar supplemented dichloran, streptomycin, and pentachloronitrobenzene (PDAspd): PDA ingredients, 2 mg dichloran, 0.5 g streptomycin, 0.2 g penta chloronitrobenzene (PCNB), distilled water to 1 L.

- Dichloran chloramphenicol peptone agar (DCPA): 15 g peptone, 1 g KH_2PO_4, 0.5 g $MgSO_4 \cdot 7H_2O$, 20 g agar, 2 mg dichloran (2,6-dichloro-nitroanilin), 0.2 g chloramphenicol, 0.05 g violet crystal, distilled water to 1 L.
- Komada's medium: 1 g KH_2PO_4, 0.5 g KCl, 0.5 g $MgSO_4 \cdot 7H_2O$, 10 mg Fe-Na-EDTA, 2 g L-asparagine, 20 g D-galactose, 15 g agar, 0.5 g PCNB, 0.25 g oxgall, 0.2 g chloramphenicol (or 0.15 g streptomycin), distilled water to 1 L (pH 3.8).
- *Fusarium* selective medium (FSM): 1 g KH_2PO_4, 0.5 g $MgSO_4 \cdot 7H_2O$, 0.2 g PCNB, 10 mL of a streptomycin sulfate stock solution (30 mg/mL), 15 g peptone, 20 g agar, distilled water to 1 L.
- Modified *Fusarium* selective medium (Modified FSM): The same as FSM, but containing 3 g streptomycin, 50 mg Rose Bengal, 1 g PCNB, and 1 g neomycin.
- Water agar (WA): Dissolve 20 g agar in 1 L distilled water.
- Spezieller-Nährstoffarmer agar (SNA) medium: 1 g KH_2PO_4, 1 g KNO_3, 0.5 g $MgSO_4 \cdot 7H_2O$, 0.5 g KCl, 0.3 g D-glucose, 0.2 g sucrose, 20 g agar, distilled water to 1 L.
- Complete medium (CM): 30 g sucrose, 2 g $NaNO_3$, 2.5 g N-ZAmine, 1 g yeast extract, 10 mL "vitamin stock" solution, 1 L distilled water. Vitamin stock: 4 g inositol, 200 mg Ca pantothenate, 200 mg chlorine-Cl, 100 mg thiamine, 75 mg pyridoxine, 75 mg nicotinamide, 50 mg ascorbic acid, 30 mg riboflavin, 5 mg *p*-aminobenzoic acid, 5 mg folic acid, 5 mg biotin, 50:50 ethanol:H_2O to 1 L.
- Carrot agar (CA): 400 g peeled and diced carrots, 20 g agar, distilled water to 1 L.
- Malt extract agar (MEA): 33 g malt extract, 20 g agar, distilled water to 1 L.

2.6 DNA Extraction from Pure Culture and Spores

1. Petri plates with cellophane membrane-covered PDA medium.
2. Fresh *F. circinatum* mycelium or spores.
3. Quartz sand.
4. Mixer Mill (e.g., Retsch MM400).
5. Vortexer.
6. Heat block.
7. Microcentrifuge.
8. Disposable polypropylene microcentrifuge tubes: 1.5 and 2 mL.

9. Lysis buffer (made freshly approximately every 2 weeks): 50 mM Tris/HCl, pH 7.2; 50 mM EDTA; 3% SDS; 1% β-mercaptoethanol.
10. Phenol:chloroform:isoamyl alcohol (25:24:1).
11. Chloroform:isoamyl alcohol (24:1).
12. Tris–EDTA (TE) buffer: 10 mM Tris–HCl pH 7.5 (25 °C), 0.1 mM EDTA.
13. Polyethylene glycol (PEG) solution: 0.6 vol. of a solution containing 20% (w/v) PEG 6000 and 2.5 M NaCl. Store at room temperature.
14. Ice-cold ethanol (70%).
15. 1.5-mL microcentrifuge tubes.
16. Micropipettes and pipet tips.
17. Disposable gloves.

2.7 Multiplex Conventional PCR

1. Disposable polypropylene DNase/RNase-free microcentrifuge tubes: 0.2 mL.
2. Molecular water.
3. dNTPs—a mixture of dATP, dCTP, dGTP, and dTTP (5 mM), stored at −20 °C.
4. Oligonucleotide primer pairs based on Ramsfield et al. [11] 10Fcirc (forward primer: 5′-AGGCCGCTTAAACCTGCAAG-3′; reverse primer: 5′- TAGGCCGCTTAAGAATATTGC -3′; 234 bp amplicon size) and 30Fcirc (forward primer: 5′-AGA GAAGTTGCCCGACAAGTTG-3′; reverse primer: 5′-TAGC TTCCAGCGTTCGTCCAC -3′; 422 bp amplicon size), re-suspended each primer to a concentration of 100 μM using sterile distilled water, and stored at −20 °C.
5. DNA polymerase (e.g., *Taq* DNA polymerase Roche Applied Science) and reaction buffer supplied by manufacturer (Roche buffer: 10 mm Tris–HCl, 1.5 mm $MgCl_2$, 50 mm KCl, pH 8.3). These reagents are stored at −20 °C.
6. *F. circinatum* DNA.
7. Thermal cycler (e.g., Applied Biosystems Veriti™).
8. Standard equipment for agarose gel electrophoresis.
9. U.V. transilluminator and camera suitable for photographing agarose gels.
10. Tris–acetate–EDTA (TAE) buffer: 40 mM, Tris–acetate pH 8.0, 1 mM EDTA.
11. Agarose D1 Low EEo.
12. Safe fluorescent nucleic acid dye.

13. Gel-loading dye 6×—30% (w/v) glycerol, 50 mM EDTA, 2.5 mg/mL bromophenol blue.
14. Markers for nucleic acid size determination.
15. Micropipettes and pipet tips.
16. Disposable gloves.

2.8 Real-Time PCR

1. Disposable polypropylene 96-well reaction plates (e.g., Applied Biosystems™ MicroAmp™ Optical plate).
2. Molecular water.
3. SYBR Green master mix (e.g., Applied Biosystems™ PowerUp™ SYBR™ Green Master Mix, cat. no. A25779: SYBR Green dye, Dual-Lock *Taq* DNA Polymerase, dNTPs with dUTP/dTTP blend, heat-labile UDG, ROX passive reference dye, and optimized buffer components). This reagent is stored at 2–8 °C.
4. Oligonucleotide primer pairs based on Schweigkofler et al. [12] CIRC1A (5′-CTTGGCTCGAGAAGGG-3′) and CIRC4A (5′-ACCTACCCTACACCTCTCACT-3′) with 360 bp of amplicon size, re-suspended to a concentration of 100 μM using sterile distilled water, and stored at −20 °C.
5. DNA from *F. circinatum*.
6. Real-time PCR instrument (e.g., Applied Biosystems™ QuantStudio™ 6 Flex).
7. Standard 96-well plate centrifuge.
8. QuantStudio™ Real-Time PCR Software.
9. Micropipettes and pipet tips.
10. Disposable gloves.

2.9 Dual-Labelled Probe Real-Time PCR

1. Disposable polypropylene 96-well reaction plates (e.g., Applied Biosystems™ MicroAmp™ Optical plate).
2. Molecular water.
3. Hot Start *Taq* DNA Polymerase (e.g., New England Biolabs, cat. no. M0495) and reaction buffer supplied by manufacturer (10× Standard *Taq* Reaction Buffer: Mg^{++} concentration of 15 mM). These reagents are stored at −20 °C.
4. dNTPs—a mixture of dATP, dCTP, dGTP, and dTTP (5 mM), stored at −20 °C.

5. Oligonucleotide primer pairs, based on Ioos et al. [13], FCIR-F (5′-TCGATGTGTCGTCTCTGGAC-3′) and FCIR-R (5-′-CGATCCTCAAATCGACCAAGA-3′), and probe FCIR-P (5′-[6-FAM]-CGAGTCTGGCGGGACTTTGTGC-[BHQ1]-3′) with 146 bp of amplicon size, re-suspended to a concentration of 100 μM using sterile distilled water, and stored at −20 °C.
6. Same Materials of Subheading 2.8 from **items 5** to **8**.

2.10 Sanger Sequencing and Interpretation

1. Arranged Sanger sequencing service for PCR products.
2. Freeware programs such as BioEdit can be used for bioinformatics analysis.

2.11 EF1 Gene Partial Sequencing

1. Disposable polypropylene DNase/RNase-free microcentrifuge tubes: 0.2 mL conical.
2. Molecular water.
3. Oligonucleotide primer pair, based on the Standard on phytosanitary measures (PM) of *F. circinatum* [14], EF1 (5-′-ATGGGTAAGGARGACAAGAC-3′) and EF2 (5′-GGARGTACCAGTSATCATGTT-3′) with 640 bp of amplicon size, re-suspended to a concentration of 100 μM using sterile distilled water and stored at −20 °C.
4. DNA polymerase with proofreading ability (e.g., Phusion® High-Fidelity DNA Polymerase New England Biolabs, cat. no. M0530) and reaction buffer supplied by manufacturer (5× Phusion HF buffer: Mg^{++} concentration of 15 mM). These reagents are stored at −20 °C.
5. Same Materials of Subheading 2.7 from **items 6** to **16**.

2.12 Maintenance and Conservation of F. circinatum *Isolates*

1. Freeze dryer.
2. Liquid nitrogen.
3. Carnation leaf agar (CLA): 20 g agar, distilled water to 1 L overlaid with sterile pieces of carnation leaves.
4. 90-mm Petri dishes.
5. Sterile vials.
6. The following materials can also be used to preserve *F. circinatum* isolates: sterile distilled water, mineral oil, silica gel, and filter or cellulose paper.

3 Methods and Protocols

3.1 Sampling in Nurseries

1. Disinfect your hands or gloves between samplings.
2. If collecting seedlings, sample the entire plants, including roots.
3. Transfer the seedlings in paper bags to the laboratory and process them immediately.
4. In the case of potted plants, include the substrate with roots. From older plants, take only portions from the stem and root system, including some surrounding substrate.
5. Label the plants, pack them in paper envelopes, insert the envelopes in sealed plastic bags, and keep them cool until being sent to the laboratory.
6. Attach to each sample a document including a collection data, and any phytosanitary treatments that were carried out (*see* **Note 1**).

3.2 Sampling in Forests and Plantations

3.2.1 Soil

1. Take the sample from the plant rhizosphere or the first 30 cm below the soil surface [15].
2. Collect the samples in paper bags, or plastic bags if the soil is too wet (*see* **Note 2**).

3.2.2 Cones and Seeds

1. Collect mature, fully developed cones, with ripe seeds. Take the cones directly from the tree (not the ground) using disinfected tools. Do not sample serotinous cones.
2. Group cones into lots (homogeneous units).
3. If sampling seeds do it randomly, as no symptoms will be visible (*see* **Note 3**).
4. Pack and label samples separately in sealed plastic bags.

3.2.3 Twigs and Branches

1. Collect twigs, shoots, or branches showing visual symptoms (*see* **Note 4**). Sample the lesion edge plus a few centimeters of healthy-looking tissue.
2. Wrap the pieces of tissue in sheets of paper and place them in a sealed plastic bag.

3.2.4 Cankers

1. Sample cankers from trunks or branches. Cut the inner bark in the area directly around the lesion using a sterile blade until the canker margin is observed (*see* **Note 5**).
2. Remove pieces of tissue, including phloem and xylem, to collect portions of the lesion edge.
3. Wrap the pieces of tissue in sheets of paper and place them in a sealed plastic bag.

3.2.5 Roots

1. Sample roots at about 1 m from the collar region of the tree, from the surface to 20–30 cm depth.
2. Take roots from 0.2 to 5 mm diameter. Roots may be apparently healthy.
3. Keep samples in paper bags during transport and storage (*see* **Note 6**).

3.2.6 Spore Trapping

1. Place the rotating arm spore trap at the selected location (*see* **Note 7**).
2. Attach vertically two wooden rods wrapped with double-sided tape to each end of the arm.
3. After 48 h, the exposed rods must be stored in 2-mL microtubes and transferred to the laboratory for keeping at −20 °C before further processing.

3.2.7 Insect Vectors of F. circinatum

1. Place baiting logs at selected locations [16, 17].
2. After captures, transport to the laboratory the potential vectors using plastic sealed bags (*see* **Note 8**).

3.3 Sample Processing in the Laboratory

The pitch canker fungus is capable of infecting all organs of the host plant, and its conidial spores are able to survive for several months in the soil [18, 19]. Keep the samples in a refrigerator until analysis, which must be performed within 8 days of arrival [14]. Optimal processing of the samples in the laboratory is crucial for successful isolation or molecular detection of the fungus.

3.3.1 Isolation of F. circinatum from Plant Tissue

1. Disinfect the surface of the plant tissue (branches, twigs, cankers, and roots) by successive immersing it in ethanol (50–70%) and/or sodium hypochlorite (1.5–7%) for 1–2 min, followed by two rinses in sterile distilled water.
2. Cut the sample into small pieces after removing borders necrotized by the aseptic compounds used for surface disinfection. The xylem surrounding the canker is the best part from which to isolate the fungus, although it can also be isolated from the phloem. The size of the wood xylem sample should be a lamina of 4–5 mm in width.
3. Place the small pieces in Petri plates with isolation media.
4. Incubate the plates at 20–25 °C in the dark for approximately 1 week, after which growth from suspected *Fusarium* colonies must be transferred to WA media to obtain pure cultures originating from single conidia.
5. Transfer the resulting cultures to the appropriate growth medium used for identification of *F. circinatum* (*see* **Note 9**).

3.3.2 Isolation of F. circinatum *from Seedlings*

1. Wash thoroughly the entire plant with water to remove soil before isolation.
2. If plants are symptomatic, isolation must be made from the edge of lesions found beneath the affected bark, mostly from stem bases or from the roots. In asymptomatic plants, different parts of the root, root collar, and stem base must be plated randomly, including roots of different diameters, from main root to fine roots.
3. Follow the steps in Subheading 3.3.1.

3.3.3 Isolation of F. circinatum *from Seeds*

1. Place the seeds directly on PDAS or Komada's medium (*see* **Note 10**).
2. Follow **steps 4** and **5** in Subheading 3.3.1.

3.3.4 Isolation of F. circinatum *from Soil Samples*

1. Air-dry the soil sample in a laminar flow hood for 24–48 h.
2. Grind the sample in a mortar in order to generate small soil particles.
3. Prepare a soil suspension ranging from 1:50 to 1:2000, depending on the relative number of *Fusarium* isolates.
4. Transfer the suspension with a pipette and spread it evenly across the surface of 0.05% WA agar medium. Use 1 mL of the soil suspension for a 100-mm-diameter Petri dish.
5. Incubate the plates under light to increase color formation.
6. If isolation media results somewhat toxic to the fungus, subculture the colonies before identification [15].

3.3.5 Isolation of F. circinatum *from Insect Samples*

Squashing Specimens in Selective Media

1. Squash trapped insects individually onto FSM and incubate them for 2 weeks at 25 °C in the dark.
2. Transfer hyphal tips from the edges of the colonies to fresh PDA in order to obtain pure cultures [20].
3. *F. circinatum* colonies can be identified following the methods described hereafter.

Sonication of Specimens

1. Each bark beetle must be transferred to a single sterilized 1.5-mL microtube with 200 μL of Tween-80 at 1% v/v [21].
2. The microtubes must be sonicated for 5 s at 40 kHz in order to obtain spore suspensions [22].
3. Culture 50 μL aliquots of spore suspensions in PDAS by extending them with a sterile loop.
4. Then, subculture each colony onto MEA in order to get pure cultures [23].
5. *F. circinatum* colonies can be identified following the methods described hereafter.

3.4 Morphological Identification of F. circinatum

Grow isolates on PDA and incubate them at 22 °C ± 6 °C under UV light (daily monitored), or in daylight for 10 days, for the examination of pigmentation and colony morphology [24]. Characters (i.e., perithecia, asci, ascospores) associated with sexual and asexual stages of *F. circinatum* will be used to identify the fungus.

3.4.1 Sexual Stage of F. circinatum

1. Make a cross between an isolate presumed to be *F. circinatum* with a standard tester strain for the species (*see* **Note 11**).
2. Use the spores produced by the presumptive strain grown on CM to fertilize a CA medium plate.
3. The perithecia produced by *F. circinatum* are ovoidal to obpyriform, non-papillate, and slightly warted around the apex. They can be superficial or immersed, nonstromatic, and have a dark purple color in water and 3% KOH, becoming red in 100% lactic acid.
4. The cylindrical asci have an apex with a shallow, refractive ring. The ascospores are ellipsoidal to fusiform, smooth, hyaline, and septate, with additional septa developing after discharge when they turn pale brown [24].

3.4.2 Asexual Stage of F. circinatum

1. Make subcultures of colonies on SNA medium [14] to produce phialides, microconidia, and sterile hyphae, or use CLA to produce uniform macroconidia [15]. These structures are produced in SNA after ca. 10 days and on CLA after ca. 20 days, at 25 °C in the dark and are visible using bright-field microscopy at 400× magnification (*see* **Note 12**).
2. Examine and compare the isolates to the *F. circinatum* descriptions of Leslie and Summerell [15], Nirenberg and O'Donnell [25], and Britz et al. [26].
3. On SNA, sclerotia are absent. Microconidia are aggregated in false heads (not in chains), with branched and erect conidiophores that can end in one or two phialides (Fig. 1). Microconidia are usually obovoid, mostly non-septate. Macroconidia are cylindrical, mostly three-septate, with slightly curved walls, an apical cell that narrows to an inwardly curved tip, and a foot-shaped basal cell. *F. circinatum* produces coiled sterile hyphae during incubation in the dark [25] that are easily observed on SNA [1].
4. On CLA, macroconidia are slender, thin-walled, usually three-septate and typical of those produced by most species in the *Fusarium fujikuroi* species complex (FFSC). Microconidia are abundant in the aerial mycelia, being obovoid, oval, or allantoid in shape, non-septate, and borne on mono- and Polyphialides that may proliferate extensively [15].
5. On PDA, *F. circinatum* grows quite rapidly [25], ranging from 2.5 to 9.0 mm/day [26, 27]. The mycelium has entire margins,

and white cottony or whitish aerial mycelium, with sometimes a salmon-colored tinge in the middle, often producing a pigment in the agar that could range from purple to dark violet or even yellow [15, 24].

3.5 Molecular Methods for *F. circinatum* Detection

The symptoms caused by *F. circinatum*, particularly in seedlings, stems, and shoots may be similar to those caused by several other pathogenic fungi (*see* **Note 13**). Therefore, it is highly important to confirm the agent causing symptoms resembling those of *F. circinatum* by molecular analysis.

After isolation of a fungus suspected to be *F. circinatum*, their identity can be determined by species-specific PCR or by sequence and phylogenetic analysis of the translation elongation factor 1 alpha (Tef1) gene [14]. However, conventional PCR (*see* **Note 14**) and real-time or quantitative PCR (qPCR) using SYBR Green or hydrolysis probes (*see* **Note 15**) can also detect *F. circinatum* directly *in planta*. In any case, the results of any PCR-based method used for pathogen detection must be confirmed with amplicon Sanger sequencing [14].

3.5.1 DNA Extraction from Pure Culture

1. Tissue disruption and cell lysing: recover 10–50 mg of fresh fungal mycelium from a cellophane-membrane-covered PDA plate to a 2-mL tube with 500 μL of quartz sand. Add 300 μL of lysis buffer and collocate the tubes in the mixer mill with frequency of 300 Hz for 4 min.
2. Incubate the tubes at 65 °C for 1 h.
3. Separating nucleic acids from proteins: add 350 μL phenol: chloroform:isoamyl alcohol (25:24:1) to each tube, vortex 1 min, and centrifuge 2 min at top speed (14,000 rpm) at room temperature.
4. Carefully transfer the upper aqueous layer to a clean tube, avoiding the white interphase as much as possible.
5. Repeat the extraction with 250 μL phenol:chloroform:isoamyl alcohol (25:24:1) until no interphase can be seen. If the final volume is low, 30–50 μL of TE can be added to act as a DNA protector.
6. Repeat the process by adding 100 μL chloroform: isoamyl alcohol (24:1), vortex 2 min, and centrifuge 3 min (14,000 rpm) at room temperature.
7. DNA precipitation: for each 100 μL of the upper layer, add 60 μL of PEG solution and mix properly. Leave tubes on ice for 20 min.
8. Centrifuge for 20 min (14,000 rpm) and discard the supernatant.

9. Wash pellet in 1 mL ice-cold 70% ethanol, centrifuge (14,000 rpm) for 5 min, and pour off supernatant carefully.
10. Leave the pellet to dry or place on 50 °C heating block with open caps for 10–15 min (until the ethanol has evaporated) and re-suspend it in 50–200 μL TE.
11. Incubate the tubes 1 h at 50 °C or overnight at room temperature (the second choice is preferable).

3.5.2 Multiplex Conventional PCR

1. Thaw required reagents and DNA samples (*see* **Note 16**). Vortex all reagents, except DNA samples, before opening and keep everything on ice.
2. Prepare a reaction mixture or "master mix" sufficient for the number of samples to be tested including a positive sample (1.0 ng DNA from *F. circinatum*) and a NTC sample (*see* **Note 17**). For an individual 25 μL PCR, calculate the volume needed according to the final concentration: 1× (2 mM $MgCl_2$) of Roche buffer, 200 μM of dNTPs, 1 μM of each primer described in Subheading 2.7 and 2 U/μL of *Taq* DNA Polymerase (e.g., Roche Applied Science).
3. Add molecular water to a volume of 24 μL of master mix per reaction, vortex, and centrifuge the mixture briefly.
4. Dispense 24 μL of the master mix to each 0.2-mL PCR tube together with 1 μL of the template DNA (containing 1.0 ng DNA).
5. Briefly centrifuge the PCR tubes, place them into the Thermocycler, and start the cycling program following the PCR conditions:
   - Run an initial denaturation step for 2 min at 94 °C.
   - Run 30 cycles of 30 s of denaturation step at 94 °C, 30 s of annealing step at 60 °C, and 1 min of elongation step at 72 °C.
   - Run a final elongation step of 7 min at 72 °C.
6. Analyze 5 μL of PCR products of each sample by agarose gel electrophoresis: to prepare 100 mL gel, add 1.5% agarose to 100 mL 1× TAE buffer. Boil the mixture in a microwave until the agarose powder is completely dissolve. Add 2 μL staining dye and shake carefully. Immediately pour the mix to the prepared tray with comb and wait until the agarose gel is solid.
7. Remove the comb and place the gel in a horizontal electrophoresis tank containing 1× TAE. Gel should be completely dipped.
8. Add gel-loading mixture to 5 μL of the PCR product samples until a concentration of 6× is reached, transfer to the wells of the agarose gel, and electrophorese.

9. Visualize using UV transilluminator and photograph. If the positive control produces both amplicons with the expected length (bands 234 and 422 bp) and the NTC does not produce amplicons, the PCRs will be considered valid. An unknown sample will be considered negative if it produces no bands or bands of a different size.

3.5.3 Real-Time PCR

1. Thaw required reagents and DNA samples (*see* **Note 18**). Vortex all reagents, except DNA samples, before opening and keep everything on ice.
2. Prepare a reaction mixture or "master mix" sufficient for the number of samples to be tested including a positive sample (10 ng DNA from *F. circinatum*) and a NTC sample. For an individual 10 μL PCR, calculate the volume needed according to the final concentration: 1× PowerUp™ SYBR™ Green Master Mix (Applied Biosystems, cat. no. A25779) and 0.5 μM of each primer as described in Subheading 2.8.
3. Add molecular water to a volume of 9 μL of master mix per reaction, vortex, and centrifuge the mixture briefly.
4. Dispense 9 μL of the master mix to each well together with 1 μL of the template DNA (containing 10 ng DNA).
5. Seal the 96-well plate with an optical adhesive cover, briefly centrifuge the plate, place them into the Thermocycler, and start the cycling program following the PCR conditions:
   - Run an UDG activation step for 2 min at 50 °C.
   - Run an initial denaturation step for 2 min at 95 °C.
   - Run 40 cycles of 15 s of denaturation step at 95 °C, and 1 min of elongation and extension step at 66 °C.
6. Analyze the results using the manufacturer's software (QuantStudio™ Real-Time PCR Software). The assay should be first validated by review of the included controls: the positive control should show an exponential amplification curve and the melting temperature value should be as expected (88.0 ± 0.5 °C), and the NTC should not amplify. If this occurs, the threshold line will be set above the NTC, intersecting the middle of the logarithmic phase of the positive amplification curve. Then, assay samples that produce an amplification curve intersecting the threshold line, generating a *Ct* value lower than 40 (*see* **Note 19**), and show the expected melting temperature value (as positive control) will be considered as positives. The samples without a *Ct* value associated or *Ct* higher than 40, producing a no-exponential curve or an unexpected melting temperature value will be deemed negatives. All unclear results should be repeated diluting the samples to reduce inhibitors that may be interfering with the reaction.

3.5.4 Dual-Labelled Probe Real-Time PCR

1. Thaw required reagents and DNA samples (*see* **Note 20**). Vortex all reagents, except DNA samples, before opening and keep everything on ice.
2. Prepare a reaction mixture or "master mix" sufficient for the number of samples to be tested including a positive sample (1.0 ng DNA from *F. circinatum*) and a NTC sample. For an individual 10 μL PCR, calculate the volume needed according to the final concentration: 1× (1.5 mM $MgCl_2$) of Standard *Taq* Reaction Buffer, 200 μM of dNTPs, 0.5 μM of each primer and probe described in Subheading 2.9 and 0.25 U of Hot Start *Taq* DNA Polymerase (New England Biolabs, cat. no. M0495).
3. Add molecular water to a volume of 9 μL of master mix per reaction, vortex, and centrifuge the mixture briefly.
4. Dispense 9 μL of the master mix to each well together with 1 μL of the template DNA (containing 1.0 ng DNA).
5. Seal the 96-well plate with an optical adhesive cover, briefly centrifuge the plate, place them into the Thermocycler, and start the cycling program following the PCR conditions:
   - Run an initial denaturation step for 10 min at 95 °C.
   - Run 40 cycles of 15 s of denaturation step at 95 °C, and 55 s of elongation and extension step at 70 °C.
6. Analyze the results using the manufacturer's software (QuantStudio™ Real-Time PCR Software). The assay should be first validated by review of the included controls: The positive control should show an exponential amplification curve and the NTC should not amplify. If this occurs, the threshold line will be set above the NTC, intersecting the middle of the logarithmic phase of the positive amplification curve. Then, assay samples that produce an amplification curve intersecting the threshold line and generate a *Ct* value lower than 32 will be regarded as positives. The samples without a *Ct* value associated or *Ct* higher than 40, or producing a no-exponential curve will be deemed negatives. All unclear results should be repeated diluting the samples to reduce inhibitors that may be interfering with the reaction.

3.5.5 F. circinatum *Detection by Sanger Sequencing of PCR Products*

1. After PCR-based methods, the assay samples with results considered positives must be sent for two-way Sanger sequencing (sequenced in both directions generating at least two sequences per fragment).
2. The chromatogram obtained from the sequencer may require visual inspection to remove regions of poorer quality, such as the first positions after the primer or at the end of the sequence. Both sequences from each fragment must be assembled into a

consensus sequence using a bioinformatics program such as BioEdit. The final consensus sequence should have the PCR primer sequences removed.

3. Once a consensus sequence is obtained, use a sequence-based search engine, such as BLAST (https://blast.ncbi.nlm.nih.gov/Blast.cgi), to look for similar sequences deposited in GenBank.
4. For species confirmation, the level of identity between the query sequence and a reference strain of *F. circinatum* (e.g., GenBank accession number AF160295) should be between 99% and 100%.

3.5.6 EF1 Gene Partial Sequencing

1. Thaw required reagents and DNA samples (*see* **Note 21**). Vortex all reagents, except DNA samples, before opening and keep everything on ice.
2. Prepare a reaction mixture or "master mix" sufficient for the number of samples to be tested. For an individual 20 μL PCR, calculate the volume needed according to the final concentration: 1× (1.5 mM $MgCl_2$) of Phusion HF Buffer, 200 μM of dNTPs, 0.5 μM of each primer described in Subheading 2.11 and 0.4 U of Phusion High-Fidelity DNA Polymerase (New England Biolabs, cat. no. M0530).
3. Add molecular water to a volume of 19 μL of master mix per reaction, vortex, and centrifuge the mixture briefly.
4. Dispense 19 μL of the master mix to each 0.2-mL PCR tube together with 1 μL of the template DNA (containing 1.0 ng DNA).
5. Briefly centrifuge the PCR tubes, place them into the Thermocycler, and start the cycling program following the PCR conditions:
   - Run an initial denaturation step for 30 s at 98 °C.
   - Run 35 cycles of 10 s of denaturation step at 98 °C, 30 s of annealing step at 55 °C, and 60 s of elongation step at 72 °C.
   - Run a final elongation step of 6 min at 72 °C.
6. Analyze 5 μL of PCR products of each sample by 1% agarose gel electrophoresis following the **steps 5**, **6** and 7 from Subheading 3.5.2. After that, visualize using UV transilluminator in order to check the presence of 640 bp band.
7. If the 640 bp band is present after visualization using UV transilluminator, send an appropriate PCR product for two-way Sanger sequencing (sequenced in both directions generating at least two sequences per fragment).

3.6 Maintenance and Conservation of F. circinatum Isolates

Although a variety of methods can be used to store *Fusarium* spp., cryopreservation in liquid nitrogen and freeze-drying are the most recognized methods [28]. The storage temperature should be 5 °C or lower. *F. circinatum* can be maintained on artificial media as a short-term preservation method, but losses in virulence, pathogenicity, and sporulation may occur with repetitive subculturing [29, 30].

For long-term preservation of fungal cultures, immersion of the mycelium or spores of the fungus in sterile distilled water is one of the simplest methods [31]. However, since the stability of fungal cells was not ensured by this simple procedure [28], the maintenance of cultures under a layer of mineral oil [32] is a good option for fungi that do not tolerate freeze-drying [30].

To increase the viability of the cultures after preservation by lyophilization, isolates should be grown on CLA in Petri dishes for 7–10 days. Several colonized carnation leaf pieces are then transferred to sterile vials, and 0.5 mL of sterile skim milk can be added to each vial as an additional cryoprotective substance before lyophilization [33] (*see* **Note 22**).

4 Notes

1. In a production nursery, both symptomatic and asymptomatic seedlings should be collected. This is because seedlings of *Pinus* spp. are commonly infected with *F. circinatum*, yet they display not symptoms. Such apparently healthy but infected seedlings often only induce disease in their hosts once they are planted in the field, where the primary symptoms are root and collar rot.
2. Regular sampling of soil is not currently required. First of all, because long-term survival of *F. circinatum* conidia in soil has not been reported, and in artificially inoculated soil, the fungus could not be detected after 8 months at temperatures of 20 and 30 °C [19]. And, as most EPPO countries prohibit the import of soil and restrict the import of plants with soil from other continents [14], regular sampling of soil to detect *F. circinatum* is not required from the plant health officials in Europe.
3. *Fusarium circinatum* can be present in cones and seeds without visual signs and symptoms, both externally and internally [34, 35], and seeds from affected areas are an important pathway for long-distance dispersal of the pathogen. There are no strict rules for sampling cones or seeds [36]. Nevertheless, as they may be asymptomatic, cones and seeds should be sampled in areas with and without symptoms. Because non-symptomatic areas may represent a huge extension to surveyed areas, sampling areas can be chosen on the basis of

different criteria, such as their location with respect to possible pathogen sources and the susceptibility of *Pinus* species. Serotinous cones may turn out to be problematic during laboratory processing because the temperatures needed for them to open may damage any present *F. circinatum* propagules, leading to false-negative results when molecular detection is not used.

4. As *F. circinatum* is a hemibiotroph capable of living in dead tissues, isolation of this species from dead seedlings or wood is usually feasible even from totally necrotic tissue and seedlings that have died weeks before.
5. The frequency of isolation is usually higher from samples extracted using an axe or chainsaw than from those extracted using a Pressler borer. The reason for the difference may be the subsequent surface disinfection of the sample in the laboratory. Samples collected with the Pressler borer are typically thinner than those collected with an axe or chainsaw, leaving the vessels open, and therefore, making it easier for the alcohol and the bleach used in the surface disinfection process to penetrate deeper into the tissues and kill the fungal community, including the target pathogen.
6. All plant material should be processed as fast as possible or refrigerated to avoid colonization by other fungal species (i.e., secondary pathogens and saprophytes) [14, 15, 24].
7. Although temporal patterns of *F. circinatum* spore dispersal have not been investigated in detail, it is known that they are dispersed by air, rain, and insect vectors [2] and have a limited natural dispersal distance [12, 37, 38] to detect and quantify the presence of inoculum in the air.
8. One of the possible pathways of *F. circinatum* spread is by insects such as bark beetles, wood borers, shoot and foliage-feeders, cone insects, predatory, and sap-feeding insects [2, 3, 39]. In Europe, only *Tomicus piniperda* has been shown to be an effective vector of *F. circinatum* [21].
9. The most commonly used media for the isolation of *F. circinatum* are PDA supplemented with streptomycin (PDAS) [14], PDAspd (PDA supplemented with dichloran, streptomycin, and pentachloronitrobenzene), dichloran chloramphenicol peptone agar (DCPA) [18, 40], Komada's medium [41], *Fusarium* Selective Medium (FSM) [42], or modified FSM [33].
10. Seeds are not surface disinfected because they can be colonized by *F. circinatum* both internally and/or externally. Hence, there is also no standard protocol on how to process seed in the laboratory [13].

11. *F. circinatum* is heterothallic and require isolates of opposite mating type for completing the sexual stage [43]. All isolates of the fungus harbor either the MAT1-1 or MAT1-2- "idiomorphs" at the mating type locus; the mating type of an isolate can thus be diagnosed using PCR-based approaches [44, 45]. A fertile cross between an isolate presumed to be *F. circinatum* with a standard tester strain for the species further provides strong evidence of conspecificity [15]. Because these tester strains are hermaphrodites, they may be used routinely as the female-fertile partner in laboratory crosses with field isolates that are commonly male-fertile. It should be noted that the sexual stage of *F. circinatum* has never been observed under field conditions.
12. Asexual structures are produced by *F. circinatum* both in culture and on plant tissues. Some isolates of the fungus are not distinctively coiled or absent [27, 46]; thus, it is recommended that the morphological identification of *F. circinatum* is verified with molecular methods [12].
13. Symptoms may be similar to those of *Sphaeropsis sapinea* (Fr.) Dyko & B. Sutton, *Gremmeniella abietina* (Lagerb.) M. Morelet, *Pachyramichloridium pini* (de Hoog & Rahman) C. Nakash., Videira & Crous, *Crumenulopsis sororia* (P. Karst.) J. W. Groves, *Cenangium ferruginosum* Fr., and *Caliciopsis pinea* Peck [8]. Damping off is also caused by agents such as *Phytophthora* spp., *Pythium* spp., *Botrytis* spp., or *Rhizoctonia* spp. In addition, needle symptoms may resemble those caused by species of *Dothistroma*, *Cyclaneusma*, *Lecanosticta*, *Lophodermium*, and *Lophodermella* [8].
14. The conventional PCR tests are easy to set up and need less expensive equipment than qPCR methods, but require an additional electrophoresis step, and thus, more time to obtain a result [47]. Another conventional PCR assay developed by Ramsfield et al. [11] may be used to detect *F. circinatum* in plant, soil, or potting mix, and it has been suggested that it brings additional reliability.
15. Real-time PCR reduces turnaround time for analysis. The test requires a high-cost equipment, but it involves less costly reagents than hydrolysis probe technology.
16. This test is recommended for use on plant tissue and soil samples. If the sample is positive in *F. circinatum* presence, two loci of unknown function are targeted to be co-amplified.
17. No template control (NTC), a sample type containing all the reaction components except the DNA template. Instead of DNA, 1 μL of molecular water is added to the NTC reactions wells to examine any contamination in the reactive mixture (primers and/or master mix).

18. This test can be used for *F. circinatum* identification from pure culture, trapped airborne spores, plant tissue, and seeds. The analysis of seeds must incorporate a previous step of biological enrichment [13]. The targeted fragment is a specific region of the rDNA Inter Genic Spacer (IGS) region of *F. circinatum* (GenBank accessions: AY249397 to AY249403).
19. The cycle threshold (*Ct*) value marks the cycle at which the fluorescence significantly exceeds the baseline signal.
20. This test is recommended for use on plant tissue, pure culture, trapped airborne spores, and seeds. The analysis of seeds must incorporate a previous step of biological enrichment [13]. The targeted fragment is a specific region of the rDNA Inter Genic Spacer (IGS) region of *F. circinatum* (GenBank accessions: AY249397 to AY249403).
21. This test is recommended for identification of pure culture. The targeted fragment is a specific region of the translation elongation factor 1 alpha (TEF1) gene.
22. The colder the storage temperature and the higher the temperature stability, the greater the likelihood for long-term viability [30, 48]. Higher viability of the isolates can be obtained when preserving *Fusarium* spp. on soil or cereal grains at temperature of 4.5 °C rather than room temperature [49]. However, one of the inconveniences is that the fungus can colonize the substrate, with the consequent chance of cultural variation and risk of physiological and genetic change [50]. For this reason, using silica gel as substrate is a useful alternative for laboratories requiring a simple technique [50]. Another simple and less expensive alternative is the modified filter paper method [51]. It also works using cellulose paper (observation of Diez J.J.).

References

1. Dwinell LD, Barrows-Broaddus JB, Kuhlman G (1985) Pitch canker: a disease complex of southern pines. Plant Dis 69:270–276. https://doi.org/10.1094/PD-69-270
2. Gordon TR, Storer AJ, Wood DL (2001) The pitch canker epidemic in California. Plant Dis 85:1128–1139. https://doi.org/10.1094/PDIS.2001.85.11.1128
3. Martínez-Álvarez P, Pando V, Diez JJ (2014) Alternative species to replace Monterey pine plantations affected by pitch canker caused by *Fusarium circinatum* in Northern Spain. Plant Pathol 63:1086–1094. https://doi.org/10.1111/ppa.12187
4. Gordon TR, Swett CL, Wingfield MJ (2015) Management of *Fusarium* diseases affecting conifers. Crop Prot 73:28–39. https://doi.org/10.1016/j.cropro.2015.02.018
5. Wingfield MJ, Hammerbacher A, Ganley RJ et al (2008) Pitch canker caused by *Fusarium circinatum* – a growing threat to pine plantations and forests worldwide. Australas Plant Pathol 37:319–334. https://doi.org/10.1071/AP08036
6. Vainio EJ, Bezos D, Bragança H et al (2019) Sampling and detection strategies for the pine pitch canker (PPC) disease pathogen *Fusarium circinatum* in Europe. Forests 10:1–27. https://doi.org/10.3390/f10090723
7. Swett CL, Kirkpatrick SC, Gordon TR (2016) Evidence for a hemibiotrophic association of the pitch canker pathogen *Fusarium*

circinatum with *Pinus radiata*. Plant Dis 100: 79–84. https://doi.org/10.1094/PDIS-03-15-0270-RE

8. Mullett M, Tubby K, Pérez-Sierra A (2017) EU template *Fusarium* field guide. COST Action FP1406 PINESTRENGTH
9. Bezos D, Martínez-Alvarez P, Fernández M, Diez JJ (2017) Epidemiology and management of pine pitch canker disease in Europe – a review. Balt For 23:279–293
10. O'Donnell K, Cigelnik E, Nirenberg HI (1998) Molecular systematics and phylogeography of the *Gibberella fujikuroi* species complex. Mycologia 90:465. https://doi.org/10.2307/3761407
11. Ramsfield TD, Dobbie K, Dick MA, Ball RD (2008) Polymerase chain reaction-based detection of *Fusarium circinatum*, the causal agent of pitch canker disease. Mol Ecol Resour 8: 1270–1273. https://doi.org/10.1111/j.1471-8286.2008.02188.x
12. Schweigkofler W, O'Donnell K, Garbelotto M (2004) Detection and quantification of airborne conidia of *Fusarium circinatum*, the causal agent of pine pitch canker, from two California sites by using a real-time PCR approach combined with a simple spore trapping method. Appl Environ Microbiol 70: 3512–3520. https://doi.org/10.1128/AEM.70.6.3512-3520.2004
13. Ioos R, Fourrier C, Iancu G, Gordon TR (2009) Sensitive detection of *Fusarium circinatum* in pine seed by combining an enrichment procedure with a real-time polymerase chain reaction using dual-labeled probe chemistry. Phytopathology 99:582–590. https://doi.org/10.1094/PHYTO-99-5-0582
14. EPPO (2019) PM 7/91 *Fusarium circinatum* (formerly *Gibberella circinata*). EPPO Bull 49: 228–247. https://doi.org/10.1111/epp.12587
15. Leslie JF, Summerell BA (2007) The *Fusarium* laboratory manual. Blackwell Publishing
16. Bezos D, Martínez-Álvarez P, Diez JJ et al (2016) Association levels between *Pityophthorus pubescens* and *Fusarium circinatum* in pitch canker disease affected plantations in Northern Spain. Entomol Gen 36:43–54. https://doi.org/10.1127/entomologia/2016/0314
17. Storer AJ, Wood DL, Gordon TR (2004) Twig beetles, *Pityophthorus* spp. (Coleoptera: Scolytidae), as vectors of the pitch canker pathogen in California. Can Entomol 136:685–693
18. Andrews S, Pitt JI (1986) Selective medium for isolation of *Fusarium* species and dematiaceous hyphomycetes from cereals. Appl Environ Microbiol 51:1235–1238. https://doi.org/10.1128/aem.51.6.1235-1238.1986
19. Serrano Y, Iturritxa E, Elvira-Recuenco M, Raposo R (2017) Survival of *Fusarium circinatum* in soil and *Pinus radiata* needle and branch segments. Plant Pathol 66:934–940. https://doi.org/10.1111/ppa.12648
20. Romon P, Troya M, de Gamarra ME et al (2008) Fungal communities associated with pitch canker disease of *Pinus radiata* caused by *Fusarium circinatum* in Northern Spain: association with insects and pathogen-saprophyte antagonistic interactions. Can J Plant Pathol 30:241–253. https://doi.org/10.1080/07060661.2008.10540539
21. Bezos D, Martinez-Alvarez P, Diez JJ, Fernandez MM (2015) The pine shoot beetle *Tomicus piniperda* as a plausible vector of *Fusarium circinatum* in Northern Spain. Ann For Sci 72:1079–1088. https://doi.org/10.1007/s13595-015-0515-4
22. Ambourn AK, Juzwik J, Eggers JE (2006) Flight periodicities, phoresy rates, and levels of *Pseudopityophthorus minutissimus* branch colonization in oak wilt centers. For Sci 52: 243–250. https://doi.org/10.1093/FORESTSCIENCE/52.3.243
23. Muñoz-Adalia EJ, Sanz-Ros AV, Flores-Pacheco JA et al (2017) *Sydowia polyspora* dominates fungal communities carried by two *Tomicus* species in pine plantations threatened by *Fusarium circinatum*. Forests 8:1–16. https://doi.org/10.3390/f8040127
24. IPPC (2017) ISPM 27 diagnostic protocols for regulated pests DP 22: *Fusarium circinatum*. Rome, Italy
25. Nirenberg HI, O'Donnell K (1998) New *Fusarium* species and combinations within the *Gibberella fujikuroi* species complex. Mycologia 90:434. https://doi.org/10.2307/3761403
26. Britz H, Coutinho TA, Wingfield MJ, Marasas WFO (2002) Validation of the description of *Gibberella circinata* and morphological differentiation of the anamorph *Fusarium circinatum*. Sydowia 54:9–22
27. Mullett M, Pérez-Sierra A, Armengol J, Berbegal M (2017) Phenotypical and molecular characterisation of *Fusarium circinatum*: correlation with virulence and fungicide sensitivity. Forests 8:1–22. https://doi.org/10.3390/f8110458
28. Espinel-Ingroff A, Montero D, Martin-Mazuelos E (2004) Long-term preservation of fungal isolates in commercially prepared cryogenic microbank vials. J Clin Microbiol 42:1257–1259. https://doi.org/10.1128/JCM.42.3.1257-1259.2004

29. Kitamoto Y, Suzuki A, Shimada S et al (2002) A new method for the preservation of fungus stock cultures by deep-freezing. Mycoscience 43:143–149
30. Humber RA (1997) Fungi: preservation of cultures. In: Manual of techniques in insect pathology. Academic Press, London, pp 269–279
31. McGinnis MR, Padhye AA, Ajello L (1974) Storage of stock cultures of filamentous fungi, yeasts, and some aerobic actinomycetes in sterile distilled water. Appl Microbiol 28:218–222. https://doi.org/10.1128/am.28.2.218-222.1974
32. Smith D, Onions AHS (1983) A comparison of some preservation techniques for fungi. Trans Br Mycol Soc 81:535–540. https://doi.org/10.1016/s0007-1536(83)80122-3
33. Hernandez-Escribano L, Iturritxa E, Aragonés A et al (2018) Root infection of canker pathogens, *Fusarium circinatum* and *Diplodia sapinea*, in asymptomatic trees in *Pinus radiata* and *Pinus pinaster* plantations. Forests 9:128. https://doi.org/10.3390/f9030128
34. Dwinell LD, Barrows-Broaddus JB (1985) Pitch canker in seed orchards. Phytopathology 72:979
35. Storer AJ, Gordon TR, Clark SL (1998) Association of the pitch canker fungus, *Fusarium subglutinans* f.sp. *pini*, with Monterey pine seeds and seedlings in California. Plant Pathol 47:649–656. https://doi.org/10.1046/j.1365-3059.1998.00288.x
36. ISTA (2002) Detection of *Fusarium moniliforme* var. *subglutinans* Wollenw & Reinke on *Pinus taeda* and *P. elliotii* (Pine). Basseldorf, Switzerland
37. Martínez-Álvarez P, Alves-Santos FM, Diez JJ (2012) *In vitro* and *in vivo* interactions between *Trichoderma viride* and *Fusarium circinatum*. Silva Fenn 46:303–316. https://doi.org/10.14214/sf.42
38. Garbelotto M, Smith T, Schweigkofler W (2008) Variation in rates of spore deposition of *Fusarium circinatum*, the causal agent of pine pitch canker, over a 12-month-period at two locations in Northern California. Phytopathology 98:137–143. https://doi.org/10.1094/PHYTO-98-1-0137
39. Brockerhoff EG, Dick M, Ganley R et al (2016) Role of insect vectors in epidemiology and invasion risk of *Fusarium circinatum*, and risk assessment of biological control of invasive *Pinus contorta*. Biol Invasions 18:1177–1190. https://doi.org/10.1007/s10530-016-1059-8
40. Ioos R, Belhadj A, Menez M (2004) Occurrence and distribution of *Microdochium nivale* and *Fusarium* species isolated from barley, durum and soft weat grains in France from 2000 to 2002. Mycopathologia 158: 351–362. https://doi.org/10.1007/s11046-004-2228-3
41. Komada H (1975) Development of a selective medium for quantitative isolation of *Fusarium oxysporum* from natural soil. Rev Plant Prot Res 8:114–124
42. Aegerter BJ, Gordon TR (2006) Rates of pitch canker induced seedling mortality among *Pinus radiata* families varying in levels of genetic resistance to *Gibberella circinata* (anamorph *Fusarium circinatum*). For Ecol Manag 235:14–17. https://doi.org/10.1016/j.foreco.2006.07.011
43. Martin SH, Wingfield BD, Wingfield MJ, Steenkamp ET (2011) Structure and evolution of the *Fusarium* mating type locus: new insights from the *Gibberella fujikuroi* complex. Fungal Genet Biol 48:731–740
44. Steenkamp ET, Wingfield BD, Teresa A et al (2000) PCR-based identification of MAT-1 and MAT-2 in the *Gibberella fujikuroi* species complex. Society 66:4378–4382. https://doi.org/10.1128/AEM.66.10.4378-4382.2000
45. Kerényi Z, Moretti A, Waalwijk C et al (2004) Mating type sequences in asexually reproducing *Fusarium* species. Appl Environ Microbiol 70:4419–4423. https://doi.org/10.1128/AEM.70.8.4419-4423.2004
46. Pérez-Sierra A, Landeras E, León M et al (2007) Characterization of *Fusarium circinatum* from *Pinus* spp. in Northern Spain. Mycol Res 111:832–839. https://doi.org/10.1016/j.mycres.2007.05.009
47. Ioos R, Aloi F, Piškur B et al (2019) Transferability of PCR-based diagnostic protocols: an international collaborative case study assessing protocols targeting the quarantine pine pathogen *Fusarium circinatum*. Sci Rep 9:8195. https://doi.org/10.1038/S41598-019-44672-8
48. MacFarlane DR, Forsyth M, Barton CA (1992) Vitrification and devitrification in cryopreservation. JAI Press, London
49. Windels CE, Burnes M, Kommedahl T (1988) Five-year preservation of *Fusarium* species on silica gel and soil. Phytopathology 78:107–109
50. Windels CE, Burnes PM, Kommedahl T (1993) *Fusarium* species stored on silica gel and soil for ten years. Mycologia 85:21. https://doi.org/10.2307/3760472
51. Fong YK, Anuar S, Lim HP et al (2000) A modified filter paper technique for long-term preservation of some fungal cultures. Mycologist 14:127–130. https://doi.org/10.1016/S0269-915X(00)80090-7

Part II

DNA Extraction Protocols

Chapter 4

Rapid Extraction of Plant Nucleic Acids by Microneedle Patch for In-Field Detection of Plant Pathogens

Rajesh Paul, Emily Ostermann, and Qingshan Wei

Abstract

Plant diseases pose a significant threat to global food security. Molecular diagnosis currently plays a crucial role in mitigating the negative impacts of plant diseases by accurately identifying the disease-causing pathogens and revealing their genotypes. However, current molecular assays are constrained to the laboratory because of the cumbersome protocols involved in plant nucleic acid extraction. To streamline this, we have developed a polymeric microneedle (MN) patch-based nucleic acid extraction method, which can be applied to various plant tissues and easily performed in field settings without using bulky laboratory equipment. The MN patch instantly isolates both host and pathogen's DNA and RNA from plant leaves by two simple steps: press and rinse with a buffer solution or nuclease-free water. The MN-extracted DNA and RNA are purification-free and directly applicable to downstream molecular assays such as polymerase chain reaction (PCR), reverse transcription-polymerase chain reaction (RT-PCR), loop-mediated isothermal amplification (LAMP), and reverse transcription loop-mediated isothermal amplification (RT-LAMP). Here, we describe the fabrication procedures of the MN patch and demonstrate the application of the MN method by extracting *Phytophthora infestans* DNA and tomato spotted wilt virus (TSWV) RNA from infected tomato leaves. After MN extraction, we directly utilize the MN-extracted nucleic acid samples to run PCR, RT-PCR, LAMP, or RT-LAMP reactions to amplify various biomarker genes, such as the ribulose-bisphosphate carboxylase (*rbcL*) gene of host tomato DNA, internal transcribed spacer (ITS) region of *P. infestans* DNA, and nucleocapsid (N) gene of TSWV RNA. Furthermore, this simple and rapid nucleic acid method can be integrated with portable nucleic acid amplification platforms such as smartphone-based microscopy devices to achieve "sample-to-answer" detection of plant pathogens directly in the field.

Key words Microneedle (MN) patch, Nucleic acid extraction, Plant pathogen, Polymerase chain reaction (PCR), Loop-mediated isothermal amplification (LAMP)

1 Introduction

In-field detection of plant pathogens is crucial for protecting crops from disease. Currently, growers rely mostly on visible symptoms to identify the plant pathogens in their field. However, diseased plants often show similar signs for different pathogens. Therefore, visual symptoms may fail to recognize the causal agent in the field. To

Nicola Luchi (ed.), *Plant Pathology: Methods and Protocols*, Methods in Molecular Biology, vol. 2536,
https://doi.org/10.1007/978-1-0716-2517-0_4,

overcome this issue, several immunoassay methods (e.g., ELISA and LFA) have been developed for in-field detection of plant pathogens [1, 2]. However, antibody-based assays usually suffer from low sensitivity and specificity. Moreover, many antibody tests cannot distinguish between different strains of pathogens, which is crucial to determine the proper treatment plans. For example, some clonal lineages of *Phytophthora infestans*, such as US-8 and US-11, are resistant to mefenoxam, a commonly used fungicide [3]. In addition, *P. infestans* has been known to undergo population shifts, in which an emerging lineage dominates and replaces previously extant lineages [4, 5]. As a result, a rapid field-portable nucleic acid detection platform is immediately needed to maximize the diagnostic capability of growers and field workers to more efficiently protect the crops. A nucleic acid assay usually consists of three steps: (1) nucleic acid extraction, (2) nucleic acid amplification, and (3) amplicon detection. Although portable PCR systems [6, 7], isothermal nucleic acid amplification assays (e.g., LAMP and RPA) [1, 8], and colorimetric [9] lateral flow-based amplicon detection systems [10] have been presented in the literature to perform nucleic acid amplification in the field settings, the nucleic acid extraction step is still posing a major challenge for developing a portable "sample-to-answer" molecular detection platforms for plant pathogens.

Isolation of nucleic acids from infected plant specimens is a complicated multistep process because the plant cells are surrounded by rigid polysaccharide cell walls [11]. Usually, plant sample preparation relies on bulky instruments for mechanical grinding and purification, and skilled personnel to operate them [12]. The representative laboratory plant nucleic acid extraction methods for DNA and RNA are cetyltrimethylammonium bromide (CTAB) extraction method [13] and TRIzol extraction method [14], respectively, which are both time-consuming and require toxic chemicals (e.g., phenol and chloroform).

Alternative to these conventional methods, we have recently developed a polymeric microneedle (MN) patch-based extraction technology to rapidly extract DNA and RNA molecules from plant leaves [11, 15, 16]. We demonstrate that the MN patch can effectively isolate both DNA and RNA molecules from hard-to-lyse plant tissues by a simple compression and retraction process on the infected leaf. During compression, the conical MNs break the plant cells and release DNA, RNA, and other intracellular molecules. As the MN patch is made of polyvinyl alcohol (PVA), a highly water-swellable polymer, the MNs instantly absorb intracellular water molecules and accumulate the released DNA, RNA, and other molecules on the surfaces of needle tips. Compared with conventional laboratory-based sample preparation protocols, the MN patch-based nucleic acid extraction method provides several advantages: (a) it is rapid (extract nucleic acids <1 min without using laboratory equipment); (b) it is low-cost (approximately 1 cent per MN patch); (c) it extracts both DNA and RNA and can potentially isolate protein biomarkers or small molecules as well; (d) the method is minimally invasive, and the extracted sample can be used for various molecular analysis without further purification; (e) the MN method can be applied to various plant species (i.e., potato, tomato, pepper [11]); (f) MN-extracted DNA and RNA are stable on the patch for minimally 1 week and 3 days, respectively, when stored at room temperature [16]; (g) single microneedle tip extracts sufficient nucleic acid to detect a plant pathogen [16]; (h) the needle array format can be easily reconfigured or patterned for multiplex detection [16]; (i) the surface chemistry of the MN patch can be rationally tuned to enable selective target extraction in future; (j) MN extraction also has the potential to be combined with DNA sequencing analysis in the field. In this chapter, we describe the detailed protocol for the MN patch-based nucleic acid extraction to isolate *P. infestans* DNA and TSWV RNA from infected tomato leaves. Moreover, we discuss the fabrication process of the MN patch, as well as the steps to perform PCR, RT-PCR, LAMP, and RT-LAMP reactions with MN-extracted samples to amplify the ribulose-bisphosphate carboxylase (*rbcL*) gene of host tomato DNA, internal transcribed spacer (ITS) region 2 of *P. infestans DNA*, and nucleoprotein (N) gene of TSWV RNA.

2 Materials

Prepare all solutions using DNase-/RNase-free distilled water; store all reagents and solutions at room temperature (unless indicated otherwise).

2.1 Microneedle Patch Fabrication

1. Polydimethylsiloxane (PDMS) MN mold (*see* **Note 1**).
2. PVA, fully hydrolyzed, molecular weight approximately 31,000–50,000 (*see* **Note 2**).
3. 10% PVA solution: Add 85-g deionized water to a glass beaker. Weigh 15-g PVA. Heat water on a magnetic stirrer hot plate to 80–90 °C. Keep the temperature constant. Transfer PVA to water gradually and stir continuously until PVA is fully dissolved. Cool and store at room temperature for up to 6 months.
4. Kimwipes.
5. Petri dish.
6. Parafilm.

2.2 MN Patch-Based Nucleic Acid Extraction

1. Healthy, *P. infestans*- and TSWV-infected tomato leaves (*see* **Note 3**).
2. MN patch.
3. 1× Tris–EDTA (TE) buffer: 10-mM Tris–HCl, 0.1-mM EDTA, pH 8.
4. Deionized (DI) water.
5. 200-μL micropipette.
6. 200-μL micropipette tips.
7. 1.5-mL microcentrifuge tube.
8. Parafilm.

2.3 Real-Time PCR and RT-PCR Amplification of MN-Extracted Nucleic Acid Samples

1. cDNA synthesis kit: 5× cDNA synthesis buffer, random hexamer, RT enzyme mix, RT enhancer; store at −20 °C.
2. 10× PCR buffer; store at −20 °C.
3. 50-mM magnesium chloride; store at −20 °C.
4. dNTPs (5 mM each); store at −20 °C.
5. 20× EvaGreen dye; store at −20 °C.
6. 50-mg/mL BSA; store at −20 °C.
7. Taq DNA polymerase; store at −20 °C.
8. Primers for RT and PCR reactions [11, 17, 18]:
   - RT reaction

     RNA primer (TSWV-N (cDNA)) [17]:

     5′-ATGTCTAAGGTTAAGCTCACTAAGG-3′.
   - PCR reactions
     (i) *rbcL* gene detection [11] (amplification product = 103 bp):

     Forward primer (*rbcl*_Rf5′- GTAACTCCT CAACCTGGAGTTC-3′).

     Reverse primer (*rbcl*_Rb): 5′- GTAAGTC CATCGGTCCATACAG-3′.

     (ii) *P. infestans* detection [18] (amplification product = 100 bp):

     Forward primer (PINF2): 5′-CTCGCTACAATAG CAGCGTC-3′.

     Reverse primer (HERB2): 5′-CGGACCGACTGC GAGTCC-3′.

     (iii) TSWV detection [17] (amplification product = 140 bp):

     Forward primer (TSWV-N_For): 5′-GCTTCCCACCCTTTGATTC-3′.

     Reverse primer (TSWV-N_Rev): 5′- ATAGCCAA GACAACACTGATC-3′.

9. Deionized (DI) water.
10. PCR tube.
11. Real-time PCR instrument.

2.4 Real-Time LAMP and RT-LAMP Amplification of MN-Extracted Nucleic Acid Samples

1. 10× isothermal amplification buffer; store at −20 °C.
2. 100-mM magnesium sulfate; store at −20 °C.
3. dNTPs (10 mM each); store at −20 °C.
4. 5-M betaine; store at 4 °C.
5. 20× EvaGreen; store at −20 °C.
6. 2.5-mM hydroxynaphthol blue (HNB).
7. Bst 2 WarmStart DNA polymerase; store at −20 °C.
8. WarmStart RTx Reverse Transcriptase; store at −20 °C.
9. LAMP and RT-LAMP primers [8, 16] (*see* **Note 4**).
   - LAMP amplification
     (i) *rbcL* gene detection [16]
     F3: 5′-GAGAGATCGTTTCTTATTTTGTG-3′.
     B3: 5′-CCAAGGTAGTATTTGCGGTAA-3′.
     FIP: 5′-TGCAGTAGCATTCAAGTAATGCCCG AAGCACTTTTT AAAGCACA-3′.
     BIP: 5′-ATGCGAAGAAATGATCAAAAGAG CCCCCCGTTAAG TAGTCATG-3′.
     Loop B: 5′-TGTATTTGCTAGAGAATTGGGCG-3′ (*see* **Note 4**).
     (ii) *P. infestans* detection [8]
     F3: 5′-CTCCAAAAGTGGTGGCATTG-3′.
     B3: 5′-GCAACAGCAAAGCCGATTC-3′.
     FIP: 5′-TCTCCATTAACGCCGCAGCAGTGG ACGCTGCTATTG.
     TAGC-3′.
     BIP: 5′-CGTGGTATGGTTGGCTTCGGCATGG TTCACCAGTCC
     ATCAC-3′.
     Loop F: 5′-ACAAACCGGTCGCCAACTC-3′.
     LoopB: 5′-ATGCGCTTATTGGGTGATTTTCCTG-3′.
   - RT-LAMP amplification [16]
     (i) TSWV detection
     F3: 5′-TCAAGCCTATGGATTACCTCT-3′.
     B3: 5′-TCTCACTGTAATGTTCCATAGC-3′.
     FIP: 5′-GGTCGATCCCGAGATCCTTGTAGCTT CAGTTGATAGCTTTGAG-3′.

BIP: 5′-ACACCAGGGAAGCCTTAGGAACCTT CTTCACCTGATCTT.
CATT-3′.
Loop F: 5′-AGCCAAGACAACACTGATCAT-3′.
Loop B: 5′-AAGTTTGCACTGTGCTGAAA-3′.

10. Deionized (DI) water.
11. PCR tube.
12. Real-time PCR instrument.

3 Methods

Wear gloves during all procedures, use RNase- and DNase-free filter tips and plastics, and carry out all experimental steps at room temperature unless otherwise specified.

3.1 Microneedle Patch Fabrication

1. Sonicate PDMS mold for 5 min to dislodge PVA fragments in needle cavities from previous MN fabrication (*see* **Note 5**).
2. Gently dry PDMS mold with a Kimwipes tissue.
3. Place PDMS mold in the vacuum chamber. Ensure the cavity side of the mold is facing up. This can be checked by gently scraping the surface of the mold with a pipette tip.
4. Pipette 500-μL 10% PVA solution onto the mold (*see* **Note 6**).
5. Seal the vacuum chamber, select −100-kPa vacuum pressure, and turn on the vacuum pump for 10 min for PVA to be drawn into cavities of the MN mold (*see* **Notes 7** and **8**).
6. Open the vacuum chamber and transfer the mold to a chemical hood (*see* **Notes 9** and **10**).
7. Store mold overnight at 25 °C in the hood to dry the PVA solution (*see* **Notes 11** and **12**).
8. Gently peel the MN patch from mold by lifting it at the corner (*see* **Note 13**). Observe the surface integrity of the MNs. A representative image of a MN patch, where all MNs are formed properly, is showed in Fig. 1a. The MN patch has 225 conical MNs, and these MNs are arranged in an array of 15 × 15. The height of each needle is 800 μm, and the diameters at the tip and base are 10 μm and 300 μm, respectively.
9. Store MN patch at room temperature in a Petri dish sealed with Parafilm for future use (*see* **Note 14**).

3.2 MN Patch-Based Nucleic Acid Extraction

1. Obtain a fresh plant leaf and determine a suitable area for nucleic acid extraction. If testing for pathogen DNA/RNA, choose a location of plant leaf that displays a visible sign of infection (e.g., water-soaked area for *P. infestans* infection and yellow spots for TSWV infection).

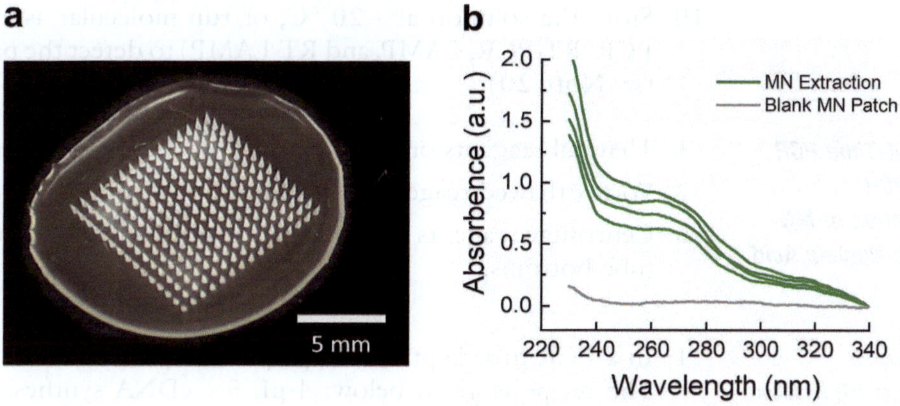

Fig. 1 (**a**) A photograph of a microneedle (MN) patch made of polyvinyl alcohol (PVA). (**b**) The UV–vis absorption spectra of MN-extracted nucleic acid solutions (green curves). The gray line represents the rinsing solution of a blank MN patch without leaf pressing

2. Cut a piece of Parafilm appropriately sized for a work surface to contain the leaf and MN patch. Place the Parafilm on a hard flat surface, such as a table, hardcover book, or the back of a phone.
3. Place the leaf on top of the Parafilm.
4. Place a MN patch on top of the leaf area selected for DNA/RNA extraction. Ensure needle tips are facing downward.
5. Gently press the MN patch into the leaf using pressure from the finger. Hold for approximately 10 s to capture plant materials (e.g., DNA and RNA) on microneedle tips (*see* **Notes 15** and **16**).
6. Gently lift the MN patch from the leaf and place it on Parafilm. Ensure needles are facing upward.
7. Rinse the MN patch eight to ten times using 100-μL TE buffer (for plant and *P. infestans* DNA isolation) or DI water (for TSWV RNA isolation) to obtain the extracted nucleic acid samples in solution. Pipette TE buffer or DI water onto the MN patch surface, and draw back into the pipette repeatedly (*see* **Note 17**).
8. Transfer the rinsing solution to a 1.5-mL microcentrifuge tube.
9. Measure the UV–vis absorbance spectrum of the rinsing solution with a NanoDrop spectrophotometer (*see* **Notes 18** and **19**). Typical absorption spectra for MN-extracted solutions are showed in Fig. 1b. The absorbance at 260-nm wavelength (A260 values) indicates that MN patches extracted significant amounts of nucleic acids from plant leaves (Fig. 1b, green curves). In contrast, no absorption at 260 nm was observed for rinsing solution from a blank MN patch without leaf pressing (Fig. 1b, gray curve).

10. Store the solution at −20 °C, or run molecular assays (e.g., PCR, RT-PCR, LAMP, and RT-LAMP) to detect the pathogen (*see* **Note 20**).

3.3 Real-Time PCR and RT-PCR Amplification of MN-Extracted Nucleic Acid Samples

1. Thaw all reagents on ice.
2. Vortex thawed reagents (*see* **Note 21**).
3. Centrifuge reagents at 100 × g for 5–10 s to collect reagents at tube bottoms.

3.3.1 Reverse Transcription (RT) Reaction (Required Only for RNA Viruses Such as TSWV)

1. In a PCR tube kept on ice, prepare 15 μL of RT master mix. The recipe is given below: 4-μL 5× cDNA synthesis buffer; 2-μL dNTPs (5 mM each); 0.5-μL random hexamer; 0.5-μL 10-μM gene-specific RNA primer; 1-μL RT enhancer; 1-μL RT enzyme mix; 6-μL DI water.
2. Add 5-μL MN-extracted solution.
3. Mix everything by pipetting up and down.
4. Close the cap and place the PCR tube in a thermal cycler set at 42 °C for 30 min for cDNA synthesis.
5. Set the temperature to 95 °C for 2 min to deactivate the RT enzyme.
6. Store cDNA solution on ice.

3.3.2 PCR Amplification

1. In a PCR tube kept on ice, prepare 23 μL of PCR master mix. Add DNA polymerase last. The recipe for PCR master mix is given below: 2.5-μL 10× PCR buffer; 0.5-μL dNTPs (5 mM each); 0.5-μL 10-μM forward primer; 0.5-μL 10-μM reverse primer; 1.25-μL 20× EvaGreen; 1.25-μL 50-mM magnesium chloride; 0.05-μL 50-mg/mL BSA; 16.35-μL deionized water; 0.1-μL 5-U/μL *Taq* DNA polymerase.
2. Add 2-μL MN-extracted solution (for *rbcL* and *P. infestans* detection) or cDNA sample (for TSWV detection).
3. Mix everything by pipetting up and down.
4. Close the cap and place PCR tubes in a thermal cycler to perform real-time PCR amplification. For *rbcL* gene amplification, perform 1 cycle of 94 °C for 2 min (initial denaturation); 35 cycles of 94 °C for 30 s (denaturation), 56 °C for 30 s (annealing), 72 °C for 30 s (extension), and fluorescence detection; and 1 cycle of 72 °C for 5 min (final extension). For *P. infestans* detection, perform 1 cycle of 94 °C for 2 min (initial denaturation); 35 cycles of 94 °C for 15 s (denaturation), 56 °C for 15 s (annealing), 72 °C for 15 s (extension), and fluorescence detection; and 1 cycle of 72 °C for 5 min (final extension). For TSWV detection, perform 1 cycle of 95 °C for 30 s (initial denaturation); 35 cycles of 95 °C for 10 s

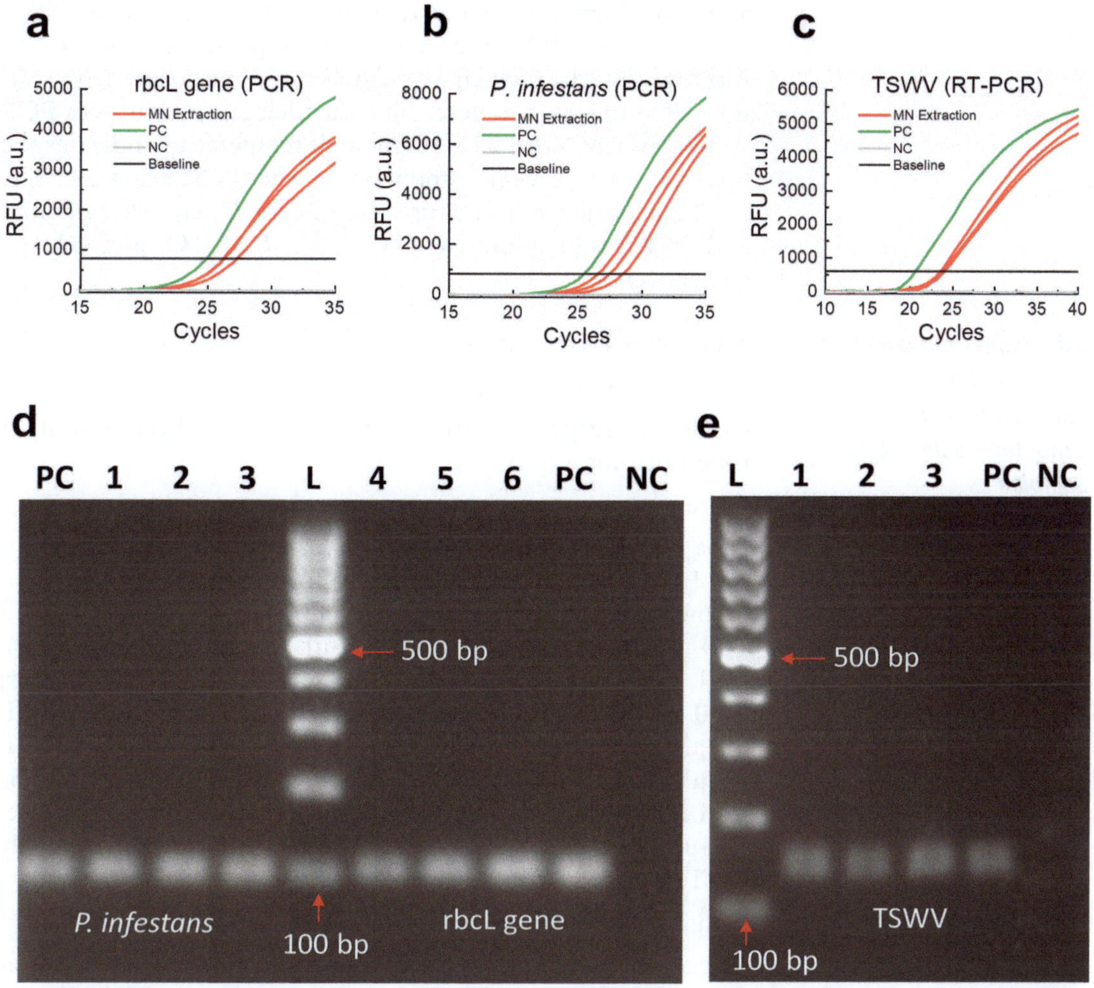

Fig. 2 PCR and RT-PCR amplification of MN-extracted nucleic acid samples. Real-time amplification curves for (**a**) *rbcL* gene, (**b**) *P. infestans*, and (**c**) TSWV detection. Gel electrophoresis showing amplified PCR products for (**d**) *P. infestans*, *rbcL* gene, and (**e**) TSWV detection. The amplicon lengths for the *rbcL* gene, *P. infestans*, and TSWV were 102 bp, 100 bp, and 140 bp, respectively. Lanes 1–3 in (**d**), three different amplified bands of *P. infestans* ITS DNA; lanes 4–6 in (**d**), three different amplified bands of *rbcL* gene; and lanes 1–3 in (**e**), three different amplified bands of TSWV N gene. L, 100-bp ladder; PC, positive control; NC, negative control (no DNA, blank rinsing solutions)

(denaturation), 55 °C for 15 s (combined annealing and extension), and fluorescence detection; and 1 cycle of 72 °C for 5 min (final extension). During PCR, EvaGreen dye binds to the amplified PCR products and emits fluorescence signals, which are recorded by a real-time PCR machine for monitoring the progress of the amplification reaction. Real-time amplification curves for the *rbcL* gene, *P. infestans,* and TSWV are reported in Fig. 2a–c.

5. Confirm the presence of a specific amplicon by gel electrophoresis in 2% agarose gel running at 120 volts for 45 min. The

amplicons' lengths for the *rbcL* gene, *P. infestans*, and TSWV were 103 bp, 100 bp, and 140 bp, respectively (Fig. 2d, e). Alternatively, the melt curve analysis can be performed in a real-time PCR machine to determine the melt temperatures of PCR amplicons (*see* **Note 22**). Single melt temperature indicates the presence of a specific amplicon in the PCR reaction. The amplicons' melt temperatures for the *rbcL* gene, *P. infestans*, and TSWV detection were 84.5 °C, 82.5 °C, and 82 °C, respectively.

3.4 Real-Time LAMP and RT-LAMP Amplification of MN-Extracted Nucleic Acid Samples

1. Thaw all reagents on ice.
2. Vortex thawed reagents (*see* **Note 21**).
3. Centrifuge reagents at 100 × g for 5–10 s to collect reagents at tube bottoms.
4. Store all reagents on ice.
5. In a PCR tube kept on ice, prepare 23 μL of LAMP or RT-LAMP master mix. The recipe for the master mix is given below: 2.5-μL 10× isothermal amplification buffer; 1.5-μL 100-mM magnesium sulfate; 3.5-μL dNTPs (10 mM each); 4-μL 5-M betaine (*see* **Note 23**); 0.50-μL 10-μM F3 primer; 0.50-μL 10-μM B3 primer; 0.8-μL 50-μM FIP primer; 0.8-μL 50-μM BIP primer; 0.2-μL 50-μM loop F primer; 0.2-μL 50-μM loop B primer; 1.25-μL 20× EvaGreen; 1.2-μL 2.5-mM HNB; 1-μL 8-U/μL Bst 2 WarmStart DNA polymerase; 0.5-μL 15-U/μL WarmStart RTx Reverse Transcriptase (only for TSWV detection); 4.55-μL nuclease-free water.
6. Add 2-μL MN rinsing solution.
7. Place the PCR tube in a real-time PCR thermal cycler set at 65 °C for 60 min, and capture fluorescence signals at 1-min intervals. Real-time LAMP amplification curves for detecting the *rbcL* gene, *P. infestans* ITS DNA, and TSWV N gene are showed in Fig. 3a, c.
8. Set the temperature at 80 °C for 5 min to deactivate Bst 2.0 WarmStart DNA Polymerase and WarmStart RTx Reverse Transcriptase.
9. Inspect the color of the reaction tube. For a positive sample, color changes from violet to sky blue due to the incorporation of HNB in the LAMP assay (Fig. 3d).
10. Confirm the presence of a specific target amplicon either by melt curve analysis or gel electrophoresis in 2% agarose gel running at 120 volts for 60 min (*see* **Note 20**). Gel images of LAMP amplicons for detecting the *rbcL* gene, *P. infestans*, and TSWV are reported in Fig. 3e, g. The amplicons' melt temperatures for the *rbcL* gene, *P infestans* ITS DNA, and TSWV N gene were 81.5 °C, 86.5 °C, and 83 °C, respectively.

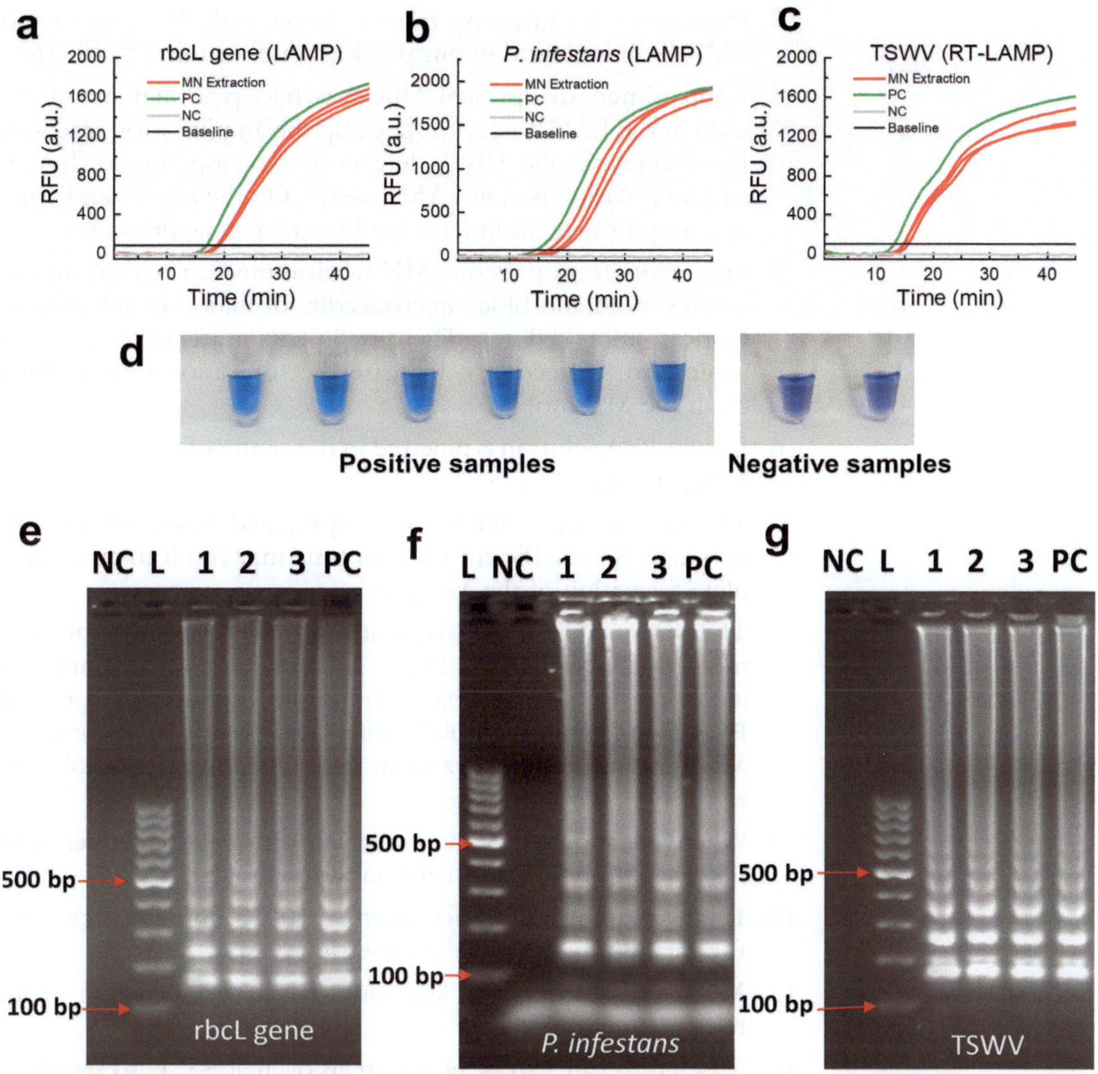

Fig. 3 LAMP and RT-LAMP amplification of MN-extracted nucleic acid samples. Real-time LAMP amplification curves for (**a**) *rbcL* gene, (**b**) *P. infestans* ITS DNA, and (**c**) TSWV N gene detection. (**d**) Visualization of the positive LAMP amplification from the color change of HNB dye. Gel electrophoresis showing LAMP amplicons for (**e**) *rbcL* gene, (**f**) *P. infestans* ITS DNA, (**g**) TSWV N gene detection. Lanes 1–3 in (**e**), LAMP amplicons for three different MN extractions from healthy leaves; lanes 1–3 in (**f**), LAMP amplicons for three different MN extractions from *P. infestans*-infected tomato leaves; and lanes 1–3 in (**g**), LAMP amplicons for three different MN extractions from TSWV-infected tomato leaves. L, 100 bp ladder; PC, positive control; NC, negative control

4 Notes

1. PDMS MN molds are purchased from Blueacre Technology, Ireland, and Micropoint Technologies Pte Ltd, Singapore. In addition, a user can also make their own MN mold with a metal MN patch [19].
2. Please select a fully hydrolyzed PVA to dissolve it entirely in DI water for MN patch fabrication.

3. Procedures for infecting tomato leaves with *P. infestans* and TSWV can be found in our previous publications [8, 15, 16].
4. Loop primers are optional. One or two loop primers are usually used in the LAMP assay to speed up the amplification reaction. For *P. infestans* and TSWV detection, two loop primers (loop F and loop B) are used in LAMP assays. On the other hand, only one loop primer (loop B) is used for *rbcL* gene detection.
5. Fragments from previous MN fabrications can remain on the PDMS mold and block microneedle formation in subsequent fabrication procedures. These fragments must be removed by sonication. If necessary, increase the sonication time to fully clean the MN mold.
6. Ensure PVA solution is pipetted in the center of mold above the MN pattern.
7. The vacuum time needs to be optimized based on vacuum pressure. An insufficient vacuum time may result in an incomplete formation of the MN patch.
8. The vacuum time also depends on the concentration and molecular weight of PVA. Increasing these parameters increases the solution viscosity. High viscosity will prevent PVA from entering mold cavities and result in incomplete MN formation. Adjust vacuum time according to the solution viscosity.
9. When releasing the vacuum, let air enter the chamber very slowly to prevent the disruption of the mold.
10. It may require waiting for another 10–15 min after releasing the vacuum for air bubbles to settle.
11. MN molds can also be kept on a laboratory bench at room temperature for drying.
12. MN molds can also be placed in an oven at 37 °C to speed up the drying process.
13. Microneedles may break when the PVA patch is peeled from the PDMS mold. Peel slowly and gently to prevent breakage. An adhesive tape can be used instead of removing patches by hand. Breakage may also occur due to insufficient drying time.
14. Ensure the storage container for MN patches is sealed from air to prevent loss of mechanical strength. The absorption of moisture can weaken the integrity of the MNs.
15. A MN patch can also be cut into two or multiple smaller pieces for nucleic acid extraction. If a smaller piece of MN patch is used, then adjust the rinsing buffer or DI water volume in the next step, accordingly.
16. The MN patch must be pressed into the plant leaf long enough (e.g., 10 s) to obtain sufficient DNA/RNA yield. If the yield is low, try to hold the pressure for a longer time.

17. The volume of washing buffer or DI water can be reduced to increase the concentration of MN-extracted nucleic acid sample.
18. If the MN patch cannot isolate enough nucleic acids for the amplification reaction, it may be due to broken needle tips or improper storage and age. Use newly fabricated MN patches with intact needles for the best results. If newly fabricated patches are mechanically weak to penetrate the leaf, use more PVA solutions or a higher concentration of PVA when fabricating.
19. Increasing the rinsing times with TE buffer may also improve the extraction yield.
20. Perform the DNA extraction, PCR, and LAMP amplification at different lab benches to prevent cross-contamination and false positives.
21. Do not vortex RT enzyme, *Taq* DNA polymerase, Bst 2.0 WarmStart DNA Polymerase, and WarmStart RTx Reverse Transcriptase, which would cause deactivation.
22. For melt curve analysis, the temperature of a reaction tube is raised in an increment of 0.5 °C from 72 °C to 95 °C. For each increment, the temperature is held constant for 5 s before measuring the fluorescence signal.
23. Betaine is optional for the LAMP assay. It is usually used to reduce false-positive amplification.

Acknowledgments

The authors thank Dr. Jean Ristaino and Dr. Anna Whitfield (Department of Entomology and Plant Pathology, NC State University) for providing *P. infestans*- and TSWV-infected tomato leaves. The authors also gratefully acknowledge the funding support from the NCSU Chancellor's Faculty Excellence Program, the Kenan Institute for Engineering, Technology & Science (KIETS), NCSU Game-Changing Research Incentive Program for the Plant Science Initiative (GRIP4PSI), USDA (No. 2019-67030-29311), USDA APHIS Farm Bill grant (No. 3.0096), and BASF.

References

1. McCoy AG, Miles TD, Bilodeau GJ et al (2020) Validation of a preformulated, field deployable, recombinase polymerase amplification assay for *Phytophthora* species. Plan Theory 9:466
2. Paul R, Ostermann E, Wei Q (2020) Advances in point-of-care nucleic acid extraction technologies for rapid diagnosis of human and plant diseases. Biosens Bioelectron 169: 112592–112592
3. Saville A, Graham K, Grünwald N et al (2015) Fungicide sensitivity of U.S. genotypes of *Phytophthora infestans* to six oomycete-targeted compounds. Plant Dis 99:659–666

4. Saville A, Ristaino JB (2019) Genetic structure and subclonal variation of extant and recent U.S. lineages of *Phytophthora infestans*. Phytopathology 109:1614–1627
5. Fry WE, Birch PRJ, Judelson HS et al (2015) Five reasons to consider *Phytophthora infestans* a reemerging pathogen. Phytopathology 105: 966–981
6. Koo C, Malapi-Wight M, Kim HS et al (2013) Development of a real-time microchip PCR system for portable plant disease diagnosis. PLoS One 8:e82704
7. Julich S, Riedel M, Kielpinski M et al (2011) Development of a lab-on-a-chip device for diagnosis of plant pathogens. Biosens Bioelectron 26:4070–4075
8. Ristaino JB, Saville AC, Paul R et al (2020) Detection of *Phytophthora infestans* by loop-mediated isothermal amplification, real-time LAMP, and droplet digital PCR. Plant Dis 104:708–716
9. Hansen ZR, Knaus BJ, Tabima JF et al (2016) Loop-mediated isothermal amplification for detection of the tomato and potato late blight pathogen, *Phytophthora infestans*. J Appl Microbiol 120:1010–1020
10. Lu X, Zheng Y, Zhang F et al (2020) A rapid, equipment-free method for detecting *Phytophthora infestans* in the field using a lateral flow strip-based recombinase polymerase amplification assay. Plant Dis 104:2774–2778
11. Paul R, Ostermann E, Gu Z et al (2020) DNA extraction from plant leaves using a microneedle patch. Curr Protoc plant Biol 5:e20104
12. Yadav S, Ram C, Singh S, Rana M (2018) Plant genomic DNA isolation-the past and the present, a review. Indian J Plant Genet Resour 31: 315–327
13. Murray MG, Thompson WF (1980) Rapid isolation of high molecular weight plant DNA. Nucleic Acids Res 8:4321–4326
14. Chomczynski P (1993) A reagent for the single-step simultaneous isolation of RNA, DNA and proteins from cell and tissue samples. BioTechniques 15:532–534
15. Paul R, Saville AC, Hansel JC et al (2019) Extraction of plant DNA by microneedle patch for rapid detection of plant diseases. ACS Nano 13:6540–6549
16. Paul R, Ostermann E, Chen Y et al (2021) Integrated microneedle-smartphone nucleic acid amplification platform for in-field diagnosis of plant diseases. Biosens Bioelectron 187: 113312
17. Nachappa P, Margolies DC, Nechols JR et al (2013) Tomato spotted wilt virus benefits a non-vector arthropod, Tetranychus urticae, by modulating different plant responses in tomato. PLoS One 8:1–14
18. Ristaino JB, Hu CH, Fitt BDL (2013) Evidence for presence of the founder Ia mtDNA haplotype of *Phytophthora infestans* in 19th century potato tubers from the Rothamsted archives. Plant Pathol 62:492–500
19. Caffarel-Salvador E, Kim S, Soares V et al (2021) A microneedle platform for buccal macromolecule delivery. Sci Adv 7:1–12

Chapter 5

DNA Extraction Methods to Obtain High DNA Quality from Different Plant Tissues

Domenico Rizzo, Daniele Da Lio, Linda Bartolini, Cristina Francia, Antonio Aronadio, Nicola Luchi, Sara Campigli, Guido Marchi, and Elisabetta Rossi

Abstract

DNA extraction from plant samples is very important for a good performance of diagnostic molecular assays in phytopathology. The variety of matrices (such as leaves, roots, and twigs) requires a differentiated approach to DNA extraction. Here we describe three categories of matrices: (a) symptomatic bark/wood tissue; (b) residues of frass resulting from insect woody trophic activities, portions of the galleries produced in the wood, and tissues surrounding exit holes; and (c) leaves of different plant species. To improve the performances of diagnostic assays, we here describe DNA extraction procedures that have been optimized for each matrix type.

Key words Quarantine pests, Invasive pathogen, Molecular identification, Diagnostic tools, Xylophagous insects, Plant pathogens

1 Introduction

DNA extraction and purification is the first crucial step to develop a reliable diagnostic molecular assay from plant matrices, since its efficiency will strongly condition the reliability and quality of downstream processes.

In general, the steps of DNA extraction include (a) lysis, in which cell envelopes are usually mechanically, physically, and/or chemically broken down; (b) nucleic acid capture step by means of buffers containing (usually) detergents; and (c) DNA elution, where DNA is eluted from the substrate. The main goal of the extraction procedures is to obtain as much DNA as possible, even from small samples, keeping in mind that the final quality will largely depend on its integrity and purity [1–4]. An ideal extraction method should also be manageable and require a relatively simple equipment and minimal amounts of toxic reagents [5].

Nicola Luchi (ed.), *Plant Pathology: Methods and Protocols*, Methods in Molecular Biology, vol. 2536, https://doi.org/10.1007/978-1-0716-2517-0_5,

In many plant matrices, the high concentration of primary and secondary metabolites such as polysaccharides, carbohydrates, polyphenols, tannins, aromatic substances, and other compounds, some of which are yet to be identified, can interfere with DNA purification [6–8] and thus with downstream biomolecular reactions (PCR, qPCR, LAMP, RFLP, etc.). For routine plant pathogens, diagnosis and nucleic acid extraction protocols are also required to be practical (high-throughput sample preparation, ease of use, etc.), with a reliable performance (high quality and integrity and relative quantity of nucleic acids extracted), and reproducible. Also, non-secondary aspects such as relative cheapness ought to be considered.

In this study we describe a two-step method for DNA extraction from different types of plant matrices, consisting of (i) the manual preparation of the sample and partial nucleic acid purification and (ii) further DNA purification and concentration with a commercial extraction kit that uses a technology based on magnetic spheres to capture and purify DNA. This protocol has been applied with excellent results in terms of performance to various recalcitrant or difficult plant matrices [9]. The various cases will be analyzed below with the appropriate modifications and additions according to the type of plant matrix taken into consideration. In particular, the matrices evaluated were (a) symptomatic bark/wood tissue for the diagnosis of canker fungi, *Geosmithia morbida* and *Ceratocystis platani*, and xylem-inhabiting bacteria, *Xylella fastidiosa*; (b) residues of frass [9] resulting from insect trophic activities (e.g., *Aromia bungii*, *Anoplophora chinensis*, *An. glabripennis*, *Xylosandrus* spp.), galleries produced in the wood (*Pityophthorus juglandis*, *Xylosandrus* spp.), or tissues surrounding exit holes (*Xylosandrus* spp.); and (c) leaves and twigs of different plant species (*Xylella fastidiosa*).

2 Materials

2.1 Sample Preparation

1. Pruning shears.
2. Plastic bags.
3. Knife.
4. Scalpel.
5. Scissor.
6. Disinfectants for tools.
7. Permanent marker.
8. Analytical balance.
9. 15-mL plastic tubes.
10. Disposable gloves.

2.2 DNA Extraction and Purification

1. CTAB 2% extraction buffer: CTAB 2%, NaCl 1.4 M, 20-mM disodium EDTA, Tris–HCl 0.1 M, PVP 40K 1%, sodium metabisulfite 1% [4] (*see* **Notes 1–3**).
2. Maxwell® RSC PureFood GMO and Authentication Kit (Promega Italia, Milan, Italy) (*see* **Note 4**).
3. Proteinase K (PK) solution.
4. RNase A solution.
5. Lysis Buffer (Promega).
6. Maxwell® RSC Cartridges (RSCI) (*see* **Note 5**).
7. Maxwell® RSC Plungers.
8. 0.5-mL elution tubes.
9. TE elution buffer: Tris–HCl 10 mM and EDTA 1 mM (*see* **Note 6**).
10. Refrigerated microcentrifuge.
11. 1.5-mL microcentrifuge tubes.
12. Micropipettes and pipette tips.
13. Vortexer.
14. Disposable gloves.

3 Methods

3.1 Sample Collection

1. In some cases, the matrices available for DNA extraction can be the mix of wood residues ejected by the larval stages inside the galleries or extruded outside them and wood fragments not digested (frass), or segments of galleries themselves, or wood shavings obtained around the exit hole of the wood borer.
2. Woody parts (twigs and leaves) from challenging matrices. Collect xylem tissues of infected woody parts, leaves and leaf petioles or sections of woody twigs. The following plant species were tested: *Arbutus unedo, Asparagus officinalis, Cercis siliquastrum, Cistus monspeliensis, Hedera helix, Erica multiflora, Eucalyptus* sp.*, Ficus carica, Laurus nobilis, Lavandula angustifolia, L. dentata, L. stoechas, Pistacia lentiscus*, and *Rosmarinus* sp. (*see* **Note 8**).

3.2 DNA Extraction in CTAB Buffer

3.2.1 Bark Woody Flakes, Woody Frass, and Xylem Shavings (See Note 7)

1. Label 10-mL steel jars for Mixer Mill MM 400 (Retsch GmbH, Haan, Germany) homogenizer with the corresponding sample code. Turn on a Thermomixer comfort (Eppendorf, Hamburg, Germany) at 65 °C.
2. For matrices consisting of (i) woody frass produced by xylophagous insects, (ii) woody parts adjacent to flickering holes, and (iii) wood shavings in correspondence with nutrition galleries

Fig. 1 *Anoplophora chinensis*. (**a**) Environmental frass at the tree foot; (**b**) frass collected from a lab rearing (photo by courtesy of Dr. F. Pennacchio)

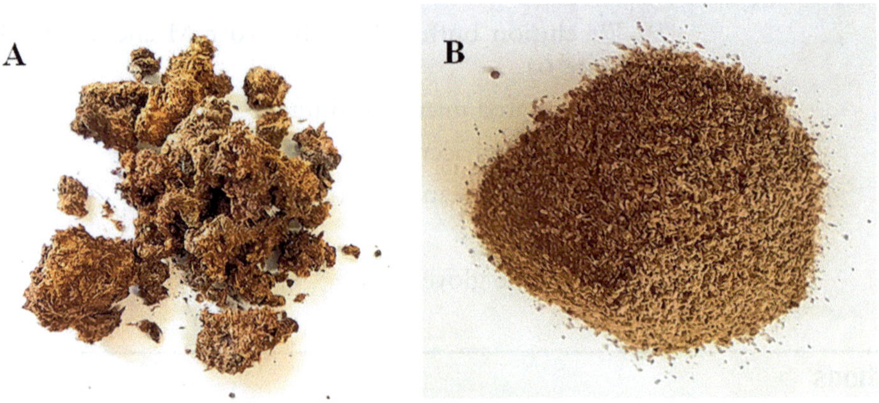

Fig. 2 *Aromia bungii*. (**a**) Environmental frass collected in the field; (**b**) frass produced in a laboratory rearing

of wood borer insects or cortical woody flakes, weigh approximately 1-g aliquots of plant material with an analytical balance (Figs. 1, 2, 3, and 4).

3. Set up a rack with 2.0-mL sterile microcentrifuge tubes labeled according to sample numbers.
4. Homogenize the individual samples at 2000 oscillations per minute for 20 s.
5. To each sample add 10 mL of 2% CTAB buffer (to respect the ratio 1 g of tissue – 10 mL of 2% CTAB buffer is mandatory) directly in the steel jar (*see* **Notes 9–11**).
6. Scrape the homogenate from the walls of the steel jars with sterile steel or plastic spatula.
7. Transfer 0.8 mL of each lysate to labeled 2-mL tubes.
8. Incubate at 65 °C for 10 min (or at 60 °C for 30 min) in the Thermomixer comfort with 30-s shaking at 1400 rpm every 5 min (*see* **Note 12**).

Fig. 3 (**a**) *Xylosandrus compactus*; (**b**) *X. crassiusculus*; (**c**) *X. germanus*. On the left, the damage caused by the three xylophagous species on twigs or trunks; on the right, a detail of the samples used for DNA extraction

Fig. 4 Black walnut branches infested by *Pityophthorus juglandis*. The frass site is indicated by the arrows

9. Under a chemical hood, add 1.2 mL of chloroform to each sample (for each Eppendorf tube), and mix well by inversion (or by vortexing for 20 s) (*see* **Note 13**).
10. Centrifuge at 13000 rpm for 10 min at room temperature (*see* **Note 14**).
11. Transfer the aqueous phase (supernatant) containing the nucleic acids (approx. 0.6 mL) to new, previously labeled sterile 2-mL/1.5-mL tubes (pre-cooled).

3.2.2 Woody Parts (Twigs and Leaves) from Challenging Matrices

1. Label nylon mesh universal bags (Bioreba, Reinach, Switzerland) (Fig. 5) with the corresponding sample number. Turn on the Thermomixer comfort at 65 °C.
2. Weigh approximately 1-g aliquots of plant material.
3. Set up a rack with 2.0-mL sterile microcentrifuge tubes labeled according to sample numbers (Fig. 6).
4. Homogenize the individual samples with the aid of a steel hammer.
5. To each sample add 2% CTAB buffer, respecting the ratio 1 g of tissue: 10 mL of 2% CTAB buffer (*see* **Notes 9–11**).
6. Transfer the bags to the Homex 6 (Bioreba, Reinach, Switzerland), and make a short homogenization step with the buffer, to allow all the cytoplasmic material to escape from the lysed plant cells.
7. Transfer 1.0 mL of each lysate to new 2-mL sterile microcentrifuge tubes.
8. Incubate at 65 °C for 10 min in the Thermomixer comfort with 30-s shaking at 1400 rpm every 5 min (*see* **Note 12**). Under a chemical hood, add 1.0 mL of chloroform to each sample (for each microtube), and mix well by inversion (or by vortexing for 2 s) (*see* **Note 13**).
9. Centrifuge at 13000 rpm for 10 min at room temperature (*see* **Note 14**).
10. Transfer the aqueous phase (supernatant) containing the nucleic acids (approx. 0.6 mL) to new, previously labeled sterile 2-mL/1.5-mL tubes (pre-cooled).

3.3 DNA Extraction by Using Maxwell® RSC PureFood GMO and Authentication Kit

1. The cartridges to be used in the Maxwell® RSC Cartridge Rack must be positioned with the labeled side facing away from the elution tube.
2. A pression on the cartridge allows it to snap it into position. Gently remove the seal so that all the plastic comes out from the top of the cartridge (*see* **Note 15**).
3. Place a CSC/RSC Plunger in well #8 of each cartridge. Well #8 is the measure closest to the elution tube (Fig. 6). Use only the

Fig. 5 Samples of leaves in Bioreba nylon mesh universal bags

plungers provided in the Maxwell® RSC PureFood GMO and Authentication Kit.

4. Place 0.5-mL elution tubes in the front of the Maxwell® RSC Cartridge Rack, adding 100 μL of the elution buffer in each elution tube.
5. Transfer 600 μL of the clear lysate sample into well #1 of the reagent cartridges, carefully avoiding to pipette any solid material from the bottom of the tube or on the surface of the liquid (*see* **Note 16**).
6. DNA purification is obtained by running the method on the Maxwell® RSC Instrument (Cat.# AS4500) (*see* **Note 17**).
7. The DNA can be eluted in 100 μL of nuclease-free water and either used in end-point or real-time PCR immediately or stored at −20 °C until use. To check the amplifiability of the DNA extracted, 1:20 DNA/ddsH_2O can be tested in a real-time PCR using a dual-labeled probe targeting a highly conserved region of the 18S rDNA [10]. The presence of PCR inhibitors in the DNA extracts can be determined by calculating the Cq and the slope of the relative amplification curves.
8. The amount of DNA (ng/μL) and the A260:280 ratios are routinely assessed for each extract using the QIAxpert spectrophotometer (Qiagen, Hilden, Germany).

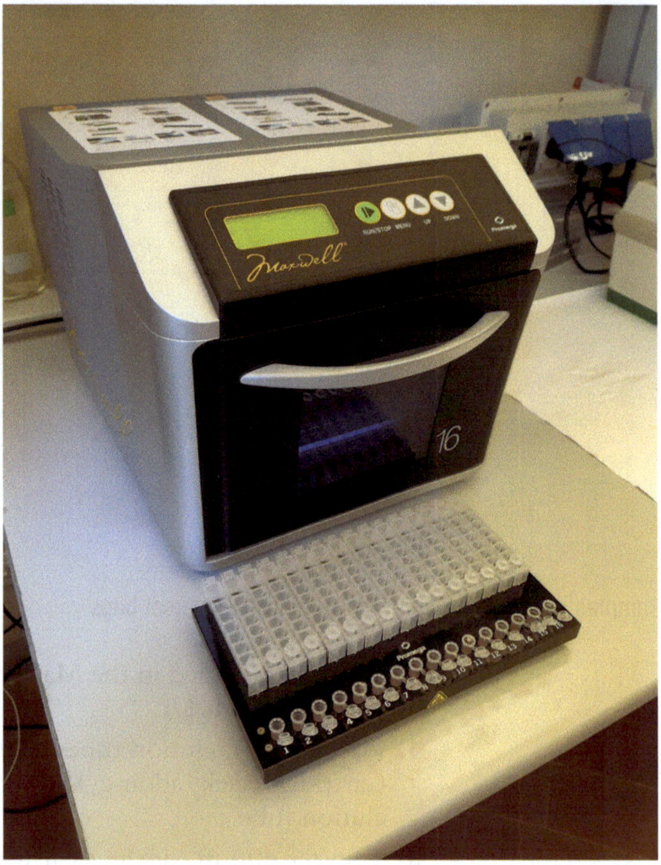

Fig. 6 Sterile microcentrifuge tubes are labeled according to the sample numbers. In the deck tray, elution buffer is added to the elution tubes. The plungers are in well #8 of the cartridge

4 Notes

1. Due to its mechanism of action, the working concentration of CTAB is variable according to the supposed amount of polysaccharides present in the sample: 1% can be fine in a simple matrix, while for complex matrices it can be as high as 5%.
2. CTAB buffer 2× is necessary in the presence of fresh tissues where the cells contain water which dilutes the buffer. When dehydrated samples are used, the concentration of the CTAB solution must therefore be reduced by half (CTAB 1× can be used or the sample can be rehydrated before buffer addition).
3. The CTAB solution is quite stable and, therefore, it can be stored at room temperature up to 3 months.
4. All solutions must be prepared using ultrapure water, which is obtained by purifying deionized water up to a resistivity of 18 MΩ-cm at 25 °C, and analytical grade reagents. All reagents

are stored at room temperature (unless otherwise indicated). All disposable products should be treated according to the local waste disposal regulations.

5. Put the reagents in a sterile beaker of suitable size (500–1000 mL) according to the volume of buffer to be prepared, and add the volumes of water required according to the final quantity of buffer to be obtained (100-, 200-, 500-, or 1000-mL H_2O ultrapure sterile). Use a heating stirrer setting it to approximately 65 °C to allow the CTAB dissolution. Once a clear solution has been obtained, bring to the desired final volume (100, 200, 500, 1000 mL) with ultrapure sterile water. Sterilize the solution by autoclaving it at 121 °C and 1 atm for 20 min. The solution can be stored at room temperature for up to 3 months.
6. The CTAB buffer has detergent properties, and, although it is irritating and corrosive, its use is essential because it binds to the polysaccharide molecules removing them from the solution, thus avoiding their binding to DNA. However, it can be found in solution as long as the NaCl concentration is greater than 0.7 M. The protocol can be used for different plant tissues as seeds, leaves, and embryos. The tissues should be grinded. The buffer must be added before DNA became in contact with the polysaccharides. The polysaccharides and CTAB buffer will be then removed from the solution using chloroform.
7. The described protocol is sufficient for 48 automated nucleic acid isolations from plant matrices or plant lysates. The recommended storage conditions of the kit are in the range of 15–30 °C.
8. The reagent cartridges contain ethanol and isopropanol, which are flammable; the guanidine hydrochloride contained in the lysis buffer is harmful and irritating. Wear gloves and follow standard safety procedures when handling these reagents. The cartridge must be handled with care because the edges of the gasket can be sharp. Detailed safety information can be found in the safety data sheets.
9. Verify that the sealing tape and any residue of adhesives are removed before positioning the cartridges in the instrument. If fewer than 16 samples are processed, the cartridges must be placed in the middle of the rack.
10. No oily traces should be present on the surface: the accidental transfer of all these materials could cause the inhibition of the downstream assays. If necessary, cleared lysate can be transferred to a new tube and centrifuged again, to avoid oily residues and particulate.

11. DNA segments obtained by this method have a length of approximately 50 kb which is acceptable for most of the subsequent applications.
12. A solution containing 30 mg of Tris base, 9 mg of disodium EDTA, and 200 mL of sterile H_2O (ultrapure) should be prepared in a sterile cylinder and stirred until the reagents are completely dissolved. The solution is brought to a final volume of 250 ml with further addition of sterile H_2O. Sterilize the solution in an autoclave at 121 °C at 1 atm for 15 min and store at room temperature.
13. Xylophagous insect pests produce aspecific damages to the host. In absence of the pest, the identification of the species can rely only on gallery size and shape, and on the host plant, but these often are not a sufficient criterion. At EU entry points, where a rapid and unambiguous identification is mandatory to avoid the introduction of invasive alien species, there is the strong need for reliable and rapid molecular tests.
14. During *Xylella fastidiosa* field survey carried out in Tuscany (Italy), DNA extraction from the mentioned plant species has proven to be difficult in terms of final integrity and purity. The DNA extraction protocol here presented gave excellent results when applied on these plant matrices.
15. This step allows the complete rupture of the cells and the denaturation of proteins [11]. To increase final DNA yield, the incubation time can be extended up to 60 min.
16. Chloroform is used to remove the CTAB that has bound to plant polysaccharides, protein, and lipids.
17. The polysaccharides, aromatic substances, and CTAB will remain in the organic phases.

References

1. Linacre AMT, Tobe SS (2013) Wildlife DNA analysis: applications in forensic science. Wiley, p 334
2. Sambrook J, Russell DW (2001) Molecular cloning – a laboratory manual, vol 3, 3rd edn. Cold Spring Harbor Laboratory Press, New York
3. Rezadoost MH, Kordrostami M, Kumleh HH (2016) An efficient protocol for isolation of inhibitor-free nucleic acids even from recalcitrant plants. 3 Biotech 6:61. https://doi.org/10.1007/s13205-016-0375-0
4. Akkurt M (2012) Comparison between modified DNA extraction protocols and commercial isolation kits in grapevine (*Vitis vinifera* L.). Genet Mol Res 11:2343–2351
5. Doyle JJ, Doyle JL (1987) A rapid DNA isolation procedure for small quantities of fresh leaf tissue. Phytochem Bull 19:11–15
6. Steiner JJ, Poklemba CJ, Fjellstrom R et al (1995) A rapid one-tube genomic DNA extraction process for PCR and RAPD analyses. Nucleic Acids Res 23:2569–2570
7. Heidari Japelaghi R, Haddad R, Garoosi GA (2011) Rapid and efficient isolation of high quality nucleic acids from plant tissues rich in polyphenols and polysaccharides. Mol Biotechnol 49:129–137
8. Csaikl UM, Bastian H, Brettschneider R et al (1998) Comparative analysis of different DNA extraction protocols: a fast, universal maxi-preparation of high quality plant DNA for

genetic evaluation and phylogenetic studies. Plant Mol Biol Rep 16:69–86

9. Li R, Mocka R, Huangb Q et al (2008) A reliable and inexpensive method of nucleic acid extraction for the PCR-based detection of diverse plant pathogens. J Virol Methods 154:48–55
10. Rizzo D, Da Lio D, Bartolini L et al (2021) The rapid identification of *Anoplophora chinensis* (Coleoptera: Cerambycidae) from adult, larval, and frass samples using TaqMan Probe assay. J Econ Entomol 114(5):2229–2235. https://doi.org/10.1093/jee/toab138
11. Ioos R, Fourrier C, Iancu G et al (2009) Sensitive detection of *Fusarium circinatum* in pine seed by combining an enrichment procedure with a real-time polymerase chain reaction using dual-labeled probe chemistry. Phytopathology 99:582–590. https://doi.org/10.1094/PHYTO-99-5-0582

Chapter 6

A Rapid Method for Extracting High-Quality DNA from Soils

Anna Maria Vettraino and Nicola Luchi

Abstract

Metagenomics offers the possibility to study the microbial community of soils at large scale, allowing the characterization also of unculturable microorganisms. Availability of high-quality DNA is a prerequisite of this approach. However, since soils have highly heterogeneous physicochemical properties, several parameters need to be considered for the development of the best molecular practices. A method for isolation of high-molecular-weight and good-quality metagenomic DNA from different soil samples is described here. The protocol combines physical and chemical strategies to ensure efficient cell lysis and precipitation of humic impurities-free DNA suitable for downstream processing for metagenomics study.

Key words Diverse soils, DNA isolation, Improved method, Metagenomic DNA, Biodiversity

1 Introduction

Microbial biodiversity within soil is significant for the balance of cycles of C, N, and P, along with maintaining plant health, structure, and fertility of soils. It is reported that 99% of microbes cannot be cultured in the laboratory using traditional culture techniques [1]. Thus, studies of microbial diversity can be incomplete if based only on isolation strategies. Metagenomics, which study the genetic material recovered directly from environmental samples, provide new insights into microbiome diversity. However, metagenomic DNA extraction remains a challenging task since soils are a heterogeneous mixture of many chemicals which affect the quality of DNA and inhibit PCR [2, 3].

Several procedures have already been developed for extraction of metagenomic DNA from soils with good quality and purity to fit metagenomics requirements [4]. In addition, commercial kits have also been specifically commercialized for the extraction of DNA from soils. Although some of these methods are shown to be suitable for the metagenomics downstream processing, like PCR amplification or restriction, they could be time-consuming, expensive, and not applicable to all types of soil samples.

Nicola Luchi (ed.), *Plant Pathology: Methods and Protocols*, Methods in Molecular Biology, vol. 2536,
https://doi.org/10.1007/978-1-0716-2517-0_6,

The protocol described below is simple, fast, and cost-effective. It ensures the yield of DNA, with no humic impurity, suitable for the preparation of metagenomic libraries. The protocol, in fact, includes chemical lysis in association with a physical method, which allows a great access to the whole microbial community [5].

2 Materials

2.1 Sample Collection

1. Plastic bags.
2. Shovel.
3. Ethanol (75–100%).
4. Sterile water.
5. Paper.
6. Disposable gloves.

2.2 DNA Extraction from Soil and Analysis

1. Bead beating tubes.
2. Bead-beater cell disruptor.
3. Glass grinding beads.
4. 1.5-ml Eppendorf tubes.
5. Disposables gloves.
6. Warm bath.
7. Centrifuge.
8. Freezer.
9. Micropipettes.
10. Qubit fluorometer.
11. Quant-iT dsDNA HS Assay.
12. Horizontal gel electrophoresis units.
13. UV lamp.

2.3 Solutions

1. Extraction buffer: 100-mM Tris–HCl (pH 8.0), 100-mM EDTA (pH 8.0), 100-mM NaCl, and 2% (wt/vol) sodium dodecyl sulfate (SDS) (*see* **Note 1**).
2. Agarose gel: Dilute GelRed® to 1× concentration using concentrated electrophoresis buffer. Add agarose powder and heat to dissolve (*see* **Note 1**).
3. TBE buffer 10× stock recipe: 108-g Tris–HCl base, 55-g boric acid, 900-mL double-distilled water; 40-mL 0.5-M EDTA solution (pH 8.0). Adjust volume to 1 L. 1× TBE preparation: Dilute 10× concentrated TBE buffer tenfold with ultrapure water (*see* **Note 1**).

3 Methods

3.1 Sample Collection

A representative number of soil subsamples should be randomly collected in the field and thoroughly bulked. The number and the weight of subsamples vary according to the field size and uniformity. Soil samples should be immediately processed. If they are wet, they should be air-dried at room temperature for 2 to 3 days. Shovels used for soil collection should be sterilized each sampling time (*see* **Notes 2** and **3**).

3.2 DNA Extraction

1. Add 0.300-g soil samples to 2-mL bead beating tube containing 0.1 g of 2-mm-diameter glass beads.
2. Add 1-mL extraction buffer and homogenized sample for 30 s at 1600 rpm in a mini-bead-beater cell disruptor.
3. Keep incubation tubes for 30 min at 65 ° C with invert mixing after every 10–15 min.
4. Centrifuge tubes at 7000 g for 20 min at 4 °C to collect the supernatant.
5. Extract the soil pellets by adding 1 mL of extraction buffer, followed by incubation at 65 °C for 15 min (*see* **Note 4**).
6. Pool the supernatants of two extractions, and mix with equal volume of chloroform/isoamyl alcohol (24:1, v/v) (*see* **Note 5**).
7. Add to the aqueous phase 0.1-volume 3-M sodium acetate along with 0.4-volume 30% (w/v) PEG 8000 (polyethylene glycol), and incubate tubes at −20 °C for 20–60 min.
8. Centrifuge the tubes at 14,000 g for 30 min at 4 °C, wash the pellet with 70% (v/v) ethanol (room temperature), and air-dry (*see* **Note 6**).
9. Dissolve the pellet in 1 ml of 1× TE buffer, mix with equal volume chloroform/isoamyl alcohol (24:1, v/v), and centrifuge at 14,000 g for 10 min at 4 °C (*see* **Note 6**).
10. Mix the aqueous phase with one volume of ice-cold isopropanol, turn the tubes upside down for 1 min to precipitate DNA, or incubate overnight at −20 °C.
11. Centrifuge the tubes at 10,000 rpm for 2 min to recover the pellet, and wash with 200 μL of absolute ethanol; then centrifuge at 10,000 rpm for 1 min.
12. Air-dry DNA and resuspend in 200-μL deionized water or TE buffer for further use and store at −20 °C.

3.3 DNA Quality and Quantity

1. Each sample is quantified with the Qubit fluorometer. The instrument is calibrated with the Quant-iT dsDNA HS Assay (declared assay range between 0.2 and 100 ng; sample starting

concentration ranged between 10 ng/μL and 100 pg/μL) according to the manufacturer's instructions.

2. The extracted DNA was evaluated by loading 100 ng of DNA based on Qubit measurements on a 0.6% agarose gel electrophoresis.
3. Red-stained agarose gels were acquired with an UV lamp and a 520-nm low-pass filter.

4 Notes

1. Prepare all solutions using ultrapure water. Use a freshly prepared solution. Diligently follow all waste disposal regulations when disposing waste materials. Use lab gloves and lab coats when performing DNA extraction.
2. Soils are characterized by a complex structural and physicochemical heterogeneity which influences the dispersion of microbial populations. Therefore, soil sample size can affect measures of community structure and comparison among experiments. It is crucial for researchers to optimize high-throughput sequencing-based experiments. However, no specific sample sizes can be defined for the analysis of different soils. Several studies agree that large soil samples are most appropriate for the description of the overall soil community but large numbers of small samples are more suitable for a determination of local microbial diversity [6, 7]. We suggest to bulk soil subsamples and to perform DNA extraction on different replicates.
3. The structure of the microbial communities is highly sensitive to DNA extraction procedures, and soil community varies with respect to soil matrix, organic and inorganic compounds, and biotic factors. Thus, we recommend carrying out preliminary tests of different DNA extraction methods according to the microbial domain or parameter (abundance, diversity, etc.) of interest in studies related to the soil microbiome characterization.
4. The extraction buffer can be added with 100-mM $CaCl_2$ to enhance the removal of humic acid whose presence inhibits the activity of some enzymes [8].
5. Steps **5** and **6** can be avoided according to the type of samples.
6. Steps **8** and **9** can be avoided if soils are not particularly rich with clay or humic acid and fulvic acid.

References

1. Burgmann H, Pesaro M, Widmer F et al (2001) A strategy for optimizing quality and quantity of DNA extracted from soil. J Microbiol Methods 45:7–2001
2. Amorim JH, Macena TNS, Lacerda-Junior GV et al (2008) An improved extraction protocol for metagenomic DNA from a soil of the Brazilian Atlantic rainforest. Genet Mol Res 7:1226–1232
3. Verma D, Satyanarayana T (2011) An improved protocol for DNA extraction from alkaline soil and sediment samples for constructing metagenomic libraries. Appl Biochem Biotechnol 165: 454–464
4. Verma SK, Singh H, Sharma PC (2017) An improved method suitable for isolation of high-quality metagenomic DNA from diverse soils. 3 Biotech 7:171
5. Robe P, Nalin R, Capellano C et al (2003) Extraction of DNA from soil. Eur J Soil Biol 39:183–190
6. Ranjard L, Lejon DP, Mougel C et al (2003) Sampling strategy in molecular microbial ecology: influence of soil sample size on DNA fingerprinting analysis of fungal and bacterial communities. Environ Microbiol 5:1111–1120
7. Nicol GW, Glover LA, Prosser JI (2003) Spatial analysis of archaeal community structure in grassland soil. Appl Environ Microbiol 69: 7420–7428
8. Dong D, Yan A, Liu H et al (2006) Removal of humic substances from soil DNA using aluminium sulphate. J Microbiol Methods 66:217–222

Part III

Molecular Methods to Detect Plant Pathogens

Chapter 7

Invasive Alien Plant Pathogens: The Need of New Detection Methods

Alberto Santini and Duccio Migliorini

Abstract

Invasive alien species are a major threat to natural and anthropogenic ecosystems and to the economy. Many invasive fungal species have severely impacted ecology and human lifestyle in the past. Most of them express a pathogenic lifestyle following introduction into a new region and hosts. They are usually cryptic during the introduction phase and hard to be identified, classified, and monitored.

The increasing number of new alien pests coincide with the rapid increase in the volume and diversity of intercontinental trade in plants for planting, underlying the need to reduce the risk of their introduction with the development of molecular-based, inexpensive, rapid, accurate, and reliable methods that can identify and intercept plant pathogens even before symptoms occur in the new environment of diffusion. Applicative aerobiology, for instance, represents a challenging research line for the implementation of pest detection protocols during the early stage of fungal introduction, being capable to target aerial dispersed propagules.

In addition to this, new metabarcoding protocols based on an innovative multigene approach, although not yet tested on fungi, are able to provide an output with very high taxonomic resolution and are likely to be considered in the next-future biosurveillance of invasive fungal pathogens.

Key words Aerobiology, Biosurveillance, Detection, Plant pathogens, Multigene metabarcoding

1 Introduction

Invasive alien species (IAS), i.e., "species whose introduction and/or spread outside their natural past or present distribution threatens biodiversity" according to Convention on Biological Diversity definition, are a major threat to natural and anthropogenic ecosystems and to economy. They are the bitter fruit of the "globalization" era, and there are no signs of a possible plateau in the accumulation of alien species [1], and it is conceivable that the damage they caused will continue to increase [2]. Biological invasions are not merely an ecological issue; they have profound implications in other disciplines including economics, ethics, sociology, and inter- and transdisciplinary studies [3]. Any organism from any

Nicola Luchi (ed.), *Plant Pathology: Methods and Protocols*, Methods in Molecular Biology, vol. 2536,
https://doi.org/10.1007/978-1-0716-2517-0_7,

taxonomic kingdom, once introduced into a new environment, can become an IAS; however, to date, invasion science has focused more on plants and animals [4]. Although underrepresented, microorganisms can cause serious impacts on ecology and economy or even on human lifestyle, an invasive alien pathogen was the cause of the Irish famine (*Phytophthora infestans*), causing a one and a half million deaths from hunger and about as many to leave Ireland to emigrate, mostly to the USA, while the agent of the chestnut blight (*Cryphonectria parasitica*), an Asian species unintentionally introduced to the USA, has caused the disappearance of chestnut from North American landscapes, or Dutch elm disease, caused by some fungi of the genus *Ophiostoma*, which were able, in the turn of a century, to wipe out the elms from Europe and North America [5]. Recently a Southeast Asian ambrosia beetle, *Euwallacea fornicatus* (polyphagous shot-hole borer, PSHB), was introduced to California, Israel, and South Africa. The beetle is associated to the fungal symbiont *Fusarium euwallaceae*, and they are an emerging invasive pest causing significant and costly damage to urban and natural trees. Female beetles dig brood galleries into trees, introducing the symbiotic fungus, which degrades gallery walls, becoming a food source for larvae and young adult beetles [6].

Plants and pathogens evolved in unique regional contexts and are isolated from others by insurmountable geographic barriers (mountain ranges, oceans, etc.). When, thanks to human activity, barriers are overcome, non-indigenous pathogenic organisms are introduced into new environments, and the consequences of exposure to novel hosts in the invaded range are difficult to predict. In the new environment, they cannot find the condition to establish and then extinguish, or they may find suitable hosts lacking resistance genes and environments favoring pathogenic behavior, potentially leading to an epidemic outbreak. Many of the alien pathogens responsible for severe epidemics of disease in plants were not known to be pathogenic before becoming established on the new hosts in the new region. Similarly, many latent plant pathogens have coevolved with native hosts, in which they behave as endophytes or epiphytes, with low virulence and no evident impact on the host within the native range but expressing a pathogenic lifestyle following introduction into a new region. For example, the ash fungal pathogen *Hymenoscyphus fraxineus* (*Lambertella albida*) was known to be nonpathogenic in its native range, Far East, Asia [7], but its introduction into Europe led to widespread dieback of the ash species *Fraxinus excelsior* and *Fraxinus angustifolia*, with which it had not coevolved, threatening the existence of this species in some parts of their native range in Europe.

2 Key Challenges for Invasive Species

Recently Paap et al. [8] have listed several key challenges to studying microorganisms as invasive species. First, microorganisms are a major component of biodiversity on Earth, and their diversity is wide, but being inconspicuous organisms, they have been far less studied than mammals or vascular plants: for example, the number of fungal species has been estimated to be at least 1.5 million, but less than 10% have been described [9]. In addition, their taxonomy is complex and it is not always easy to distinguish one species from another.

Second, even when a microorganism is known, it is hard to know its geographical origin. About 30% of European invasive pathogens are of unknown origin [10]. For example, *Ophiostoma ulmi* s.l. (causal agents of one of the most devastating tree pandemics, Dutch elm disease) is supposed to be originated in Eastern Asia and introduced in Europe and North America in the early 1900s, but there are no evidences [11]. This makes impossible to search for potential biological control agents [12].

Third, the functional diversity of microorganisms is wide, many different life-history strategies exist, and many different traits are involved.

Detection is the fourth key challenge. Invasive microorganisms are often only perceived when their impacts become palpable (Fig. 1).

Microorganisms are cryptic, and, as previously mentioned, they are hard to be identified and classified, making difficult any effort to monitor their presence and determine their function seriously affecting our ability to proactively manage their invasion pathways and to reduce the chances of further epidemic.

The increasing number of new alien pests in many parts of the world [10, 13, 14] coincides with the rapid increase in the volume and diversity of intercontinental trade in plants for planting [10, 13, 15, 16]. The large number of plants traded and the inconspicuousness of microorganisms emphasize the importance of limiting the risk associated with this pathway [17]. Due to the vast number of imported plants, commodity border controls are limited to visual inspections that will not completely prevent the introduction of new pests because they may be difficult to see or they can remain latent for a long part of their life [18–20]. Research has shown that many plant pathogens experience a lag between establishment and detection, which is generally associated with an outbreak [21]. This detection time lag further limits the chance of a successful eradication.

Fig. 1 Examples of impacts caused by invasive forest pathogens: (**a**) ash dieback, caused by *Hymenoscyphus fraxineus*; (**b**) Dutch elm disease caused by *Ophiostoma novo-ulmi*; (**c**) pine pitch canker caused by *Fusarium circinatum*; (**d**) canker stain disease, caused by *Ceratocystis platani*; and (**e**) *Euwallacea fornicatus*, polyphagous shot-hole borer

3 Innovative Methods for Biosurveillance

A prerequisite for setting up measures to reduce the risk of introduction of potentially harmful organisms in the frame of international and national phytosanitary regulations is that the identity of the harmful organism is clear. Classical detection methods are time-consuming and not precise enough and may lead to an inaccurate diagnosis: false negative may have significant economic and ecological consequences on native species; false positive, in case of quarantine organisms, may lead to an economic impact for the measures put in act. It has also been noted that there are marked differences between interception and establishment data [22]. This may be also due to the fact that there is a very short time available for the phytosanitary inspection and, given the difficulty of reliably identifying pathogens only basing on symptoms or signs of their presence, many pathogens may pass borders unspotted. A harmonized pool of new technical, methodological, and conceptual solutions for biosurveillance is needed. Rapid and accurate identification of biological samples collected during border inspections is, therefore, an essential key point in preventing the introduction of alien pathogens [23]. New molecular-based, inexpensive, rapid, accurate, and reliable methods that can identify and intercept plant pathogens even before symptoms occur in the host are therefore needed.

4 Airborne Spore Trapping

A pathway even more complex to control and manage is that of airborne plant pathogens, even if the means of dispersal is more common at continental level; however, there are some groups able to travel over very long distances keeping their viability and pathogenicity [24], and there are reports of transoceanic movement, occurred as a consequence of extreme and rare meteorological events, that has led to epidemic events [25]. Sampling fungal spores in the air with the purposes of biosurveillance and monitoring to detect the introduction and to study the dispersion of invasive plant pathogens [26] is a field of application of aerobiology (Fig. 2). To this aim spore sampling and quantitative DNA analyses methods, specifically quantitative real-time PCR, have been combined resulting in efficient early detection of pathogens [26–29]. The main limitation of this method is that it can only analyze organisms known to science and only one or a few species at a time [30]. In order to remove this limitation, the use of semiquantitative HTS is increasingly resorted to [31]. This scenario is also becoming a reality in plant pathology and fungal ecology [32].

Fig. 2 Examples of spore traps. (**a**) Volumetric cyclone sampler (Burkard Manufacturing Co Ltd); (**b**) volumetric VPPS 2010 (Lanzoni s.r.l.); (**c**) rotating arm spore trap ROTTRAP 68 (Miloň Dvořák, Boršov nad Vltavou, Czech Republic)

5 Molecular Promising Approaches

A very promising method is represented by a mixed technique, where eDNA samples are processed either by quantification (qPCR) of target organisms at the species level or by metabarcoding [33]. This double approach received so far little attention in the study of aerial trapped fungi [34, 35]; however, as underlined in [36] for the clinic sector, a precise estimation of the abundance of certain taxa is a key outcome when important allergens and pathogens are detected by metabarcoding, for which there is the need to define species- and dose-effect relationships.

Furthermore, given the speed at which metabarcoding has evolved in recent years, the applied research in plant pathology will adopt innovative HTS techniques probably already in the near future. It is highly possible that deep sequencing of environmental DNA samples will be carried out using a multigene approach, with sequencing instruments and bioinformatics pipelines capable of obtaining and analyzing several target genes simultaneously. This research approach has not been explored yet on fungi, including in the field of fungal ecology [37], but it has already been utilized as a population genetic tool in exosystemic studies [38], and it definitely represents the next challenge of plant pathology.

New methods and tools are extremely needed to unmask the hidden identity of new pathogens, thus preventing their introduction possibly threatening the survival of a single plant species up to an ecosystem.

References

1. Seebens H, Blackburn TM, Dyer EE et al (2017) No saturation in the accumulation of alien species worldwide. Nat Commun 8:1–9. https://doi.org/10.1038/ncomms14435
2. Pyšek P, Hulme PE, Simberloff D et al (2020) Scientists' warning on invasive alien species. Biol Rev 95:1511–1534. https://doi.org/10.1111/brv.12627
3. Hui C, Richardson DM (2017) Invasion dynamics. Oxford University Press
4. Wilson JRU, Datta A, Mbobo T et al (2020) Is invasion science moving towards agreed standards? The influence of selected frameworks. NeoBiota 62:569–590. https://doi.org/10.3897/neobiota.62.53243
5. Paap T, Wingfield MJ, Burgess TI et al (2020) Harmonising the fields of invasion science and forest pathology. NeoBiota 62:301–332. https://doi.org/10.3897/neobiota.62.52991
6. Paap T, de Beer ZW, Migliorini D et al (2018) The polyphagous shot hole borer (PSHB) and its fungal symbiont *Fusarium euwallaceae*: a new invasion in South Africa. Australas Plant Pathol 47:231–237. https://doi.org/10.1007/s13313-018-0545-0
7. Gross A, Hosoya T, Queloz V (2014) Population structure of the invasive forest pathogen *Hymenoscyphus pseudoalbidus*. Mol Ecol 23:2943–2960. https://doi.org/10.1111/mec.12792
8. Paap T, Wingfield MJ, Burgess TI et al (2022) Invasion frameworks: a forest pathogen perspective. Curr For Reports 8:74–89. https://doi.org/10.1007/s40725-021-00157-4

9. Hawksworth DL (2001) The magnitude of fungal diversity: the 1.5 million. Mycol Res 105:1422–1432. https://doi.org/10.1017/S0953756201004725
10. Santini A, Ghelardini L, De Pace C et al (2013) Biogeographical patterns and determinants of invasion by forest pathogens in Europe. New Phytol 197:238–250. https://doi.org/10.1111/j.1469-8137.2012.04364.x
11. Brasier CM, Buck KW (2001) Rapid evolutionary changes in a globally invading fungal pathogen (Dutch elm disease). Biol Invasions 3: 223–233. https://doi.org/10.1023/A:1015248819864
12. Prospero S, Botella L, Santini A, Robin C (2021) Biological control of emerging forest diseases: how can we move from dreams to reality? For Ecol Manag 496:119377. https://doi.org/10.1016/j.foreco.2021.119377
13. Liebhold AM, Brockerhoff EG, Garrett LJ et al (2012) Live plant imports: the major pathway for forest insect and pathogen invasions of the US. Front Ecol Environ 10:135–143. https://doi.org/10.1890/110198
14. Nahrung HF, Carnegie AJ (2020) Non-native forest insects and pathogens in Australia: establishment, spread, and impact. Front For Glob Change 3:1–12. https://doi.org/10.3389/ffgc.2020.00037
15. Kenis M, Rabitsch W, Auger-Rozenberg MA, Roques A (2007) How can alien species inventories and interception data help us prevent insect invasions? Bull Entomol Res 97: 489–502. https://doi.org/10.1017/S0007485307005184
16. Roques A, Rabitsch W, Rasplus JY et al (2009) Alien terrestrial invertebrates of Europe. In: Handbook of alien species in Europe. Springer, Dordrecht, pp 63–79
17. Eschen R, O'Hanlon R, Santini A et al (2018) Safeguarding global plant health: the rise of sentinels. J Pest Sci 92:29–36. https://doi.org/10.1007/s10340-018-1041-6
18. Migliorini D, Ghelardini L, Tondini E et al (2015) The potential of symptomless potted plants for carrying invasive soilborne plant pathogens. Divers Distrib 21:1218–1229. https://doi.org/10.1111/ddi.12347
19. Meurisse N, Rassati D, Hurley BP et al (2019) Common pathways by which non - native forest insects move internationally and domestically. J Pest Sci 92:13–27. https://doi.org/10.1007/s10340-018-0990-0
20. Vettraino AM, Santini A (2021) Forest health in Italy: learning from the Xylella incursion. Front For Glob Chang 4:1–8. https://doi.org/10.3389/ffgc.2021.699393
21. Chapple DG, Simmonds SM, Wong BBM (2012) Can behavioral and personality traits influence the success of unintentional species introductions? Trends Ecol Evol 27:57–64. https://doi.org/10.1016/j.tree.2011.09.010
22. Eschen R, Roques A, Santini A (2015) Taxonomic dissimilarity in patterns of interception and establishment of alien arthropods, nematodes and pathogens affecting woody plants in Europe. Divers Distrib 21:36–45. https://doi.org/10.1111/ddi.12267
23. Hamelin RC, Roe AD (2020) Genomic biosurveillance of forest invasive alien enemies: a story written in code. Evol Appl 13:95–115. https://doi.org/10.1111/eva.12853
24. Stakman EC, Christensen CM (1946) Aerobiology in relation to plant disease. Bot Rev 12: 205–253
25. Brown JKM, Hovmøller MS (2002) Aerial dispersal of fungi on the global and continental scales and its consequences for plant disease. Science 297:537–541. https://doi.org/10.1126/science.1072678
26. Chandelier A, Helson M, Dvorak M, Gischer F (2014) Detection and quantification of airborne inoculum of *Hymenoscyphus pseudoalbidus* using real-time PCR assays. Plant Pathol 63:1296–1305. https://doi.org/10.1111/ppa.12218
27. Dvořák M, Janoš P, Botella L, et al (2017) Spore dispersal pattern of *Fusarium circinatum* on an infected Monterey pine forest in North-Western Spain. In: Invasive forest pathogens & implications for biology & policy, IUFRO Working Party 7.02.02 May 7-11, 2017 Niagara Falls, Ontario (USA). p 23
28. Grosdidier M, Ioos R, Husson C et al (2018) Tracking the invasion: dispersal of Hymenoscyphus fraxineus airborne inoculum at different scales. FEMS Microbiol Ecol 94:1–11. https://doi.org/10.1093/femsec/fiy049
29. Quesada T, Hughes J, Smith K et al (2018) A low-cost spore trap allows collection and real-time PCR quantification of airborne *Fusarium circinatum* spores. Forests 9:6–10. https://doi.org/10.3390/f9100586
30. Luchi N, Pepori AL, Bartolini P et al (2018) Duplex real-time PCR assay for the simultaneous detection of *Caliciopsis pinea* and *Fusarium circinatum* in pine samples. Appl Microbiol Biotechnol 102:7135–7146. https://doi.org/10.1007/s00253-018-9184-1

31. Castaño C, Berlin A, Brandström Durling M et al (2020) Optimized metabarcoding with Pacific biosciences enables semi-quantitative analysis of fungal communities. New Phytol 228:1149–1158. https://doi.org/10.1111/nph.16731
32. Banchi E, Pallavicini A, Muggia L (2020) Relevance of plant and fungal DNA metabarcoding in aerobiology. Aerobiologia (Bologna) 36: 9–23. https://doi.org/10.1007/s10453-019-09574-2
33. Chandelier A, Hulin J, Martin GS et al (2021) Comparison of qPCR and metabarcoding methods as tools for the detection of airborne inoculum of forest fungal pathogens. Phytopathology 111:570–581. https://doi.org/10.1094/PHYTO-02-20-0034-R
34. Dannemiller KC, Lang-Yona N, Yamamoto N et al (2014) Combining real-time PCR and next-generation DNA sequencing to provide quantitative comparisons of fungal aerosol populations. Atmos Environ 84:113–121. https://doi.org/10.1016/j.atmosenv.2013.11.036
35. Nicolaisen M, West JS, Sapkota R et al (2017) Fungal communities including plant pathogens in near surface air are similar across northwestern Europe. Front Microbiol 8:1–11. https://doi.org/10.3389/fmicb.2017.01729
36. Rocchi S, Valot B, Reboux G, Millon L (2017) DNA metabarcoding to assess indoor fungal communities: electrostatic dust collectors and Illumina sequencing. J Microbiol Methods 139:107–112. https://doi.org/10.1016/j.mimet.2017.05.014
37. Adams CIM, Knapp M, Gemmell NJ et al (2019) Beyond biodiversity: can environmental DNA (eDNA) cut it as a population genetics tool? Genes (Basel) 10(3):192. https://doi.org/10.3390/genes10030192
38. Weitemier K, Penaluna BE, Hauck LL et al (2021) Estimating the genetic diversity of Pacific salmon and trout using multigene eDNA metabarcoding. Mol Ecol 30(20): 4970–4990. https://doi.org/10.1111/mec.15811

Chapter 8

Detection of Airborne Inoculum of *Hymenoscyphus fraxineus:* The Causal Agent of Ash Dieback

Miloň Dvořák

Abstract

The inoculum of *H. fraxineus* consists mainly of ascospores released from apothecia which are growing on fallen leaves infected during the previous year. The ascospores can be detected in various manners due to their high concentration in the air during the main sporulation season, which corresponds to astronomic summer. This methodology is focused on one of the methods which have been successfully used. It employs a cheap, but highly efficient rotating arm air sampler and a specific quantitative real-time PCR method for the quantification of the air samples. The methodology is accompanied by lots of detailed theoretical and practical notes for its smooth application, including mentioning other alternatives.

Key words Rotorod, ROTTRAP, Rotating arm air sampler, Spore trap, Real-time PCR, qPCR, Spore suspension, Spore dispersal pattern, Sampling rate

1 Introduction

Hymenoscyphus fraxineus (T. Kowalski) Baral, Queloz & Hosoya is an alien invasive fungal pathogen currently severely destructing ash trees and their ecosystems throughout Europe [1–3]. It is an ascomycete fungus infecting leaves, shoots, branches, stems, root collars, and seeds of ashes [4–9]. Spatial spread of this pathogen is provided mainly by sexual propagules = ascospores released from apothecia [5] with a contribution of the asexual conidia [5, 6]. The apothecia are commonly grown on leaf litter (Fig. 1) under infected trees during summer [5, 6, 10–13]. Detection of ascospores as infective propagules has considerable importance for the management of ash dieback disease. It can help in detection of the disease without following the symptoms, neither even the host trees—for example landscape-scale surveillance [14]. Data about the detected inoculum can be used for modeling of the disease spread in a particular area, or can help to describe a particular seasonal pattern

Nicola Luchi (ed.), *Plant Pathology: Methods and Protocols*, Methods in Molecular Biology, vol. 2536,
https://doi.org/10.1007/978-1-0716-2517-0_8,

Fig. 1 Matured apothecia growing from the pseudosclerotium on ash leave litter

of spread. Such outputs can provide valuable support for effective forest management, like, for example, curative treatment of the disease in forest nurseries.

Detection of the *H. fraxineus* propagules in the air can be performed by various air samplers based on various principles. The simplest air samplers which have been used for this task are passive spore traps. Spores are randomly trapped on a sticky [12] or wet [15] surface without any activity of the trap to attract them. Active air samplers are more complicated and are equipped with a mechanism to concentrate the air particles to enhance the probability of the positive result of the sampling. Some types of active spore traps called volumetric air samplers sample a measurable amount of air and usually enable sampling in real time to estimate exactly the time particle occurrence [11, 16–18]. They are mostly constructed based on the principle of Hirst [17] and have been used for seasonal and diurnal dispersal patterns studies of *H. fraxineus* [11, 18]. Although they are perfect for the studies describing the dynamics of spore dispersal, there are other active air samplers which can be used, preferably due to their lower price and weight or higher sampling rates, like for example rotating arm spore traps (rotorods) [19, 20] described by Perkins [21] and later modified [12, 22]. Rotorods have been successfully employed for the detection and quantification of *H. fraxineus* inoculum in Belgium [12] and the Czech Republic [18, 23].

This chapter is focused on the detection and quantification of *H. fraxineus* by rotorods. As a model case of rotorod sampler, the ROTTRAP 104 (Fig. 2; Miloň Dvořák, Boršov nad Vltavou, Czech Republic) is presented, but the general principles for the rotorod construction and choice are properly explained. For sample

Fig. 2 ROTTRAP 104 without the sampling tape

processing and quantification, a modified protocol of Chandelier et al. [12, 24] is introduced by using the qPCR LightCycler® 480 Instrument II (Roche Diagnostics, Basel, Switzerland). The detection process is divided into three main steps:

1. Field sampling with rotorods;
2. DNA extraction from field samples;
3. Detection and quantification of the inoculum by real-time quantitative PCR (qPCR).

2 Materials

2.1 Field Sampling

1. Rotorods: ROTTRAP (Miloň Dvořák, Boršov nad Vltavou, Czech Republic).
2. Power supply: adaptor providing DC 9 V of at least 500 mA or an accumulator 9 V/20 Ah.
3. Double-sided non-woven tape (Tesa SE, Norderstedt, Germany).
4. 2-mL screw cap microtubes.
5. Laboratory tweezers.
6. 75% Ethanol.

2.2 DNA Extraction

1. 0.1- and 0.5-mm glass beads.
2. 0.1% 4-Nonylphenyl-polyethylene glycol (= Nonidet P40—substitute; AppliChem, Darmstadt, Germany).
3. DNEasy Plant Mini Kit (Qiagen, Valencia, CA, USA).
4. 1.5-mL microcentrifuge tubes.

5. Pipettes and pipette tips.
6. Homogenizer (Mixer Mill MM400; Retsch, Haan, Germany).
7. Microcentrifuge.
8. Vortex.
9. Disposable gloves.

2.3 Real-Time Quantitative PCR (qPCR)

1. qPCR instrument (LightCycler® 480 II; Roche Diagnostics, Basel, Switzerland).
2. LightCycler® 480 Probes Master (Roche Diagnostics, Basel, Switzerland).
3. Forward primer Cf-F (5′-3′): CCCTTGTGTATATTA TATTGTTGCTTTAGC.
4. Reverse primer Cf-R (5′-3′): GGGTCCTCTAGCAGGCA CAGT.
5. Dual-labeled probe Cf-S (5′-3′): 6FAM-TCTGGGCGTCGGCCTCGG-BHQ1.
6. Deionized sterile water (DSW).
7. Plates, strips, or microtubes specific for qPCR instrument.
8. Pipettes and pipette tips.
9. Disposable gloves.

3 Methods

3.1 Ascospore Sampling by Rotorods

1. Sampling of *H. fraxineus* ascospores is carried out by rotorods (ROTTRAP) (Figs. 2 and 3) (*see* **Notes 1** and **2**).
2. The trapping medium of the rotorods—double-sided tape—should cover only the area of the vertical rods. The best way how to apply the tape is to:
   (a) clean vertical rods with 96% ethanol;
   (b) cut 1 cm of the tape (width 5 cm);
   (c) fix it on the clean vertical rods' impaction side (5 cm long) having the rod approximately in the middle of the tape and remove the yellow cover strip with forceps (Fig. 4);
   (d) cut the flanking parts with a very sharp razor or scalpel following the edge of the vertical rod (Fig. 5) (*see* **Note 3**).
3. ROTTRAP is normally mounted at a chosen height on a bar (Fig. 3) using a worldwide available shelve console Element system (DIY Element System GmbH, Rottenacker, Germany) (*see* **Notes 4** and **5**). Sampling targeted to a vertical dispersal pattern or broader scale surveillance can be easily proceeded with high objects in the infected areas. In Fig. 6, the rotorods are mounted in different heights of an agricultural silo (*see* **Note 6**).

Fig. 3 ROTTRAP 52 mounted on the "Element system" bar

Fig. 4 Fixing the double-sided tape to the rods

4. Power supply is provided by commonly available accumulators providing DC of 9 V (or 12 V for other spore traps) (*see* **Note 7**). The voltage directly influences the sampling rate—volume of air sampled per unit of time (*see* **Note 8**).
5. The duration of the sampling is normally 48 h [12, 18, 23], but may be shortened due to higher concentrations of other particles (dust or pollen) causing an overload of the trapping medium to even 24 h [25].
6. After sampling, the exposed tapes are taken out by a clean forceps and stuck on the inner wall of a microtube (1.5 or 2.0 ml) and stored at −20 °C (*see* **Note 9**).

Fig. 5 Cutting the flanking edges of the tape by razor

3.2 DNA Extraction of H. fraxineus *Spores from Tapes*

1. Proper spore disruption (*see* **Note 10**) should be processed with 0.3 g of 0.1-mm and 0.1 g of 0.5-mm glass beads (Pkit, Prague, Czech Republic) added to the microtubes with the samples together with 250 μL of 0.1% Nonidet P40—substitute (AppliChem, Darmstadt, Germany) and disrupted for 10 min by 30 Hz in a homogenizer (Mixer Mill MM400—Retsch, Haan, Germany).
2. For further processing, the DNEasy Plant Mini Kit (or other extraction kit) (*see* **Note 11**) is used according to the manufacturer's instructions.
3. The final elution is reduced to only one-step elution with 100 μL of buffer AE. Extracted DNA is stored at −20 °C (*see* **Note 12**).

3.3 Ascospore Suspension

1. Ascospore suspension must be prepared from freshly collected apothecia (*see* **Note 13**).
2. Add 1 ml of Nonidet into 1.5-mL microtubes and go to the ash dieback focus early in the morning.
3. Find a rachis with high abundance of clean (dust and debris free), fresh, and matured apothecia (Fig. 7).
4. Detach one apothecium with forceps without touching the receptacle. Open the microtube and touch a few times (10–12×) Nonidet level with the upper surface of the receptaculum (*see* **Note 14**).
5. Repeat with more microtubes (to have at least three repetitions).

Fig. 6 ROTTRAP 52 mounted in different heights (red arrows) on an old agricultural silo

6. Back in the laboratory check the content of each microtube after 5 min of still standing. Transfer the clean part of the suspension (avoid sedimenting pollution) into a new microtube and vortex it properly.
7. Drop a droplet of the suspension (50 μL) onto the hemocytometer (Bürker chamber) and cover it with a cover glass.
8. Estimate the concentration of spores (*see* **Note 15**).
9. Vortex properly again and prepare the serial ten-fold dilutions of the suspension to get 10^5 to 10 ascospores into each microtube with clean (unexposed) double-sided tape to simulate the standard content of an exposed sample.
10. Extract the DNA the same way like the field samples.

Fig. 7 Matured apothecia are visible on the dark and moist soil surface under the infected ashes

3.4 qPCR Detection and Quantification

1. Specific quantitative real-time PCR is performed using specific primers and dual-labeled probe (*see* **Note 16**) mixed in the master mix previously described (*see* **Note 17**).
2. The qPCR is performed in microtubes (*see* **Note 18**) in a final volume of 20 μL of a mixture composed of:
   (a) 10 μL Light Cycler 480 Probes master mix;
   (b) Forward primer Cf-F: final concentration 400 nmol/L (0.8 μl of a working solution 10 pmol/μL);
   (c) Reverse primer Cf-R: final concentration 400 nmol/L (0.8 μL of a working solution 10 pmol/μL);
   (d) Dual-labeled probe Cf-S: final concentration 200 nmol/l (0.4 μL of a working solution 10 pmol/μL) (*see* **Note 19**);
   (e) 3 μL template DNA;
   (f) 5 μL deionized sterile water (DSW).

   Every reaction is performed in three technical repetitions together with a negative control containing DSW instead of template DNA.
3. Setting of the real-time PCR cycler is in accordance with the manufacturer's recommendations (*see* **Note 20**). The Roche's Light Cycler 480 Probes Master indicates the following setting of the Light Cycler 480's run program 1 cycle of 10 min at 95 °C (pre-incubation); 40 cycles of 95 °C for 10 s (denaturation), 60 °C for 30 s (annealing), 72 °C for 1 s with single acquisition mode (extension).

4. The amount of ascospores of *H. fraxineus* in every sample and consequently in the air is determined by the absolute quantification which can be performed in two ways:
   (a) Serial dilutions of 10 to 10^5 ascospore suspension's extracts (*see* the ascospore suspension preparation method above and **Note 21**).
   (b) As a reference standard, a cloned plasmid of *Escherichia coli* can be used. It is a more reliable option due to the stability of the plasmid's concentration for longer-term storage and handling. The *H. fraxineus*-specific amplicon resulting from the qPCR mentioned above is inserted into a commercially available plasmid; for example, pCR™ 2.1 TOPO® TA vector (Invitrogen, Carlsbad, California, USA). Plasmid with the inserted amplicon is then quantified to calculate the exact number of its copies. The reference standard it becomes after its quantification throughout the above-mentioned qPCR absolute quantification by ascospore suspension dilution series.

4 Notes

1. Sampling with rotorods is easy. However, every time it must be decided what kind of information about the inoculum is needed. Rotorods are normally used for studies focused on the spatial spread of the inoculum [12, 26]. It is an economical solution for higher numbers or repetitions of simultaneous sampling. ROTTRAP is not the only rotorod device you can use. Rotorods are very simple devices and instructions how to build them are accessible even in youtube.com [27]. More elaborated rotorods are commercialized as well, for example, Gripst-2009 Pollen and Spore Sampler (Aerobiology research laboratories, Nepean, Ontario, Canada). However, the principle is every time the same.
2. Effectivity of the sampling of *H. fraxineus* ascospores depends strongly on the construction of the rotorods. Specific disadvantage of rotorods is their low sensitivity for particles smaller than a certain limit [12, 21, 22, 28]. This insensitivity is caused by aerodynamical reasons and must be taken into consideration during choosing the most appropriate construction of the sampler. Running square-profiled rod is pushing a certain amount of resisting air mass in front of the impaction side. Therefore, too small particles which are not inertial enough are blown around the impaction rod and do not get trapped on the sticky sampling surface. The critical size of a particle is negatively influenced by the speed of rotation and positively by the width of the impaction surface. In practice, it means that

the faster and narrower the rod is, the smaller particle is likely to be impacted. Approximate calculations for the critical particle with 85–100% probability of being impacted can be calculated according to the following equation [28]:

$$d = \sqrt{H \times \frac{W}{v}}$$

where
d = particle diameter (cm)
H = constant (=0.0162)
W = width of impaction rod (cm)
v = impaction rod velocity (cm/s)
Velocity of the rod is calculated from the horizontal rotating arm diameter and the number of rotations per minute as follows:

$$v = \pi \times D \times \frac{\text{rpm}}{60}$$

where
v = impaction rod velocity (cm/s)
π = constant (= 3.1416)
D = rotating diameter; length of the horizontal rotating arm (cm)
rpm = number of rotorod rotations per minute
Example for ROTTRAP 104 (9 V DC):

$$v = 3.1416 \times 20 \times \frac{2067}{60} = 2164 \text{ cm/s}$$

$$d = \sqrt{0.0162 \times \frac{0.08}{2164}} = 7.739 \times 10^{-4} \text{ cm} = 7.739 \text{ μm}$$

which means that particles with 7.74 μm diameter and bigger should be impacted with a probability 85–100%. The size of *H. fraxineus* ascospores is known to be 13–17 (−21) × 3.5–5 μm [29]. The mean size is therefore 15 × 4.25 μm oscillating around the critical size of an ideally rounded particle.

3. Any flanking edges may affect the efficiency causing turbulences; a wider impact area of the rod decreases the efficiency as well due to the negative "air pillow" effect. This effect may disable small particles to reach the trapping surface [19, 20, 28].
4. The sampling height is an essential factor for a correct detection of *H. fraxineus* spores. The source of the inoculum (apothecia) is positioned on the soil surface and the occurrence of the spores in the air is totally dependent on the air currents and therefore varies a lot. It was found that with the height of sampling the inoculum load decreases exponentially [12], but can be detected very high [14]; even in 60 m above ground (publication in preparation).

5. To avoid any deformation of the mounting sockets, the bar should be mounted (with bolts) on a more robust metal profile hammered into the soil. If mounted on any other object, the shortest variant of the bar can be used and screwed onto its surface (Fig. 2).
6. Maintenance is feasible in the windows and there are no problems with power supply. However, sampling on silos (or other high buildings) may have specific disadvantages: (a) *Overloading of samples by dust* produced by the stored material, which can act as a mechanical barrier for catching spores (decreasing the sticky surface of the rods) and/or chemical inhibition for further molecular processing. Appropriate shortening of the sampling period may be helpful; (b) *Unusual air currents* originating in a ventilation system of the silo and in the heat radiation of the wall exposed to the sun. To prevent this, it is better to sample on the shaded side of the silo without any ventilation outlets.
7. The power supply can be a critical factor in the field. From the point of view of the precision of the results, the connection to the electrical network with constant voltage is always the best option. In such a case, commonly available adaptors providing DC of 9 V (or 12 V for other spore traps) are the optimal solution. It is always necessary to protect them against rain or excessive humidity, for example, in an upside-down installed bucket. It is recommendable to check the adaptor during the first sampling to avoid its overheating and possible fire. Cables should have thick skin or sleeves to protect it against rodents if they are laid down on the soil surface. Alternative power supply can be an accumulator, but with the sampling time the voltage is decreasing, which means a certain decline of the sampling rate. To avoid this, a solar panel with a proper regulator can be employed. Designing the solar charging, the length of the expected daylight, and the ratio of sunshine must be considered. For a sufficient supply, a 120 W diffuse light solar cell has been used for this purpose (Fig. 8) [18].
8. The sampling rate (a certain amount of air sampled per certain time) of a rotorod is an important value for a final absolute quantification of the trapped spores related to the amount of air. The producer is normally providing this information. However, in case of doubt or homemade rotorods, it can be calculated as follows:

$$R = 0.12 \times v \times W \times L$$

where
R = Sampling rate (l/min)
v = impaction rod velocity (cm/s)
W = width of impaction rod (cm)

Fig. 8 Solar cell (120 W) for a proper power supply of rotorods and volumetric air samplers

L = length of impaction rod (cm)
Example for ROTTRAP 104:

$$R = 0.12 \times 2164 \times 0.08 \times 5 = 103.872 \text{ l/min}$$

It is not recommended to use the values of rpm stated by the producer of the electric motor used for the homemade rotorods. In practice, that value is decreased by the air resistance of the rotating arms and rods during operation. The rpm can be experimentally estimated by taking a photograph by a camera of good quality (Fig. 9). For that purpose, the camera should be mounted on a tripod, set on a fixed shutter time (in Fig. 9 it is $1/200$ s = 0.005 s), the axis of the camera's lens

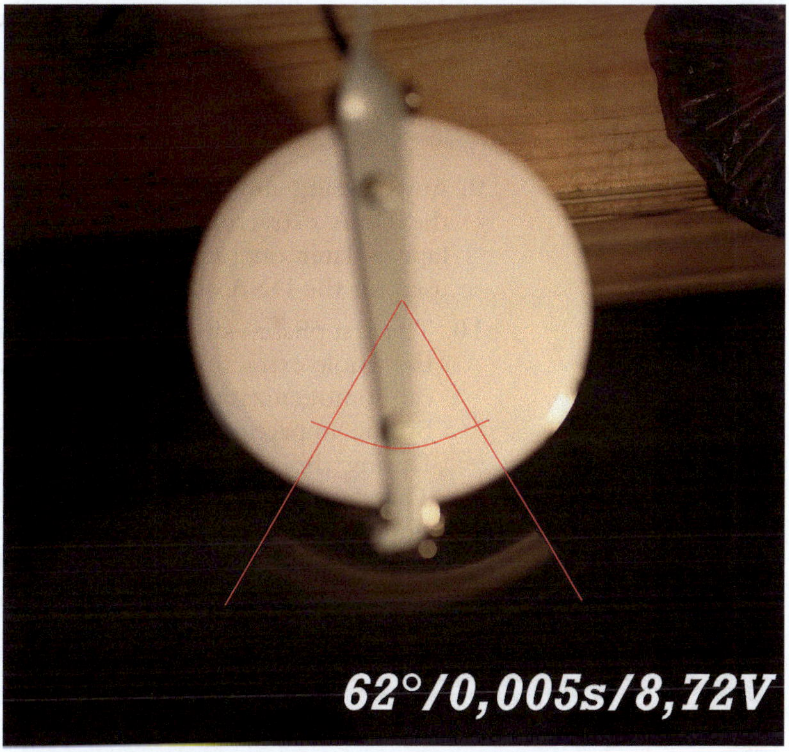

Fig. 9 Experimental estimation of the rotorod's speed of rotation with DSLR camera

must be identical with the axis of the rotating motor, the camera must be focused on the horizontal arm of the rotorod (or slightly further to display sharply any point of the vertical rod), the rotating arm must be strongly illuminated by tungsten beam (continuous light), and the picture should be taken against neutrally dark background. Normally, more pictures should be taken to catch a well-illuminated rotating arm. The aperture value of the camera must be chosen experimentally. It depends on the lens used and the luminosity and distance of the light source. Generally, low aperture values (f/4, f/5,6) are better to homogenize the unfocused background. The next step is to measure the angle drawn by the arm (or rod) on the clearest picture and calculate the number of rotations per minute as follows:

$$\text{rpm} = \propto \times T \times 0.1667$$

where

rpm = number of rotations per min

α = angle drawn by the arm (rod) in the photograph (°)

T = shutter time selected on the camera ($1/T$ s; for $1/200$ s: $T = 200$)

The example in Fig. 9 would be calculated:

$$\text{rpm} = 62 \times 200 \times 0.1667 = 2067$$

9. Any crossing and overlapping of the tape itself are decreasing the DNA extraction efficiency. Microtube with the exposed tape is later on (after possible storage at −20 °C) directly used for the DNA extraction.
10. The first phase—disruption of the sample, is a crucial point of the whole extraction and it combines mechanical and chemical homogenization. The mechanical disruption is provided by the impaction of hard beads on the sampled material. During this process, the sample can get unexpectedly hot. Therefore, the addition of any liquid is appropriate to buffer the increase of temperature. Apart from this role, the detergent Nonidet helps to make the cell walls of the spores permeable for the following step. The chemical disruption is intermediated by incubation in 65 °C and addition of the AP1 buffer. It is recommendable to use microtubes with screw caps. Microtubes with standard caps containing warmed-up, expanding samples are normally contaminating the whole thermal block or water in the bath. Sometimes there is even a danger of cross-contamination between samples by chaotically self-opening tubes. The microtubes should be inverted a few times during the thermal incubation, because it happens very often that not all material in the sample is immersed in the buffer. Permanent movement of the tubes would probably improve the DNA yield.
11. There are other useful protocols for DNA extraction from *H. fraxineus* spores in air samples published [10–12, 15]. The method described above is a compromise of these methods regarding DNA yield, price, duration, repeatability, feasibility, availability of materials, and safety of work. Good results were obtained with DNEasy Plant Mini Kit (Qiagen, Valencia, California, USA) [23, 30, 31]. Every extraction run should be started with one empty microtube as a negative control of extraction.
12. In the last step, DNA is eluted only once with 100 μL of preheated elution buffer to obtain a solution of higher DNA concentration. The suggested 2 × 100 μL probably counts with higher DNA content, higher than usual low amounts of air samples. The yield of elution may be probably enhanced by AE buffer preheated to 65 °C or/and a few minutes' incubation before its last centrifugation.

13. The ideal source of apothecia is the expected source of the air inoculum—rachises of last-year leaves on the moist soil surface. The peak of the fruiting period in Europe is considered in July and August. During that time, apothecia are actively releasing ascospores almost every day, early in the morning after sunrise, and therefore, the spore suspension can be easily prepared.
14. Repeat with ca. 20 apothecia. Do not immerse the apothecia completely = avoid any other fungal parts or debris release.
15. Estimate the number of spores per volume using a microscope. Count the number of spores in ten randomly chosen 0.2 × 0.2 mm squares (Fig. 10) and calculate the number of spores per 1 ml, following the equation:

$$C_e = \frac{c_c}{4 \times 10^{-5}}$$

where

C_e = estimated concentration of spores in the suspension (number of spores/1 mL)

c_c = number of spores counted in ten squares (number of spores)

Example from Fig. 10:

$$C_e = \frac{15}{4 \times 10^{-5}} = 3.75 \times 10^5$$

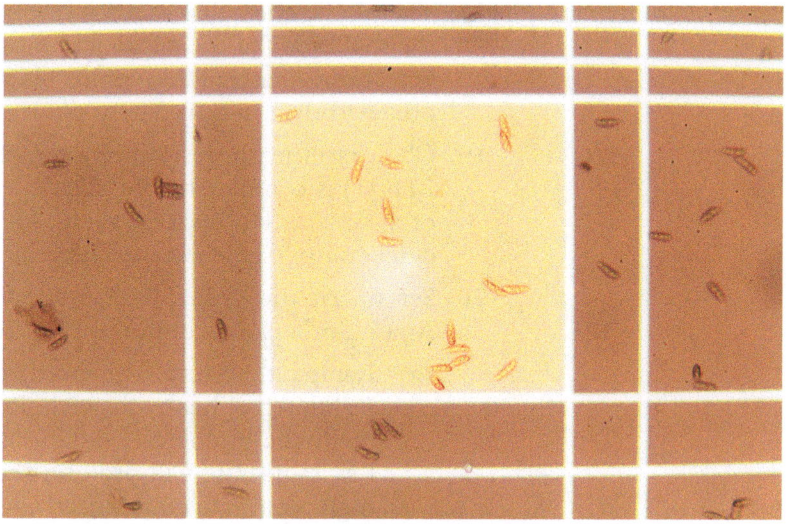

Fig. 10 Microscopic view on the Bürker chamber grid. The elucidated square with ascospores is mentioned in the example in the text

16. Dual-labeled probes are light sensitive. Furthermore, they keep their best fluorescence with the least number of freezing–thawing cycles. Therefore, it is advisable to prepare working solutions in aliquots, for example, as follows:
    (a) Add the prescribed (by provider) amount of DSW (or buffer—depends on provider) to the delivered lyophilized probe in the original microtube to achieve the 100 pmol/μL stock solution. Vortex one minute.
    (b) Prepare and mark ten black 1.5-mL microtubes, pipette DSW and then add the 100 pmol/μL stock solution in each of them to achieve 10 pmol/μL working solution.

        Example: 350 μL of 100 pmol/μL stock solution: 315 μL of DSW + 35 μL of stock solution in each black microtube.
    (c) Working solutions are kept in −20 °C; defrosted and properly vortexed directly before use. As an example, 350 μL of the working solution in each microtube will be sufficient for max. 875 reactions. Using the full 96 plates for each cycler run, it means nine freezing–thawing cycles.
17. The suggested master mix is chosen due to two reasons: (i) it has been designed for LightCycler 480 which is used as the model machine throughout this protocol and (ii) it has been one of the cheapest options. However, other master mixes can be used for LightCycler 480 without any effect on the results. If using other real-time PCR platforms, be aware of their particular requirements (for example, ROX content as a passive reference dye).
18. The qPCR microtubes (strips or plates) must match with the thermal block of the cycler and must be approved for optical detection. Generally, in white plastic higher fluorescence is obtained than in transparent plastic.
19. Choice of the fluorescent reporter dye (in this case FAM and BHQ1) is completely free and depends on the configuration of the cycler. The cycler must be equipped with a filter for the detection of particular wavelengths.
20. Set up of the cycler is very individual for each machine. Among others, it is important not to forget:
    (a) Ramp rate setting. The speed of temperature change may be limited either by the technical capabilities of the cycler or by the biological limitations of the polymerase. Very low ramp rate will prolong the reaction run, while too high ramp rate could inactivate the polymerase;
    (b) To set the reporter dye (FAM in this case);
    (c) To set the point of acquisition = the point of the cycle when the fluorescence is measured (single acquisition mode in the extension phase).

21. For more accuracy, preparation of the spore suspension should be repeated (three times) and run together in one qPCR assigning each concentration as a standard. Resulting standard curve should be then saved in the thermocycler software and later used as the external standard curve for qPCRs with field samples, together with a reference standard. This reference standard can be any of the ascospore serial dilutions used for the standard curve construction. Concentration of this reference standard should be assigned following its absolute quantification during the standard curve construction. The number of ascospores in each field sample is then quantified by the thermocycler software after each reaction. As a final result, the theoretical average ascospore concentration in the air for a particular sampling period is calculated as follows:

$$C = \frac{c_q \times 1000}{T \times R}$$

where

C = theoretical average ascospore concentration (number of spores/m^3)

c_q = number of spores in one sample quantified by qPCR (number of spores)

T = duration of the sampling period (min)

R = sampling rate of the spore trap (l/min)

Example for ROTTRAP 104, $c_q = 25{,}000$, and sampling period 48 h:

$$C = \frac{25{,}000 \times 1000}{2880 \times 103.872} = 83.570 \text{ spores/m}^3$$

Acknowledgment

This work was supported by using the laboratories of Mendel University in Brno, Phytophthora Research Centre (funded by project CZ.02.1.01/0.0/0.0/15_003/0000453), and laboratories of the University of South Bohemia in České Budějovice, Faculty of Agriculture and Technologies, Department of Genetics and Agricultural Biotechnology. Furthermore, this work was financed from a HORIZON 2020 project (agreement no. 771271): Holistic Management of Emerging Forest Pests and Diseases (HOMED).

References

1. Kowalski T (2006) *Chalara fraxinea* sp. nov. associated with dieback of ash (*Fraxinus excelsior*) in Poland. For Pathol 36:264–270. https://doi.org/10.1111/j.1439-0329.2006.00453.x
2. Pautasso M, Aas G, Queloz V, Holdenrieder O (2013) European ash (*Fraxinus excelsior*) dieback – a conservation biology challenge. Biol Conserv 158:37–49. https://doi.org/10.1016/j.biocon.2012.08.026
3. Vasaitis R, Enderle R (2017) Dieback of European ash (*Fraxinus* spp.) -consequences and guidelines for sustainable management the report on European Cooperation in Science & Technology (COST) Action FP1103 FRAXBACK
4. Kowalski T, Holdenrieder O (2009) Pathogenicity of *Chalara fraxinea*. For Pathol 39:1–7. https://doi.org/10.1111/j.1439-0329.2008.00565.x
5. Gross A, Zaffarano PL, Duo A, Grünig CR (2012) Reproductive mode and life cycle of the ash dieback pathogen *Hymenoscyphus pseudoalbidus*. Fungal Genet Biol 49:977–986. https://doi.org/10.1016/j.fgb.2012.08.008
6. Fones HN, Mardon C, Gurr SJ (2016) A role for the asexual spores in infection of *Fraxinus excelsior* by the ash-dieback fungus *Hymenoscyphus fraxineus*. Sci Rep 6:1–10. https://doi.org/10.1038/srep34638
7. Cleary MR, Arhipova N, Gaitnieks T et al (2013) Natural infection of *Fraxinus excelsior* seeds by *Chalara fraxinea*. For Pathol 43:83–85. https://doi.org/10.1111/efp.12012
8. Chandelier A, Gerarts F, San Martin G et al (2016) Temporal evolution of collar lesions associated with ash dieback and the occurrence of *Armillaria* in Belgian forests. For Pathol 46:289–297. https://doi.org/10.1111/efp.12258
9. Kräutler K, Treitler R, Kirisits T (2015) *Hymenoscyphus fraxineus* can directly infect intact current-year shoots of *Fraxinus excelsior* and artificially exposed leaf scars. For Pathol 45:274–280. https://doi.org/10.1111/efp.12168
10. Hietala AM, Timmermann V, BØrja I et al (2013) The invasive ash dieback pathogen *Hymenoscyphus pseudoalbidus* exerts maximal infection pressure prior to the onset of host leaf senescence. Fungal Ecol 6:302–308. https://doi.org/10.1016/j.funeco.2013.03.008
11. Timmermann V, Børja I, Hietala AM et al (2011) Ash dieback: pathogen spread and diurnal patterns of ascospore dispersal, with special emphasis on Norway. EPPO Bull 41:14–20. https://doi.org/10.1111/j.1365-2338.2010.02429.x
12. Chandelier A, Helson M, Dvorak M et al (2014) Detection and quantification of airborne inoculum of *Hymenoscyphus pseudoalbidus* using real-time PCR assays. Plant Pathol 63:1296–1305. https://doi.org/10.1111/ppa.12218
13. Husson C, Caël O, Grandjean JP et al (2012) Occurrence of *Hymenoscyphus pseudoalbidus* on infected ash logs. Plant Pathol 61:889–895. https://doi.org/10.1111/j.1365-3059.2011.02578.x
14. Aguayo J, Husson C, Chancerel E et al (2021) Combining permanent aerobiological networks and molecular analyses for large-scale surveillance of forest fungal pathogens: a proof-of-concept. Plant Pathol 70:181–194. https://doi.org/10.1111/ppa.13265
15. Grosdidier M, Aguayo J, Marçais B et al (2017) Detection of plant pathogens using real-time-PCR: how reliable are late Ct values? Plant Pathol 66:359–367. https://doi.org/10.1111/ppa.12591
16. D'Amato G, Lobefalo G (1989) Allergenic pollens in the southern Mediterranean area. J Allergy Clin Immunol 83:116–122
17. Hirst JM (1952) An automatic volumetric spore trap. Ann Appl Biol 39:257–265
18. Dvořák M, Rotková G, Botella L (2016) Detection of airborne inoculum of *Hymenoscyphus fraxineus* and *H. albidus* during seasonal fluctuations associated with absence of apothecia. Forests 7:1–13. https://doi.org/10.3390/f7010001
19. Dhingra OD, Sinclair JB (1995) Basic plant pathology methods, 2nd edn. CRC Press
20. Lacey ME, West JS (2006) The air Spora – a manual for catching and identifying airborne biological particles. Springer
21. Perkins WA, Leighton PA (1957) The rotorod sampler. Second semi-annual report no. CML 186, aerosol laboratory
22. McCartney HA, Fitt BDL, Schmechel D (1997) Sampling bioaerosols in plant pathology. J Aerosol Sci 28:349–364
23. Čermáková V, Kudláček T, Rotková G et al (2017) *Hymenoscyphus fraxineus* mitovirus 1 naturally disperses through the airborne inoculum of its host, *Hymenoscyphus fraxineus*, in the Czech Republic. Biocontrol Sci Tech 27:992–1008. https://doi.org/10.1080/09583157.2017.1368455

24. Chandelier A, André F, Laurent F (2010) Detection of *Chalara fraxinea* in common ash (*Fraxinus excelsior*) using real time PCR. For Pathol 40:87–95. https://doi.org/10.1111/j.1439-0329.2009.00610.x
25. Botella L, Bačová A, Dvořák M et al (2019) Detection and quantification of the air inoculum of *Caliciopsis pinea* in a plantation of *Pinus radiata* in Italy. iForest - Biogeosc For 12: 193–198. https://doi.org/10.3832/ifor2866-012
26. Dvořák M, Janoš P, Botella L et al (2017) Spore dispersal patterns of *fusarium circinatum* on an infested Monterey pine forest in North-Western Spain. Forests 8:432. https://doi.org/10.3390/f8110432
27. UCCE Napa Viticulture Advisor (2020) How to build a rotating arm spore trap for powdery mildew. https://www.youtube.com/watch?v=D0kJZbNw7tc
28. Noll KE (1970) A rotary inertial impactor for sampling giant particles in the atmosphere. Atmos Environ 4:9–19
29. Kowalski T, Holdenrieder O (2009) The teleomorph of *Chalara fraxinea*, the causal agent of ash dieback. For Pathol 39:304–308. https://doi.org/10.1111/j.1439-0329.2008.00589.x
30. Dvořák M, Rotková G (2015) Spore dispersal of *Hymenoscyphus fraxineus* in the Czech Republic. In: Joint IUFRO Working Party Meetings 7–12 June, 2015 in Uppsala, Sweden. Swedish University of Agricultural Sciences, p 30
31. Koukol O, Haňáčková Z, Dvořák M, Havrdová L (2016) Unseen, but still present in Czechia: *Hymenoscyphus albidus* detected by real-time PCR, but not by intensive sampling. Mycol Prog 15:1–9. https://doi.org/10.1007/s11557-015-1149-5

Chapter 9

Molecular Detection of Wheat Blast Pathogen in Seeds

Renaud Ioos

Abstract

Pyricularia oryzae is a fungal plant pathogen causing blast disease in several species of the Poaceae family. It encompasses several genetic lineages, including one that is pathogenic on wheat and belongs to the *Triticum* lineage of *P. oryzae*. The fungus spreads at short distances by its airborne and rain-splash dispersed spores, and at longer distances via cryptically infected wheat seeds, through trade. Here, we describe a practical method to detect *P. oryzae Triticum* lineage in wheat seeds, after a biological enrichment step, with various options for molecular testing involving several DNA-based technologies: polymerase chain reaction (PCR), real-time PCR, and loop-mediated isothermal amplification (LAMP). The array of available molecular assays is presented in this protocol, each of them targeting specific regions of the *P. oryzae Triticum* lineage and offering different levels in terms of sensitivity and specificity.

Key words Diagnosis, *Magnaporthe oryzae*, *Pyricularia oryzae*, Quarantine, Lineage, Emerging plant disease, Biological enrichment

1 Introduction

Wheat blast is an emerging disease jeopardizing wheat production at the global scale. The disease induces necrotic lesions on the leaves, stems, and grains, and bleaching of the spikes, which contain empty or missing grains. It was first reported in 1985 and spread within Brazil [1], before moving to other South American countries such as Bolivia, Paraguay, and Argentina [2]. This devastating disease was recently reported on the Asian continent, reaching Bangladesh in 2016 [3], and this leap to such a distant region of the globe was attributed to the introduction of infected seed from South America [4]. The causal agent, a specific lineage of the fungus named *Pyricularia oryzae Triticum* lineage, disperses at short distances by its airborne two-celled spores called conidia, by the wind, or by rain splashes during the vegetative growth of wheat. Infection of wheat by *P. oryzae Triticum* lineage has been demonstrated experimentally [5], and the example of the Bangladesh outbreaks illustrates how the fungus can spread further through

Nicola Luchi (ed.), *Plant Pathology: Methods and Protocols*, Methods in Molecular Biology, vol. 2536,
https://doi.org/10.1007/978-1-0716-2517-0_9,

trade of contaminated wheat seeds that are moved globally and may travel from infected areas to disease-free countries. It is therefore paramount that wheat seed lots are tested before trading to prevent long-distance spread of the disease. Importantly, the disease has proven very difficult to eradicate once introduced [6, 7]. Symptoms of wheat blast on grains are sometimes difficult to observe and may be confused with shriveled grains caused by *Fusarium* head blight. In addition, seeds collected from healthy-looking spikes of some wheat cultivars may have a similar degree of infection and show no symptoms at all [8]. Laboratory analysis of seeds is therefore required to test for the presence of *Pyricularia oryzae Triticum* lineage. Although planting the seeds on an artificial medium to isolate the fungus may be a method to check for the presence of Pyric*ularia oryzae* sensu lato, this species actually encompasses numerous genetic lineages that are virtually undistinguishable by morphology or cultural characteristics [9]. Identification of the *Triticum* lineage, which is the only one able to cause blast epidemics on wheat, relies on pathogenicity testing on wheat cultivars, or on molecular-based assays, targeting DNA regions that are virtually exclusively present in the *P. oryzae Triticum* lineage genome. This first approach is time-consuming and requires prior isolation of the pathogen in pure culture, and facilities to grow wheat plant throughout the year, with necessary biosafety precautions. The approach we will be presenting here uses PCR, real-time PCR, or LAMP assays, carried out on DNA directly extracted from ground wheat samples, and provides a result within 1 to 5 days, depending on whether biological enrichment of the seeds is applied. Several PCR [10, 11], real-time PCR [10–12], or LAMP assays [11, 13] are currently available and were designed based on comparative genomics. However, none of them proved to be 100% specific. Of note, several studies have demonstrated that gene flow occurs between different *P. oryzae* lineages [14, 15], and cross-reactions in the described assays with DNA from nontarget lineages are reported (*see* **Notes 5–12**). However, a combination of the tests described here, each targeting different loci in the *P. oryzae Triticum* lineage, will help to secure the diagnosis.

2 Materials

2.1 Biological Enrichment

1. Plastic container: 9-cm-diameter sterile Petri dishes can be used, or any other sterile plastic or glass container of similar volume that can accommodate incubation of 400 wheat seeds as a single layer.
2. Potato dextrose broth (PDB): prepared as per the manufacturer's instructions, and sterilized before use by autoclaving for 15 min at 120 °C, and kept in a vial in a refrigerator before use.

2.2 Grinding of Seeds

1. Laboratory mixer: A Microtron™ MB 550 homogenizer was used by Thierry et al. [12] and Thierry et al. [11], equipped with a sterilized 125-mL glass grinding bowl and stainless steel cutting blades assembly (*see* **Note 1**).
2. Micropipette tips: 1000-μL sterile filter tips are required. The plastic cone is truncated approximately 1 cm from its tip using sterile scissors or a scalpel blade, so that the opening of the tip enables pipetting of a thick liquid, such as the seed and PDB macerate.

2.3 Extraction of Total DNA

1. Plant DNA extraction kit (*see* **Note 2**).
2. Heating bath for incubation of the microtubes at a specified temperature recommended for lysis by the plant DNA extraction kit manufacturer (65 °C for most of them).
3. Pipette filter tips of various volumes (1000 μL, 200 μL, 20 μL, 10 μL). Tips must be sterile.
4. Two-mL microtubes, sterile.

2.4 Amplification of DNA by Conventional or Real-Time PCR or by LAMP

1. For conventional PCR: reagents for conventional PCR (PCR buffer, $MgCl_2$, dNTP, DNA polymerase) or ready-to-use master mix used in Pieck et al. [10] or Thierry et al. [11]. Agarose for gel electrophoresis (1% final concentration). Electrophoresis cell and electrophoresis buffer (Tris–borate–EDTA 0.5×). DNA ladder for gel electrophoresis.
2. For real-time PCR: reagents for real-time PCR using a hydrolysis probe (PCR buffer, $MgCl_2$, dNTP, DNA polymerase) or ready-to-use real-time PCR master mix used in Pieck et al. [10], Thierry et al. [12], or Thierry et al. [11].
3. For LAMP: reagents for LAMP (isothermal amplification master mix) used in Yasuhara-Bell et al. [13], or Thierry et al. [11].
4. Thermal cycler, real-time thermal cycler, or LAMP device equipped with a fluorescence measurement system.
5. Conventional PCR primers of Pieck et al. [10] and Thierry et al. [11] (*see* Table 1).
6. Real-time primers and hydrolysis probe of Pieck et al. [10] or Thierry et al. [11, 12]. Table 1 shows the primer and hydrolysis probe sequences, including the fluorescent dye and quencher labelling the probes, according to the original description. Alternative dyes may be used, provided that they are suitable for the real-time equipment used.
7. LAMP primers of Yasuhara-Bell et al. [13] and Thierry et al. [11]. *See* Table 1.
8. DNA from a reference strain of *Pyricularia oryzae Triticum* lineage, which will be used as a positive control in conventional PCR, real-time PCR, or LAMP reactions.

Table 1
Primer and probe sequences and modifications used in this protocol

Assay	Name	Sequence (5′ to 3′)	References
Conventional PCR	MoT3F	GTCGTCATCAACGTGACCAG	[10]
	MoT3R	ACTTGACCCAAGCCTCGAAT	[10]
Conventional PCR	C45-forward	TCTTTCACTCCTCCGAAAGAC[a]	[11]
	C45-reverse	GTATAGCTGGGTATCTTGGTAGAC	[11]
Real-time PCR	MoT3_1F	AGGATGTATGCCCTGACTGG	[10]
	MoT3_1R	CTCGGCGATGCAAAAGTGAA	[10]
	MoT3_FAM2	[6FAM]-TTCTAACGGTTTGCAATTGCACAAAACAAC-[BHQ1]	[10]
Real-time PCR	C17-F	CGA TAG AAA CTT GAG GAA GAT CAA GTA AG	[12]
	C17-R	TCA CCG AGA GAT GTG CCA	[12]
	C17-P	[6FAM]-TCG CTA ACA ATG TCC ACC CCG CC-[BHQ1]	[12]
Real-time PCR	C45-forward	TCTTTCACTCCTCCGAAAGAC	[11]
	C45-reverse	GTATAGCTGGGTATCTTGGTAGAC	[11]
	C45-probe	[6FAM]-TGCCCTCATCAAAACCTGCAGCCAT-[BHQ1]	[11]
LAMP	MoT3-F3	CCAAGGATGTATGCCCTGAC	[13]
	MoT3-B3	GTTGGGGGCTTCGTATGC	[13]
	MoT3-FIP	CAAACCGTTAGAAAGCTGTACTTCAGTTGTTACAACATCACGAATCG	[13]
	MoT3-BIP	TTGCACAAAACAACCCAGTTCAGAGACGTTGTTTATGAATTCCG	[13]
	MoT3-LoopB	CTTTTGCATCGCCGAGCC	[13]
	MoT3-Probe	[5Cy5]-ACGCTGAGGACCCGGATGCGAATGCGGA TGCGGATGCCGACTTTTGCATCGCCGAGCC	[13]

	Quencher probe	TCGGCATCCGCATCCGCATTCGCATCCGGGTCCTCAGCGT-[BHQ1]	[20]
LAMP	No.5-F3	TGCGTGATCAACGAATGGC	[11]
	No.5-B3	CGGAAGCAAACTCTGCGATT	[11]
	No.5-FIP	GCAAGATGCCTACCGTGGGGTTTTCTGGGTTCCCCTCACCAT	[11]
	No.5-BIP	TCGTAAGGAGCATGAAGGGCTGTTTTGCGGACGGACATACGTAGT	[11]

[a]Should be highly purified when prepared by the manufacturer to ensure integrity of the primer at the 3′ end (e.g., HPLC purification)

3 Methods

Seed sampling, biological enrichment, and the first steps of DNA extraction (steps **3.1-1** to **3.3-1**) should be carried out under an appropriate level of biosafety. In regions where wheat blast does not occur, the highest precautions should be taken in order to prevent any spread of the pathogen if present (e.g., work under a class II safety cabinet and biohazard level 3 room).

Different methods may be used, each of them offering different sensitivity and specificity levels. They may be chosen based on the availability of the equipment and reagents, or skills of the staff. Some of them have been assessed with wheat seeds, but for only three of them [11], an optimized DNA extraction protocol from seeds has been described and will be presented here. It is important to emphasize that, given the genetic closeness between the different *P. oryzae* lineages, false-positive results might be obtained with DNA from a few strains of nontarget *P. oryzae* lineages [10–13, 16]. Positive cases using one assay should be confirmed by another complimentary test, targeting another region of the genome. This is particularly important when positive cases are obtained with seeds originating from regions where wheat blast has not been reported so far.

3.1 Seed Sampling and Biological Enrichment

1. As virtually no typical symptoms can be observed on seeds, the lot should be sampled randomly after manual mixing. Take a subsample of 400 seeds per seed lot using a disinfected spoon or spatula (*see* **Note 3**).
2. Transfer the 400 seeds to a sterile Petri dish, spread them in a single grain layer, and cover them with 15 mL of potato dextrose broth (PDB). Label the dish with sample ID, and indicate the date and time of preparation.
3. Transfer 15 mL of PDB to another Petri dish. Label the dish as a negative control, and indicate the date and time of preparation.
4. Incubate the seeds and the negative control for 72 h on a bench at room temperature (ca 23–25 °C) under natural light (day/-night). This 72-h incubation will serve as a biological enrichment step and stimulate the growth of any fungal species present on/in the seeds, thus improving the detection threshold of *P. oryzae Triticum* lineage when present [11]. This step may be skipped, and the seeds may be directly ground for DNA extraction and testing, thus reducing the turnaround time for analysis to 1 day. However, Thierry et al. [11] showed that a 72-h biological enrichment step greatly improved the sensitivity of the molecular assay.

3.2 Grinding of Seeds and DNA Extraction

1. After incubation, transfer aseptically the content of the Petri dish (seeds + PDB) to a laboratory mixer, and add an additional 30-mL PDB.
2. Grind the mixture at maximum speed until a homogenous, milky macerate is obtained (*see* **Note 4**). This may take between 30 and 60 s depending on the type of blender used.
3. Collect three subsamples of 500 μL each using a truncated 1-mL cone, and transfer to individual 2-mL sterile microtubes.
4. Collect one sample of 500 μL of PDB from the negative control and transfer to a 2-mL microtube ("negative extraction control").
5. Each of the four 500-μL macerates can be stored in a freezer or used subsequently for total DNA extraction.

3.3 Total DNA Extraction

1. For each seed sample, extract the total DNA of each of the three 500-μL macerates, as well as the DNA from the negative control (72-h incubated PDB broth), following the manufacturer's instructions, and elute the total DNA in 100 μL of the provided elution buffer. Store the four total DNA extracts in a freezer until testing.
2. Test each DNA extract individually using one of the molecular methods presented below. A seed sample will therefore be rated as positive if at least one out of the three subsamples is positive after testing.

3.4 Testing by Conventional PCR [10]

1. Perform the PCR assays in a 50-μL reaction mixture containing 1× PCR buffer including $MgCl_2$, 5 U of DNA polymerase, 0.2-mM dNTP, 0.4 μM of each MoT3F and MoT3R primer (*see* Table 1), and 2.5 μL of DNA template.
2. The PCR parameters are initial denaturation at 94 °C for 90 s, followed by 30 cycles at 94 °C for 30 s, 62 °C for 30 s, and 72 °C for 60 s and a final extension at 72 °C for 2 min.
3. In each run, include a positive DNA control (DNA from a reference strain of *P. oryzae Triticum* lineage) and a negative water control (molecular grade water), and include the negative extraction control.
4. Visualize amplified products by electrophoresis on a 2% agarose gel, after staining with ethidium bromide. Add a 100-bp ladder to a well to determine the size of the amplicons.
5. Samples showing a 361-bp amplicon should be considered positive provided that the negative water control and negative extraction control show no amplification. Samples not showing a 361-bp amplicon should be considered negative at the limit of detection of the test, provided that the positive DNA control shows a 361-bp amplicon and provided that the DNA extract

was proven to be amplifiable. The limit of detection of this test is 100 pg of target DNA. However, Pieck et al. [10] reported the occurrence of *P. oryzae* strains whose DNA yielded false-positive or false-negative results (*see* **Note 5**).

3.5 Testing by Conventional PCR [11]

1. Perform PCR assays in a 20-μL reaction mixture containing 1× PCR buffer, 5-mM $MgCl_2$, 0.5 U of DNA polymerase, 0.25-mM dNTP, 0.3 μM of each C45-forward and C45-reverse primer (*see* Table 1), and 2 μL of DNA template.
2. The PCR parameters are initial denaturation at 95 °C for 10 min, followed by 40 cycles at 95 °C for 30 s, 65 °C for 30 s, and 72 °C for 45 s and a final extension at 65 °C for 10 min.
3. In each run, include a positive DNA control (DNA from a reference strain of *P. oryzae Triticum* lineage) and a negative water control (molecular grade water), and include the negative extraction control.
4. Visualize amplified products by electrophoresis on a 1.5% agarose gel, after staining with ethidium bromide. Add a 100-bp ladder to a well to determine the size of the amplicons.
5. Samples showing a 102-bp amplicon should be considered positive, provided that the negative water control and negative extraction control show no amplification. Samples not showing a 102-bp amplicon should be considered negative at the limit of detection of the test, provided that the positive DNA control shows a 102-bp amplicon and provided that the DNA extract was proven to be amplifiable. The limit of detection of this test is 5 pg of target DNA per PCR tube. However, although no false-positive or false-negative results were observed during their study, Thierry et al. [11] recommend the use of a confirmation assay in the case of a positive result (*see* **Note 6**).

3.6 Testing by Real-Time PCR [10]

1. Perform real-time PCR assays in a 20-μL reaction mixture containing 1× real-time PCR master mix containing dNTP, $MgCl_2$ and DNA polymerase, 0.4 μM of each MoT3_1F and MoT3_1R primer (*see* Table 1), 0.1-μM MoT3_FAM2 probe, and 1 μL of DNA template. Test each DNA extract in triplicate.
2. The PCR parameters are initial denaturation at 95 °C for 10 min, followed by 40 cycles of 95 °C for 15 s and 60 °C for 60 s. Monitor fluorescence at the end of each annealing/extension step at 60 °C.
3. In each run, include a positive DNA control (DNA from a reference strain of *P. oryzae Triticum* lineage) and a negative water control (molecular grade water), and include the negative extraction control.

4. Record the threshold value as the cycle at which the fluorescent signal, associated with exponential amplification of the PCR product, intersected the threshold set manually by the operator or automatically by the analysis software.
5. Samples showing a Ct value should be considered positive provided that the negative water control and negative extraction control show no amplification. Samples not showing a Ct value should be considered negative at the limit of detection of the test, provided that the positive DNA control shows a Ct value and provided that the DNA extract was proven to be amplifiable. The limit of detection of this test is 10 pg of target DNA per tube. However, Pieck et al. [10] reported the occurrence of *P. oryzae* strains whose DNA yielded false-positive or false-negative results (*see* **Note** 7).

3.7 Testing by Real-Time PCR [12]

1. Perform real-time PCR assays in a 20-µL reaction mixture containing 1× real-time PCR buffer, 0.2-mM dNTP, 5-mM $MgCl_2$, 0.5 U DNA polymerase, 0.3 µM of each C17-F and C17-R primer, 0.1-µM C17-P probe (*see* Table 1), and 2 µL of DNA template. Test each DNA extract in duplicate.
2. The PCR parameters are initial denaturation at 95 °C for 10 min, followed by 40 cycles at 95 °C for 15 s and 65 °C for 60 s. Monitor fluorescence at the end of each annealing/extension step at 65 °C.
3. In each run, include a positive DNA control (DNA from a reference strain of *P. oryzae Triticum* lineage) and a negative water control (molecular grade water), and include the negative extraction control.
4. Record the threshold value as the cycle at which the fluorescent signal, associated with exponential amplification of the PCR product, intersected the threshold set manually by the operator or automatically by the analysis software.
5. Samples showing a Ct value should be considered positive provided that the negative water control and negative extraction control show no amplification. Samples not showing a Ct value should be considered negative at the limit of detection of the test, provided that the positive DNA control shows a Ct value and provided that the DNA extract was proven to be amplifiable. The limit of detection of this test is 1 pg of target DNA per tube. However, Thierry et al. [12] reported the occurrence of *P. oryzae* strains whose DNA yielded false-positive or false-negative results and recommended the use of a confirmation assay in the case of a positive result (*see* **Note 8**).

3.8 Testing by Real-Time PCR [11]

1. Perform real-time PCR assays in a 20-µL reaction mixture containing 1× real-time PCR master mix, 0.3 µM of each C45-F and C45-R primer, 0.1-µM C45-P probe (*see*

Table 1), and 2 μL of DNA template. Test each DNA extract in duplicate.

2. The PCR parameters are initial denaturation at 95 °C for 10 min, followed by 40 cycles at 95 °C for 15 s and 65 °C for 55 s. Fluorescence is monitored at the end of each annealing/extension step at 65 °C.
3. In each run, include a positive DNA control (DNA from a reference strain of *P. oryzae Triticum* lineage) and a negative water control (molecular grade water), and include the negative extraction control.
4. Record the threshold value as the cycle at which the fluorescent signal, associated with exponential amplification of the PCR product, intersected the threshold set manually by the operator or automatically by the analysis software.
5. Samples showing a Ct value should be considered positive provided that the water negative control and the negative extraction control show no amplification. Samples not showing a Ct value should be considered negative at the limit of detection of the test, provided that the positive DNA control shows a Ct value and provided that the DNA extract was proven to be amplifiable. The limit of detection of this test is 5 pg of target DNA per tube. However, Thierry et al. [11] reported the occurrence of *P. oryzae* strains whose DNA yielded false-positive results and recommended the use of a confirmation assay in the case of a positive result (*see* **Note 9**).

3.9 Testing by LAMP [13]

1. Perform LAMP reactions in a 30-μL reaction mixture containing 15 μL of isothermal master mix, 5 μL of primer master mix (1 μM MoT3-F3, 1 μM MoT3-B3, 8 μM MoT3-FIP, 8 μM MoT3-BIP, 4 μM MoT3-LoopB, 0.4 μM MoT3-Probe, and 0.8 μM quencher probe) (*see* Table 1), and 5 μL of DNA template. Test each DNA extract in triplicate (*see* **Note 10**).
2. Run and analyze LAMP reactions using real-time PCR equipment or a dedicated LAMP instrument set at 65 °C for 30 min, with fluorescence readings taken at 1-min intervals.
3. In each run, include a positive DNA control (DNA from a reference strain of *P. oryzae Triticum* lineage) and a negative water control (molecular grade water), and include the negative extraction control.
4. The threshold value is recorded as the time at which the fluorescent signal, associated with exponential amplification of the LAMP product, intersected the threshold set manually by the operator or automatically by the analysis software.
5. Samples showing a threshold fluorescence value should be considered positive provided that the negative water control

and negative extraction control show no amplification. Samples not showing a threshold fluorescence value should be considered negative at the limit of detection of the test, provided that the positive DNA control shows a threshold fluorescence value and provided that the DNA extract was proven to be amplifiable. The limit of detection of this test is 5 pg of target DNA per tube. However, Yasuhara-Bell et al. [13] reported the occurrence of *P. oryzae* strains whose DNA yielded false-positive or false-negative results and recommended the use of a confirmation assay in the case of a positive result (*see* **Note 11**).

3.10 Testing by LAMP [11]

1. Perform LAMP reactions in a 25-μL reaction containing 1× isothermal master mix, 0.2 μM of each external primer No.5-F3 and No.5-B3, 0.8 μM of each internal primer No.5-FIP and No.5-BIP) (*see* Table 1), and 2 μL of DNA template. Test each DNA extract in duplicate.
2. Run and analyze LAMP reactions using real-time PCR equipment or a dedicated LAMP instrument set at 65 °C for 30 min, with fluorescence readings taken at 1-min intervals.
3. In each run, include a positive DNA control (DNA from a reference strain of *P. oryzae Triticum* lineage) and a negative water control (molecular grade water), and include the negative extraction control.
4. Record the threshold value as the time at which the fluorescent signal, associated with exponential amplification of the LAMP product, intersected the threshold set manually by the operator or automatically by the analysis software.
5. Samples showing a threshold fluorescence value should be considered positive provided that the negative water control and negative extraction control show no amplification. Samples not showing a threshold fluorescence value should be considered negative at the limit of detection of the test, provided that the positive DNA control shows a threshold fluorescence value and provided that the DNA extract was proven to be amplifiable. The limit of detection of this test is 5 pg of target DNA per tube. However, Thierry et al. [11] reported the occurrence of *P. oryzae* strains whose DNA yielded false-positive or false-negative results and recommended the use of a confirmation assay in the case of a positive result (*see* **Note 12**).

4 Notes

1. Alternative grinding systems are possible. Crushing incubated seeds with PDB is feasible using any other similar grinding device. Grinding manually in a mortar using a pestle is

discouraged because it would not produce a sufficiently homogenous macerate.

2. The commercial kit used by Thierry et al. [11] and Thierry et al. [12] is NucleoSpin Plant II (Macherey-Nagel), but other brands of plant DNA extraction kits may be used.
3. As counting of seeds may be laborious, the sampled seeds may be weighed instead of counted. For instance, weigh 3 replicates of 10 seeds, calculate the mean weight, and deduce from this the weight corresponding to 400 seeds. Depending on the method chosen for the identification, the total number of seeds to be tested per lot in order to detect the pathogen at different levels of infection in the lot may be different and needs to be determined statistically (useful guidance is given in Tables 1 and 2 of ISPM 31, Methodologies for sampling of consignments) [17]. The sample size recommended by the International Seed Testing Association (ISTA) is 400 seeds for plating [18]. However, larger samples (e.g., 1000 seeds) can easily be processed by biological enrichment before DNA analysis.
4. This grinding step may take between 30 s to 1 min depending on the blender used. Shake the grinding bowl manually from time to time if needed, in order to avoid the presence of a thick agglomerate on the grinding bowl's inner walls.
5. In their study, a collection of 284 *P. oryzae* isolates was tested by Pieck et al. [10], from a large range of host plants, including 115 from *Triticum* spp. Using their MoT3 conventional PCR assay, and in their own conditions, Pieck et al. [10] reported a false-positive result with DNA from *P. oryzae* isolate P-0029 isolated from *Bromus tectorum* and a false-negative result with DNA from *P. oryzae* isolate PY 05010, from *Triticum aestivum*. Additional evidence, however, suggests that isolate P-0029 actually belongs to *P. oryzae Triticum* lineage [10].
6. In their study, a collection of 185 *P. oryzae* isolates was tested by Thierry et al. [11], from a large range of host plants, including 34 from *Triticum* spp. Using their C17 conventional PCR assay, and in their own conditions, Thierry et al. [11] reported a false-negative result with DNA from *P. oryzae* isolate AG0102, from *Triticum aestivum*. Pathogenicity tests, however, suggest that isolate AG0102 does not induce blast symptoms on wheat, and probably does not belong to *P. oryzae Triticum* lineage [11].
7. In their study, a collection of 284 *P. oryzae* isolates was tested by Pieck et al. [10], from a large range of host plants, including 115 from *Triticum* spp. Using their MoT3 real-time PCR assay, and in their own conditions, Pieck et al. [10] reported a false-positive result with DNA from *P. oryzae* isolate P-0029

isolated from *Bromus tectorum* and a false-negative result with DNA from *P. oryzae* isolate PY 05010, from *Triticum aestivum*. Additional evidence, however, suggests that isolate P-0029 actually belongs to *P. oryzae Triticum* lineage [10]. In their work, Thierry et al. [12] showed that in their own condition, the MoT3 real-time assay of Pieck et al. [10] was not able to detect DNA from two strains of *P. oryzae Triticum* lineage (BR0032 and BR0043, both from *Triticum* sp.), whereas it yielded false-positive results with DNA from isolate AG0055 from *Bromus*.

8. In their study, a collection of 110 *P. oryzae* isolates was tested by Thierry et al. [12], from a large range of host plants, including 30 from *Triticum* spp. Using their C17 real-time PCR assay, and in their own conditions, Thierry et al. [12] reported false-positive results with DNA from *P. oryzae* isolates P29 from *Bromus* sp., CR0023 from *Echinochloa crus-galli*, CR0057 from *Lolium* sp., and JP0033 from *Eriochloa villosa*. The latter three isolates were unable to induce blast symptoms when artificially inoculated on wheat, whereas P29 was actually shown to belong to the *Triticum* lineage, although isolated from *Bromus* [12].
9. In their study, a collection of 185 *P. oryzae* isolates was tested by Thierry et al. [11], from a large range of host plants, including 34 from *Triticum* spp. Using their C45 real-time PCR assay, and in their own conditions, Thierry et al. [11] reported a false-negative result with DNA from *P. oryzae* isolate AG0102, from *Triticum aestivum*. Pathogenicity tests, however, suggest that isolate AG0102 does not induce blast symptoms on wheat, and probably does not belong to *P. oryzae Triticum* lineage [11]. In addition, Thierry et al. [11] reported potential false-positive results with DNA from ten non-wheat isolates of *P. oryzae*. However, all these cross-reactions sometimes yielded very late Ct values (Ct > 35.5) and were not repeatable. Additional experiments with wheat seeds artificially contaminated with one of these isolates (isolate BR0079 from *Eleusine* sp.) showed that the C45 real-time PCR assay did not yield false-positive results with this nontarget isolate, when present on seeds at the level of 50 contaminated grains/400.
10. The MoT3 LAMP assay of Yasuhara-Bell et al. [13] can be run individually, or in a duplex reaction using Nad5 LAMP. The mitochondrial NADH dehydrogenase (*nad5*) gene [19] is used as an internal control for plant DNA. The master mix composition for the duplex MoT3/Nad5 reaction is described in Yasuhara-Bell et al. [13].
11. In their study, a collection of 159 *P. oryzae* isolates was tested by Yasuhara-Bell et al. [13], from a large range of host plants,

including 19 from *Triticum* spp. Using their MoT3 LAMP assay, and in their own conditions, Yasuhara-Bell et al. [13] reported three potential false-negative results with DNA of *P. oryzae Triticum* lineage isolates BR32, PY5010, and PY86.1 from *Triticum* spp. and false-negative results with DNA of *P. oryzae* isolates FPH-2015-44 from *Hakonechloa macra* and Br35 from *Brachiaria* sp.

12. In their study, a collection of 185 *P. oryzae* isolates was tested by Thierry et al. [11], from a large range of host plants, including 34 from *Triticum* spp. Using their No5 LAMP assay, and in their own conditions, Thierry et al. [11] reported a false-negative result with DNA from *P. oryzae* isolate AG0102, from *Triticum aestivum*. Pathogenicity tests, however, suggest that isolate AG0102 does not induce blast symptoms on wheat, and probably does not belong to *P. oryzae Triticum* lineage [11]. In addition, Thierry et al. [11] reported false-positive results with DNA from the *P. oryzae* isolates CD88125 from *Echinochloa colona*, CR0023 from *Echinochloa crus-galli*, CR0057 from *Lolium* sp., JP0031 from *Panicum coloratum*, JP0033 from *Eriochloa villosa*, and AG0067 from *Phalaris canariensis*. Other cross-reactions were sometimes observed (non-repeatable) for 22 other nontarget lineages, but shortening the LAMP reaction time to 8 min helped to remove these late false positives [11].

References

1. Anjos J, Da Silva DB, D'Avila Charchar MJ et al (1996) Occurrence of blast fungus (*Pyricularia grisea*) on wheat and rye in the savanna region of central Brazil. Pesq Agropec Bras 31: 79–82
2. Cruz CD, Valent B (2017) Wheat blast disease: danger on the move. Trop Plant Pathol 42: 210–222
3. Malaker PK, Barma NCD, Tiwari TP et al (2016) First report of wheat blast caused by *Magnaporthe oryzae* pathotype triticum in Bangladesh. Plant Dis 100:2330–2330
4. Islam MT, Croll D, Gladieux P et al (2016) Emergence of wheat blast in Bangladesh was caused by a South American lineage of *Magnaporthe oryzae*. BMC Biol 14:84
5. Gomes DP, Rocha VS et al (2018) Potential of transmission of *Pyricularia graminis-tritici* from plant to seed and from seed to seedling in wheat genotypes with different degrees of blast resistance. J Seed Sci 40:16–24
6. Ceresini PC, Castroagudín VL, Rodrigues FÁ et al (2018) Wheat blast: past, present, and future. Annu Rev Phytopathol 56:427–456
7. Yesmin N, Jenny F, Abdullah HM et al (2020) A review on South Asian wheat blast: the present status and future perspective. Plant Pathol 69:1618–1629
8. Urashima AS, Grosso C, Stabili A et al (2009) Effect of *Magnaporthe grisea* on seed germination, yield and quality of wheat. In: Wang GLVB (ed) Advances in genetic, genomics and control of rice blast disease. Springer, pp 267–277
9. Castroagudín VL, Moreira SI, Pereira DAS et al (2016) *Pyricularia graminis-tritici*, a new Pyricularia species causing wheat blast. Persoonia – Mol Phyl Evol Fungi 37:199–216
10. Pieck ML, Ruck A, Farman ML et al (2017) Genomics-based marker discovery and diagnostic assay development for wheat blast. Plant Dis 101:103–109
11. Thierry M, Chatet A, Fournier E et al (2020) A PCR, qPCR, and LAMP toolkit for the detection of the wheat blast pathogen in seeds. Plan Theory 9:277
12. Thierry M, Gladieux P, Fournier E et al (2019) A genomic approach to develop a new qPCR

test enabling detection of the *Pyricularia oryzae* lineage causing wheat blast. Plant Dis 104: 60–70
13. Yasuhara-Bell J, Pedley KF, Farman M et al (2018) Specific detection of the wheat blast pathogen (*Magnaporthe oryzae Triticum*) by loop-mediated isothermal amplification. Plant Dis 102:2550–2559
14. Gladieux P, Condon B, Ravel S et al (2018) Gene flow between divergent cereal- and grass-specific lineages of the rice blast fungus *Magnaporthe oryzae*. mBio 9(1). https://doi.org/10.1128/mBio.01219-17
15. Inoue Y, Vy TTP, Yoshida K et al (2017) Evolution of the wheat blast fungus through functional losses in a host specificity determinant. Science 357:80–83
16. Yasuhara-Bell J, Pieck ML, Ruck A et al (2019) Regarding the MoT3 wheat blast diagnostic assay. Phytopathology 109:509–511
17. ISPM (2008) ISPM No. 31 methodologies for sampling of consignments. FAO, Rome
18. ISTA (2007) ISTA method validation for seed testing. ISTA, Basseldorf
19. Ecke W, Schmitz U, Michaelis G (1990) The mitochondrial nad5 gene of sugar beet (Beta vulgaris) encoding a subunit of the respiratory NADH dehydrogenase. Curr Genet 18: 133–139
20. Kubota RM, Alvarez A, Su WW et al (2011) FRET-based assimilating probe for sequence-specific real-time monitoring of loop-mediated isothermal amplification (LAMP). Biol Eng Trans 4:81–100

Chapter 10

Detection and Identification of the Causal Agents of Dothistroma Needle Blight

Funda Oskay, Asko Lehtijärvi, and Hatice Tuğba Doğmuş Lehtijärvi

Abstract

Dothistroma needle blight (DNB) is one of the most damaging foliage diseases of pine in plantations and natural forests worldwide and is caused by two closely related fungi: *Dothistroma septosporum* and *D. pini*, which are virtually impossible to differentiate from each other based on morphology. Although diagnosis of DNB based on symptoms is relatively reliable in the later stages of the disease when fruit bodies (conidiomata) are formed, for diagnosis in the early stages, as well as identification of the causal agent at species level, molecular methods are required. In addition, reliable and sensitive diagnostics before sporulation is a prerequisite for early detection to minimize accidental introductions of disease through movement of infected plant materials, especially seedlings. While amplification and sequencing of the ITS region of the rDNA alone is not reliable to differentiate the two species, conventional PCR (cPCR) using species-specific primers or mating type-specific primers and quantitative PCR (qPCR) are widely used and accepted molecular methods to identify and differentiate the DNB pathogens, either from cultures or directly from needles.

Key words *Dothistroma septosporum*, *Dothistroma pini*, ITS, Species-specific primers, cPCR, Mating type

1 Introduction

Dothistroma needle blight (DNB) is one of the most damaging foliage diseases of pine worldwide [1]. Symptoms of infection include needle necrosis, defoliation, reduction in radial increment and height growth, and, in severe cases, tree death. The disease was thought to be caused by a single pathogen species until Barnes et al. [2] in 2004 revealed the presence of two distinct species: *Dothistroma septosporum* (Dorog.) Morelet and *D. pini* Hulbary. *Dothistroma septosporum* and *D. pini* produce identical symptoms on host trees, and differentiation of the two species based solely on morphology is virtually impossible [2, 3]. However, a range of molecular methods can be used to identify and differentiate the DNB pathogens, either from cultures or directly from needles [2, 4–6].

Nicola Luchi (ed.), *Plant Pathology: Methods and Protocols*, Methods in Molecular Biology, vol. 2536,
https://doi.org/10.1007/978-1-0716-2517-0_10,

The known geographic and host range of the *Dothistroma* species has increased from 61 countries and 89 host taxa [7] to 76 countries and 109 host taxa [1] over the last decade, with continuing contributions documenting new geographical and host ranges within only the last few years [8–10]. While *D. septosporum* is reported to have a worldwide distribution and a broad host range, *D. pini* is assumed to have a more limited distribution and host range [1], although recent studies utilizing molecular detection tools have suggested that it is more common than previously thought [10].

Given the importance of DNB disease and its continuously increasing prevalence and impact in pine forests and plantations in Europe since the 1990s, it is critical to accurately identify the causal agent. Reliable and efficient identification and detection methods facilitate disease surveys and early detection, reducing introductions and mitigating spread and damage by enabling more efficient disease management.

Amplification and sequencing of the ITS region of the rDNA of isolates is a principal molecular identification tool for most of the fungal pathogens; however, when used alone, it is not reliable for differentiation between the two *Dothistroma* species. Therefore, identification and detection of *Dothistroma* spp. mainly relies on species-specific PCR assays. Conventional PCR (cPCR) using species-specific primers developed for *D. septosporum* and *D. pini* by Ioos et al. [5] has been successfully used for confirming the species identification of isolates. On the other hand, mating-type primer pairs for *D. septosporum* and *D. pini* [4] have been more widely used for species-specific identification of isolates as they can ensure simultaneous identification of the *Dothistroma* species and the mating type of the isolates. Nonetheless, the causal agent at species level can be detected and differentiated directly in needle samples via cPCR using the same species-specific primer sets for each species developed by Ioos et al. [5].

2 Materials

2.1 Microscopy and Fungal Isolation

1. Stereo/dissecting microscope.
2. Binocular microscope (40–400× magnifications).
3. Microscope slides and cover glasses.
4. Distilled sterile water.
5. Microcentrifuge tubes (1.5 or 2 mL).
6. Vortex.
7. Water agar (WA).
8. Pasteur pipette.
9. 2% Malt extract agar (MEA).

10. 70% Ethanol.
11. Acetone.
12. Alcohol spirit lamp for flaming.
13. 10% Glycerol.
14. Reverse osmosis (RO) water.
15. Brown paper envelopes of different sizes.

2.2 DNA Extraction

1. DNA extraction kit (Thermo Scientific GeneJET Genomic DNA Purification Kit).
2. 96–100% Ethanol.
3. Tissue homogenizer (Retsch MM400 with 2 × 10/24 microcentrifuge tube adaptors).
4. Ø 1.6–3 mm stainless steel beads.
5. Hot water bath or hot plate or oven capable of heating up to 56 °C.
6. Microcentrifuge capable of at least 12,000 × *g*.
7. 1.5 mL Microcentrifuge tubes.
8. Vortex.
9. Micropipettes and pipet tips.
10. Disposable gloves.

2.3 PCR

1. Oligonucleotide primers (Table 1).
2. Deoxyribonucleotides (dNTPs).
3. Molecular grade water.
4. GoTaq Green Master Mix (Promega, Madison, WI, USA).
5. Microcentrifuge tubes (1.5/2 mL).
6. 0.2 mL PCR tubes.
7. Pipettes and sterile pipette tips.
8. Ice (for keeping reagents cold or frozen ice bricks with wells to hold the microcentrifuge tubes/PCR tubes).
9. Positive reference sample(s).

2.4 Gel Electrophoresis

1. Gel electrophoresis system.
2. UV gel documentation system.
3. Agarose (i.e., SeaKem® LE Agarose LONZA).
4. Gel staining (EtBr (Ethidium bromide 1%)).
5. 0.5X TBE.
6. Microwave oven.
7. 100 bp DNA Ladder.

Table 1
Primer pairs for molecular identification and detection of *Dothistroma septosporum* and *D. pini*

Primer type	Target organism/sequence	Primer name	Primer sequence (5′ to 3′)	Targeting gene region	Product size (bp)	Reference
Universal primer pair	Fungi	ITS1	TCCGTAGGTGAACCTGCGG	ITS	ca 600	[11]
		ITS4	TCCTCCGCTTATTGATATGC			
Fungi-specific primer pair	Fungi	ITS1F	CTTGGTCATTTAGAGGAAGTAA	ITS		[12]
		ITS4	TCCTCCGCTTATTGATATGC			
Species-specific conventional PCR primer pairs	*D. septosporum*	DStub2-F	CGAACATGGACTGAGCAAAAC	β-tub2	ca 231	[5]
		DStub2-R	GCACGGCTCTTTCAAATGAC			
	D. pini	DPtef-F	ATTTTTCGCTGCTCGTCACT	EFI-α	ca 193	[5]
		DPtef-R	CAATGTGAGATGTTCGTCGTG			
Species-specific mating-type primer pairs	*D. septosporum MAT1*	DseptoMat1F	CGCAGTAAGTGATGCCCTGAC	*MAT1-1-1*	ca 820	[4]
		DotMat1r	TTGCCTGACCGGCTGCTGGTG			
	D. septosporum MAT2	DsetptoMat2F	GTGAGTGAACGCCGCACATGG	*MAT1-2*	ca 480	
		DotMat2r	CTGGTCGTGAAGTCCATCGTC			
	D. pini MAT1	DpiniMat1f2	AGTAAGCGACGCGCTCCCATG	*MAT1-1-1*	ca 820	[4]
		DotMat1r	TTGCCTGACCGGCTGCTGGTG			
	D. pini MAT2	DpiniMat2f	GTAAGTGATCGTTGAACATGC	*MAT1-2*	ca. 480	
		DotMat2r	CTGGTCGTGAAGTCCATCGTC			

3 Methods

3.1 Microscopical Examination

1. Select needles with *Dothistroma* sp. symptoms and signs (conidiomata) (Fig. 1).
2. Clean needles gently by wiping them with a tissue soaked in 70% ethanol (*see* **Note 1**).
3. Examine needles under a dissecting microscope for conidiomata typical of *Dothistroma* species.
4. Remove the outer surface of the conidiomata to expose the hymenium using a scalpel or needle.
5. Put a droplet of distilled water on the surface of the microscope slide and place a piece of hymenium over the droplet.
6. Cover it with a clean cover slip.
7. Gently crush the conidiomata under the cover slip.
8. View under a compound microscope to find the fusiform to short clavate, hyaline, smooth, and thin-walled 1–5 septate conidiospores of *Dothistroma*.

3.2 Single Spore (SS) Isolation From Needles

1. Clean needles bearing mature conidiomata of *Dothistroma* sp. gently by wiping with a tissue soaked in 70% ethanol (*see* **Note 2**).
2. Examine needles under a dissecting microscope for conidiomata typical of *Dothistroma* species (*see* **Note 3**).
3. Remove the single conidioma gently with a sterile scalpel or dissecting needle and place the conidiomata into a 1.5 mL microcentrifuge tube containing 1 mL of sterile distilled water.

Fig. 1 Symptoms of Dothistroma needle blight on 1-year-old needles of *Pinus brutia*

4. Vortex the tube thoroughly to release the spores.
5. Spread the resulting spore suspension onto 1% water agar in Ø 90 mm Petri dishes.
6. Incubate the Petri dishes at 20 °C (17–22 °C) for 2–5 days. Check germination daily. Do not wait too long. Spores usually germinate after 2 days.
7. Locate germinated spores under a microscope.
8. Pick a single germinating colony from the agar using a glass Pasteur pipette (*see* **Note 4, 5**).
9. Transfer to a Petri dish containing 2% malt extract agar (MEA) to obtain the single spore isolate (SSI).
10. Pick a few SSI from a single conidioma (*see* **Note 6**) and incubate the new plates at 20 °C for 2–3 weeks to obatain axenic *Dothistroma* single spore cultures.
11. Submit molecularly identified Dothistroma isolates to a culture collection (*see* **Note 7**).
12. Additionally, deposit needle samples with *Dothistroma* sp. conidiomata into a fungal herbarium (*see* **Note 8**).

3.3 Molecular Identification of Isolates

3.3.1 DNA Extraction

1. Transfer isolates onto 2% MEA covered with cellophane for 2–3 weeks at 20 °C.
2. Transfer 10–50 mg of mycelium from the colony edge to a 2.0 mL microcentrifuge tube (*see* **Note 9**).
3. Store at −20 °C until homogenization.
4. Add 5–7 sterile metal beads (Ø 1.6 mm) into microcentrifuge tubes containing the mycelia.
5. Homogenize the mycelia for 1 min at a frequency of 30 Hz using a Retsch MM400 homogenizer (Retsch GmbH, Haan, Germany).
6. Add 200 mL of digestion solution and 20 mL of Proteinaz K solution supplied with the DNA extraction kit and mix thoroughly by vortexing to obtain a uniform suspension.
7. Centrifuge at a low speed (i.e., 3300 × *g*) for 1 min to efficiently include the tissues on the lid of the microcentrifuge tube.
8. Incubate the sample at 56 °C for 4 h or longer (up to 12 h) until the tissue is completely lysed and no particles remain. Vortex the tubes a few times during incubation.
9. Add 20 μL of RNase A Solution, mix by vortexing, and then incubate for 10 min at room temperature.
10. Add 200 μL of lysis solution, mix thoroughly by vortexing for 15 s until a homogeneous mixture is obtained then Add 400 μL of 50% ethanol, and mix by pipetting or vortexing.

11. Centrifuge at >20.000 × *g* for 5 min.
12. Transfer the prepared lysate to a GeneJET Genomic DNA Purification Column inserted into a collection tube. Centrifuge the column for 1 min at 6000 × *g*.
13. Discard the collection tube containing the flow-through solution. Place the GeneJET Genomic DNA Purification Column into a new 2 mL collection tube.
14. Add 500 μL of wash buffer (WB) I, centrifuge for 1 min at 8000 × *g*, discard the flow-through, and place the purification column back into the collection tube.
15. Add 500 μL of WB II to the GeneJET Genomic DNA Purification Column, centrifuge for 3 min at maximum speed (≥12,000 × *g*), discard the collection tube containing the flow-through solution, and transfer the GeneJET Genomic DNA Purification column to a sterile 1.5 mL microcentrifuge tube.
16. Add 40 μL of Elution Buffer (EB) to the center of the GeneJET Genomic DNA Purification Column membrane to elute genomic DNA, incubate for 2 min at room temperature, and centrifuge for 1 min at 8000 × *g*. Repeat the step but this time add 30 μL EB.
17. Discard the purification column. Use the purified DNA immediately in downstream applications or store at −20 °C.

3.3.2 ITS PCR and Sequencing

1. Set up PCR reactions in a contamination-free environment using ITS1 or ITS1F and ITS4 primers (Table 1).
2. Prepare the master mix by adding 2x GoTaq Green Master Mix (i.e., 12.5 μL for a final concentration of 1x per sample), primers (10 μM) (i.e., 2.5 μL each for a final concentration of 1 μM per sample), and 5.5 μL of molecular grade water per sample (*see* **Note 10**).
3. Vortex reaction mixture briefly.
4. Aliquot 23 μL into each PCR tube or each well of a PCR plate.
5. Add 2 μL (5–20 ng/μL) of template DNA sample for each 25 μL reaction.
6. Set the thermal cycling conditions on a thermocycler: 95 °C for 2 min followed by 35 cycles of 95 °C for 20 s, 55 °C for 25 s, and 72 °C for 50 s, followed by a final extension step at 72 °C for 10 min and perform the amplification.
7. Resolve the PCR products by agarose gel electrophoresis to check the success of amplification (presence of a clear single band of 450–600 bp on gel) (*see* Subheading 3.3.5 for methodology of gel electrophoresis).

8. Send PCR products to a commercial laboratory for sequencing (Sanger sequencing) of samples at both directions using ITS1-ITS4 primers.
9. Alternatively, the ITS1F-ITS4 PCR products can be sequenced using an ITS5 primer in one direction.
10. Edit sequences manually and align (those sequenced in two directions) and obtain the consensus sequence using BioEdit version 7.2.5.
11. Subject sequences to a BLAST analysis on NCBI's databases.
12. Submit obtained sequences to GenBank.

3.3.3 Mating-Type PCR

Species-specific mating-type primers developed by Groenewald et al. [4] are widely used for both determining *Dothistroma* species and their mating type.

1. Set up PCR reactions in a contamination-free environment using mating-type primers specific for the species of interest (Table 1).
2. Test samples first by using the *D. septosporum*-specific mating-type primers.
3. Prepare the master mix by adding 2x GoTaq Green Master Mix (i.e., 12.5 μL for a final concentration of 1x per sample), *D. septosporum*-specific mating-type primers—DseptoMat1f, DotMat1r, DseptoMat2f, DotMat2r (10 μM) (i.e., 0.5 μL each for a final concentration of 0.1 μM per sample)—and 9.5 μL of molecular grade water per sample (*see* **Note 10**).
4. Vortex reaction mixture briefly.
5. Aliquot 23 μL into each PCR tube or each well of a PCR plate.
6. Add 2 μL of template DNA sample (5–20 ng/μL) for each 25 μL reaction.
7. Set the thermal cycling conditions on a thermocycler: 94 °C for 5 min followed by 40 cycles of 94 °C for 20 s, 65 °C for 20 s, and 72 °C for 40 s, followed by a final extension step at 72 °C for 5 min [4], and perform the amplification.
8. Resolve the PCR products by agarose gel electrophoresis (*see* Subheading 3.3.5).
9. Any isolates with weak or absent bands should be rerun using the *D. pini* primer set (DpiniMat1f2, DotMat1R, DpiniMat2f, DotMat2R).
10. Amplification of the *MAT1-1-1* idiomorph produces an amplicon of size c. 820 bp for both *D. septosporum* and *D. pini*. The *MAT1-2* idiomorph produces an amplicon of c. 480 bp (Fig. 2).
11. Repeat the PCR and subsequent visualization on the agarose gel at least twice to ascertain the results.

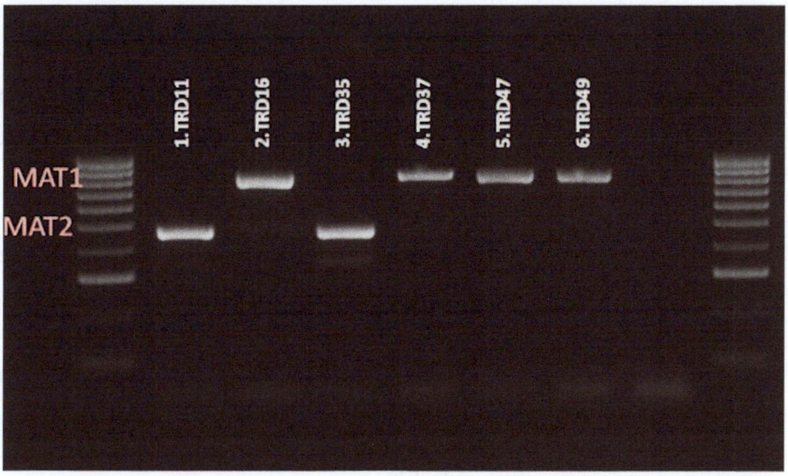

Fig. 2 Ethidium bromide-stained agarose gel showing PCR amplicons obtained using species-specific mating-type primers DseptoMat1F, DotMat1R, Dsepto-Mat2F, and DotMat2R

3.3.4 Conventional PCR: Species Identification Using Species-Specific Primers

1. Set up cPCR reactions in a contamination-free environment using species-specific primers (Table 1).
2. Test samples first by using the *D. septosporum*-specific primers.
3. Prepare the master mix by adding 2x GoTaq Green Master Mix (i.e., 12.5 μL for a final concentration of 1x per sample), *D. septosporum*-specific primer pair, DStub2-F/DStub2-R (10 μM) (i.e., 1 μL each for a final concentration of 0.4 μM per sample), and 8.5 μL of molecular grade water per sample (*see* **Note 10**).
4. Vortex reaction mixture briefly.
5. Aliquot 23 μL into each PCR tube or each well of a PCR plate.
6. Add 2 μL of template DNA sample for each 25 μL reaction.
7. Set the thermal cycling conditions on a thermocycler: 95 °C for 10 min followed by 35 cycles of 95 °C for 30 s, 60 °C for 30 s, and 72 °C for 60 s, followed by a final extension step at 72 °C for 10 min [5], and perform the amplification.
8. Resolve the PCR products by agarose gel electrophoresis (*see* Subheading 3.3.5).
9. Any isolates with weak or absent bands should be rerun using the *D. pini* primer pair (DStub2-F/ DStub2-R).
10. Amplification of the *D. septosporum* produces an amplicon of size c. 231 bp. The size of the band for *D. pini* should be ca 193 bp.
11. Repeat the PCR and subsequent visualization on the agarose gel at least twice to ascertain the results.

3.3.5 Gel Electrophoresis

1. Dissolve 100 g of agarose in TBE 0.5× by boiling in a microwave.
2. Add stain (Etbr, etc.) to the agarose when warm enough.
3. Pour agarose into the tank, insert the combs, and wait until it solidifies in a sucking hood.
4. Load 5 μL of the PCR product.
5. Include a 100 bp or 1 kb DNA ladder.
6. Run electrophoresis for 40 min at 120 V or run at 75 V for 50 min.
7. Visualize the amplicon of interest under UV light using a gel documentation system.
8. The amplicon size can be established by comparing it to a DNA ladder.

4 Notes

1. If the needle is relatively clean, no surface sterilization is necessary.
2. Avoid using excessive sterilization. The sterilant can penetrate the conidiomata and destroy the conidiospores.
3. Intend for a single conidioma for isolation. Avoid multiple or coherent conidiomata.
4. The tip of the Pasteur pipette can be modified by snapping off a small piece from the tip with a finger nail and by heating it a couple of centimeters from the tip in a small flame and bending it, e.g., 30–40 degrees to facilitate easier working under a compound microscope. A cotton plug is inserted at the other end, preventing contamination when the captured spore on an agar piece is blown out into the culture medium by forcing air out of one's mouth and through the pipette. The tip of the pipette can be cleaned by immersing and rinsing in a series of water (for rinsing by sucking in and then blowing out a small amount of water), ethanol, and acetone and thereafter flaming it briefly to remove the acetone and blown dry [13].
5. Avoid picking from germinating colonies close to each other.
6. Note each SSI from the same conidioma (i.e., SSI-1 of Conidioma-1) and label them accordingly.
7. For storage of *Dothistroma* sp. cultures, two methods described in Frazer et al. [14] can be suggested: (a) under sterile reverse osmosis (RO) water at 4 °C and (b) under sterile 10% glycerol at −80 °C. At least tree replicates of each isolate are suggested to be stored with either one of these methods:

(a) Cut a mycelial plug (i.e., 5 mm in diameter) from the growing edge of a *Dothistroma* sp. culture growing on MEA or any other growth media and place it into a 2 mL tube with screw caps.

(b) Add 1 mL sterile (otoclaved at 105 kPa, 15 min) reverse osmosis (RO) water and store at 4 °C.

(c) Add 1 mL sterile (autoclaved at 105 kPa, 15 min) 10% glycerol and store at −80 °C.

8. Needle samples with symptoms and/or signs (conidiomata) of DNB can be stored in brown paper envelopes (17 × 25 or larger). A collection number should be given to the sample, and envelopes should be labeled accordingly. The label should include at least the collection date, collection location, and host name. Name of collector, coordinates and elevation of collection location, type of site—urban area (i.e., arboretum, park), natural forest, or plantation forest—age or age class of host, health status of host, and *Dothistroma* species (if determined) can be written on the label or in a separate file for the sample.

Select needles with *Dothistroma* sp. symptoms and signs (conidiomata) and clean needles gently by wiping with a tissue soaked in 70% ethanol. Then let the needles completely dry and then place the needle samples in paper envelopes and label the envelopes.

9. The mycelium of *Dothistroma* sp. can be scratched easily from the colony edges of cultures grown on cellophane using a sterile scalpel.

(a) Avoid scratching the solid medium together with the mycelia.

(b) Thermo Scientific GeneJET Genomic DNA Purification Kit requires comparatively small amounts of tissue (10–20 mg is suggested in the protocol). Thus avoid using excess material (maximum 40–50 mg of mycelia).

(c) The younger the mycelia, the greater the yield and quality of DNA; thus avoid incubating cultures for more than three weeks.

10. At least one negative amplification control (NAC; i.e., PCR grade water used in the reaction mix) and one positive amplification control (PAC; DNA from *D. septosporum* and/or *D. pini* isolate whose identity and mating type is determined and confirmed) should be included for each series of amplification of the target nucleic acid. The master mix should be sufficient for all samples and the NAC and PAC samples.

References

1. Drenkhan R, Tomešová-Haataja V, Fraser S et al (2016) Global geographic distribution and host range of Dothistroma species: a comprehensive review. For Pathol 46:408–442
2. Barnes I, Crous PW, Wingfield BD et al (2004) Multigene phylogenies reveal that red band needle blight of Pinus is caused by two distinct species of Dothistroma, *D. septosporum* and *D. pini*. Stud Mycol 50:551–565
3. Barnes I, Kirisits T, Wingfield MJ et al (2011) Needle blight of pine caused by two species of Dothistroma in Hungary. For Pathol 41(5): 361–369
4. Groenewald M, Barnes I, Bradshaw RE et al (2007) Characterization and distribution of mating type genes in the Dothistroma needle blight pathogens. Phytopathology 97(7): 825–834
5. Ioos R, Fabre B, Saurat C et al (2010) Development, comparison, and validation of real-time and conventional PCR tools for the detection of the fungal pathogens causing brown spot and red band needle blights of pine. Phytopathology 100(1):105–114
6. Schneider S, Jung E, Queloz V et al (2019) Detection of pine needle diseases caused by *Dothistroma septosporum*, *Dothistroma pini* and *Lecanosticta acicola* using different methodologies. For Pathol 49(2):e12495
7. Watt MS, Kriticos DJ, Alcaraz S et al (2009) The hosts and potential geographic range of Dothistroma needle blight. Forest Ecol Manag 257(6):1505–1519
8. Janosikova-Heckova Z, Ondruskova E, Barta M et al (2018) The hosts and geographic range of Dothistroma needle blight in Slovakia. For Pathol 48(3):e12421
9. Matsiakh I, Dogmus-Lehtijarvi HT, Kramarets V et al (2018) *Dothistroma* spp. in Western Ukraine and Georgia. For Pathol 48(2): e12409
10. Mullett MS, Adamson K, Bragança H et al (2018) New country and regional records of the pine needle blight pathogens *Lecanosticta acicola*, *Dothistroma septosporum* and *Dothistroma pini*. For Pathol 48(5):e12440
11. White TJ, Bruns T, Lee S et al (1990) Amplification and direct sequencing of fungal ribosomal RNA genes for phylogenetics. In: Innis MA, Gelfand DH, Sninsky J, White TJ (eds) PCR protocols: a guide to methods and applications. Academic Press, San Diego
12. Gardes M, Bruns TD (1993) ITS primers with enhanced specificity for basidiomycetes-application to the identification of mycorrhizae and rusts. Mol Ecol 2(2):113–118
13. Korhonen K, Hintikka V (1980) Simple isolation and inoculation methods for fungal cultures. Karstenia 20(1):19–22
14. Fraser S, Weitz HJ, Brown AV et al (2016) Storage of *Dothistroma septosporum* cultures. For Pathol 46(5):547–550

Chapter 11

The Use of qPCR to Detect *Cryphonectria parasitica* in Plants

Anne Chandelier

Abstract

Cryphonectria parasitica is a fungal pathogen that causes lethal bark necrosis in chestnut. A duplex qPCR allowing detection of the pathogen and its host, *Castanea sativa*, is described. The method can be used for early detection of the pathogen in chestnut bark tissues with an internal control of false-negative results caused by PCR inhibitors and/or DNA extraction failure. A positive amplification control of qPCR that allows detection of any deviation from a normal qPCR run based on a control chart is also described. As *C. parasitica* is a regulated pathogen in Europe, the protocol also provides information on the way to collect and handle bark samples to fulfil biosecurity rules.

Key words Chestnut blight, Endophyte, Fungal pathogen, Internal amplification control, Positive amplification control, Real-time PCR

1 Introduction

Chestnut blight, caused by the fungus *Cryphonectria parasitica* (Murrill) Barr, is a very damaging disease. Native to eastern Asia, the fungus was first introduced to North America, where it decimated the American chestnut trees (*Castanea dentata*) at the beginning of the twentieth century. The first infections on European chestnuts (*Castanea sativa*) were observed in Italy in the 1930s [1]. Since then, the disease has spread throughout Europe, where the European chestnut is present [2]. *Cryphonectria parasitica* is a bark-inhabiting fungus that causes lethal necrosis on the bark of trunks and branches. It can be present as an endophyte in the phellem [3]. Rare infections have also been reported on oaks (*Quercus* spp.) in areas with high disease pressure [4–6]. Strains infected by a virus (CHV1) reducing the virulence of the fungus [7] have been detected in the population of *C. parasitica*. This phenomenon is called hypovirulence. The virus is transmitted by hyphal anastomosis between vegetative compatible strains of the pathogen and can be used as a biocontrol agent [8, 9].

Nicola Luchi (ed.), *Plant Pathology: Methods and Protocols*, Methods in Molecular Biology, vol. 2536,
https://doi.org/10.1007/978-1-0716-2517-0_11,

The disease in Europe has been less severe than in North America. This can be explained by the higher susceptibility of American chestnut to the pathogen and by a larger number of vegetative compatibility groups in North America, limiting the spread of the virus [10]. Introductions of new strains of the pathogen from North America and Asia into Europe could modify this situation because new strains of the pathogen could be more aggressive on European chestnut. Moreover, a greater diversity of vegetative compatibility groups in Europe could jeopardize management of the disease based on hypovirulence [11]. Although *C. parasitica* is a regulated pathogen in Europe (EU directive 2000/29/EC [12] recently replaced by regulation 2019/2072 [13]), multiple introductions have occurred since the first report of the disease [14–16]. The main introduction pathways into new areas include not only plants for planting but also wood with bark (notably chestnut posts or firewood) as the pathogen can survive and sporulate on dead wood [17]. It is therefore necessary to conduct surveillance of this pathogen and to analyze any suspect plant material with a reliable and rapid detection method. A fluorescence-based real-time quantitative PCR (qPCR) allowing early detection of *C. parasitica* in chestnut tissues has been developed for this purpose [18].

Plant tissues, and in particular woody tissues, may contain PCR inhibitory compounds that can lead to negative results while the target is present in the sample [19, 20]. The choice of suitable DNA extraction and PCR amplification kits for this analysis, as well as the use of an exogenous inhibition control, can solve these problems, as shown in [18]. However, using an internal amplification control (IAC) amplified at the same time as the target is becoming a necessary condition in diagnostic tests, as shown for the detection of foodborne pathogens [21]. Compared to exogenous controls, IACs offer the advantage of controlling both the efficiency of the extraction step and the absence of PCR-inhibiting compounds in the sample. As the two reactions are carried out in the same tube (duplex reaction), they also avoid the need to repeat a test for each sample to be processed [22]. An efficient internal control is expected to be present in all field samples. It must display a high sensitivity to inhibition (i.e., allow detection of strong and weak inhibitions). Its use should not affect the sensitivity of the main target detection method (i.e., there should be no interaction between the two sets of primers and probe).

A duplex qPCR combining the detection of *C. parasitica* using the primers and probe described in [18], with the detection of *C. sativa*, used as an endogenous control of DNA integrity and PCR inhibition, is presented. As the interpretation of the results is based on data that are linked to a qPCR instrument and specific reagents, implementation of the method in a laboratory will require preliminary tests to be carried out to define the decision criteria (acceptance or rejection of the analysis, positive or negative status of the sample).

2 Materials

2.1 Sample Acquisition and Handling

1. Wood chisel.
2. 70% Ethanol.
3. Disposable gloves.
4. Zipped bag.

2.2 DNA Extraction From Bark and Mycelium

1. Mortar and pestle.
2. 95% Ethanol.
3. 70% Ethanol.
4. 1% Bleach.
5. Sterile demineralized water.
6. Liquid nitrogen.
7. 1.5 mL microcentrifuge tubes.
8. A subculture of *C. parasitica* on potato dextrose agar (PDA).
9. Round-bottom 2 mL microcentrifuge tubes.
10. Bead beaters (for instance, Retsch MM200 mixer mill) and stainless steel beads (5 mm).
11. Dry bath/block heater.
12. TEX buffer: 100 mM Tris pH 8.0, 20 mM EDTA, 1.4 M NaCl, 2% CTAB, 2% PVP-K30 (*see* **Note 1**).
13. Commercial kit for plant DNA extraction (for instance, NucleoSpin Plant II kit, Macherey Nagel, *see* **Note 2**).
14. Autoclave

2.3 Preparation of the PCR Controls

1. A spectrophotometer (to measure the DNA concentration).
2. Tris buffer: 10 mM Tris pH 7.5 (sterilized solution, 121 °C, 20 min).
3. 1.5 mL Low binding tubes.

2.4 Duplex qPCR

1. Kit for qPCR amplification (for instance, FastGene Probe 2x qPCR Universal Mix, Nippon Genetics Europe GmbH, *see* **Note 2**).
2. qPCR instrument (for instance, the BioRad CFX96 as used in this study), allowing duplex real-time PCRs with dyes with little or no overlap in their emission spectra (for instance, FAM and Yakima Yellow, *see* **Note 3**).
3. Ultrapure water.
4. Primers and probes for duplex qPCR:

- *C. parasitica* (amplification product = 92 bp, [18]):

  Forward primer (CpF4): 5'- GATACCCTTTGTGAACT TATAA-3'.

  Reverse primer (CpR3): 5'-GGGGAGAAGGAAGAAAA TC-3'.

  TaqMan probe (CpS3): 5'-6-FAM-TTTATCGTTGCCTC GGCGCTGA-BHQ1-3'.

- *C. sativa* (amplification product = 162 bp):

  Forward primer (CastaF): 5'-ACGACAATCGGTGG TTCTTC-3'.

  Reverse primer (CastaR): 5'-TCCTCCGCTTATTG ATATGCTT-3'.

  TaqMan probe (CastaP): 5'-Yakima Yellow-GGGAC-TACCCGCTGAGTTT-BHQ1-3'.

5. Calibrated micropipets (10, 100, 1000 μL) with filtered tips.
6. PCR tubes in strips and caps (for instance, MicroAmp Fast Reaction Tubes, ThermoFisher Scientific).

3 Method

3.1 Sample Acquisition and Handling

1. *C. parasitica* is a regulated pathogen in Europe (i.e., a quarantine pathogen in some protected areas and a regulated non-quarantine pest in other areas, Regulation (EU) 2019/2072, [13]). Moreover, it produces huge amounts of spores. It is therefore necessary to manipulate infected plant tissues with disposable gloves in a laminar flow with a HEPA filter in a quarantine facility. Plant tissues and fungal cultures will be inactivated by autoclaving after use.
2. Bark samples (inner and outer bark, ~2 × 2 cm) are collected on the trunk or on branches of chestnut displaying canker (Fig. 1a) with a wood chisel. If fructifications of the fungus are visible (Fig. 1b), the sample is taken preferably in this area (Fig. 1c). Otherwise, the sample is taken at the margin between infected and healthy tissue (Fig. 1d, e).
3. The sample is placed in a zipped plastic bag and stored at 2–8 °C until DNA extraction (which should take place ideally within 2–3 days).
4. The wood chisel is disinfected with ethanol 70% between two successive samplings (*see* **Note 4**).

3.2 DNA Extraction From Chestnut Bark and Mycelium

To avoid cross-contamination, it is necessary to ensure that the DNA extraction and PCR steps are carried out in separate rooms.

Fig. 1 Sampling of a piece of chestnut bark (~2 × 2 cm) for the detection of *C. parasitica* by qPCR. (**a**) canker on a chestnut tree; (**b**, **c**) canker area with orange fructifications of the fungus. (**d**) slight depression in the outer bark and (**e**) necrotic front visible in the inner bark at the margin between healthy and infected bark. (Photos: A. Chandelier, CRAW)

3.2.1 Chestnut Bark

1. The bark sample (*see* **Note 5**) is surface disinfected for 10 s in 95% ethanol and 4 min in 1% bleach, rinsed with sterile water, and dried.
2. The internal part (inner + outer bark, ~2 cm^2) is cut into small pieces that are ground with liquid nitrogen in a mortar.
3. A quantity of ground bark corresponding to ~100 mg is put into four 1.5 mL microcentrifuge tubes (*see* **Note 6**).
4. DNA extractions are carried out on two of the four tubes (*see* **Note 7**) with the NucleoSpin Plant II kit (Machery Nagel). The procedure follows the user's instructions provided by the manufacturer except for the lysis step which is performed at 65 °C for 1 h with 400 μL of TEX buffer (in house buffer) and 10 μL of RNAse A (provided with the Nucleospin Plant II kit at 10 mg mL^{-1}). A negative extraction control (sterile water, NEC) is used at each DNA extraction run. The extracted DNA from plant tissues is stored at −20 °C until qPCR.

3.2.2 Mycelium

1. A positive amplification control (PAC) is prepared from a subculture of *C. parasitica* on potato dextrose agar (PDA).
2. The mycelium is collected in a 2 mL microfuge tube (~100 mg). It is ground with one 5 mm stainless steel bead in a bead beater (Retsch MM200, 1 min, 30 Hz).
3. DNA extraction is carried out with the NucleoSpin Plant II kit (Macherey Nagel) using PL1 as the lysis buffer.
4. Extracted DNA is stored at −20 °C until further processing (*see* Subheading 3.3).

3.3 Preparation of the PCR Controls

If a calibration curve is not performed in each PCR run, it is essential to use, as positive amplification control (PAC), a quantification calibrator at the same concentration in each run. The use of a calibrator allows results to be compared between qPCR runs and highlights any deviation from a normal run [24]. To prepare this control (*see* **Note 8**):

1. Quantify the DNA extracted from a pure culture of *C. parasitica* with a spectrophotometer.
2. Adjust the DNA concentration to 50 pg per μL. To minimize DNA loss with time, the dilutions will be prepared in 10 mM Tris pH 7.5.
3. Store the diluted DNA at −20 °C in 1.5 mL low binding tubes, considering a maximum volume of 6 μL per tube to avoid thawing PAC solutions more than twice.

A negative amplification control (NAC) is made of ultrapure water. It will be the last sample in the series to be prepared for qPCR.

3.4 Duplex qPCR

The use of a PCR workstation for the preparation of PCR master mix is highly recommended to protect PCR runs against contamination.

1. Duplex qPCRs for the amplification of *C. parasitica* and *C. sativa* are carried out in 20 μL.
2. The PCR master mix for the analysis of samples and controls (NEC, PAC, NAC) is prepared according to Table 1.
3. A volume of 18 μL is placed in each PCR tube.
4. The DNA samples or controls are then added (2 μL).
5. The PCR tubes are properly closed and loaded onto the thermal cycler (in this study, the BioRad CFX96), adjusting the program to allow for duplex analysis (two fluorophores, FAM and VIC– *see* **Note 3**) and including a 10 min step at 95 °C followed by 40 cycles of 95 °C, 15 s and 60 °C, 1 min.

Table 1
Master mix for the detection of *Cryphonectria parasitica* and *Castanea sativa* by duplex qPCR (volume to multiply according to the number of reactions to prepare)

Component	Stock	Final conc	Vol(μL)
FastGene Buffer	2x	1x	10
CastaF	5 μM	50 nM	0.2
CastaR	5 μM	50 nM	0.2
CastaP	5 μM	50 nM	0.2
CpF4	5 μM	250 nM	1
CpR3	5 μM	250 nM	1
CpS3	5 μM	250 nM	1
Sterile water	–	–	4.4
Total volume			18

3.5 Results Interpretation

The interpretation of the results is based on the Cq value (quantification cycle) corresponding to the number of PCR cycles at which the emitted fluorescence increases above a threshold. In order for the results to be comparable from one PCR run to another (qualitative analysis, presence or absence of the pathogen in the sample), the fluorescence thresholds for the two targets are manually set after the PCR run (*see* **Note 9**).

3.5.1 Experimental Controls

1. There should be no amplification of the *C. parasitica* target in the negative controls (NEC and NAC).
2. The PAC for *C. parasitica* should give a Cq corresponding to the amount of *C. parasitica* DNA present in the sample, taking account of the expected variation under routine operating conditions (*see* **Note 10**).
3. The internal control (*C. sativa*) must give a clear amplification curve in each sample reaction. Figure 2 illustrates the Cq values obtained on 54 chestnut bark samples in the conditions of this study (*see* **Note 11**).

3.5.2 Samples

1. If the controls give the expected results, the positive or negative status of the samples is determined.
2. Based on standard curves established for duplex (*C. parasitica* and *C. sativa*) qPCRs, the limit of detection (LOD) of the method is determined (*see* **Note 12**).
3. In the conditions of this study, it has been established that 10 fg *C. parasitica* DNA per PCR is sufficient. This corresponds to a Cq value of around 36 ($Cq_{LOD} = 36$; Fig. 3).

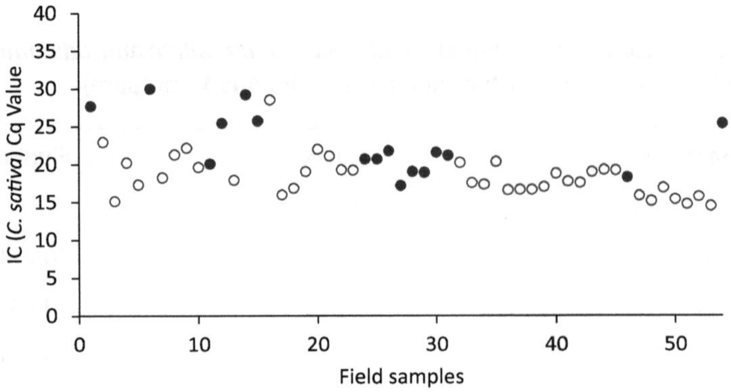

Fig. 2 Cq value of the internal control (IC) in 54 field samples (bark collected from chestnuts in forests, parks, and along roads in Belgium) analyzed by duplex qPCRs after DNA extraction with the NucleoSpin Plant II kit and the TEX lysis buffer. Black dots indicate *C. parasitica* positive samples. (All samples are described in [18])

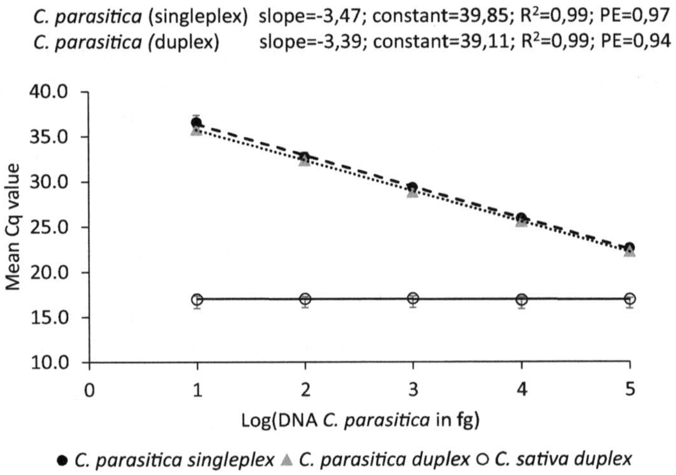

Fig. 3 Standard curves based on tenfold dilutions of *C. parasitica* DNA (from 10 fg to 100 pg/PCR, two replicates per fungal isolate and per concentration level, and two isolates) diluted in *C. sativa* DNA (5 ng/PCR) in singleplex (*C. parasitica*) and duplex (*C. parasitica/ C. sativa*) qPCRs. The slope, constant, correlation coefficient R^2, and PCR efficiencies are indicated for the detection of *C. parasitica* with the two qPCRs (singleplex /duplex)

4. Samples with clear amplification curves and a $Cq < Cq_{LOD}$ for both replicates are positive. Samples with a $Cq \geq Cq_{LOD}$ for one or two replicates (i.e., below the limit of detection) are doubtful.
5. In this case, a new extraction is performed from tubes stored at −20 °C (*see* Subheading 3.2). If the results are again inconclusive, the sample is considered negative.

4 Notes

1. A solution of 100 mM Tris pH 8.0, 20 mM EDTA, and 1.4 M NaCl is autoclaved (121 °C, 20 min). The solution is then aliquoted by 10 mL into plastic tubes stored at 2–8 °C. CTAB 2% and PVP-K30 2% are added at the time of extraction (10 mL lysis buffer allows the extraction of 24 samples). An incubation of the suspension at 65 °C for a few minutes allows the solubilization of CTAB and PVP-K30.
2. Commercial kits described in this chapter have provided reliable results. Other kits can be selected. However, bark tissues contain compounds that can inhibit the PCR. It is therefore recommended to use DNA extraction kits and PCR amplification kits adapted to this type of material (intra-laboratory validation required).
3. Yakima Yellow is a freedom dye and an alternative to VIC. Users should always check if the selected dye is compatible with the PCR instrument used. The channel and filter settings for VIC on the BioRad CFX96 can also be used for Yakima Yellow.
4. It is advisable to wear nitrile/latex gloves that are also disinfected between two samplings.
5. The test can also be done on the bark from oak.
6. After use, the mortar and pestle are disinfected in 1% bleach for 2–3 h, rinsed under tap water, and sprayed with ethanol 70%. The unused plant material is inactivated by autoclave sterilization (121 °C, 30 min [23]).
7. Two tubes are stored at −20 °C to make new extractions in case of doubtful results.
8. Another method consists of using as control the *C. parasitica* amplification product cloned in plasmid DNA, used at 1000 copies per μL as described in [18].
9. The fluorescence thresholds for the two targets depend on the PCR instrument and the PCR conditions. They should be determined by intra-laboratory validation. Using the BioRad CFX96 instrument in the conditions described in this study, the fluorescence thresholds were set at 300 (FAM) and 15 (VIC) for *C. parasitica* and *C. sativa*, respectively.
10. To determine this variation (part of the intra-laboratory validation), thirty replicates of the PAC are analyzed in at least ten duplex qPCR runs corresponding to different routine conditions (different batches of the qPCR kit, different preparations of primers and probes, different operators). The mean and the standard deviation are calculated. An analysis is rejected if the Cq value of the PAC is outside the control limits corresponding to the mean ± 3 standard deviations. This procedure is based on the Shewhart chart [24].

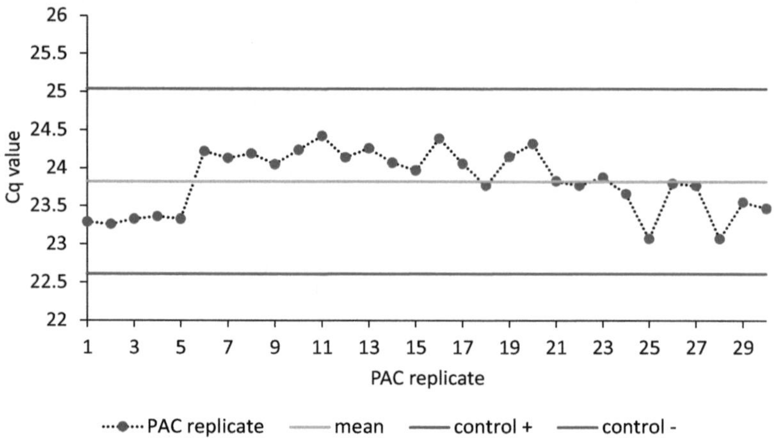

Fig. 4 Variation of the positive amplification control (PAC) Cq value in 30 replicates analyzed in 10 duplex qPCRs. Definition of the upper and lower control limits (mean ± 3 SD)

Figure 4 illustrates the variation of the Cq values of 30 analyses on the BioRad CFX96 using a PAC at 50 pg μL^{-1} and the reagents used in this study.

11. In the case of PCR inhibition, it is recommended to carry out a purification of the DNA extract on a PVPP spin column, as described in [18], and to repeat the qPCR. If the results are still inconclusive, a new DNA extraction can be carried out on the two tubes of ground bark stored at −20 °C.

12. Each time a new qPCR method is used as a diagnostic test, the linear dynamic range, the PCR efficiency, and the correlation coefficient should be determined by means of standard curves [25]. This step of the validation can also be used to determine the limit of detection of the method (LOD) corresponding to the smallest concentration of spiked target DNA detected accurately in 95% of replicates [26]. Based on a standard curve with the logarithm of the initial template concentration plotted on the x-axis, the PCR efficiency = $(10^{(-1/\text{slope})} - 1)$ [26].

References

1. Rigling D, Prospero S (2018) *Cryphonectria parasitica*, the causal agent of chestnut blight: invasion history, population biology and disease control. Mol Plant Pathol 19:7–20. https://doi.org/10.1111/mpp.12542
2. EFSA PLH Panel (EFSA Panel on Plant Health) (2014) Scientific opinion on the pest categorization of *Cryphonectria parasitica* (Murrill) Barr. EFSA J 12(10):3859, 42. https://doi.org/10.2903/j.efsa.2014.3859
3. Bisseger M, Sieber T (1994) Assemblages of endophytic fungi in coppice shoots of *Castanea sativa*. Mycologia 86:648–655
4. Adamčikova K, Kobza M, Juhásová G (2010) Characteristics of the *Cryphonectria parasitica* isolated from *Quercus* in Slovakia. Forest Pathol 40:443–449
5. Haltofová P, Jankovský L, Palovčíková D (2005) New finds of *Cryphonectria parasitica* and the first record of chestnut blight on red

oak *Quercus rubra* L. in the Czech Republic. J For Sci 51:256–258

6. Tziros GT, Nakopoulou ZG, Diamandis S (2015) *Cryphonectria parasitica*, the chestnut blight fungus, causes cankers on *Quercus frainetto* in Greece. Aust Plant Dis Notes 10:19. https://doi.org/10.1007/s13314-015-0171-4
7. Mlinarec J, Nuskem L, Ježić M et al (2018) Molecular evolution and invasion pattern of *Cryphonectria* hypovirus 1 in Europe: mutation rate, and selection pressure differ between genome domains. Virology 514:156–164
8. Heiniger U, Rigling D (1994) Biological control of chestnut blight in Europe. Annu Rev Phytopathol 32:581–599
9. Kunova A, Pizzatti C, Cerea M et al (2017) New formulation and delivery method of *Cryphonectria parasitica* for biological control. J Appl Microbiol 122:180–187
10. Dennert F, Rigling D, Meyer JB et al (2020) Testing the pathogenic potential of *Cryphonectria parasitica* and related species on three common European Fagaceae. Front For Glob Change 3:52. https://doi.org/10.3389/ffgc.2020.00052
11. EFSA PLH Panel (EFSA Panel on Plant Health) (2016) Scientific opinion on the risk assessment and reduction options for *Cryphonectria parasitica* in the EU. EFSA J 14(12): 4641, 54. https://doi.org/10.2903/j.efsa.2016.4641
12. Council of the European Union (2000) Directive 2000/29/EC of 8 May 2000 on protective measures against the introduction into the Community of organisms harmful to plants or plant products and against their spread within the Community. Off J Eur Union L169:1–112
13. European Commission (2019) Commission Implementing Regulation (EU) 2019/2072 of 28 November 2019 establishing uniform conditions for the implementation of Regulation (EU) 2016/2031 of the European Parliament and the Council, as regards protective measures against pests of plants, and repealing Commission Regulation (EC) No 690/2008 and amending Commission Implementing Regulation (EU) 2018/2019. Off J Eur Union L319:1–279
14. Dutech C, Fabreguettes O, Capdevielle X et al (2010) Multiple introductions of divergent genetic lineages in an invasive fungal pathogen, *Cryphonectria parasitica*, in France. Heredity 105:220–228
15. Peters FS, Bußkamp J, Prospero S et al (2014) Genetic diversification of the chestnut blight fungus *Cryphonectria parasitica* and its associated hypovirus in Germany. Fungal Biol 118:193–210
16. Prospero S, Cleary M (2017) Effects of host variability on the spread of invasive forest diseases. Forests 8:80. https://doi.org/10.3390/f8030080
17. Prospero S, Conedera M, Heiniger U et al (2006) Saprophytic activity and sporulation of *Cryphonectria parasitica* on dead chestnut wood in forests with naturally established hypovirulence. Phytopathology 96:1337–1344. https://doi.org/10.1094/phyto-96-1337
18. Chandelier A, Massot M, Fabreguettes O et al (2019) Early detection of *Cryphonectria parasitica* by real-time PCR. Eur J Plant Pathol 153:29–46. https://doi.org/10.1007/s10658-018-1538-0
19. Barbier FF, Chabikwa TG, Ahsan MU et al (2019) A phenol/chloroform-free method to extract nucleic acids from recalcitrant, woody tropical species for gene expression and sequencing. Plant Methods 15:62. https://doi.org/10.1186/s13007-019-0447-3
20. Gouker FE, Guo Y, Pooler MR (2020) Using acetone for rapid PCR-amplifiable DNA extraction from recalcitrant woody plant taxa. Appl Plant Sci 8(12):e11403. https://doi.org/10.1002/aps3.11403
21. Hoorfar J, Malorny B, Abdulmawjood A et al (2004) Practical considerations in design of internal amplification controls for diagnostic PCR assay. J Clin Microbiol 42:1863–1868
22. Mittelberger C, Obkircher L, Oberkofler V et al (2020) Development of a universal endogenous qPCR control for eukaryotic DNA samples. Plant Methods 16:53. https://doi.org/10.1186/s13007-020-00597-2
23. Emmanuel J, Hrdinka C, Gluszynski P et al (2004) Non-incineration medical waste treatment technologies in Europe. Health Care Without Harm Europe (ed) https://www.env-health.org/IMG/pdf/altech_Europe_updated_version_10_12_2004.pdf
24. Westgard JO, Barry PL, Hunt MR (1981) A multi-rule Shewhart Chart for quality control in clinical chemistry. Clin Chem 27:493–501
25. Bustin SA, Benes V, Garson JA et al (2009) The MIQE guidelines: minimum information for publication of quantitative real-time PCR experiments. Clin Chem 55:611–622
26. Kralik P, Ricchi M (2017) A basic guide to real time PCR in microbial diagnostics: definitions, parameters, and everything. Front Microbiol 8: 108. https://doi.org/10.3389/fmicb.2017.00108

Chapter 12

Rapid Molecular Diagnostics in the Field and Laboratory to Detect Plant Pathogen DNA in Potential Insect Vectors

Karolina Pusz-Bochenska, Edel Pérez-López, Tim J. Dumonceaux, Chrystel Olivier, and Tyler J. Wist

Abstract

A variety of sensitive and specific molecular diagnostic assays has been described for detecting nucleic acids in biological samples that may harbor pathogens of interest. These methods include very rapid, isothermal nucleic acid amplification methods that can be deployed outside of the laboratory environment, such as loop-mediated isothermal DNA amplification (LAMP) and recombinase-polymerase amplification (RPA). However, all molecular diagnostic assays must be preceded by nucleic acid extraction from the biological samples of interest, which provides suitable template molecules for the assays. To exploit the features of the amplification assays and be utilized outside of the lab, these methods must be rapid and avoid the need for typical laboratory chemicals and equipment. We describe a protocol for the extraction of DNA from field-collected insects that can be implemented at the point of collection and used to detect the presence of DNA sequences from potential plant pathogens that may be vectored by the insects. This protocol provides template DNA that is suitable for PCR, LAMP, and RPA. The FTA PlantSaver card-based DNA extraction product was also confirmed to amplify the mitochondrial cytochrome oxidase 1 (CO1) universal barcode that could later be sequenced to identify any insect. Lastly, we provide an example using field-collected insects, *Neokolla (Graphocephala) heiroglyphica*, and demonstrate the detection of the plant pathogen *Xylella fastidiosa* in carrier insects using PCR, RPA, and LAMP.

Key words DNA extraction, Aster yellows phytoplasma, *Xyllela fastidiosa*, LAMP, RPA, Field-adaptable molecular methods

1 Introduction

Phytophagous insects cause significant yield losses in crop plants of all types, both through direct feeding damage [1] and as a result of plant pathogenic bacteria and viruses that can be vectored by insect feeding and impact productivity throughout the growing season [2, 3]. The lack of treatment options for many of these insect-vectored diseases, coupled with the fact that the appearance of symptoms in host plants can lag feeding activity by several weeks, means that surveillance for insect-vectored diseases must include

Nicola Luchi (ed.), *Plant Pathology: Methods and Protocols*, Methods in Molecular Biology, vol. 2536,
https://doi.org/10.1007/978-1-0716-2517-0_12,

both the host plants and the insects that spread the infection through their feeding activity. Rapid, inexpensive, and accurate methods to detect the appearance of vectors harboring the pathogen of concern will help producers implement appropriate mitigative measures and provide regulators with data on the appearance and spread of diseases of concern.

Xylella fastidiosa [4] is a xylem-limited, leafhopper-vectored pathogen responsible for several diseases depending upon the plant that is infected. In citrus crops, *X. fastidiosa* causes citrus variegated chlorosis disease [5] and, in coffee, coffee leaf scorch disease [6]. In grapevine, *X. fastidiosa* threatens the viticulture industry by causing a lethal disease, named after the pathologist who first described the symptoms in the 1800s, called "Pierce's disease" [4].

Very rapid and accurate detection of the DNA of pathogens such as *X. fastidiosa* in biological samples is provided using molecular diagnostic methods. For some pathogens that are difficult or impossible to culture in the laboratory, such as "*Candidatus* Phytoplasma" spp., such methods are the most commonly employed means of detection and strain typing [7]. In other cases, for example, *X. fastidiosa* [3] bacteria may be culturable, but the time required for strain growth precludes the use of culture for surveillance activities [3]. Methods such as PCR [8], quantitative PCR [9], and droplet digital PCR [10] offer the possibility of highly accurate detection and typing of pathogen DNA in insect and host plant tissues with a relatively rapid turnaround. However, these methods are laboratory-based, which means that plant or insect samples must be collected in the field, then transported to the laboratory for processing, including both DNA extraction and the application of the molecular diagnostic procedure. These steps add considerable time between the collection of the sample and the reporting of the result, opening the door to further transmission of the diseases.

Alternative molecular diagnostic methods have been described that are very rapid and do not require the extensive use of laboratory equipment such as thermocyclers, centrifuges, or gel electrophoresis tanks and power supplies. These methods are typically isothermal, meaning that they can be run at a single temperature with portable equipment. Isothermal methods can therefore be applied more conveniently outside the laboratory, and the steps of sample collection and pathogen detection can be combined into a single step. This combination of steps will greatly reduce the turnaround time for the generation of a result, especially given the typical speed of these diagnostic methods. For example, insulated isothermal PCR (iiPCR) has been applied to the on-site detection of equid alphaherpesvirus 3 in genital swabs of animals, providing detection in under an hour [11]. Another commonly used isothermal method that is suitable for on-site plant pathogen detection is loop-mediated isothermal DNA amplification (LAMP) [12, 13]. LAMP provides an array of product detection methods including colorimetric, fluorometric, and turbidometric and can generate a positive result in well under 30 min for bacteria, viruses, or fungi [14–16]. Recombinase-polymerase amplification (RPA) [17] is an alternative isothermal method that offers very rapid, sensitive, and specific detection of target pathogen nucleic acid and can be adapted to on-site or point-of-care applications [18, 19].

Irrespective of the molecular diagnostic method employed, however, the basic challenge of extracting nucleic acids from the sample that are suitable for the intended purpose must be overcome. Typical laboratory-based methods for plants and insects may use commercial kits with steps including sample disruption using liquid nitrogen and multiple centrifugation steps to purify nucleic acids that are relatively pure and free of inhibitors of PCR amplification or may use harsh chemicals such as phenol-chloroform or quaternary ammonium surfactants such as cetyltrimethylammonium bromide [20]. Such methods are clearly not suited for application outside of the laboratory environment, since centrifuges and hazardous chemical disposal protocols are not readily available. However, one of the advantages of the isothermal amplification methods that have been mentioned is that they typically do not require nucleic acid that is as free of contaminants as PCR-based methods. Accordingly, a very rapid nucleic acid extraction method (30 s) has been developed for plant tissue [21] that involves sample disruption and the application of a paper dipstick that adsorbs nucleic acid, facilitating its transfer from the sample to the amplification master mix. Alternatively, rapid and inexpensive detection of SARS-Cov2 in human nasal swabs and saliva samples has been reported using LAMP and an inactivation and purification protocol using glass microbeads [14]. However, no nucleic acid extraction method suitable for application outside of the laboratory environment has been described for insects. Dickey et al. [22] described a protocol for elution of DNA from Flinders Technology Associates (FTA) cards from freshly collected whiteflies (*Bemisia tabaci*), but the protocol described involves a vacuum centrifugation step and was not shown to be applicable to other insects. To overcome the challenges associated with the field-based extraction of amplification-ready DNA from insects, we applied FTA cards with an extraction and elution protocol that was applicable to a wide range of insects in various matrices, including live caught, sticky cards, and ethanol-stored samples [23]. The nucleic acid extracted from the insects was demonstrated to be suitable for the detection of the insect-vectored plant pathogen "*Candidatus* Phytoplasma *asteris*" using LAMP, as well as for insect identification using PCR

Fig. 1 (**a**) (dorsal), (**b**) (ventral view of frons). *Neokolla* (*Graphocephala*) *hieroglyphica* insects used for DNA extraction. (Insects were collected in blueberry fields in Quebec, Canada, in 2018)

and sequencing of cytochrome oxidase I [23]. In the present work, we describe this method in detail and demonstrate its utility for LAMP, RPA, and conventional PCR to detect the bacterial plant pathogen *X. fastidiosa* [24], which is vectored by several insects, including the xylem-feeding sharpshooters of the genus *Neokolla* (formerly of the genus *Graphocephala)*, including *N. heiroglyphica* (Hemiptera: Cicadellidae) [3, 25] (Fig. 1).

2 Materials

2.1 Sample Collection

1. 70% Ethanol.
2. Sweep net.
3. Yellow sticky card traps (e.g., Catchmaster, U.S., catalog no. 980-y, or AlphaScents SKU YC4X6).
4. Scissors.

2.2 Nucleic Acid Extraction Using FTA Matrix Cards

1. FTA PlantSaver cards (Whatman).
2. AP1 buffer from the DNeasy Plant Mini Kit (Qiagen).
3. Rubber mallet (can be obtained at a local hardware store).
4. Hair dryer.
5. A fabric punch with a variable punch size (can be obtained at a fabric store).
6. Forceps.
7. Pipettes and pipette tips.
8. Thermal block (available from a variety of suppliers).
9. 1.5 mL Centrifuge tubes (e.g., Eppendorf).
10. Buffer $TE_{0.1}$ (10 mM Tris-Cl pH 8.0; 0.1 mM EDTA) (10 mL): Mix 100 μL of a stock solution of 1 M Tris-Cl pH 8.0 and 2 μL of a stock solution of 0.5 M EDTA in a final volume of 10 mL water. Store it at room temperature.
11. Ethanol 70% (10 mL): Add 3 mL of 95–100% ethanol to 7 mL of water. Store at room temperature – discard solution and make fresh after 1 week.
12. Disposable gloves.

2.3 Nucleic Acid Amplification Using PCR

1. QIAquick PCR purification kit (Qiagen),
2. PCR primers amplifying insect CO1, *X. fastidiosa rpoD*, or other targets (available from a variety of suppliers),

   CO1 primer sequences [26]:

   - HCO2198: 5'-TAAACTTCAGGGTGACCAAAAAATCA-3',
   - LCO1490: 5'-GCTCAACAAATCATAAAGATATTGG-3',

   rpoD primer sequences [27]:

   - rst31: 5'- GCGTTAATTTTCGAAGTGATTCGATTGC - 3',
   - rst33: 5'-CACCATTCGTATCCCGGTG-3',
3. 10x PCR buffer (Invitrogen),
4. 100 mM dNTP set (Invitrogen).

5. 50 mM $MgCl_2$ (Invitrogen).
6. *Taq* polymerase (Invitrogen).
7. Molecular biology grade or ultraviolet light-treated H_2O.
8. 1 kb DNA ladder (Invitrogen).
9. Agarose powder (e.g., UltraPure Agarose – Invitrogen).
10. TBE buffer.
11. 1.5 mL Centrifuge tubes (Eppendorf).
12. PCR strip tubes with caps (available from a variety of suppliers).
13. Gel electrophoresis system (e.g., Wide Mini-Sub Cell GT system, Bio-Rad + Power Pac HC, Bio-Rad).
14. Gel imager (e.g., Gel Doc XR+, Bio-Rad).
15. Mini-centrifuge accommodating 1.5 mL tubes, with an adaptor that will enable the centrifugation of PCR/LAMP strip tubes (e.g., VWR® Mini Centrifuge).
16. Thermal cycler (e.g., C1000 Touch Thermal Cycler, Bio-Rad)
17. Pipettes and pipette tips.
18. Disposable gloves.

2.4 DNA Amplification Using RPA

2.4.1 RPA Product Detection Using Agarose Gel Electrophoresis

1. TwistAmp Basic kit (TwistDx, UK).
2. Oligonucleotide primers for amplification (RPA product size = 236 bp). (See Subheading 2.4.2 for primer sequences).
3. QIAquick PCR purification kit (Qiagen).
4. Molecular weight marker (available from a variety of suppliers).
5. Research grade water H_2O.
6. E-Gel 1.2% agarose gel with SYBR Safe (Invitrogen).
7. E-Gel Power Snap electrophoresis device with camera (Invitrogen).
8. 1.5 mL centrifuge tubes (Eppendorf).
9. Mini-centrifuge accommodating 1.5 mL tubes, with an adaptor that will enable the centrifugation of PCR/LAMP strip tubes (e.g., VWR® Mini Centrifuge).
10. Heating block or thermocycler.
11. Pipettes and pipette tips.
12. Disposable gloves.

2.4.2 RPA Product Detection Using a Real-Time PCR Instrument

1. TwistAmp Exo kit (TwistDx, UK).
2. Oligonucleotide primers for amplification and exo-probe for product detection (RPA product size = 236 bp):
   - Xf-170-31: 5'- CAACCATCTCTGCTAACTCAGACGAGTCTAT-3'.

- Xf-405-30: 5'-TTTCTTGTCGTGAAGGAGGATGTAGG GATT-3'.
- Xf-2p: 5'- CAGTTTGACCGTGGTTACTCTTCGCCA-TATT-tF-H-a-tQ-CAACAACCAGCAGTC -3'. (The RPA probe contains the following modifications: tF, deoxythymidine residue coupled to fluorophore FAM (dT-FAM); H, tetrahydrofuran; tQ, deoxythymidine residue coupled to quencher BHQ-1 (dT-BHQ-1). Probe ordered from Biosearch Technologies).

3. Real-time PCR instrument for detection of RPA-exo products (*see* **Note 1**) (e.g., CFX96 Real-Time System, Bio-Rad).
4. PCR strip tubes, 0.2 mL volume (available from a variety of suppliers).
5. 1.5 mL Centrifuge tubes (Eppendorf).
6. Mini-centrifuge accommodating 1.5 mL tubes, with an adaptor that will enable the centrifugation of PCR/LAMP strip tubes (VWR® Mini Centrifuge).
7. Electrical equipment can be connected to a portable power supply in a field setting (e.g., MotoMaster Nautilus Recreational Power Pack (800 A; 400 W; 33 Ah) which is available from the Canadian Tire store in Canada).
8. Pipettes and pipette tips.
9. Disposable gloves.

2.5 DNA Amplification Using LAMP

1. Master mix for LAMP product generation and detection (e.g., WarmStart LAMP Kit, New England Biolabs). Amplification primers targeting *X. fastidiosa cpn60*:
   - Xf-cpn60-F3: 5'-CAATGGATCTTAAGCGTGGTA-3'.
   - Xf-cpn60-B3: 5'-AATATGGCGAAGAGTAACCAC-3'.
   - Xf-cpn60-FIP: 5'- -GTTAGCAGAGATGGTTGCGACTGGTTACC-GAACTGAAGAAGAT-3'.
   - Xf-cpn60-BIP: 5'-CAGACGAGTCTATCGGCAACATGT GCCTTCCTCAATGGTAAT-3'.
   - Xf-cpn60-loopF: 5'-TTATCGTCGCTGGTTGGC-3'.
   - Xf-cpn60-loopB: 5'-AAGCAATGAAGAAGGTCGGT-3'.
2. Aluminum sticky foil seals for PCR (e.g., VWR catalog number 60941-124).
3. 1.5 mL Centrifuge tubes (Eppendorf).
4. Genie apparatus and Genie LAMP tube strips (0.2 mL volume) (OptiGene, UK).

5. Mini-centrifuge accommodating 1.5 mL tubes, with an adaptor that will enable the centrifugation of PCR/LAMP strip tubes (e.g., VWR® Mini Centrifuge).
6. Portable power supply.
7. Pipettes and pipette tips.
8. Disposable gloves.

3 Methods

3.1 Sample Collection

We have previously demonstrated [23] the utility of this method for extracting DNA from insects in various matrices that are typically employed in field collection, including sticky cards, live caught, and ethanol preserved. The current example will employ insects that have been preserved in ethanol for over a year, but the method would not be any different for live collected insects. Four *Neokolla (Graphocephala) heiroglyphica* (Hemiptera: Cicadellidae) used in this assay were collected using sweep netting from blueberry fields in Quebec, Canada, as part of a project on blueberry-stunt disease (Fig. 1). Because the *X. fastidiosa* carrier status of the individual insects could not be known prior to DNA extraction, we added 5 ul of *X. fastidiosa* genomic DNA to two of the individuals prior to extraction (added after sample homogenization) and added the same amount of $TE_{0.1}$ buffer to the other two individuals. Control non-carrier insects (*Macrosteles quadrilineatus*) (Hemiptera: Cicadellidae) were collected from a laboratory colony maintained at 24 degrees Celsius (16:8 day-night) on barley (*Hordeum vulgare*) seedlings at the Agriculture and Agri-Food Canada Research Saskatoon Research and Development Centre.

3.2 Nucleic Acid Extraction Using FTA Matrix Cards

Nucleic acids are extracted from whole insects by homogenization onto the FTA matrix cards, followed by washing. Nucleic acids are then eluted from the paper matrix to provide a suitable template for the amplification methods described (*see* **Note 2**).

1. Place insect (up to 5 depending on size) onto the matrix card. Put the cover sheet on top (cover sheet can be marked with a pen to identify samples).
2. Hit the paper with the hard rubber mallet 5–10 times. Check to see that the matrix is slightly wet.
3. Dry the matrix using a hair dryer for 2 min (low setting) OR air dry for at least 1 hour (*see* **Note 3**).
4. Using the fabric punch, remove a disc from the center of the insect spot. A fabric punch features a variable punch size that can be adjusted to accommodate the size of the insect (*see* **Note 4**).

5. Transfer the punch to a 1.7 mL Eppendorf tube. To clean the punch tool, punch into clean filter paper and/or wipe the DNA away between samples.
6. Add to the punch 200 μL of cell lysis buffer AP1 (from the Qiagen plant DNA kit). Invert the tube 1–2x and incubate 2–3 min at room temperature (ensure the punch remains in the liquid and not caught up in the lid). Remove buffer AP1 and discard.
7. Add 200 μL of fresh AP1 buffer and repeat **step 6** (total of 2 washes with AP1).
8. Add 200 μL of $TE_{0.1}$ (10 mM Tris pH 8.0, 0.1 mM EDTA), invert tube 1–2x, and incubate 1 min at ambient temperature. Remove buffer and discard.
9. Repeat **step 8** twice (three washes).
10. Rinse the punch once with 200 μL of 70% ethanol. Discard wash and allow punch to dry for 5 min with the lid open at ambient temperature.
11. Elute DNA from the punch using $TE_{0.1}$ (20 μL for 1.2 mm punch, 150 μL for 5 mm punch, 200 μL for 7 mm punch). Incubate the tube at 90 °C for 20 min.
12. Quickly spin the tube to collect all the condensation. (Optional: transfer liquid to a fresh tube or remove paper with forceps.)

3.3 Conventional PCR Targeting Insect CO1 and *X. fastidiosa rpoD*

DNA extracted from the FTA matrix card can be used as a template in a PCR with universal primers targeting the mitochondrial cytochrome oxidase 1 (CO1) gene [26]. To determine the taxonomic affiliation of the insect, amplicons are purified from the amplification mixture using QIAquick columns (Qiagen) and sequenced using the amplification primers (Eurofins Genomics). Insects are identified based on the CO1 amplicon sequence obtained by comparison to the Barcode of Life database [28, 29] and/or comparison to GenBank using the Basic Local Alignment Search Tool (BLAST) (www.ncbi.nlm.nih.gov). The current example yielded the sequence deposited in GenBank under accession number MZ396076. The same CO1 results were obtained from all four insects.

1. Prepare an amplification master mix to provide the following final concentrations: 1x PCR buffer (Invitrogen), 1.5 mM $MgCl_2$, 0.2 mM of each dNTP, 0.4 μM of each primer, and 1 U Taq polymerase (Invitrogen). The final PCR volume, including template, is 50 μL.
2. Add 48 μL of the amplification master mix from **step 1** to a PCR tube.

3. Add 2 μL of template or control DNA to each tube. Snap the lid on the tube immediately after adding the template.
4. Centrifuge the strip tubes briefly to bring the template DNA into the mixture.
5. Place the strip tube into the thermal cycler and close the lid.
6. Start the CO1 PCR program, which includes 1 min at 94 °C; 5 cycles at 94 °C for 40 s, 45 °C for 40 s, and 72 °C for 1 min; 35 cycles at 94 °C for 40 s, 51 °C for 40 s, and 72 °C for 1 min; 1 cycle at 72 °C for 1 min; and a final hold at 12 °C.
7. Analyze the results on a 1% agarose gel with TBE or TAE (Bio-Rad, Wide Mini-Sub Cell GT system). The expected amplicon size is approximately 650 bp [26].
8. Add 1.2 μL of 10x loading buffer (Invitrogen) to 10 μL of the amplicon, and load the samples into the wells of the gel. Include a molecular weight marker that provides sizing in the 100–1000 bp range (e.g., Invitrogen 1 kb ladder).
9. Run the gel at 80–150 V for about 1–1.5 hours.
10. Visualize the results.
11. Before sending the PCR product for sequencing, purify the reactions using QIAquick columns.
    (a) Transfer the reactions to a new 1.5 mL Eppendorf tube.
    (b) Add 5 volumes (250 μL) of buffer PB to each reaction.
    (c) Mix and transfer to a QIAquick column in a supplied 2 mL tube.
    (d) Centrifuge for 1 min at 14000 g.
    (e) Discard eluate; add 750 μL of buffer PE to the column.
    (f) Centrifuge as in d.
    (g) Discard eluate; spin column dry for 1 min at 14000 g.
    (h) Add 30 μL elution buffer EB to each column; incubate for 1 min at room temperature. Transfer the column to a fresh 1.5 mL Eppendorf tube.
    (i) Centrifuge at 8000 g or higher (caps tend to snap off at high g) for 1 min.
12. Prepare the samples according to sample submission guidelines and send them out to the company that provides sequencing service (we have used Eurofins Genomics; eurofinsgenomics.com).
13. Amplification of *rpoD* from *X. fastidiosa* is performed using the primers described in Subheading 2.3. The PCR protocol is generally the same as for CO1 (**steps 1–9**), except that an amplification master mix is prepared containing 1x PCR buffer (Invitrogen), 2.5 mM $MgCl_2$, 0.5 mM of each dNTP, 0.4 μM of each primer, and 1 U Taq polymerase (Invitrogen). The final PCR volume, including template, is 50 μL.

14. *rpoD* Amplification program: 1 cycle at 95 °C for 3 min; 40 cycles at 95 °C for 30 sec, 55 °C for 30 sec, and 72 °C for 30 sec and 1 cycle at 72 °C for 5 min. The expected PCR product size is 733 bp.

3.4 Nucleic Acid Amplification Using RPA

3.4.1 RPA Product Detection Using Agarose Gel Electrophoresis

RPA is performed using a commercially available, proprietary master mix, many of the key ingredients of which are provided in a lyophilized format. This is especially useful for in-field applications since it improves the stability of the reagents. The sequences of the primers used to amplify *X. fastidiosa cpn60* are provided in Subheading 2.4.2.

1. Prepare the reaction mixture according to the instructions below. Prepare a master mix including extra volume to account for pipetting errors:

Component	μL/reaction	Final concentration
Xf-170-31 (10 μM)	2.4	480 nM
Xf-405-30 (10 μM)	2.4	480 nM
Primer-free rehydration buffer	29.5	–
Water (*see* **Note 5**, **Note 6**)	11.2	–
Total	*45.5*	–

2. Add 42.5–45.5 μL of the reaction mixture from **step 1** to a tube containing lyophilized reagents. Mix to rehydrate.
3. Add 2–5 μL of template or control DNA to each tube (*see* **Note 6**). Do not close the lids on the tubes after adding the templates; rather, place it upside down beside the strip tube.
4. Add 2.5 μL of 280 mM magnesium acetate (MgOAc – supplied in the kit) to the tube lids. Snap the caps onto the tubes; then centrifuge the strip. The reaction starts upon the addition of MgOAc (14 mM final concentration), so it is important to work quickly.
5. Incubate the strip tube at 39 °C for 20 min.
6. Purify the reactions using QIAquick columns. The reactions contain too much salt to be directly analyzed by agarose gel electrophoresis.
   (a) Transfer the reactions to a new 1.5 mL Eppendorf tube.
   (b) Add 5 volumes (250 μL) of buffer PB to each reaction.
   (c) Mix and transfer to a QIAquick column in a supplied 2 mL tube.
   (d) Centrifuge for 1 min at 14000 g.
   (e) Discard eluate; add 750 μL buffer PE to the column.

(f) Centrifuge as in d.

(g) Discard eluate; spin column dry for 1 min at 14000 g.

(h) Add 30 μL of elution buffer (EB) to each column; incubate for 1 min at room temperature. Transfer the column to a fresh 1.5 mL Eppendorf tube.

(i) Centrifuge at 8000 g or higher (caps tend to snap off at high g) for 1 min.

7. Analyze the results of the RPA using agarose gel electrophoresis on a 1.2% E-Gel.

   (a) Add 20 μL of molecular weight marker to well 1.

   (b) Add 20 μL of purified RPA products to subsequent wells. Add 20 μL of 0.5x TBE buffer to any empty wells.

   (c) Run the gel for 10 min.

   (d) Observe the results using the back light. Record the results using the camera.

8. When the gel has finished, note the results in terms of positive/ negative (*see* **Note 7**).

3.4.2 RPA Product Detection Using a Real-Time PCR Instrument

The RPA exo kit is a suitable choice for product detection in real time. The procedure is described here with detection using a real-time PCR instrument (Bio-Rad CFX96), which is not suitable for in-field applications (*see* **Note 1**). The sequences of the primers and exo probe are provided in Subheading 2.4.2.

1. Prepare the reaction mixture according to the instructions below. Prepare a master mix including extra volume to account for pipetting errors:

Component	μL/Reaction	Final concentration
Xf-170-31 (10 μM)	2.1	420 nM
Xf-405-30 (10 μM)	2.1	420 nM
Xf-2p (10 μM)	0.6	120 nM
Primer-free rehydration buffer	29.5	–
Water (*see* **Note 5, Note 6**)	11.2	–
Total	*45.5*	–

2. Add 42.5–45.5 μL of the reaction mixture from **step 1** to a tube containing lyophilized reagents. Mix to rehydrate.

3. Add 2–5 μL of template or control DNA to each tube (*see* **Note 6**). Do not close the lids on the tubes after adding the templates; rather, place it upside down beside the strip tube.

4. Add 2.5 μL of 280 mM magnesium acetate (MgOAc – supplied in the kit) to the tube lids. Snap the caps onto the tubes; then centrifuge the strip. The reaction starts upon the addition of MgOAc (14 mM final concentration), so it is important to work quickly.
5. Use the following amplification program on the Bio-Rad CFX96 or equivalent thermocycler: 40 cycles of 39 °C for 30 s (plate read; FAM channel).
6. Results are provided by the CFX96 in terms of quantification cycle (C_q). At 30 s per cycle, the time to positive (T_p, minutes) can be derived by $C_q/2$ (*see* **Note 8**).

3.5 Nucleic Acid Amplification Using LAMP

This nucleic acid extraction method was previously shown to provide LAMP-ready DNA from *M. quadriliniatus* harboring the plant pathogen "*Candidatus* Phytoplasma *asteris*" [23]. Similar methods, detailed below, are used to determine if *N. heiroglyphica* individuals carry *X. fastidiosa*. Here we use a set of LAMP primers we designed to amplify a fragment of *cpn60* from the target bacterium (Subheading 2.5). The protocol provided here uses a WarmStart LAMP Master Mix with fluorometric detection (New England Biolabs – NEB). This master mix uses a proprietary fluorescent dye that interacts with DNA to generate a fluorescent signal. Products are detected using a Genie apparatus (OptiGene, UK) – *see* **Note 9**).

1. Prepare the LAMP primer mixture according the instructions below. Prepare a master mix sufficient for 100–500 reactions (*see* **Note 10**):

Component	μL/ assay	[stock], μM	Final concentration in LAMP, μM
Forward outer primer (F3)	0.05	100	0.2
Reverse outer primer (B3)	0.05	100	0.2
Forward inner primer (FIP)	0.4	100	1.6
Backward inner primer (BIP)	0.4	100	1.6
Loop inner primer (loopF)	0.2	100	0.8
Loop outer primer (loopB)	0.2	100	0.8
Water	3.7		
Total	*5*		

2. Prepare the reaction mixture according to the instructions below, based on the number of reactions to be performed. Prepare extra master mix to account for pipetting errors (*see* **Note 11**). The reaction mix and fluorescent dye are from New England Biolabs (Subheading 2.5):

Component	μL/assay
2x Reaction mix	12.5
Fluorescent dye (*see* **Note 12**)	0.5
Water	7–10
Template DNA	2–5
Total	*25*

3. Add 20–23 μL of the reaction mixture to each well of a LAMP strip tube. Puncture the foil seal using the pipette tip.
4. Add 2–5 μL of template or control DNA to each tube. Snap the lid on the tube immediately after adding the template.
5. Centrifuge the strip tubes briefly to bring the template DNA into the LAMP mixture.
6. Place the strip tube into the Genie apparatus and close the lid.
7. Start the LAMP program, which includes an annealing curve: 40 °C for 1 min, then 63 °C for 60 min; 98 °C–80 °C at 0.05 °C/second.
8. Observe the fluorescent traces in real time and note the results in terms of time to positive (T_p, minutes) and annealing temperature of the products generated (T_a, °C).
9. When the program has finished, note the results in time to positive, and observe the results of the annealing curve that occurs at the end of the program (*see* **Note 13**).

4 Notes

1. A portable apparatus is available for RPA exo applications in-field (e.g., Axxin T16-ISO instrument, catalog no P002921).
2. A video has been prepared demonstrating this method. It can be found here: https://www.youtube.com/watch?v=Y251r_amzFA (or search "Rapid DNA extraction method using filter paper cards in 10 simple steps").
3. At this point, the sample can be stored at room temperature for long periods of time prior to starting the DNA extraction. Typically, however, nucleic acid extraction would proceed immediately in the field.

4. The size of the punch should be selected according to the size of the homogenized insect after hitting with the hammer (Fig. 2). Try to get as little extra filter paper as possible while capturing the entire insect. Some insects, such as *M. quadrilineatus*, are so small that they do not leave the FTA matrix visibly wet. This does not appear to impact DNA extraction efficiency.

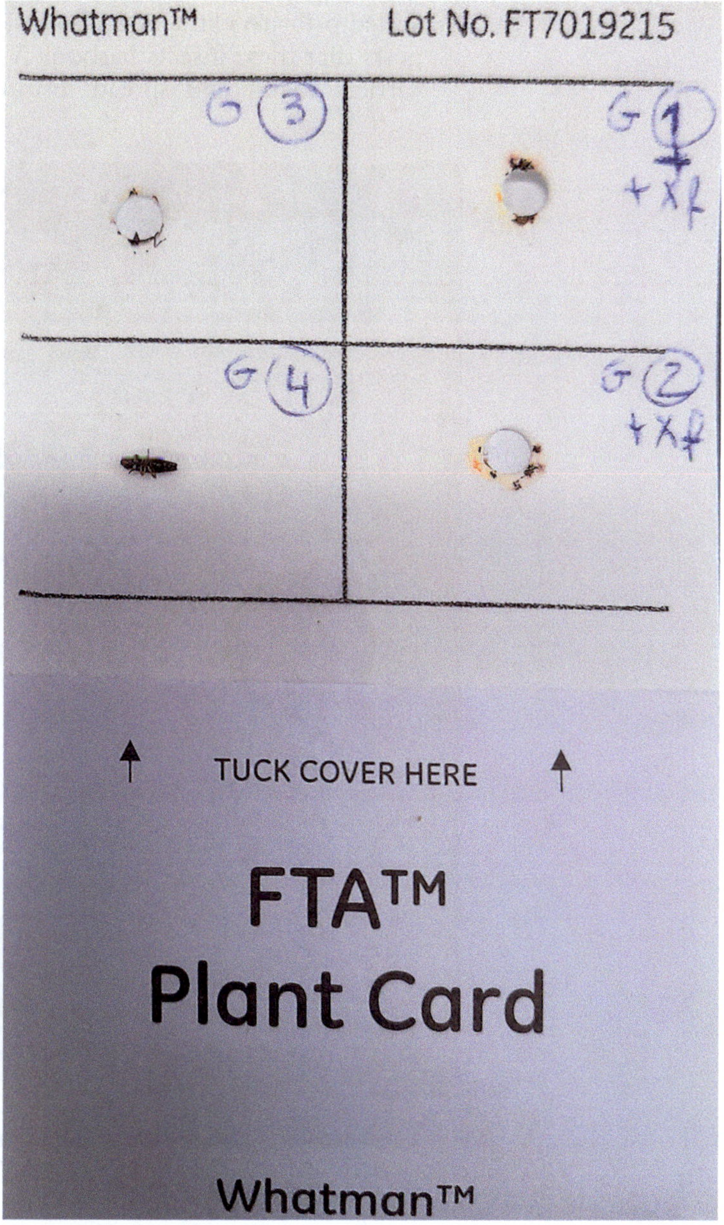

Fig. 2 FTA matrix cards with *Neokolla* (*Graphocephala*) *heiroglyphica* after homogenization and sample punch removal. One insect (quadrant 4) is shown prior to homogenization

5. After the addition of water, the mixture is used to rehydrate the remaining reaction components, which are supplied in lyophilized format.
6. Up to 5 μL of template can be added, in which case the volume of water should be adjusted accordingly.
7. Using this method on the *N. hieroglyphica* insects examined in this example, it is clear that all of the insects were positive, including the two insects that did not have *X. fastidiosa* DNA added prior to extraction, "spiked insects" (Fig. 3). This suggests that these insects harbour *X. fastidiosa* and are potential carriers that could spread disease to plants. This result is

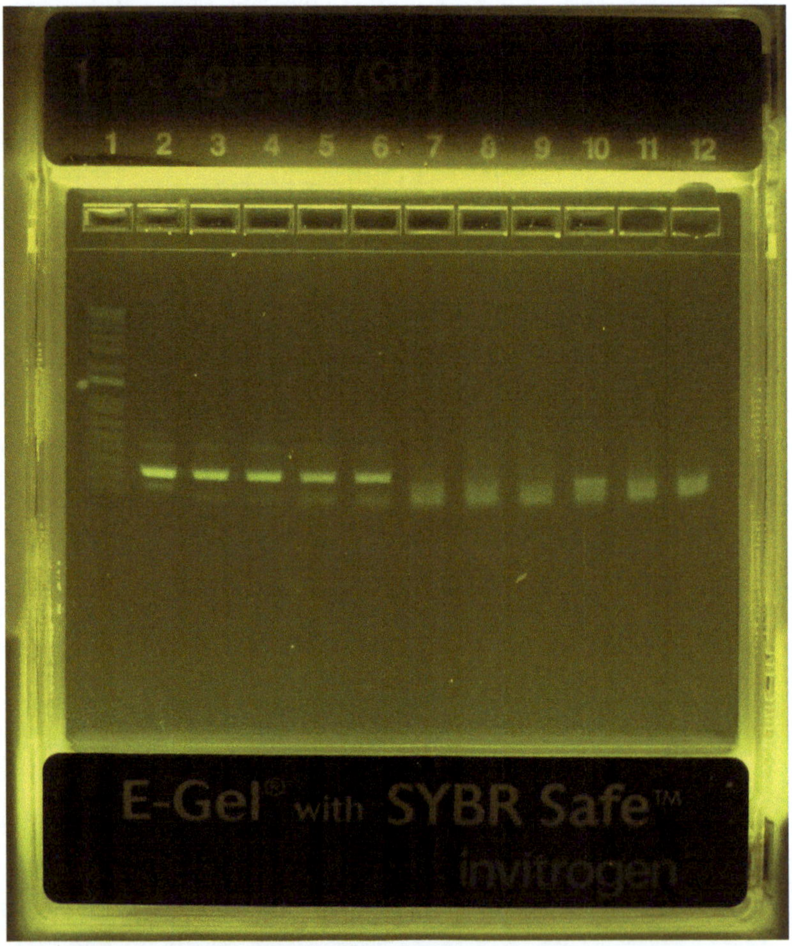

Fig. 3 Detection of *Xylella fastidiosa* DNA in *Neokolla* (*Graphocephala*) *heiroglyphica* insects extracted using the FTA matrix cards using RPA basic. Lane 1, molecular weight marker; lane 2, positive control template containing 10^7 copies of *X. fastidiosa cpn60* plasmid; lanes 3–6, *N. heiroglyphica* samples 1–4; lanes 7–10, *Macrosteles quadrilineatus* non-carrier samples; lanes 11–12, no template controls. RPA products of the expected size (236 bp) are observed in samples carrying *X. fastidiosa*

consistent with the faint *rpoD* PCR products observed in the "unspiked insects" (Fig. 4). While the RPA basic method does not appear to be quantitative, it does provide rapid and unequivocal positive results in a short time. Control non-carrier insects were negative (Fig. 3), suggesting that the method does not generate a result from insects that do not harbour the pathogen. However, the RPA basic method is not particularly well suited to field applications due to the requirement for column purification and gel electrophoresis of the products.

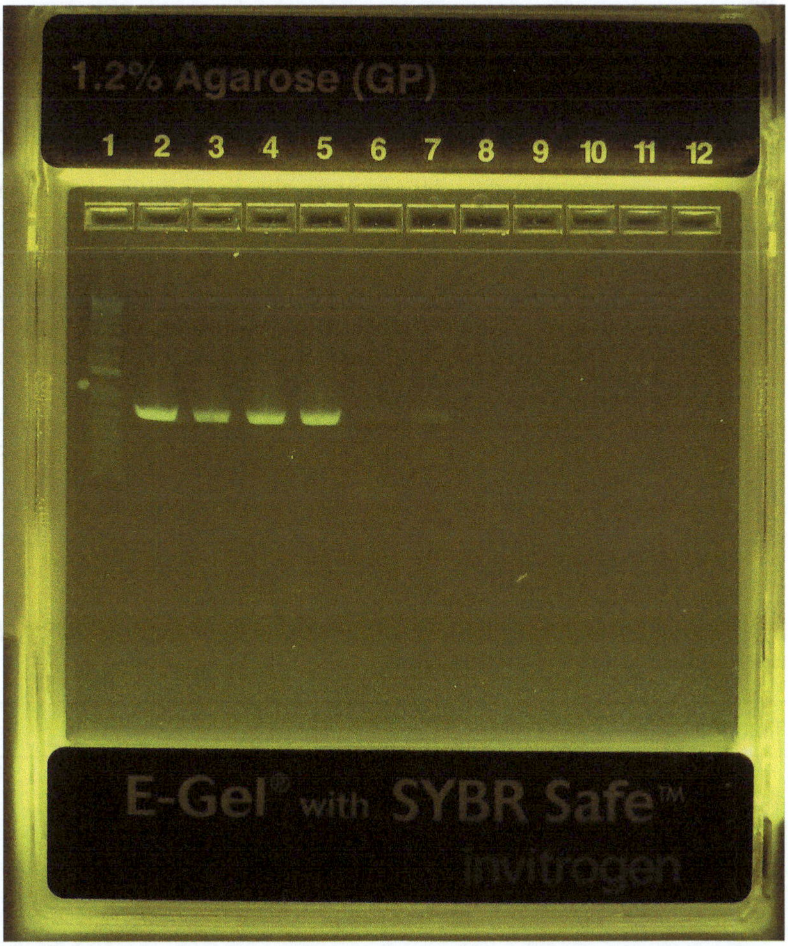

Fig. 4 Conventional PCR for *Xylella fastidiosa rpoD* using insect DNA extracted from FTA matrix cards as a template. Lane 1, molecular weight marker; lanes 2–3, *X. fastidiosa* genomic DNA (positive controls); lanes 4–5, *Neokolla* (*Graphocephala*) *hieroglyphica* insects spiked with *X. fastidiosa* DNA; lanes 6–7, *N. hieroglyphica* insects without exogenous *X. fastidiosa* DNA added; lanes 8–12, *Macrosteles quadrilineatus* non-carrier control insects with no added DNA

8. In the current example, the two *N. hieroglyphica* insects with *X. fastidiosa* DNA added prior to extraction (spiked insects) yielded positive results very quickly, while the unspiked insects were also positive but with a longer T_p, indicating lower levels of *X. fastidiosa* DNA compared to the spiked individuals (Table 1). Control non-carrier insects were negative (not shown). This result suggests that, like qPCR vs. conventional PCR, RPA exo is more quantitative compared to RPA basic.
9. Like the RPA exo reaction, LAMP product detection could also proceed using a real-time PCR instrument with detection on the FAM channel. However, this approach does not lend itself to product detection outside of the laboratory. Many alternative methods have been described for the detection of LAMP products, including calcein [12], hydroxyl naphthol blue [30], and colorimetric detection using a commercial WarmStart master mix (available from NEB). Another fluorometric alternative is offered by OptiGene (Isothermal Mastermix) that provides very rapid detection in a proprietary formulation. The choice of detection method will depend on a variety of factors, especially portability, but also cost, detection limits required, and the value of additional information on product annealing temperature that is provided by some of the fluorometric methods. An annealing temperature (T_a) is provided using the NEB WarmStart fluorescent and Isothermal Mastermixes that can help identify the formation of the correct LAMP product, which is analogous to the use of annealing or melt curves to identify the formation of the correct product in SYBR Green-based quantitative PCR.

Table 1
Real-time RPA exo detection of *Xylella fastidiosa* in *Neokolla* (*Graphocephala*) *hieroglyphica* insects extracted using the FTA matrix cards

Sample	T_p, minutes
X. fastidiosa cpn60 plasmid DNA, 10^6 copies	2.84
X. fastidiosa cpn60 plasmid DNA, 10^5 copies	3.66
N. heiroglyphica extract 1, *X. fastidiosa* DNA added	3.69
N. heiroglyphica extract 2, *X. fastidiosa* DNA added	3.81
N. heiroglyphica extract 3, no *X. fastidiosa* DNA added	13.73
N. heiroglyphica extract 4, no *X. fastidiosa* DNA added	17.91
No template	0.00

Data were collected using a real-time PCR apparatus (Bio-Rad CFX96) as described in the text, and time to positive (T_p) is reported as $C_q/2$ due to the use of 30-second cycles. Control non-carrier insects (*Macrosteles quadrilineatus*) were negative by real-time RPA (not shown)

10. The primer mixture can be used to prepare LAMP tubes with lyophilized primers, which will minimize the number of pipetting steps required in the field and will decrease the chance of cross contamination of samples. To do this, distribute 5 μL of the primer mix into each well of the Genie LAMP tube strip and lyophilize the strip for about 30 min. Once the primers are dried, cut a strip of sticky foil (VWR catalog number 60941-124) to the appropriate size and place it over the open wells. The foil can be punctured with a pipette tip and the lid closed on top of the foil with no effect on amplification or detection. The LAMP tube strips so prepared are very stable at room temperature – we have successfully generated products using strips stored at ambient temperature for at least six months.
11. The protocol is described here assuming that the primers are lyophilized in the tubes according to the note. If this is the case, it is important to centrifuge the LAMP tube strip with lyophilized primers briefly to ensure that the primer pellet is at the bottom of each tube.
12. The fluorescent dye is dissolved in dimethyl sulfoxide (DMSO). This has two implications: first, it is very important to wear gloves as DMSO is absorbed readily through the skin. Second, DMSO freezes at 18 °C, so if the field conditions are colder than that, it may be necessary to thaw the dye, for example, by placing the tube between two fingers.
13. Like RPA, LAMP generates rapid results, with the *N. hieroglyphica* samples examined in the current example generating a positive signal within 10.5–24.5 min (Table 2). Because the instrument used to generate this result (Genie) is field-deployable, these results could be easily generated within a few minutes of completing the DNA extraction according to

Table 2
LAMP detection of *Xylella fastidiosa* in *Neokolla* (*Graphocephala*) *heiroglyphica* insects extracted using the FTA matrix cards

Sample	T_p, minutes	T_a, °C
X. fastidiosa cpn60 plasmid DNA, 10^7 copies	8.5	86.8
X. fastidiosa cpn60 plasmid DNA, 10^6 copies	11.5	86.6
X. fastidiosa cpn60 plasmid DNA, 10^5 copies	12.5	86.6
N. heiroglyphica extract 1, *X. fastidiosa* DNA added	23.5	86.3
N. heiroglyphica extract 2, *X. fastidiosa* DNA added	10.5	86.4
N. heiroglyphica extract 3, no *X. fastidiosa* DNA added	24.5	86.2
N. heiroglyphica extract 4, no *X. fastidiosa* DNA added	No call	86.4
No template	Neg	–

the protocol described here. However, for lab-based LAMP detection, a real-time PCR instrument (e.g., Bio-Rad CFX96) can also be used to generate results, with a program similar to that used for real-time RPA (e.g., 60 cycles of 30 s each at 63 °C).

References

1. Bushing RW, Burton VE (1974) Leafhopper damage to Silage Corn in California 13. J Econ Entomol 67(5):656–658
2. Maejima K, Oshima K, Namba S (2014) Exploring the phytoplasmas, plant pathogenic bacteria. J Gen Plant Pathol 80(3):210–221
3. Overall LM, Rebek EJ (2017) Insect vectors and current management strategies for diseases caused by *Xylella fastidiosa* in the Southern United States. J Integr Pest Manag 8(1):1
4. Hopkins DL, Purcell AH (2002) *Xylella fastidiosa*: cause of Pierce's disease of Grapevine and other emergent diseases. Plant Dis 86(10): 1056–1066
5. Coletta-Filho HD, Castillo AI, Laranjeira FF et al (2020) Citrus variegated chlorosis: an overview of 30 years of research and disease management. Trop Plant Pathol 45(3): 175–191
6. de Lima JEO, Miranda VS, Hartung JS et al (1998) Coffee leaf Scorch bacterium: axenic culture, pathogenicity, and comparison with *Xylella fastidiosa* of citrus. Plant Dis 82(1): 94–97
7. Bertaccini A, Duduk B, Paltrinieri S et al (2014) Phytoplasmas and phytoplasma diseases: a severe threat to agriculture. Am J Plant Sci 5:1763–1788
8. Martini M, Bottner-Parker KD, Lee IM (2019) PCR-based sequence analysis on multiple genes other than 16s rRNA gene for differentiation of phytoplasmas. Methods Mol Biol 1875:97
9. Ikten C, Ustun R, Catal M et al (2016) Multiplex real-time qPCR assay for simultaneous and sensitive detection of phytoplasmas in sesame plants and insect vectors. PLoS One 11(5): e0155891
10. Pérez-López E, Rodríguez-Martínez D, Olivier CY et al (2017) Molecular diagnostic assays based on cpn60 UT sequences reveal the geographic distribution of subgroup 16SrXIII-(A/I)I phytoplasma in Mexico. Sci Rep 7(1): 950
11. Vissani MA, Tordoya MS, Tsai YL et al (2018) On-site detection of equid alphaherpesvirus 3 in perineal and genital swabs of mares and stallions. J Virol Meth 257:29–32
12. Tomita N, Mori Y, Kanda H et al (2008) Loop-mediated isothermal amplification (LAMP) of gene sequences and simple visual detection of products. Nat Protocols 3(5):877–882
13. Notomi T, Okayama H, Masubuchi H et al (2000) Loop-mediated isothermal amplification of DNA. Nucleic Acids Res 28(12):e63
14. Rabe BA, Cepko C (2020) SARS-CoV-2 detection using isothermal amplification and a rapid, inexpensive protocol for sample inactivation and purification. Proc Natl Acad of Sci USA 117(39):24450–24458
15. Ocenar J, Arizala D, Boluk G et al (2019) Development of a robust, field-deployable loop-mediated isothermal amplification (LAMP) assay for specific detection of potato pathogen *Dickeya dianthicola* targeting a unique genomic region. PLoS One 14(6): e0218868
16. Manjunatha C, Sharma S, Kulshreshtha D et al (2018) Rapid detection of *Puccinia triticina* causing leaf rust of wheat by PCR and loop mediated isothermal amplification. PLoS One 13(4):e0196409
17. Lobato IM, O'Sullivan CK (2018) Recombinase polymerase amplification: basics, applications and recent advances. TRAC -Trend Anal Chem 98:19–35
18. Conrad CC, Daher RK, Stanford K et al (2020) A sensitive and accurate recombinase polymerase amplification assay for detection of the primary bacterial pathogens causing Bovine respiratory disease. Front Vet Sci 7:208
19. Zhao G, Hou P, Huan Y et al (2018) Development of a recombinase polymerase amplification combined with a lateral flow dipstick assay for rapid detection of the Mycoplasma bovis. BMC Vet Res 14(1):412
20. Kalendar R, Boronnikova S, Seppänen M (2021) Isolation and purification of DNA from complicated biological samples. Meth Mol Biol 2222:57
21. Zou Y, Mason MG, Wang Y et al (2017) Nucleic acid purification from plants, animals and microbes in under 30 seconds. PLoS Biol 15(11):e2003916
22. Dickey AM, Shatters RG, McKenzie CL (2012) A comparison of two methods of

eluting insect DNA from Flinders Technology Associates cards. Fla Entomol 95(3):790–793
23. Pusz-Bochenska K, Perez-Lopez E, Dumonceaux TJ et al (2020) A rapid, simple, laboratory and field-adaptable DNA extraction and diagnostic method suitable for insect-transmitted plant pathogen and insect identification. Plant Health Progr 21(1):63–68
24. Harper SJ, Ward LI, Clover GR (2010) Development of LAMP and real-time PCR methods for the rapid detection of *Xylella fastidiosa* for quarantine and field applications. Phytopathology 100(12):1282–1288
25. Frazier NW, Freitag JH (1946) Ten additional leafhopper vectors of the virus causing Pierce's disease of grapes. Phytopathology 36:634–637
26. Hebert PDN, Cywinska A, Ball SL et al (2003) Biological identifications through DNA barcodes. Proc R Soc Lond [Biol] 270(1512):313–321
27. Minsavage GV, Thompson CM, Hopkins DL et al (1994) Development of a polymerase chain reaction protocol for detection of Xylella fastidiosa in plant tissue. Phytopathology 84:456–461
28. Ratnasingham S, Hebert PDN (2013) A DNA-based registry for all animal species: the Barcode Index Number (BIN) System. PLoS One 8(7):e66213
29. Ratnasingham S, Hebert PDN (2007) BOLD: the Barcode of Life Data System. (http://www.barcodinglife.org). Mol Ecol Notes 7 (3):355–364
30. Goto M, Honda E, Ogura A et al (2009) Colorimetric detection of loop-mediated isothermal amplification reaction by using hydroxy naphthol blue. BioTechniques 46(3):167–172

Chapter 13

A Panel of Real-Time PCR Assays for the Direct Detection of All of the *Xylella fastidiosa* Subspecies

Jennifer Hodgetts

Abstract

The bacterial plant pathogen *Xylella fastidiosa* causes disease in hundreds of plant species worldwide including many crops, and as such accurate determination of the subspecies of the bacteria is vital to control, containment, and eradication measures. Conventional methods to determine the subspecies of *X. fastidiosa* rely on time consuming multilocus sequence typing (MLST), a laborious multistage process. This chapter provides a rapid alternative to MLST utilizing real-time PCR assays to provide highly specific and sensitive detection of the pathogen subspecies. Here we describe the methodology for sampling plant material, performing the DNA extraction and undertaking the real-time PCR assays. This method allows straightforward, robust, reliable, high-throughput, and rapid determination of the *X. fastidiosa* subspecies.

Key words Diagnostics, Molecular detection, Identification, qPCR, MLST

1 Introduction

Xylella fastidiosa (Wells et al., [1]) is a Gram-negative bacterium which is a regulated (quarantine) pathogen within the EU (Annex I/A/II, EPPO A2 list). The pathogen is endemic in the Americas where it is associated with numerous major plant diseases such as Pierce's disease of grapevine, almond leaf scorch, and citrus variegated chlorosis [2, 3]. The first confirmed European finding of *X. fastidiosa* was in 2013 in Puglia, Italy [4], causing serious damage to olive groves, with subsequent findings throughout Europe in the following years in numerous host species [5].

Xylella fastidiosa is a challenging pathogen to diagnose due to a lack of distinctive characteristic symptoms and very large host range of nearly 600 species worldwide (at the time of writing). Symptoms of the pathogen are easily confused with biotic and abiotic factors and typically include leaf scorch, wilting, tip dieback, and subsequent plant death. The pathogen is also associated with a high prevalence of asymptomatic infection [6].

Nicola Luchi (ed.), *Plant Pathology: Methods and Protocols*, Methods in Molecular Biology, vol. 2536,
https://doi.org/10.1007/978-1-0716-2517-0_13,

The taxonomy of *X. fastidiosa* is complex, utilizing subspecies level descriptions for differentiation. There are three formally described subspecies: *X. fastidiosa* subsp. *fastidiosa*, *X. fastidiosa* subsp. *multiplex*, and *X. fastidiosa* subsp. *pauca*, along with two further subspecies *X. fastidiosa* subsp. *sandyi* and *X. fastidiosa* subsp. *morus*, which are widely used within the scientific community but have an uncertain taxonomic standing due to a lack of formal subspecies descriptions within the literature [7–9]. The various subspecies can be associated with different host ranges, vectors, climatic growth zones, and geographic distributions. Therefore, it is important that the subspecies of a given sample is accurately determined.

Real-time PCR (qPCR) is a widely used molecular biology technique which provides rapid, sensitive, and highly specific detection of targets. qPCR has advantages compared to conventional PCR of greater sensitivity, rapidity, a lack of post PCR manipulations, reduced laboratory contamination, and amenability to high throughput testing [10]. The assays used here exploit the TaqMan™ fluorophore probe-based system of qPCR, which provides greater specificity than those using intercalating dyes such as SYBR® Green. To provide robust detection, the presence of potential inhibitors within the DNA extract is checked for by testing two concentrations of the DNA. In comparison to conventional PCR and DNA sequencing (MLST, the alternative method to determine *X. fastidiosa* subspecies), these qPCR assays are substantially faster and provide greater sensitivity.

Five assays are described, which together allow direct detection of all of the subspecies: *X. fastidiosa* subsp. *fastidiosa*, *X. fastidiosa* subsp. *multiplex*, *X. fastidiosa* subsp. *pauca*, *X. fastidiosa* subsp. *morus*, and *X. fastidiosa* subsp. *sandyi* [11]. These are most effectively used as an alternative to MLST for subspecies determination of a sample which is already identified as containing *X. fastidiosa*. They can also be used for initial identification and subspecies determination in a single step if desired. Protocols for using a *X. fastidiosa* "universal" assay [12] which detects to the species level (all of the subspecies together) and the plant cytochrome oxidase I (COX) gene [13, 14] are also described. The Harper et al. [12] assay can be used as an initial screen to identify positive samples with further identification to the subspecies then achieved using the subspecies assays. The COX gene assay is used as a control to demonstrate success of the DNA extraction protocol, allowing the unambiguous interpretation of negative results.

A CTAB DNA extraction method (modified from Doyle and Doyle, [15]) is suitable for a wide range of plant species and provides a high yield of good quality DNA. This is the method of choice for *X. fastidiosa* samples, where the pathogen can be present in low levels. However, some hosts (particularly *Quercus* sp. (oak) and *Cistus* sp.) can fail to extract DNA of suitable quality using the CTAB method. In these instances, an alternative extraction method is more appropriate, and the QIAGEN Mericon® method should be used in preference for these hosts. While a CTAB DNA extraction protocol can have a lower throughput than some other common DNA extraction protocols, due to the intrinsic nature of *X. fastidiosa*, it is important a rigorous method such as the CTAB protocol is used to prevent false negative results.

This chapter describes protocols for testing *in planta* samples. Fundamental to the success of the test is correct sampling of the suspect plant, for which information is provided. The real-time PCR assays described can also be used for subspecies identification of *X. fastidiosa* bacterial cultures if desired.

2 Materials

All reagents and plasticware should be sterile and certified DNase-, RNase-, and pyrogen-free.

2.1 Sampling Plant Tissue

1. Ethanol, 70% in water (or industrial methylated spirits, IMS) (*see* **Note 1**).
2. Distel high-level (medical surface/laboratory) disinfectant (Distel, UK), 1% solution diluted in water (optional) (*see* **Note 2**).
3. Tray, plastic, smooth surfaced, approximately 30 × 20 cm.
4. Cut-resistant/proof gloves.
5. Scalpel or razor blades.
6. Disposable weigh boats (plastic or paper).
7. Balance, two decimal places.
8. Polythene bags, approximately 500 gauge, approximately 4″ × 6″ (e.g., Polybags Ltd., catalogue number 46500) (*see* **Note 3**).

2.2 Sampling From Woody Hosts

1. As per Subheading 2.1.
2. Ziploc (sandwich) bags, slightly larger than the polythene bags (Subheading 2.1, **item 8**), for example, 6″ × 8″.

2.3 DNA Extraction

1. Suspected *X. fastidiosa*-infected plant sample or bacterial culture.
2. Healthy plant material, for DNA extraction control (*see* **Note 4**).
3. CTAB buffer: weigh 20 g CTAB (hexadecyltrimethylammonium bromide). Add 280 mL 5 M NaCl (sodium chloride), 40 mL 0.5 M EDTA (ethylenediaminetetraacetic acid), and 100 mL of 1 M Tris-HCl (Trizma® hydrochloride) and make up to 1 L with molecular biology grade water. Store at room temperature (*see* **Notes 5–7**).

4. Sodium diethyldithiocarbamate trihydrate (e.g., Sigma Aldrich catalogue number 71480).
5. Polyvinylpyrrolidone average mol wt 40,000 (PVP-40), (e.g., Sigma Aldrich catalogue number PVP-40).
6. Chloroform-isoamyl alcohol mixture (24:1 v/v) (e.g., Sigma Aldrich catalogue number 25666).
7. Ethanol, 100%, molecular biology grade.
8. Ethanol, 70%, molecular biology grade diluted in molecular biology grade water.
9. Isopropanol (also known as 2-propanol), 100%, molecular biology grade, stored at 2 to 5 °C.
10. 5M NaCl (sodium chloride) solution, molecular biology grade.
11. TE buffer, 1x, pH 8, molecular biology grade. Add 1 mL of 1 M Tris-HCl and 200 μL of 0.5 M EDTA to 98.8 mL molecular biology grade water. Store at room temperature in small aliquots, for example, 5 mL (*see* **Note 8**).
12. QIAamp® DNA Mini Kit (QIAGEN, catalogue number 51304) (optional, for bacterial cultures only).
13. DNeasy® Mericon® Food Kit (QIAGEN, catalogue number 69514) (optional, for certain plant species).
14. 50 mL tube, leak proof, for example, centrifuge tube.
15. Water bath (static) set at 65 °C and floating rack for 2 mL microcentrifuge tubes.
16. Shaking incubator set at 50 °C (optional) (*see* **Note 9**).
17. Balance, two decimal places.
18. Homogenizer, for example, BIOREBA Homex hand model (*see* **Note 10**).
19. Mallet with rubber head (optional).
20. Fume hood.
21. Freezer, −20 to −30 °C.
22. Ice and ice bucket.
23. Gloves, disposable, chloroprene (for when handling chloroform).

2.4 DNA Extraction From Woody Hosts

1. As per Subheading 2.3.
2. Fridge, 2–5 °C.

2.5 DNA Purification

1. Ethanol, ice-cold 70%, molecular biology grade.
2. Ethanol, ice-cold 100%, molecular biology grade.
3. 3M NaAc (sodium acetate), pH 5.2, molecular biology grade.
4. Freezer, −20 to −30 °C.

2.6 qPCR

1. DNA extracts for each *X. fastidiosa* subspecies (positive control DNA).
2. DNA extract of a plant (positive control for COX assay) (optional).
3. Oligonucleotides (primers) for *X. fastidiosa* subspecies (and *X. fastidiosa* and plant control, optional) assays (Table 1) (*see* **Notes 11,12**). Stocks prepared to 100 μM in molecular biology grade water and stored at −20 °C. Prepare working solutions at 7.5 μM by diluting 7.5 μL of 100 μM stock in 92.5 μL molecular biology grade. Store at −20 °C.
4. Fluorescently labeled probes, labeled with FAM-BHQ1, FAM-MGB-NFQ, and VIC-TAMRA (Table 1) (*see* **Notes 12–16**). Stocks prepared to 100 μM in molecular biology grade water and stored at −20 °C. Prepare working solutions at 5 μM by diluting 5 μl of 100 μM stock in 95 μL molecular biology grade. Store at −20 °C.
5. Applied Biosystems™ TaqMan™ Fast Universal PCR Master Mix (2X), No AmpErase™ UNG (ThermoFisher Scientific, catalogue number 10575385).
6. Applied Biosystems™ AmpliTaq Gold™ DNA Polymerase with 10x Buffer A, 25 mM $MgCl_2$ and 2.5 mM each dNTP mix (catalog number 4304441) (optional, for use with the plant COX control gene assay only).
7. UltraPure™ BSA (non-acetylated bovine serum albumin) 50 mg mL^{-1} (e.g., catalogue number AM2616, Ambion).
8. Molecular biology grade water.
9. Real-time PCR platform.
10. 96-Well plates and lids (self-adhesive, heat sealed or flat cap strips) compatible with available PCR platform (*see* **Notes 17–18**).
11. Fridge (2 to 5 °C) and freezer (−20 to −30 °C) (for storage of reagents).

2.7 Equipment and General Lab Supplies

1. 1.5 mL and 2 mL microcentrifuge tubes.
2. Pipette tips with filters, various sizes.
3. Vortex.
4. Centrifuge, for microcentrifuge tubes (1.5 mL and 2 mL) and 96-well plates.
5. Pipettes, 1 μL to 1 mL size range, multiple sets for each stage of the process (*see* **Note 19**).
6. UV workstation cabinet (optional) (*see* **Note 20**).
7. Gloves, disposable (e.g., latex or nitrile).

Table 1
Primers and probes for the *Xylella fastidiosa* subspecies specific assays and the optional *X. fastidiosa* universal and plant COX control assays

Target	Element	Sequence (5′-3′)	Reporter-quencher	Reference
X. f. subsp. *fastidiosa*	Xff_3-Fw-a_122365	TCG TTG TGG GAT TAC CGT TAA A		[11]
	Xff_3-Rv_122365	ACC TGA GAA TTG CCC TTA ATC G		
	Xff_3-Pr_122365	TCG AAA ACA CCG GAC TTG CCA ACA	FAM-BHQ1	
X. f. subsp. *multiplex*	Xfmu_3-Fw_15747	CAA TCG CTT TTG AGG TCA TCC		[11]
	Xfmu_3-Rv_15747	GCG ATT GTT TCT TCT CTA CAC CAA G		
	Xfmu_3-Pr_15747	TCT GCA AAC GCT TTA AAA ACT GCT CGC C	FAM-BHQ1	
X. f. subsp. *pauca*	Xfp_1-Fw_28775	GCA TCC TCA CCA CCG AAG G		[11]
	Xfp_1-Rw_28775	TCC ACA TCC AGC AAG GTG AC		
	Xfp_1-Pr_28775	CCT TGG ACG CGG ATA CCC GCA	FAM-BHQ1	
X. f. subsp. *morus*	Xfmo_7_Fw_67218	CCA CCT CGC TTT AGT TAC GTG ATT		[11]
	Xfmo_7_Rva_67218	GGA GTT TAT TTG GCT GAA CTG AGT G		
	Xfmo_7_Pr_67218	AAG CGT GAT ACT ACT CC	FAM-MGB-NFQ	
X. f. subsp. *sandyi*	Xfs_4_Fwb_112076	CCC CGC TGT GGC AGA A		[11]
	Xfs_4_Rv_112076	GGT CCG AGC CAT ACG GC		
	Xfs_4_Pr_112076	CAG CGC CTT CAA TC	FAM-MGB-NFQ	
X. fastidiosa (all subspecies)	XF-F	CAC GGC TGG TAA CGG AAG A		[12]
	XF-R	GGG TTG CGT GGT GAA ATC AAG		
	XF-P	TCG CAT CCC GTG GCT CAG TCC	FAM-BHQ1	
Plant cytochrome oxidase I (COX) gene	COX F	CGT CGC ATT CCA GAT TAT CCA		[13, 14]
	COX R_W	CAA CTA CGG ATA TAT AAG RRC CRR AAC TG		
	COXSOL 1511T	AGG GCA TTC CAT CCA GCG TAA GCA	VIC-TAMRA	

3 Methods

Ensure compliance with all relevant local plant health regulations.

Standard laboratory safety protocols should be followed at all times. Staff should be familiar with relevant COSHH assessments, emergency response protocols, and appropriate waste disposal routes for hazardous chemicals. All laboratory protocols should undergo risk assessment following the laboratories protocols before first use.

Carry out all procedures at room temperature unless otherwise specified.

3.1 Sampling Plant Tissue

The most common symptoms of *X. fastidiosa*, which are generally non-specific and can vary depending upon the host plant, include tip dieback, scorching, yellowing (chlorosis) or bronzing, and crown dieback (*see* **Note 21**) and can easily be confused with other abiotic or biotic factors.

Laboratories should ensure they have a system for assigning unique identifying numbers to samples for tracking and reporting purposes.

Samples will typically be harvested from the plant and posted to a laboratory for testing (*see* **Note 22**). Samples can be stored at room temperature for a short period of time after collection from the plant (e.g., during transit and before sampling), although refrigerated storage minimizes sample degradation, especially after shipping to the laboratory. Freeze-dried material cannot be assessed for symptoms and should be sampled at random (*see* **Note 23**). Samples other than leaves and stems (e.g., root, flower, or fruit) should be sampled following the principles described below.

1. Disinfect the work surfaces, balance, and tray by wiping with paper towel and 70% ethanol or 1% Distel. Allow to air dry before use (*see* **Note 24**).
2. A new blade and weigh boat and a fresh pair of gloves should be worn for handling each sample, and care should be taken to prevent contamination of other equipment or work surfaces.
3. Remove the sample from its packaging and spread out on a tray to assess for symptoms (*see* **Notes 25–28**).
4. Decaying samples with visible fungal growth may be unfit for testing and should be avoided if possible (*see* **Note 29**).
5. Identify any symptoms present which are consistent with *X. fastidiosa* and gather this material together for sampling. If possible collect leaves from multiple stems within the sample.
6. If the leaves to be sampled are still attached to a stem, then carefully remove the leaf from the stem leaving the petiole intact with the leaf.

7. A sample should be comprised of small sections of leaf petiole/midribs from 10 to 25 leaves (*see* **Note 30**).
8. Use a scalpel/razor blade to make two cuts through the lower section of the leaf angled towards the midrib to excise the petiole and a section of midrib (Fig. 1). Discard the rest of the leaf (*see* **Notes 31, 32**).
9. Repeat for a minimum of 10 leaves, up to a maximum of 25 leaves (depending upon the leaf size).
10. For very small leaved plants, approximately 2 cm in length or smaller (e.g., lavender) whole leaves may be sampled.
11. Place a new disposable weigh boat onto a balance and tare (zero) the balance.
12. Transfer the excised petiole-midribs into the weigh boat and weigh. A subsample for testing should weigh 0.6 g $\pm$ 0.01 g (*see* **Notes 33, 34**).
13. If the subsample weight is less than 0.6 g, select additional petioles-midribs and add these to the subsample until the weight reaches 0.6 g.
14. If the subsample weight is greater than 0.6 g, trim the tissue along the previously cut edges until the weight is reduced to 0.6 g.
15. Transfer the subsample to a plastic bag and label with the samples unique reference number (*see* **Note 35**). The material does not need to be cut into smaller pieces.
16. If the sample contains material which is thorny or sharp, there is a risk of the extraction bag being punctured during processing. If this is the case, use the blade to cut the sharp material into smaller pieces (*see* **Note 36**).
17. If testing is not taking place immediately, the extraction bag should be placed into a sealable plastic bag, labeled with the date of sampling, and stored at -20 °C in an upright position (*see* **Note 37**).
18. Dispose of used blades into an appropriate sharps waste bin.
19. Surplus plant material should be stored at room temperature or preferably refrigerated until test results are produced.
20. Dispose of the weigh boat, disinfect the work surfaces, balance and tray by wiping with paper towel and 70% ethanol or 1% Distel, and change gloves prior to handling another sample (*see* **Notes 38–40**).

Fig. 1 Sampling plant leaves for *X. fastidiosa* testing. Two cuts are used to excise a section of leaf midrib and the petiole which is used for the DNA extraction (the tissue within the triangle), while the remaining upper leaf tissue is discarded

3.2 Sampling From Woody Hosts

1. Prepare samples as per Subheading 3.1, **steps 1 to 4**.
2. Identify any symptoms present which are consistent with *X. fastidiosa* and gather this material together for sampling (*see* **Note 26**). If possible, select stems from multiple branches within the sample.

3. Use a fresh scalpel/razor blade to shave off the outer tissue/bark layer from a 5–10 cm length of stem, exposing the vascular bundle (*see* **Note 31**).
4. Discard the trimmed outer bark tissue.
5. Prepare at least 3 stem sections, preferably from different stems of the sample.
6. Shave thin layers/slivers of the xylem tissue from the inner stem always cutting away from yourself.
7. Continue from Subheading 3.1, **step 11**.

3.3 DNA Extraction

Foliar samples (e.g., leaves, shoots, green stem, flowers, fruit) should be processed by the following method (*see* **Notes 41–45**). Freeze-dried material should be processed by this method in the first instance. For samples comprised of non-green (woody) stem, other dry/tough material (e.g., roots), or desiccated leaf samples *see* Subheading 3.4 "DNA Extraction from Woody Hosts". Bacterial cultures should be processed using a commercially available DNA extraction kit (*see* **Note 46**).

Each batch of DNA extracts must include a negative isolation control (NIC) and a positive isolation control (PIC) (*see* **Notes 47–52**).

1. Turn on a water bath to 65 °C and a shaking incubator to 50 °C and allow to heat to the set temperatures (*see* **Note 53**).
2. Prepare an aliquot of CTAB buffer containing 2% PVP-40 and 0.2% sodium diethyldithiocarbamate in a leak-proof 50 mL tube, for example, add 0.6 g PVP-40 and 0.06 g sodium diethyldithiocarbamate to 30 mL CTAB buffer (*see* **Notes 54, 55**).
3. Incubate the buffer at 50 °C with gentle shaking until the additives have dissolved (approximately 30 min) (*see* **Note 56**).
4. Once the additives have dissolved fully, preheat the buffer to 65 °C in a water bath.
5. Fully processing each sample in turn (*see* **Note 57**), homogenize the tissue by grinding in a polythene bag;
   (i) Using a hand homogenizer (Homex) and/or mallet, firmly homogenize all the plant material until the material is partially crushed, focusing on the thicker petioles and midribs of the tissue.
   (ii) Add 2 mL of the hot CTAB buffer + additives, loosely fold over the end of bag and hold shut (*see* **Notes 58–60**).
   (iii) Using a mallet/hand homogenizer, firmly homogenize all the plant material again until the tissue is completely crushed and the buffer has turned homogenously green (*see* **Notes 61–63**).

(iv) Transfer the lysate (sample-buffer slurry) by pouring approximately 0.9 mL into an appropriately labeled 2 mL microcentrifuge tube. For smaller samples where a reduced amount of tissue and CTAB buffer was used, 0.6 mL of lysate should be recovered as a minimum (*see* **Notes 64, 65**).

(v) Discard the used bag and any remaining homogenate.

6. Incubate the microcentrifuge tube at 65 °C in the water bath in a floating tube rack for 30 min (*see* **Notes 66, 67**).
7. Twice during the incubation invert and gently shake the tubes by hand to mix.
8. Remove the samples from water bath and dry the outside of tubes with paper towel.
9. Working in a fume hood, add an equal volume (e.g., 0.9 mL) of chloroform-isoamyl alcohol (24:1) (*see* **Notes 68, 69**).
10. Working in a fume hood, vortex each tube for 30 s ensuring the phases are mixing well throughout (*see* **Notes 70, 71**). Change gloves after each sample.
11. Centrifuge tubes at 13,000 rpm for 10 min.
12. Working in a fume hood, carefully remove the aqueous supernatant (upper layer) using a 200 μl pipette to a clean labeled 2 mL microcentrifuge tube. Ensure none of the interface layer is pipetted (*see* **Notes 72, 73**).
13. Working in a fume hood, add an equal volume of chloroform-isoamyl alcohol (24:1) to the aqueous layer.
14. Mix the tubes gently by inversion several times until the phases mix well (*see* **Note 74**).
15. Centrifuge at 4000 rpm for 10 min.
16. Working in a fume hood, carefully remove the aqueous supernatant (upper layer) to a clean labeled 1.5 mL microcentrifuge tube. Ensure none of the interface layer is pipetted (*see* **Notes 72, 73**).
17. Add an equal volume of fridge cold 100% isopropanol and a half volume of room-temperature 5 M NaCl to the aqueous layer (e.g., to 500 μL supernatant and 500 μL isopropanol and 250 μL NaCl) (*see* **Note 75**).
18. Invert the tubes gently by hand to mix (*see* **Notes 76, 77**).
19. Store the tubes at −20 °C for at least 2 h or overnight (*see* **Notes 78, 79**).
20. Centrifuge tubes at 13,000 rpm for 30 min (*see* **Note 80**).
21. Place a tube of 70% ethanol on ice to cool.
22. After centrifugation visually locate the pellet in the microcentrifuge tube (*see* **Note 81**).

23. Discard the supernatant by tipping into a microcentrifuge tube. Take care to ensure the pellet does not move or get discarded with the supernatant (*see* **Note 82**).
24. Add 1 mL of ice-cold 70% ethanol to the pellet. Invert the tube several times by hand to wash the pellet.
25. Centrifuge the tube for 1 min at 13,000 rpm (*see* **Note 80**).
26. Carefully tip ethanol off pellet into a microcentrifuge tube (*see* **Note 82**).
27. Add 100 μL of ice-cold 70% ethanol to the pellet. Invert the tube several times by hand to wash the pellet.
28. Centrifuge the tube for 1 min at 13,000 rpm (*see* **Note 80**).
29. Use a small volume pipette to remove all traces of ethanol from the pellet (*see* **Note 83**).
30. Resuspend the DNA pellet in 100 μL 1x TE buffer. Initially add the TE buffer, pipette up and down to release the pellet from the side of the tube, and then leave to stand for a few minutes. Then fully resuspend the pellet by gently pipetting up and down. If this cannot be achieved, add additional TE buffer in 20 μL increments mixing well after each addition (*see* **Notes 84–87)**.
31. Prepare 1:10 dilutions of each DNA extract (including the PIC and NIC controls) by adding 10 μL of each DNA to 90 μL of 1x TE buffer (or molecular grade water) in a new 1.5 mL microcentrifuge tube.
32. Store the DNA extracts and dilutions at −20 °C until required for testing.

3.4 DNA Extraction From Woody Hosts

Samples which are not comprised of fresh plant material should be subject to overnight soaking prior to DNA extraction following the method variation as detailed below. This may include samples of desiccated (dried) leaf, plant roots (woody or fibrous), and mature stem sections (*see* **Note 88**). This variation can also be used for particularly tough/dry leaf samples.

1. Prepare CTAB buffer with additives as per Subheading 3.3, **steps 1 to 3**.
2. Add 2 mL of hot CTAB buffer to the sample bag and gently massage the bag/sample using your fingers to ensure all of the sample is in contact with the buffer (*see* **Note 89**).
3. Carefully place the sample bag(s) within a sealable plastic bag (e.g., a Ziploc sandwich bag) in an upright orientation (to prevent buffer spills) and seal.
4. Place the samples upright at 4 °C overnight (*see* **Note 90**).

5. Remove the samples from the fridge and homogenize as follows:
   (i) Loosely fold the end of the bag over and hold shut.
   (ii) Using a mallet/hand homogenizer carefully but firmly homogenize all the plant material until the shavings are visually disrupted and the buffer has colored (yellow-red-brown depending on the species) (*see* **Notes 91, 92**).
   (iii) Transfer the lysate (sample-buffer slurry) by pouring approximately 0.9 mL into an appropriately labeled 2 mL microcentrifuge tube.
   (iv) Discard the used bag/remaining homogenate.
6. Continue from Subheading 3.3, **step 6**.

3.5 DNA Purification

If a DNA extract tests negative with a plant control assay (when tested with the DNA both neat and diluted 1:10), then the following DNA clean-up protocol may be used. This is typically used for when the final DNA extract is not clear in color and can be performed before testing by qPCR if desired. The method can be scaled up or down as required.

1. Add 2.5 volumes ice-cold 100% ethanol and $^{1}/_{10}$ volume 3M NaAc pH 5.2 (i.e., to 50 μL DNA extract, add 250 μL 100% ethanol and 5 μL 3M NaAc). Mix gently by inversion (*see* **Note 93**).
2. Incubate the tubes at −20 °C for at least 20 min or overnight if preferred (*see* **Note 79**).
3. Centrifuge tubes at 13,000 rpm for 30 min (*see* **Note 80**).
4. Carefully decant the supernatant into a microcentrifuge tube and discard (*see* **Notes 81, 82**).
5. Add 1 mL ice-cold 70% ethanol to the tube containing the DNA pellet. Invert gently to mix.
6. Centrifuge the tubes at 13,000 rpm for 30 s. Decant the supernatant into a microcentrifuge tube and discard (*see* **Notes 80–82**).
7. Centrifuge the pellet at 13,000 rpm for 1 min. Pipette away all traces of ethanol (*see* **Note 80 and 83**).
8. Resuspend in an appropriate volume of 1x TE buffer (e.g., the starting volume of DNA used in **step 1**) (*see* **Notes 94, 95**).
9. If the DNA cannot be fully resuspended add additional 1x TE buffer in small increments (e.g., 20 μL) (*see* **Note 86**).

3.6 Setting up qPCR Reactions and Plates

Before first use in any laboratory, the performance characteristics of any assay should be verified to ensure it is as per the published validation data (*see* **Note 96**).

The plant cytochrome oxidase I (COX) gene real-time PCR assay of Weller et al., [13] as modified by Hughes et al., [14] should be used to confirm the success of DNA extraction from plant samples. If a sample has already be shown to be positive for *X. fastidiosa* by qPCR, then the subspecies assays can be used directly without the COX assay (*see* **Notes 97–103**).

1. Prepare plate plan(s) (Fig. 2). All samples should be tested in duplicate (*see* **Note 104**), and each plate should contain a negative amplification control (NAC) and positive amplification control (PAC) for each assay tested in duplicate (*see* **Notes 105–107**). Ensure that the extraction controls (NIC and PIC) are tested with all assays.
2. Prepare PCR master mix calculations following Tables 2 and 3 using the formula $(n + 3) \times 2$, where "n" is the number of samples and extraction controls. The "+ 3" allows for the NAC, PAC, and a small excess, and the "× 2" allows for all samples to be tested in duplicate wells. For example, if there are 6 samples/extraction controls to be tested, then calculate enough master mix for 18 reactions $(6 + 3) \times 2 = 18$.
3. Thaw the primers, probes, and reagents (except for the *Taq* polymerase) in the dark at room temperature (*see* **Notes 108, 109**). Thaw the DNA samples, dilutions of the samples, and the positive amplification control DNA extracts at room temperature.
4. Briefly vortex all reagents (except the *Taq* polymerase) and briefly centrifuge to pool the contents to the bottom of the tubes.

	1	2	3	4	5	6	7	8	9	10	11	12	
A B	NIC	PIC	S1	S2	S3	S4	S5	S6	S7	S8	NAC	PAC	Assay 1, e.g. *X. f.* subsp. *fastidiosa*
C D	NIC d	PIC d	S1 d	S2 d	S3 d	S4 d	S5 d	S6 d	S7 d	S8 d			
E F	NIC	PIC	S1	S2	S3	S4	S5	S6	S7	S8	NAC	PAC	Assay 2, e.g. *X. f.* subsp. *multiplex*
G H	NIC d	PIC d	S1 d	S2 d	S3 d	S4 d	S5 d	S6 d	S7 d	S8 d			

Fig. 2 Typical plate plan for qPCR with the *Xylella fastidiosa* subspecies assays. In this example, a batch of DNA extractions comprising eight samples and two extraction controls (NIC and PIC) are assayed for two subspecies. All samples/extraction controls are tested in duplicate wells and at two concentrations (neat and 1:10 dilutions). Three plates would be prepared to allow all five subspecies assays to be tested; plate 1 testing assays 1 and 2, plate 2 testing assays 3 and 4, and plate 3 testing assay 5. Sx (x = 1-8) samples neat DNA; Sx d 1:10 dilutions of samples DNA. NIC, negative isolation control; PIC, positive extraction control; NAC, negative amplification control; and PAC, positive amplification control

Table 2
PCR master mix reaction composition using Applied Biosystems™ TaqMan™ Fast Universal reagents for the *X. fastidiosa* subspecies and universal assays

Reaction component	Stock concentration	Volume (μl) for 1 well[a]	Final concentration
TaqMan™ Fast Universal PCR Master Mix	2 x	10	1 x
Forward primer	7.5 μM	1	300 nm
Reverse primer	7.5 μM	1	300 nm
Probe	5 μM	0.5	100 nm
BSA	50 mg mL^{-1}	0.125	6.25 μg
Water, molecular biology grade	n/a	5.375	n/a

[a]final total volume 20 μL

Table 3
PCR master mix reaction composition using Applied Biosystems™ TaqMan™ Gold with Buffer A for the plant COX assay

Reaction component	Stock concentration	Volume (μl) for 1 well[a]	Final concentration
Buffer A	10 x	2.5	1 x
$MgCl_2$	25 mM	5.5	5.5 mM
dNTP mix	2.5 mM each	2	200 μM each
Forward primer	7.5 μM	1	300 nm
Reverse primer	7.5 μM	1	300 nm
Probe	5 μM	0.5	100 nm
Taq (Gold)	5 U μl^{-1}	0.125	0.625 U
BSA	50 mg mL^{-1}	0.125	6.25 μg
Water, molecular biology grade	n/a	11.25	n/a

[a]final total volume 25 μL

5. Prepare the master mix(s) in an appropriately sized microcentrifuge tube (1.5, 2 or 5 mL size), adding the reagents in the order listed in Tables 2 and 3 (*see* **Note 110**).
6. Briefly vortex the completed master mix to mix.
7. Aliquot the PCR master mix(s) into the appropriate wells of a 96-well plate according to the plate plan, pipetting 18 μL for the *X. fastidiosa* assays and 24 μL for the plant COX assay.
8. Add 2 μL of water (that used to prepare the master mix) to the NAC wells.

9. Add 2 μL of the appropriate template DNA to the PAC wells. Gently flick tube to mix before use; do not vortex.
10. Add 2 μL template DNA or 1:10 dilution of the template DNA to the appropriate wells. Gently flick the tube to mix before use; do not vortex.
11. Seal the plate (*see* **Note 111**).
12. Run the plate, immediately if possible (*see* **Note 112**).

3.7 Perform the qPCR Run

1. Briefly (3 s) centrifuge the plate at low speed to pool the contents and then load the plate into the real-time PCR machine.
2. Launch the software and select the appropriate reporter-quencher combinations and reaction volume and input the sample locations and identifying information.
3. Run the qPCR using the appropriate cycling conditions. For the *X. fastidiosa* subspecies and universal assays, with a reaction volume of 20 μL, the cycling conditions are 50 °C for 2 min, 95 °C for 10 min, followed by 40 cycles of 95 °C for 10 s, and 62 °C for 40 s. For the plant COX assay, with a reaction volume of 25 μL, the cycling conditions are 50 °C for 2 min, 95 °C for 10 min followed by 40 cycles of 95 °C for 15 s, and 60 °C for 1 min.
4. See the instruments manual for detailed information on setting up, starting, and saving the run.

3.8 Analyze Data

1. Set the baseline and threshold and analyze the data (see the instruments manual for detailed instructions) (*see* **Notes 113–115**).
2. A test will be considered positive if both replicates produce an exponential (S-shaped) amplification curve which passes the threshold value.
3. A test will be considered negative if both replicates produce no amplification curve, described as C_T 40, or if they produce amplification curves which are not exponential in shape.
4. If the two replicate wells for a given sample vary by more than 3 C_T, then the result should be considered void and repeated (*see* **Note 116**).
5. First analyze the NAC and ensure that this is negative. If this is positive, then the plate must be repeated.
6. Analyze the PAC and ensure that this is positive and the C_T value within the expected range for the DNA extract (*see* **Note 117**). If this is negative, then the plate must be repeated.

7. Analyze the NIC and ensure that this is negative. If this is positive, the batch of DNA extracts must be discarded and the extraction process repeated.
8. Analyze the PIC and ensure that this is positive and the C_T value within the expected range for the host species. If the C_T value falls outside the expected range, the batch of DNA extracts may be suboptimal any negative samples must be investigated further.
9. Analyze the sample DNA extracts and compare the C_T values obtained with the neat and diluted DNA, which should show an increase of 3.3 C_T between the neat and diluted DNA, for example, C_T 20.0 with the neat DNA extract should be C_T 23.3 with the 1:10 dilution (*see* **Note 118**).
10. Tests should be repeated if any results are contradictory or unclear (*see* **Notes 119–122**).

4 Notes

The procedures described herein are carried out in accordance with principles of the EPPO diagnostic protocol PM7/24 (4) [16] [https://onlinelibrary.wiley.com/doi/full/10.1111/epp.12575], and the subspecies assays described were validated in line with PM 7/98 (4) [17] [https://onlinelibrary.wiley.com/doi/full/10.1111/epp.12629].

1. Industrial methylated spirits (IMS) can be used for disinfection of laboratory surfaces and equipment. However it cannot be used for later DNA extraction stages. Therefore we find it convenient to use ethanol throughout.
2. Distel is our disinfecting agent of choice for cleaning the laboratory environment and equipment between samples due to its broad-spectrum activity (bactericidal, fungicidal, and virudical) and ability to degrade DNA/RNA. However other laboratory disinfectants such as Virkon (Antec International, UK) may be used.
3. Many labs perform homogenization of samples using plastic bags with a mesh insert (e.g., BIOREBA extraction bags) to aid recovery of the liquid slurry without tissue fragments. However we find that these bags can sometimes reduce slurry yield in the small buffer volumes used in this protocol and that the small tissue fragments transferred can help to maintain the integrity of the interface at the chloroform extraction phase, so we prefer bags without an insert.
4. Leaf tissue of a healthy plant is used as a control for the DNA extraction protocol. We typically use leaves from young plants

(grown from seed for this purpose) of species such as tomato, *Nicotiana*, aubergine, and Chinese cabbage, which are amenable to DNA extraction and thereby also act as a gauge of extraction efficiency.

5. Prepare the CTAB buffer aseptically to prevent the need for autoclaving. Gentle heating (up to 60 °C) may be used to help dissolve the CTAB powder. The pH does not need to be adjusted.
6. We use commercially manufactured (sterile) molecular biology grade buffer solutions and water to ensure that no DNA is present in the buffer stocks. Alternatively, the stock solutions can be prepared using standard lab protocols.
7. When handling CTAB (powder/buffer) wear disposable gloves (e.g., nitrile), and in case of contact with skin, wash immediately with large quantities of fresh water.
8. Alternatively, a 100x stock solution of TE buffer can be purchased and diluted to 1x using molecular biology grade water.
9. A shaking incubator/orbital shaker allows the easiest resuspension of additives into the base buffer; however this is not essential. Resuspension can be aided by shaking the tube vigorously by hand and by gently heating at temperatures up to 65 °C.
10. In our experience, the BIOREBA hand homogenizer (400010) provides very good homogenization of a wide range of plant tissues. The larger Homex 6 automated model (400005) allows for homogenization of up to 5 samples in parallel and is also suitable. However good homogenization can also be achieved without this equipment by using a combination of a mallet/hammer and a sturdy wall paper seam roller (available for DIY/hardware stores).
11. Primers can be purchased either lyophilized or at a range of concentrations (generally 50–200 μM) from various manufacturers. If delivered lyophilized, primers should be reconstituted following the manufactures recommendations in either molecular biology grade water or 1x TE buffer.
12. Typically, a stock solution of 100 μM is maintained along with small aliquots (e.g., 100–200 μL) of a working solution of 7.5 μM (primers) or 5 μM (probes) at −20 °C in the dark. It is wise to use aliquots of the working solution (rather than diluting all of the stock) to avoid multiple freeze-thaw cycles which can degrade the primers/probes and also to help prevent contamination of stocks. Working solutions should not be prepared in large volumes to prevent freeze-thaw cycle damage.
13. The assays have been validated with the specific reporter-quencher combinations detailed, and these should not be

substituted. If for any reason the reporter and/or quencher used is different to those specified, then the assay(s) would require complete revalidation.

14. Fluorescently labeled probes can be purchased from a range of manufacturers, although not all manufacturers supply all of the reporters and quenchers.
15. Minor groove binding (MGB) probes are available from Thermo Fisher and must be ordered in this format to ensure assay performance. MGB probes contain a MGB moiety at the 3′ end that increases the melting temperature (Tm) of the probe allowing shorter probe sequences to be utilized. These cannot be substituted for standard probes as the assay specificity would be altered.
16. Fluorescently labeled probes should always be kept in the dark to prevent degradation of the fluorophore, including while thawing.
17. Source 96-well plates and lids which are compatible with the real-time PCR machine available for use. Plates are typically clear and must be suitable for real-time PCR applications (i.e., optical readings) and must fit the block of the machine. Plates can be fully, semi-, or non-skirted, and most manufacturers provide a compatibility table.
18. Clear lids are typically used to seal the plates, which again must be compatible with the machine and plate used. They are typically either heat-sealed or self-adhesive "sticky" lids; however, with some machines, flat PCR caps (in strips) can be used.
19. Due to the highly sensitive nature of qPCR, laboratory contamination must be avoided to prevent false positive results. It is best practice to have separate work areas and dedicated equipment (especially pipettes) for (i) DNA extraction, (ii) preparing the PCR master mix, (iii) adding template DNA to the PCR plate, and (iv) running the assays.
20. If possible, it is good practice to prepare PCR master mix while working in a PCR cabinet, to reduce the likelihood of contamination of reagents, although this is not essential.
21. Images of *X. fastidiosa* symptoms in a selection of plant hosts can be found on the EPPO global database: https://gd.eppo.int/taxon/XYLEFA/photos.
22. See the EPPO standards PM 3/81 (2) [18] and PM 3/82 (2) [19] for further information on sampling procedures for consignments and places of production for *X. fastidiosa*, respectively.
23. Freeze-dried or silica-dried tissue can be tested; however the starting weight should be reduced by a factor of 5. Additional subsamples may be processed to ensure adequate

representation of the sampled plant within the subsample used for testing.

24. We use plastic trays as a working surface to contain the sample and allow easy disinfection of the working area between samples. Alternatively sheets of paper or paper towel can be used to line a work bench and changed between samples.
25. Only one sample at a time should be processed per workspace. All other samples must remain sealed and separated from the work area to prevent cross contamination.
26. Symptomatic material must take priority over asymptomatic material for inclusion in DNA extraction subsamples.
27. The highest likelihood of detection is obtained by sampling when the plant is in active growth but not new growing shoots. *Xylella fastidiosa* can have an uneven distribution in host, and correct sampling of the host plant is crucial to the success of the test.
28. The same methods should be used for both symptomatic and asymptomatic samples; however the level of confidence of detection is lower in asymptomatic samples. For asymptomatic samples, use a random selection of leaves from as many stems/branches of the plant as possible.
29. Samples with advanced decay (e.g., as a result of delays in shipping and/or secondary pathogen growth) are unlikely to yield a reliable result, and the plant should be resampled if possible. However, if this is not possible, the sample can be processed, but negative results may be considered unreliable.
30. The number of leaves included in the subsample should be between 10 and 25, with a higher number of leaves increasing the likelihood of detection by including a greater proportion of the plant within the tested sample. Only for plants with very large leaves, such as *Coffea* spp. (coffee), can less leaves be used, and in these instances, a minimum of five leaves should be sampled.
31. We would suggest the use of cut-proof gloves when using blades.
32. This sampling method removes excess leaf material to allow a greater quantity of petiole and midrib to be included in the sample. These tissues have a higher ratio of xylem vessels where *X. fastidiosa* is located.
33. The parameters used for sampling of the plant material remains the same regardless of the DNA extraction protocol being used, with only the starting mass varying. Ensure that the correct sample weight has been prepared for the extraction

method: 3 × 0.1 g for DNeasy® Plant Mini kit, 0.4 g for DNeasy® Mericon® Food Kit, and 0.6 g for CTAB.

34. Should there be inadequate sample to obtain 0.6 g, then a smaller sample can be tested, for example, 0.3 g and the CTAB buffer volume reduced proportionally. The subsample weight should be noted in the sample records as a negative result would have less confidence.
35. To prevent contamination of the outside of the bag, do not touch the outside of the extraction bag with the plant material or while wearing gloves which have handled the plant material during sampling.
36. Alternative, thicker extraction bags (e.g., 800 gauge) can be used for thorny/sharp samples.
37. Once the plant tissue has been sampled, then it should be processed quickly for DNA extraction (within 1–2 h). If this is not possible, the sample should be frozen (−20 °C minimum) to prevent degradation.
38. If desired, multiple subsamples can be tested from a single sample. This allows a greater amount of material to be tested and provides higher confidence in the results, especially in cases of a negative result. However this is not essential, and one subsample is acceptable in the majority of circumstances.
39. Dispose of all waste material in compliance with local plant health regulations.
40. It is prudent to store surplus plant material until the testing process has been successfully completed on the sample, at which point surplus material can be disposed.
41. A range of different DNA extraction methods and protocols can be used for DNA extraction of samples to be tested for *X. fastidiosa*. However, due to the extensive host range, which includes many species known to be challenging to extract nucleic acids from, it is important that a DNA extraction protocol which results in a high yield of good-quality, pure DNA is used. Crude DNA extraction protocols, such as simple, rapid lysis-based extractions (e.g., PEG buffer) often used for other molecular methods such as LAMP are typically inadequate for *in planta* detection of *X. fastidiosa*.
42. If a plant species has not been tested before, use the CTAB protocol in the first instance.
43. A widely used alternative to the CTAB protocol is based upon the QIAGEN DNeasy® Mericon® Food Kit (catalogue number 69514). The protocol is performed following the manufacturer's "standard protocol (200 mg)" with minor modifications as follows. Prepare 0.4 g sample following guidelines in Subheadings 3.1 and 3.2. Homogenize the tissue as per

protocol Subheading 3.3, **step 5** stages i to iii using 2 mL food lysis buffer in place of CTAB buffer. Proteinase K is omitted. Transfer 0.2 mL lysate to a fresh tube and then continue with the manufacturer's protocol from **step 2**. Follow the manufacturer's protocol as stated until **step 11** when the elution volume is reduced to 100 μl buffer EB.

44. If a given sample fails testing by one method (e.g., is negative with the plant COX assay), then it should be re-extracted using the alternate method in the first instance.
45. DNA can also be extracted using a QIAGEN DNeasy® Plant Mini Kit (catalogue number 69104) according to the manufacturer's instructions, although less validation data is available for this method. It is important to note that the sample size is reduced to 100 mg (0.1 g) with this method. Therefore, it is recommended that at least three subsamples are processed from each sample. If using this method, samples should be prepared in high impact 2 mL screw-cap tubes using glass beads or metal ball bearings. A Precellys bead beater or TissueLyser (or equivalent) should be used for homogenization, ensuring the tissue is evenly and thoroughly homogenized, which may require optimization based on the tissue type.
46. If testing bacterial cultures, then we would recommend the QIAamp® DNA Mini Kit (QIAGEN, catalogue number 51304) is used for DNA extraction, following the manufactures protocol for spin columns.
47. As a minimum, disposable gloves (e.g., nitrile) and safety glasses should be worn when performing the DNA extraction.
48. Ensure that all local regulations for safe disposal of hazardous waste are followed.
49. Negative isolation controls (NICs), or "extraction blanks," are used to demonstrate the absence of within batch cross-contamination during the DNA extraction protocol and to show the absence of contamination within in any of the buffers used. This is performed by processing a sample in the same manner but without any plant material present.
50. Positive isolation controls (PICs), or "extraction controls," are used to confirm the success of a batch of DNA extracts and monitor for variation in extraction efficiency between batches. Any plant species which has been shown to be successfully extracted by the method can be used. The healthy plant tissue is subject to sampling and DNA extraction in the same manner as the samples.
51. Every batch of DNA extractions should contain one NIC and one PIC alongside the samples; however, as a minimum, a NIC must always be included with every batch.

52. The size of a batch of DNA extractions can vary. In our experience, 12 samples in total is a practical number. Due to the numerous centrifugation steps, it is pragmatic to always ensure an even number of samples (so that the centrifuge can be easily balanced), and to not process batches larger than the centrifuge rotor, which typically accept 18 or 24 tubes. Also *see* **Note 66**.
53. If a shaking incubator is not available, then place the buffer tube into the 65 °C water bath to allow the additives to dissolve and periodically remove the tube and shake vigorously by hand to aid dissolving.
54. Only add the additives to the CTAB base buffer on the day of use. Do not store any remaining CTAB buffer with additives.
55. Prepare enough CTAB buffer with additives for the expected number of samples in the batch (including controls) and a small excess. For example, for a batch of 12 samples/controls, we would typically prepare 30 mL of buffer.
56. Ensure the additives have fully dissolved and the buffer is no longer viscous before use.
57. To prevent degradation of the samples, ensure that one sample is processed from homogenization to the water bath stage completely before starting to process the next sample. For example, do not homogenize all of the samples and then start to decant them.
58. If the starting weight of the samples was less than 0.6 g, reduce the volume of CTAB buffer proportionally.
59. Freeze-dried material is likely to absorb some of the buffer, and additional buffer may need to be added in 0.5 mL increments until there is free buffer to allow adequate homogenization. The sample-buffer slurry may benefit from standing at room temperature for a few minutes to allow the material to fully rehydrate.
60. Care should be taken that the buffer does not escape from the top of the bag. Hold the bag by the folded over end and elevate this slightly from the work surface while homogenizing.
61. It is important that the tissue is thoroughly disrupted. Alternate between a homogenizer, a mallet, and a roller until this is achieved. The buffer color/texture should be homogeneous.
62. Ensure that the entire sample is thoroughly homogenized and no tissue is trapped in the corners of the bag.
63. If the grinding bag breaks, discard it and immediately clean the equipment and work area and change gloves.
64. If recovery is less than the required volume, then add additional buffer (0.5 mL) to the sample bag and repeat the

homogenization to recover additional sample. Record this information with the sample records.

65. The presence of small fragments of the plant tissue in the lysate removed from the sample bag is acceptable.

66. Throughout the DNA extraction protocol, it is good practice to not place tubes in adjacent wells of a rack/centrifuge, but to always leaving a gap between them. This mitigates against the possibility of cross-contamination from tubes touching.

67. Time the 30 min from the time the last sample is added to the water bath.

68. Care must be taken that all steps using chloroform are performed in a constant air fume hood.

69. When handling chloroform, wear double gloves (2 layers) with the outer pair being chloroprene gloves. Change gloves immediately if they come into contact with chloroform.

70. Ensure that the phases are fully mixed for the duration of the vortexing step. If the phases are not mixing well, then shake the tube vigorously by hand and immediately place onto an active vortex mixer. Repeat as needed during the mixing time if needed.

71. In our experience, this stage is a prime cause of cross-contamination within a batch of samples. Therefore, we change gloves after ever sample is vortexed to prevent cross-contamination between the samples.

72. The interface phase is very delicate, and it is vital that none is transferred with the supernatant (upper layer). Use a small volume pipette (e.g., 200 μL) and remove in aliquots. Leave behind a small volume of the supernatant if in doubt or if the interface degrades/collapses. Should the interface degrade before adequate supernatant is removed, re-centrifuge the sample and then repeat the process.

73. Discard tubes containing chloroform-isoamyl alcohol waste into a designated chlorinated solvent waste container in a fume hood using the appropriate hazardous waste disposal container.

74. It is critical that the samples are treated gently at this stage to prevent shearing/fragmentation of the DNA. Therefore the tubes are not vortexed but inverted by hand 5 to 10 times until the phases can be seen to have mixed. The phases will begin to separate almost immediately, and the sample will not remain homogenous.

75. If the combined volume of isopropanol and NaCl is too large to fit into the microcentrifuge tube, then prioritise adding the isopropanol and then top up with as much NaCl as possible.

76. It is critical that the samples are treated gently at this stage to prevent shearing/fragmentation of the DNA, and the tubes should be inverted three to five times gently by hand to mix the solutions.
77. In some instances upon mixing of the tubes, DNA may become visible (as white strands). Continue with the protocol regardless of this phenomenon.
78. The duration of this incubation should be a minimum of 2 h; however overnight (e.g., up to 18 h) has no detrimental effect on results and is most often used in our lab due to providing a practical method to fit within the working day. For example, the samples will be processed up to this step on day 1 and then completed and tested by qPCR on day 2.
79. A minimum temperature of −20 °C is required, although within the range of −20 °C to −30 °C is acceptable. Do not use ultra-low temperature freezers for this stage, for example, −80 °C or colder.
80. When placing the microcentrifuge tube in the rotor, locate the hinge at the "top" of the rotor to ensure the pellet will form directly under this point which will mean locating the DNA pellet in the tube is easier.
81. The DNA pellet may be small and clear/white in color and be difficult to locate, or invisible to the naked eye. It should be at the base of the tube directly in line with the tube hinge if the hinge was at the top of the centrifuge rotor.
82. The DNA pellet may not be firmly attached to the tube after centrifugation. Before discarding the supernatant visually, locate the pellet, and then ensure it is still present after the discard. Always discard into a single use (sterile) tube so that the pellet can be recovered should it be inadvertently tipped away with the supernatant.
83. Pipette all traces away with a small sized tip, for example, 10 μL. If residue remains air dry the pellet (with the lid open on the bench) for as short a duration as possible. Over-drying of the DNA pellet can make later resuspension more challenging and should therefore be avoided.
84. Once completely resuspended, the DNA extract should not have any strands of DNA visible and should not be viscous. The DNA should be a free-flowing liquid which is clear in color. Some samples may require several rounds of gently pipetting and standing to fully resuspend, and patience may be required.
85. Do not vortex the DNA to aid resuspension which would act to fragment the DNA.

86. Add the minimum volume of TE buffer required to fully resuspend the DNA pellet to avoid over dilution of the DNA extract.
87. Some DNA extracts are not clear colored, and these are more likely to contain PCR inhibitors. These samples may benefit from DNA purification as detailed in Subheading 3.5 prior to testing. Alternatively, the samples can be tested first (neat and diluted) and subject to the purification if the results of the plant control assay show PCR inhibition.
88. While our lab does not have experience of DNA extraction from root tissue for *X. fastidiosa* testing, the evidence from the literature suggest that the methods detailed within this chapter should be successful with this matrix and that the woody tissue variation would be most likely to be successful. Soil should be removed from the roots before testing by washing in cool tap water.
89. Should the sample immediately absorb all the buffer, then add further buffer in 0.5 mL increments until there is a small volume of free buffer. Do not add excessive quantities as this may over dilute the resultant sample resulting in a decrease in test sensitivity.
90. We would typically use a minimum of 16 h and a maximum of 24 h soaking time, which fits well within the working day. Shorter durations may be acceptable, but we have not tested these routinely. Longer durations may lead to sample degradation and should be avoided.
91. Tissue fragments of this nature may remain sharp even after soaking posing a high risk of rupturing the bag during homogenization. If necessary, a second empty bag can be placed on top of the sample to reduce the likelihood of breakage.
92. Wood shavings/woody samples do not typically completely disintegrate after soaking-homogenization; however, ensure that all fragments have been physically disrupted as much as possible.
93. Either the entire DNA extract or an aliquot of the extract can be used for the purification protocol.
94. Do not resuspend the cleaned-up DNA into a larger volume than the starting volume unless absolutely necessary, to avoid diluting the resultant DNA.
95. If desired, the DNA pellet can be resuspended in a volume smaller than the starting volume to concentrate the DNA.
96. For further information on assay validation and verification, see the EPPO standard PM7/98(4), specific requirements for laboratories preparing accreditation for a plant pest diagnostic activity [17].

97. The order in which the assays are used to test a sample can be altered depending upon whether the desired outcome is the most rapid result, or the most cost-effective.
98. For a sample which has already been demonstrated to be positive for *X. fastidiosa* (e.g., by another test), then DNA extracts can be screened directly with the subspecies assays. This directly leverages the time saving compared to running an MLST analysis. While the five assays could be used sequentially until the subspecies is identified, we prefer to run them in parallel for ease.
99. For unknown samples where a rapid outcome is required, either (i) all of the assays can be run in parallel (COX, *X. fastidiosa* universal, and five *X. fastidiosa* subspecies assays) or (ii) the COX and Harper assays can be run in parallel and then followed by the five subspecies assays in parallel on any samples positive with the universal assay.
100. For unknown samples where a rapid outcome is not required or where cost-effectiveness is the priority, samples should first be screened with the COX assay (to confirm the success of the DNA extraction), then the universal assay (to identify the presence/absence of *X. fastidiosa*), and then any positive samples screened with all five *X. fastidiosa* subspecies.
101. The assay testing order may also be influenced by the nature of the samples being tested and the likelihood of *X. fastidiosa* positives. If the likelihood of positive samples is very low, then initial screening with the Harper and COX assays minimises unnecessary testing. If the likelihood of positive samples is high, then initial screening with the Harper assay may be unnecessary, and samples can be tested with the subspecies assays directly.
102. If testing DNA extracts from bacterial cultures (as opposed to plant samples), then the plant COX assay should not be used. An alternative test/assay must be used to confirm the success of the DNA extraction and ability of the extract to support PCR, for example, a bacterial control gene assay.
103. It is not uncommon for a target and plant control assays, or multiple target assays, to be multiplexed into a single reaction. However, optimization of the assays must be taken to allow multiplexing without any detrimental effect on the assay's performance characteristics. In our lab, the plant control assay and target assays are used with different reagents, and the assays are therefore ran on separate plates (due to the different cycling conditions). Furthermore, in this instance, there would be five assays to multiplex, and the likelihood of no negative effects on assay performance is low. We find it

simpler to run the assays separately to obtain optimal performance from each.

104. As a minimum, each DNA extract should be tested in duplicate wells with each assay. It is not uncommon for samples to be tested in triplicate, particularly if quantification is being undertaken. However, for these tests, duplicate is adequate.

105. Each plate must contain both positive and negative amplification controls, tested in duplicate wells for each of the assays being used. The negative amplification control (NAC), often called a "no template" control, has water added in place of the template DNA. This acts to demonstrate the absence of contamination of the PCR reagents. The positive amplification control (PAC) is a DNA extract which is known to produce a positive result with the assay in question. This acts to demonstrate the success of the PCR run for the assay.

106. If possible, a DNA extract with a known C_T value for the assay in question should be used as the positive control. By having a pre-determined C_T value, this can also be used to monitor assay performance, with an increase in C_T indicating a potential problem with the plate/run.

107. We would typically produce small aliquots (20 μL) of positive control DNA to limit degradation due to repeated freeze-thaw cycles.

108. The fluorophores of the probes are degraded by light and should always kept in the dark unless in use. We wrap the rack containing the tubes with a double layer of paper towel during thawing to maintain darkness.

109. The *Taq* polymerase should be left in the freezer until immediately before use.

110. *Taq* polymerase should not be vortexed. Gently flick the tube with a finger to mix before use.

111. Always hold plates by the edges and avoid handling the top of the plate, especially once sealed, so that the top of the plate does not get dirty, marked, or have fingerprints which may affect the machine reading of results.

112. Plates made with TaqMan™ Gold reagents can be frozen at −20 °C for a few days, thawed, and then run without any detrimental effect on results. Plates made with Fast Universal reagents cannot be frozen. Store at 4 °C prior to run and run within a few hours of preparation.

113. The terms C_T (cycle threshold) and C_Q (quantification cycle) can both be used for qPCR analysis. However, as these assays are not being used in a quantitative manner, we prefer the term C_T.

114. It is vital that the threshold and baseline values are set properly to allow correct interpretation of the results.

115. Some labs use cut-off values with assays to aid interpretation of results for samples with higher C_T values, for example, > C_T 35. However, the validation data for these assays indicated that any C_T value up to C_T 40 demonstrates the presence of *X. fastidiosa* DNA. Laboratory contamination can be manifested as late C_T value amplification (e.g., C_T 35–40), and adequate control measures to prevent this occurring and subsequently identify if this has occurred are required.

116. We would typically expect the C_T values of duplicate wells to be within 1 C_T value most often within 0.1 C_T. Anything outside of this range indicates poor replication which warrants investigation.

117. The C_T value of the PAC should be within a small range of the known value for this DNA extract, for example, within 1–2 C_T. Should the C_T vary outside of this range, this may be an indicator of inefficiency in the run which should be investigated further.

118. A tenfold dilution of a DNA extract should result in increase of 3.3 C_T. A larger increase may indicate the presence of inhibitors within the DNA extract or inaccurately prepared dilutions.

119. Should a sample test negative with all five subspecies assays, it should only be classified as negative for *X. fastidiosa* if (i) the success of the DNA extract and (ii) the ability of the extract to support qPCR have both been confirmed (e.g., by testing with the plant COX gene assay).

120. Should a sample be negative with the COX assay, with the target assay having amplification profiles which are S-shaped and C_T values <40, then the results for the target assay can still be used.

121. For a result to be provided in compliance with the EPPO guidelines, then any positive sample should be confirmed by two separate tests. Should these both be molecular tests, then they should target difference genes; however, for other diagnostic purposes, a positive result with these assays can be used to provide a *X. fastidiosa* identification.

122. If concerned about a batch of DNA extracts, then compare the C_T value with the COX assay of the extraction control extract. If the same species is used, this should be within a small range every time, and a delayed C_T may indicate an issue with the batch of DNA extracts.

References

1. Wells JM, Raju BC, Hung HY et al (1987) *Xylella fastidiosa* gen. nov., sp. nov: gram-negative, xylem-limited, fastidious plant bacteria related to *Xanthomonas* spp. Int J Syst Evol Microbiol 37(2):136–143
2. Jacques MA, Denancé N, Legendre B et al (2016) New coffee plant-infecting *Xylella fastidiosa* variants derived via homologous recombination. Appl Environ Microbiol 82(5): 1556–1568
3. Janse JD, Obradovic A (2010) *Xylella fastidiosa*: its biology, diagnosis, control and risks. J Plant Pathol 92:S35–S48
4. Saponari M, Boscia D, Nigro F et al (2013) Identification of DNA sequences related to *Xylella fastidiosa* in oleander, almond and olive trees exhibiting leaf scorch symptoms in Apulia (Southern Italy). J Plant Pathol 95(3): 668
5. European Commission (2018) Latest Developments of *Xylella fastidiosa* in the EU territory. [ONLINE] Available at: https://ec.europa.eu/food/plant/plant_health_biosecurity/legislation/emergency_measures/xylella-fastidiosa/latest-developments_en. Accessed 25 Sept 2021
6. EFSA PLH Panel (EFSA Panel on Plant Health) (2015) Scientific Opinion on the risks to plant health posed by *Xylella fastidiosa* in the EU territory, with the identification and evaluation of risk reduction options. EFSA J 13(1): 3989
7. Schaad NW, Postnikova E, Lacy G (2004) *Xylella fastidiosa* subspecies: *X. fastidiosa* subsp *piercei*, subsp. nov., *X. fastidiosa* subsp. *multiplex* subsp. nov., and *X. fastidiosa* subsp. *pauca* subsp. nov. Syst Appl Microbiol 27(3): 290-300. Erratum. Syst Appl Microbiol 27(6): 763
8. Schuenzel EL, Scally M, Stouthamer R et al (2005) A multigene phylogenetic study of clonal diversity and divergence in North American strains of the plant pathogen *Xylella fastidiosa*. Appl Environ Microbiol 71(7): 3832–3839
9. Nunney L, Schuenzel EL, Scally M et al (2014) Large scale intersubspecific recombination in the plant pathogenic bacterium *Xylella fastidiosa* is associated with the host shift to mulberry. Appl Environ Microbiol 80(10): 3025–3033
10. Rubio S, Barnes A, Webb K (2017) A real-time PCR assay for improved rapid, specific detection of *Cryphonectria parasitica*. Ann Appl Biol 171(1):52–61
11. Hodgetts J, Glover R, Cole J et al (2021) Genomics informed design of a suite of real-time PCR assays for the specific detection of each *Xylella fastidiosa* subspecies. J Appl Microbiol 131(2):855–872
12. Harper SJ, Ward LI, Clover GRG (2010) Development of LAMP and real-time PCR methods for the rapid detection of *Xylella fastidiosa* for quarantine and field applications. Phytopathology 100:1282–1288. Erratum 2013
13. Weller SA, Elphinstone JG, Smith NC et al (2000) Detection of *Ralstonia solanacearum* strains with a quantitative, multiplex, real-time, fluorogenic PCR (TaqMan) assay. Appl Environ Microbiol 66:2853–2858
14. Hughes KJ, Tomlinson JA, Griffin RL et al (2006) Development of a one-step real-time polymerase chain reaction assay for diagnosis of *Phytophthora ramorum*. Phytopathology 96:975–981
15. Doyle JJ, Doyle JL (1990) Isolation of plant DNA from fresh tissue. Focus 12:13–15
16. EPPO (European and Mediterranean Plant Protection Organization) (2019a) PM 7/24 (4) *Xylella fastidiosa*. Bulletin OEPP 49(2): 175–227
17. EPPO (European and Mediterranean Plant Protection Organization) (2019b) PM 7/98 (4) Specific requirements for laboratories preparing accreditation for a plant pest diagnostic activity. Bulletin OEPP 49(3):530–563
18. EPPO (European and Mediterranean Plant Protection Organization) (2020a) PM 3/81 (2) Inspection of consignments for *Xylella fastidiosa*. Bulletin OEPP 50(3):401–414
19. EPPO (European and Mediterranean Plant Protection Organization) (2020b) PM 3/82 (2) Inspection of places of production for *Xylella fastidiosa*. Bulletin OEPP 50(3): 415–428

Chapter 14

Selective Quantification of *Erwinia amylovora* Live Cells in Pome Fruit Tree Cankers by Viability Digital PCR

Ricardo Delgado Santander, Katarina Gašić, and Srđan Goran Aćimović

Abstract

The accurate assessment of *Erwinia amylovora* live cell populations in fire blight cankers by classic microbiology methods has major limitations. Some of them are the presence of competitive microbiota in samples that inhibit *E. amylovora*'s growth and the release of toxic compounds by plant material during sample processing, which may hamper the pathogen's ability to form colonies on solid media. Digital PCR (dPCR) combined with the photo-reactive DNA-binding dye propidium monoazide (PMA) allows selective detection and quantification of live *E. amylovora* cells in woody samples while overcoming the constraints of culture-dependent methods. This work describes a reliable viability dPCR procedure to determine *E. amylovora* live cell concentrations in fire blight cankers from pome fruit trees. This protocol can be adapted for the analysis of other types of plant material and enables investigation of ecological, epidemiological, and management significance of cankers as a relatively underexplored part of the fire blight disease cycle.

Key words Fire blight, Cankers, Bacterial survival, Bacterial viability, Propidium monoazide, dPCR, Molecular detection methods, Apple, Pear, Asian pear

1 Introduction

Most of the published fire blight diagnostic and detection protocols for *Erwinia amylovora* rely on the pathogen's ability to grow on solid media [1–4]. Culture-dependent isolation methods are usually expensive, time-consuming, and involve an anticipated preparation of materials. Besides, these procedures may fail in detecting stressed bacteria or bacteria in a nonculturable state [5, 6], and in general, their success and reproducibility are highly dependable on factors varying between samples (state of preservation, tissue damage, etc.) [7–10]. *E. amylovora*'s growth, for example, can be outcompeted by the saprophytic microbiota present in some samples [11]. Besides, the processing of plant tissues, especially those composed of bark and/or wood, may involve the release of tannins and other toxic secondary metabolites which may have an impact on the

Nicola Luchi (ed.), *Plant Pathology: Methods and Protocols*, Methods in Molecular Biology, vol. 2536,
https://doi.org/10.1007/978-1-0716-2517-0_14,

pathogen's ability to form colonies on agar plates [12, 13]. Some of these drawbacks can be circumvented by molecular detection methods. However, although molecular strategies can support or, in some cases, be an alternative to culture-dependent methods, most of them fail to determine whether the detected cells are alive or dead [14]. These drawbacks have important implications, for example, during the analysis of asymptomatic plant material, the study of potential vectors, and reservoirs of the pathogen, etc., where knowing the detected cell's viability is crucial.

To study *E. amylovora* populations in pome fruit tree cankers, we designed a sample preservation and processing method maximizing the amount of recovered *E. amylovora* cells from the lignified tissues of fire blight cankers. Then, we optimized a molecular approach for selective quantification of *E. amylovora* live cells in cankers. Briefly, after canker collection, samples are flash-frozen with liquid nitrogen and stored at −80 °C until processing. For sample analysis, woody tissues are physically processed with a hammer inside plastic bags containing an antioxidant maceration buffer. Then, samples are treated with a viability PCR reagent, PMAxx™ Dye, which is an improved version of propidium monoazide (PMA). The dye penetrates only the damaged dead-cell membranes and after photo-activation it binds covalently to their DNA and inhibits the subsequent PCR amplification. Meanwhile, live cell's DNA remains protected by the cell lipid membranes, which are impermeable to the dye. After treatment with PMAxx, sample DNA is extracted and viable cell numbers are quantified by digital PCR (dPCR), a technology that allows precise and absolute quantification of target DNA copies without the need for standard curves and which is less affected by inhibitors than other PCRs [15–17].

With this methodology, we have successfully quantified *E. amylovora* live cells in fire blight cankers on a variety of cultivars of different pome fruit host species, in tissues with different degrees of damage by fire blight necrosis. We foresee a significant potential of this procedure for evaluating the efficacy of bactericidal compounds *in planta*, fire blight diagnostics, and studies on different aspects of *E. amylovora* epidemiology such as population dynamics during infection and analysis of viable populations in insect vectors and reservoirs of the pathogen. Overall, our viability dPCR protocol can be used to investigate the ecological, epidemiological, and management significance of cankers as a relatively underexplored part of the fire blight disease cycle [18].

2 Materials

2.1 Sample Collection and Preservation

1. Pruning shears.
2. Bleach.
3. Paper towels.
4. Gloves.
5. 70% Ethanol.
6. 6 × 9″ 2 Mil polyethylene resealable bags (ULINE, Pleasant Prairie, WI, Ref. S-1296) containing a piece of paper with information about the sample (host species, cultivar, date, experiment, replicate, etc.).
7. Portable camping cooler, preferably rotomolded, dry ice-resistant, with a rubber lid gasket filled with ice.
8. Liquid nitrogen.
9. Kitchen tongs.
10. Ultralow temperature freezer set at −80 °C.

2.2 Physical Sample Processing

1. Gloves.
2. Paper towels.
3. Sterile scalpels.
4. Weighing paper.
5. Scale.
6. Graduated cylinders of different volumes.
7. 12 × 15″ 4 Mil polyethylene reclosable plastic bags (ULINE, Pleasant Prairie, WI, Ref. S-1306) (*see* **Note 1**).
8. Hammer.
9. Earplugs or any other noise cancelation system to protect hearing during sample processing.
10. Sterile distilled water: Autoclave at 121 °C for 20 min (*see* **Note 2**). Let it cool and store at room temperature.
11. pH meter.
12. Phosphate-buffered saline (PBS), pH 7.4: In a beaker, using a magnetic stirrer, dissolve in 800 mL of distilled water 80 g NaCl (*M*, 58.4 g/mol), 2 g KCl (*M*, 74.551 g/mol), 14.4 g Na_2HPO_4 (*M*, 141.96 g/mol), and 2.45 g KH_2PO_4 (*M*, 136.086 g/mol) to obtain 10X concentrated PBS (*see* **Note 3**). Adjust pH to 7.4 ± 0.1 with NaOH (*see* **Note 4**), transfer the solution to a volumetric flask, and increase the volume up to 1 L by adding distilled water. Autoclave PBS at 121 °C for 20 min, cool, and store at room temperature.
13. Sterile borosilicate glass bottles.

14. Filtration unit with 0.2-mm pore diameter filter, coupled to a vacuum system (*see* **Note 5**).
15. Antioxidant maceration buffer (AMB), pH 7.4: To prepare AMB according to the EPPO 2013 [2], mix 100 mL 10X PBS pH 7.4 with 700 mL distilled water (at room temperature, non-sterile) in a beaker (*see* **Note 6**). Then, using a magnetic stirrer dissolve 20 g polyvinylpyrrolidone PVP-10 (*M*, 10,000 g/mol) (*see* **Note** 7), 10 g D-mannitol (*M*, 182.17 g/mol), 1.76 g ascorbic acid (*M*, 176.13 g/mol), 3 g reduced glutathione (*M*, 307.33 g/mol). Adjust pH to 7.4 ± 0.1 with NaOH (*see* **Note 4**) and increase the volume up to 1 L by adding distilled water using a volumetric flask. Sterilize using a filtration unit coupled to a vacuum system (*see* **Note 8**). Store at 4 °C protected from light (*see* **Note 9**). For sample processing, AMB must be diluted with sterile 1X PBS in a ratio of 1:9 (*see* **Note 10**). For each gram of sample add 50 mL 0.1X AMB.
16. Ice.

2.3 PMAxx™ Dye Treatment

1. PMAxx™ Dye (Biotium, Fremont, CA), sold as a 20 mM aqueous solution. Each vial allows the analysis of around 20–40 samples, depending on the sample volume treated with the dye.
2. 5X PMA Enhancer™ for Gram-negative bacteria (Biotium, Fremont, CA) (*see* **Note 11**).
3. 2-mL Microcentrifuge tubes (*see* **Note 12**).
4. Pipettes and sterile pipette tips.
5. Two 500-W halogen bulbs in a fixture(s).
6. Tray filled with flake ice placed at a distance of 20 cm from the halogen bulbs.
7. Microcentrifuge.
8. Timer.
9. Orbital shaker at 25 °C, 150 rpm.
10. A container to discard the PMAxx-treated sample supernatants (*see* **Note 13**).

2.4 DNA Extraction

1. DNeasy Plant Mini Kit (Qiagen, Frederick, MD) or any other reliable DNA extraction method (*see* **Note 14**).
2. Absolute ethanol for AW1 and AW2 buffers (Qiagen, Frederick, MD).
3. Dry bath/block heater.
4. Microcentrifuge.
5. Microcentrifuge tubes.

6. Pipettes and sterile pipette tips.
7. Flake ice.
8. Freezer (−20 °C).

2.5 Digital PCR

1. 1.5-mL microcentrifuge tubes (DNase/RNase-free, non-stick, or low-binding).
2. Pipettes and sterile barrier pipette tips.
3. Powder-free gloves.
4. QuantStudio™ 3D Digital PCR Master Mix v2 (Applied Biosystems, Waltham, MA).
5. 20X primers/probe stock: In a sterile microcentrifuge tube, mix 18 μL of a 100 mM stock solution of primer Ams06KbF (5′-AAT TGG TTC CGC TAT AAC TTG CAG-3′) [12], with 18 μL of a 100 mM solution of primer Ams189R (5′-GGG TAT TTG CGC TAA TTT TAT TCG-3′) [1], 4 μL of a 100 mM solution of TaqMan probe Ams141T (3′-FAM-CCA GAA TCT GGC CCG CGT ATA CCG-TAMRA-3′) [1] and 60 μL of sterile ultra-pure water. Store the mixture in small volume aliquots at −20 °C but use at room temperature.
6. Sterile, nuclease-free water.
7. Extracted sample DNA.
8. *E. amylovora* genomic DNA to use as a positive control.
9. Vortex mixer.
10. Microcentrifuge.
11. QuantStudio 3D Digital PCR 20K Chip Kit v2, containing QuantStudio 3D Digital PCR 20K Chips v2, QuantStudio 3D Digital PCR Chip Lids v2, QuantStudio 3D Digital PCR Sample Loading Blades, QuantStudio 3D Digital PCR Immersion Fluid syringes, and QuantStudio 3D Digital PCR Master Mix v2 (Applied Biosystems, Waltham, MA). The QuantStudio 3D Digital PCR Master Mix v2 can also be purchased separately.
12. Permanent marker to label samples.
13. Kimwipes (Kimberly-Clark Professional, Irving, TX) or any other type of clean room-grade wipes.
14. Isopropanol.
15. QuantStudio™ 3D Digital PCR Chip Loader (Applied Biosystems, Waltham, MA).
16. Dual Flat Block GeneAmp® 9700 PCR System (Applied Biosystems, Waltham, MA). The ProFlex™ 2 x Flat PCR System (Applied Biosystems, Waltham, MA) is also compatible with QuantStudio 3D Digital PCR technology.

17. QuantStudio™ 3D Digital PCR Instrument for chip imaging (Applied Biosystems, Waltham, MA).
18. USB flash drive (in case the QuantStudio™ 3D Digital PCR Instrument is not connected to a working network or an online cloud-based platform).
19. QuantStudio™ 3D AnalysisSuite™ Software.

3 Methods

3.1 Canker Sample Collection and Cryopreservation

1. By using pruning shears, excise the portion of the branch containing the canker by cutting the asymptomatic wood at least 2–3 cm above and 2–3 cm below the canker (*see* **Note 15**). Between samples, wet a paper towel with bleach and wipe the pruning shear blades for surface disinfection.
2. Handle the canker with gloves and place it in a 6 × 9″ plastic resealable bag including a paper label containing information about the sample. Change gloves or use 70% ethanol to disinfect the gloves between samples.
3. Similarly, collect samples from trees without fire blight symptoms to use as negative control. These samples allow determining the lower dPCR detection and quantification limits (*see* **Note 16**).
4. Flash-freeze the collected samples in a high-density polyethylene (HDPE) plastic Dewar flask with liquid nitrogen (*see* **Note 17**).
5. When the dipped samples stop bubbling in the liquid nitrogen, transfer them carefully to an ultracold freezer at −80 °C until processing. This cryopreservation protocol for the flash-frozen samples was optimal compared to other preservation and processing conditions tested (Fig. 1).

3.2 Sample Processing

1. Take the frozen branch sample out of −80 °C and place it on the bench at room temperature for 10–15 min until it becomes easier to manipulate (Fig. 2a). Use gloves surface-disinfected with 70% ethanol or change gloves between samples.
2. With a sterile scalpel, excise the bark and vascular cambium encompassing the canker plus a 4-mm ring area of asymptomatic tissue beyond the canker margin (*see* **Note 18**) (Fig. 2b, c). Cut the sample in as many pieces as necessary to carve out the tissues from the whole canker plus the 4-mm ring area (Fig. 2d). Sterilize the scalpels between samples by flaming, after dipping them in absolute ethanol or another appropriate method.

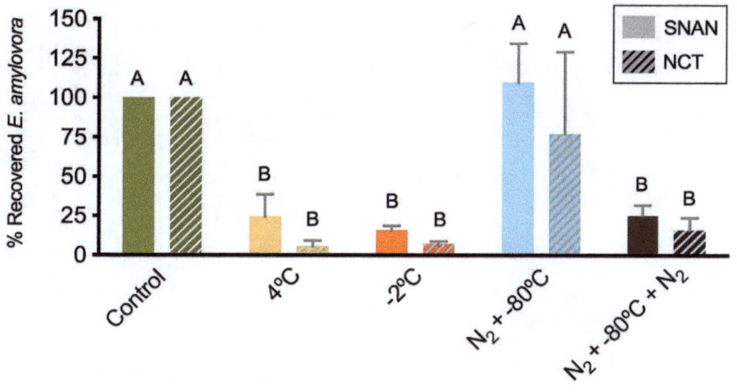

Fig. 1 Comparison of different sample storage and processing conditions. *E. amylovora* was artificially inoculated into branch pieces under the bark by making a V-shaped incision in the bark, raising the tip of the cut area, and placing a 40 μL droplet of an *E. amylovora* suspension at 10^8 CFU/mL in PBS. Afterward, the samples were processed either immediately (control) or placed under different storing conditions (4 °C, −2 °C and −80 °C after flash-freezing samples with liquid nitrogen). After one week of incubation at the temperatures mentioned above, samples were processed by removing the bark of a square area surrounding the V-shaped incision, placing it in resealable plastic bags containing 50 mL PBS for each gram of weighed sample and homogenized by smashing with a hammer. An additional group of samples stored at −80 °C after flash-freezing with liquid nitrogen was processed by pulverizing in liquid N_2 with mortar and pestle and then dissolving the powder in PBS in the same proportions as mentioned above. After sample processing, *E. amylovora* colony plate counts were performed on SNAN (Sucrose Nutrient Agar amended with Natamycin) [14] and NCT (Natamycin, Crystal violet, Thallium Nitrate agar) [19] culture media. The assay was performed in triplicate. Different letters indicate statistically significant differences between the compared methods

3. Weigh the sample on the scale and place it into a labeled 12 × 15″ resealable plastic bag.
4. Calculate the volume of ice-cold 0.1X AMB (in mL) to mix with the sample cut in pieces by multiplying the weighed grams by 50 (i.e., 50 mL 0.1X AMB for each gram of sample) (*see* **Note 19**).
5. Using graduated cylinders, add the specific volume of 0.1X AMB calculated for each sample to the bag containing the canker cut into pieces. Remove the air inside the bag and seal it (Fig. 2e).
6. Place the bag on a flat, resistant surface and using a hammer homogenize the sample by smashing the cut pieces of canker tissue against the hard surface. Stop hammering when the sample fragments are around 1–2 mm diameter and do not break despite the hammering (*see* **Note 20**) (Fig. 2f).

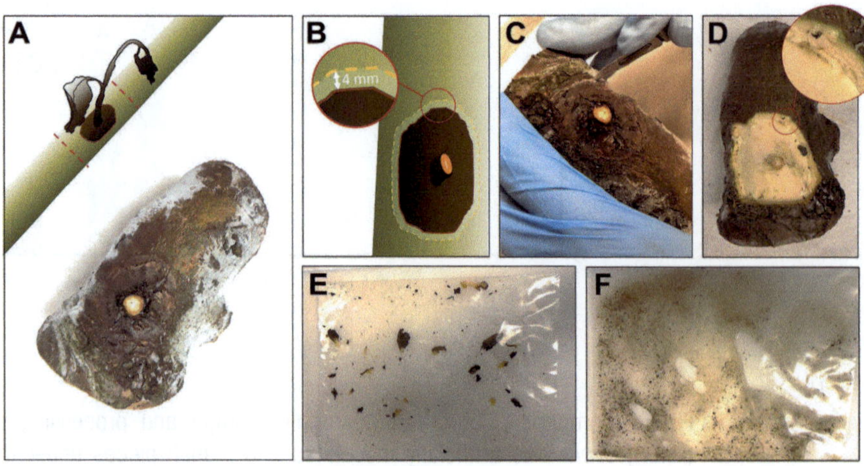

Fig. 2 Steps for canker processing. Cankers on pome fruit tree branches are cut with pruning shears (red lines), flash-frozen with liquid nitrogen, and stored at −80 °C. For sample processing, cankers are taken out of the freezer and kept at room temperature until tissues soften (**a**). A sterile scalpel is used to delimit a 4-mm-thick ring area surrounding the canker borders (**b**, **c**). The tissues within this area, which include the cork, bark, and vascular cambium, are carved out with a scalpel (**d**). After collection, the excised canker pieces are weighed on a scale and placed into a resealable plastic bag with ice-cold 0.1X antioxidant maceration buffer (AMB) in a ratio of 50 mL AMB per gram of canker tissue (**e**). Samples are processed mechanically with a hammer until the remaining fragments are 1–2 mm in diameter and challenging to homogenize (**f**). Samples are cooled on ice up to 30 min until treatments with PMAxx Dye

7. Incubate samples covered with ice up to 30 min before carrying out the PMAxx Dye treatment (*see* **Note 21**).

3.3 PMAxx Dye Treatment

1. Gently shake the bag content to homogenize the sample. Decant a small volume of the plant macerate into a sterile Petri dish for easier pipetting of the macerate.
2. Using a pipette, transfer a 0.4–1 mL aliquot of the macerate to a sterile 2-mL microcentrifuge tube (*see* **Note 22**) and add 5X PMA Enhancer for Gram-negative bacteria and 20 mM PMAxx Dye, to a final concentration of 1X PMA Enhancer and 100 mM PMAxx Dye in each tube. This step is performed at room temperature and in low-light conditions (PMAxx Dye is sensitive to light).
3. Immediately, place the closed 2-mL tubes horizontally in a Petri dish or any other container and transfer them to an orbital shaker at 25 °C, protected from light. Shake the tubes at 150 rpm for 5 min.
4. In a tray containing ice, at 20 cm distance from two 500-W halogen bulb fixtures, place the tubes horizontally, dipped halfway in the ice, with the upper half of the tube facing the light source. To photo-activate the PMAxx Dye, turn the lights on and expose the samples to light for 10 min (*see* **Note 23**).

After 5 min, without turning the lights off, gently tip the tray containing the tubes back and forth to mix the tube's content by repeatedly raising one side of the tray while lowering the opposite one, and vice versa. After 5–10 s of mixing, put the tray back under the light until the timer signals the 10-min light exposure completion.

5. Centrifuge the tubes at 15,800 × *g* for 10 min and discard the supernatants.
6. Place the tubes with the pellets at −80 °C for long-term storage or perform the DNA extraction immediately.

3.4 DNA Extraction

1. Take the tubes with sample pellets out of −80 °C and use DNeasy Plant Mini Kit (Qiagen, Frederick, MD) (*see* **Note 24**) or any other reliable DNA extraction method to isolate DNA from samples. The volume used for eluting DNA in the process of extraction must be recorded to accurately calculate the initial *E. amylovora* concentration in the original sample.
2. After extraction, sample DNA can be stored at −20 °C or −80 °C until use.

3.5 Digital PCR

1. Turn on the QuantStudio 3D Digital PCR Chip Loader, so it is ready for use when all the chips are prepared. Wear gloves while manipulating samples and QuantStudio 3D Digital PCR components and reagents.
2. Set up the dPCR by mixing the QuantStudio 3D Digital PCR Master Mix v2, primers, TaqMan probe, and sample in a DNAse/RNAse-free microcentrifuge tube. All the reaction components must be at room temperature and be mixed by inversion or low-speed vortexing before use. For each sample, the dPCR mix consists of 8 μL of QuantStudio 3D Digital PCR Master Mix v2, 0.8 μL of 20x primers/TaqMan probe mix, 6.4 μL of direct or diluted DNA sample (*see* **Note 25**), and 0.8 μL of nuclease-free water. The final reaction volume of 16 μL is over the required volume to load a chip (14.8 μL) to compensate for volume loss from pipetting. We recommend preparing enough reaction volume to load two chips per sample. *E. amylovora* genomic DNA extracted from pure cultures can be used as a positive control. In this case, the volume of sample DNA can be reduced to 1 μL, while raising the final volume of the control reaction mix to 16 μL with nuclease-free water. In negative controls, the volume of sample DNA is substituted by an equivalent volume of nuclease-free water. After adding all the reaction components to the tube, mix well by gently pipetting up and down and immediately load the chip.
3. To load the chips, first check whether the light in the chip loader is green, indicating that it reached the correct temperature to load chips. Place the chip facing up in the nest using the

nest lever and the clamp to lock it in position. Then, place a disposable loading blade on the loader head and fill the tip of the blade with 15 μL of the dPCR mix. Avoid forming air bubbles while loading the blade. Press the loading button to initiate the automatic transfer of the dPCR mix from the blade to the chip surface homogeneously. The blade can be used to load the chip manually, but for better reproducibility, we recommend using the automatic procedure.

4. Once the chip loading is complete (the status light flashes during the loading and returns to solid green when completed), immediately cover the reaction wells (a black square in the middle of the chip) with the QuantStudio 3D Digital PCR Immersion Fluid using the syringe, by placing the tip 1–2 cm above the chip surface. The Immersion Fluid must be released slowly, without touching the chip surface with the syringe tip. Use a wipe moistened with isopropanol to remove any drop of immersion oil from the chip edges (*see* **Note 26**).
5. The chip lid is a separate piece with an adhesive edge on the lower side. To seal the chip with the lid, remove the red sticker covering the adhesive on the lower edges of the lid. Then, position the lid over the chip and firmly press down the lid against the chip for 15–30 s using either your fingers or the arm of the chip loader. The lid must be handled carefully and powder-free gloves must be worn during the whole process. Grab the lid by the edges and avoid touching the lid glass with your fingers. This step can also be performed using the arm of the chip loader.
6. Remove the chip from the QuantStudio 3D Digital PCR Chip Loader and pull back the label on the top half of the lid. Hold the chip and the peeled back label with your fingers to access the fill port, which should be on the top right side of the chip. Using the syringe, fill the remaining space between the lid and the chip surface with Immersion Fluid through the fill port, avoiding bubble formation. Use the wipe moistened with isopropanol to remove any droplet of Immersion Fluid that could accidentally fall on the chip surface. Finally, carefully remove the sticker on the inner side of the pulled back label and seal the fill port by firmly pressing the top half of the label against it for 15 s. Store the prepared chip in the dark while preparing the next one. Take into consideration that only 24 samples (chips), including positive and negative controls, can be loaded in the thermocycler at the same time and that, once opened, the Immersion Fluid in the syringe is usable for 1 h (*see* **Note 27**).
7. Place the loaded chips in the flat block thermal cycler and run the following program: One step of 10 min at 95 °C, followed by 39 two-step cycles of 1 min at 60 °C and 15 s at 98 °C, and a final extension step at 60 °C for 2 min.

8. Once the PCR is completed, read the chips by loading them, one at a time, into the QuantStudio 3D Digital PCR Instrument. The imaging takes around 60 s per chip. Chips must be removed from the Instrument tray only when the indicator on the screen of the QuantStudio 3D Digital PCR Instrument reports it is safe to do so. The Instrument detects the fluorescence signals in each of the 20,000 wells of the chip and records the raw fluorescence data to a USB drive, network file server, or directly on the Connect cloud-based platform, which allows direct access to the QuantStudio 3D Digital PCR AnalysisSuite.

9. To analyze the samples, data must be imported into QuantStudio 3D Digital PCR AnalysisSuite software, a web-based application that enables chip data visualization, quality assessment (Fig. 3), and target DNA copy quantification (Table 1).

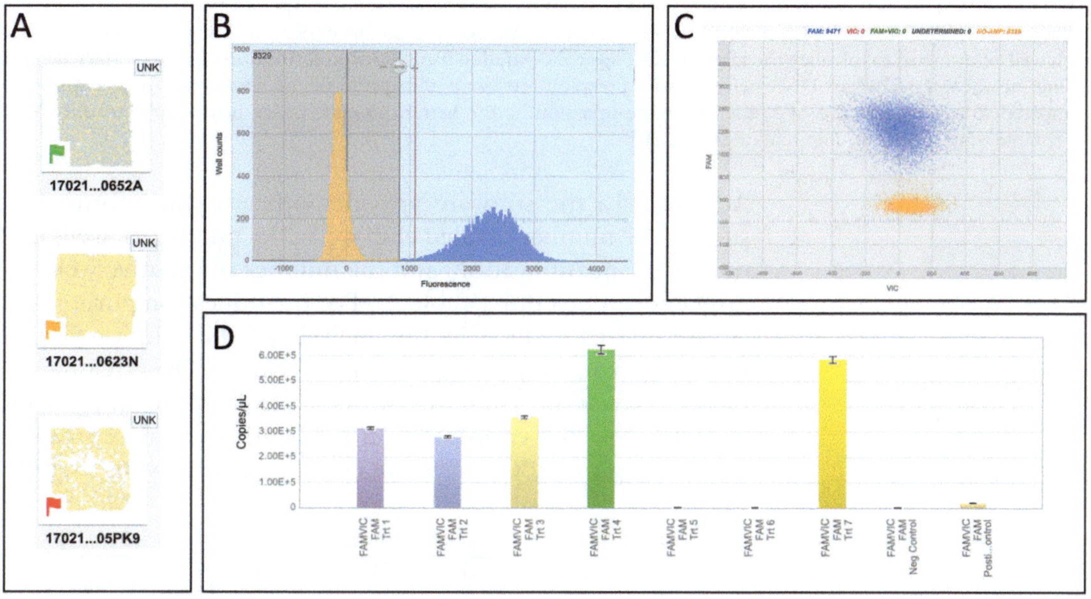

Fig. 3 QuantStudio 3D Digital PCR AnalysisSuite graphic outputs. For quality assessment, the chip fluorescence data for each sample is represented with a chip image where yellow and blue spots depict negative and positive reaction wells, respectively. Each chip (**a**) contains a unique label associated with the sample (e.g., 17021-193345-C05PK9) and a data quality flag, whose color varies depending on well quality thresholds and other data analysis characteristics. Green flags designate data meeting all the quality thresholds. Yellow flags usually indicate that the target DNA copies are outside of the optimal range (200–2000 copies/μL), which is common when analyzing natural samples with unknown bacterial content. The review of yellow flagged chips is suggested, particularly ensuring that the software's automatic threshold separates positive and negative well signals correctly. Red flags usually indicate low-quality wells or a reduced number of filled wells, usually because of the presence of air bubbles, defective loading of the chip, and/or a leak of Immersion Fluid during the thermal cycling. Rerunning red-flagged samples is frequently the best recommendation (**a**). The chip image analysis software also provides a representation of fluorescence frequency distributions as a histogram (**b**) and the fluorescence units detected in each well as a scatter plot (**c**). After reviewing the chip analysis information and implementing the dilution factors for each sample, the software also provides a column graph including the values of copies/μL automatically calculated for each sample (**d**)

Table 1
Typical "Quantification Results" table provided by QuantStudio 3D Digital PCR AnalysisSuite Software. The "Quantification Results" table summarizes the results taking into consideration the dilution factors and averaging replicate values

Assay	Target[a]	Sample[b]	Copies/μL[c]	CI Copies/μL	Precision[d]	Chips
FAMVIC	FAM	Sample 1	3.13E+05	3.09E+5–3.18E+5	1.46%	2
FAMVIC	FAM	Sample 2	2.80E+05	2.76E+5–2.84E+5	1.42%	2
FAMVIC	FAM	Sample 3	3.57E+05	3.51E+5–3.62E+5	1.54%	2

[a]Specifies the fluorophore detected by the QuantStudio 3D Digital PCR Instrument for Imaging, in this case, the probe used is labeled with fluorescein (FAM)
[b]For the software to consider different chips as replicates of the same sample, these must be introduced with the same sample name. In this example, there were two chips (see the last column) labeled as "Sample 1," which were used to calculate values in the next columns
[c]The software calculates these values considering all the dilution factors specified when importing data and averages the results of all the replicates, indicated in the last column (in this example, two replicates per sample)
[d]When dPCR chips are loaded, samples are partitioned in thousands of wells so that either 0 or 1 target DNA copies fall in each well, allowing us to calculate the target DNA copy concentration using Poisson distribution statistics. If we define λ as the mean copies of target DNA per partition, precision represents the spread of the confidence interval around λ compared to the actual value of λ; the lower the precision %, the better. Usually, values below 10% are considered acceptable when analyzing chip replicates [17, 20]

Additionally, the software provides other useful information, such as the number of filled chip wells, the number of wells that qualify for quantification, and the number of negative wells for target copy detection (Table 2). For quantification purposes, it is essential to include in the analysis all the dilution factors resulting from sample processing, PMAxx Dye treatments, DNA extractions, and dPCR. To calculate the final viable *E. amylovora* cell concentration per gram of analyzed canker tissue using the dPCR data, we can use the following formula:

$$\frac{\text{Viable Ea cells}}{g\ \text{canker}} = \frac{\text{copies}}{\mu\text{L}} \times \frac{1}{D_{\text{rxn}_{\text{tube}}}} \times \frac{1}{D_{\text{DNA}}} \times \frac{1}{E_{\text{DNA}_{\text{extr}}}} \times \frac{\text{Vol}_{\text{Elution}(\mu\text{L})}}{\text{Vol}_{\text{macerates}(\text{mL})}} \times \frac{\text{Vol}_{\text{AMB}(\text{mL})}}{\text{Weight}_{\text{canker}(\text{g})}}$$

where copies/μL is the concentration of target DNA copies in the reaction tube, provided by the QuantStudio 3D Digital PCR AnalysisSuite software; $D_{\text{rxn tube}}$ is the dilution of sample DNA in the dPCR tube (e.g., following our protocol, this factor equals to 0.4, i.e., 6.4 μL sample DNA in 16 μL of dPCR MasterMix); D_{DNA} refers to an optional dilution of sample DNA before dPCR (when no DNA dilution is performed, this value equals to 1); $E_{\text{DNA extr}}$ is the efficiency of the DNA extraction with the kit of choice (e.g., if the efficiency is 75%, the value of E_{DNAextr} is 0.75); $\text{Vol}_{\text{Elution }(\mu\text{L})}$ is the final elution volume of sample DNA after extraction with the kit (in our case, 200 μL); $\text{Vol}_{\text{macerates }(\text{mL})}$ is the volume of plant

Table 2
Typical "Replicates" table provided by QuantStudio 3D Digital PCR AnalysisSuite Software. The "Replicates" table enables visualization of specific parameters linked to the individual chips analyzed

Assay	Target	Sample[a]	Dilution[b]	Chip[c]	Copies/Rxn[d]	CI Copies/Rxn	Copies/μL[e]	CI Copies/μL	# of Neg[f]	# qualified by QT[g]	# of Filled[h]
FAM VIC	FAM	Sample 1	0.01	170311_200027_C05T6W.eds	2.511	2.46–2.563	3103.3	3040.3–3167.7	1397	16662	18319
FAM VIC	FAM	Sample 1	0.01	170311_200127_C05WZF.eds	2.561	2.509–2.615	3166.1	3101.7–3231.8	1366	16965	18325
FAM VIC	FAM	Sample 2	0.01	170311_200331_C05T7R.eds	2.371	2.323–2.419	2930.2	2871.5–2990	1576	16457	18226
FAM VIC	FAM	Sample 2	0.01	170311_200236_C05Z7T.eds	2.166	2.124–2.209	2677.8	2625.9–2730.7	1967	16797	19021
FAM VIC	FAM	Sample 3	0.01	170311_200513_C064FO.eds	3.005	2.939–3.072	3714.5	3633.3–3797.5	973	17508	18382
FAM VIC	FAM	Sample 3	0.01	170311_200424_C05XXL.eds	2.778	2.72–2.838	3434.4	3362.5–3507.9	1136	17226	18874

[a]Each replicate of the same sample must be labeled with the same name. The different replicates can be distinguished by their different chip labels

[b]Dilution factors taken into consideration to calculate the copies/μL. In this case, "0.01" corresponds to a DNA sample that was diluted ten times before the analysis, and for the dPCR, only 1.6 μL of the sample was added to a final reaction volume of 16 μL. Accordingly, the final dilution factor is: $1/10 \times 1.6/16 = 0.1 \times 10 = 0.01$

[c]Unique label of the chip associated with a sample/replicate. The label associated with each chip contains an optional prefix (absent in this example), a number indicating the date (*yymmdd*), a number indicating time (in 24 h format), and chip ID (e.g., 170311_200424_C05XXL corresponds to a sample analyzed on March 11 of 2017, at 20 h 4 min 24 sec and the Chip ID is C05XXL)

[d]Quantity of target DNA copies per reaction well in the chip

[e]Copies/μL, copies in the dPCR mixture in the tube, directly calculated using Poisson distribution statistics without considering any dilution factor. To calculate the final cell concentration, values in this column must be multiplied by all the dilution factors resulting from sample processing, PMAxx Dye treatments, DNA extraction, and reaction mix preparation. For the average values of copies/μL of all the replicates after applying all the dilution factors, the software creates the Quantitation Results table (*see* Table 1)

[f–h]#Of Neg, #Qualified for QT, and # of Filled refer to the number of wells per chip showing negative PCR amplification (i.e., no significant increase of fluorescence), the number of wells qualified for quantification (i.e., properly filled and fitting all the quality thresholds evaluated by the software), and the number of wells filled after loading the chip, respectively. The number of positive wells (i.e., showing a significant increase of fluorescence values linked to FAM) can be easily calculated by subtracting the #of Neg from the #Qualified for QT macerates used for DNA extraction (we use 0.4 mL); finally, following our protocol, the ratio $Vol_{AMB\ (mL)}/Weight_{canker\ (g)}$ is always 50, i.e., we add 50 mL 0.1xAMB for every gram of canker processed. As an example, if the target DNA concentration provided by dPCR is 200 copies/μL, to calculate the *E. amylovora* live cell concentration in the canker using the parameters mentioned above, we would multiply $200 \times 2.5 \times 1 \times 1.33 \times 500 \times 50 = 1.7 \times 10^7$ live *E. amylovora* cells/g canker.

4 Notes

1. We use 12 × 15″ 4 Mil polyethylene reclosable plastic bags of the brand Uline (Ref. S-1306). These bags allow the proper processing of most samples without leaks. If it is not possible to find the suggested bags, we recommend trying different brands to minimize leaks during the sample processing.
2. Sterile distilled water is used to prepare 0.1X Antioxidant Maceration Buffer (AMB) pH 7.4. This buffer must be cold for use, so in case it is stored at room temperature, it is recommended to prepare 0.1X AMB hours or days before the assay and cool it down in the fridge or on ice before use.
3. 10X PBS is used to prepare 1X PBS, which is required for AMB and 0.1X AMB preparation. The suggested volumes are convenient for the analysis of a high number of samples. If necessary, volumes can be scaled up or down.
4. Use 10N NaOH or a saturated NaOH solution to adjust the pH. For more precision, when approaching the desired pH, use a more diluted NaOH solution (e.g., 1N NaOH).
5. For convenience, we use Nalgene™ Rapid-Flow™ Sterile Single Use Bottle Top Filters (Nalgene, Rochester, NY, Ref. 290-4520) assembled to a sterile bottle and a vacuum system. Still, these can be substituted by any other equivalent filtration method.
6. The suggested volumes are convenient for the analysis of a high number of samples. If necessary, volumes can be scaled up or down.
7. The original AMB formulation specifies the use of PVP-10 [2]. In an assay, we compared the effect of using PVP-40 (*M*, 40,000 g/mol) instead of PVP-10 (*M*, 10,000 g/mol) for AMB preparation in *E. amylovora* total cell number quantification by dPCR. We observed a significant reduction of total cell counts by processing samples in AMB prepared with PVP-40. Hence, we suggest not substituting PVP-10 with PVPs of other molecular weights to prepare AMB.

8. Filter sterilization avoids incorporating contaminant microorganisms into the samples and increases the buffer's life span. Freshly prepared AMB is translucent with a pale-yellow color and weak smell of rotten eggs due to the sulfur present in glutathione. Turbidity, dark yellow color, and a strong smell like rotten eggs indicate AMB contamination/deterioration. Do not use contaminated AMB.
9. When prepared and used under aseptic conditions, AMB stored in the fridge is usable for around two weeks. We recommend the preparation of fresh AMB 24–72 h before use.
10. To reduce sample oxidation, 0.1X AMB must be cold when mixed with samples. We recommend preparing it in advance and store it in the refrigerator at 4 °C. To prepare 1 L sterile 0.1X AMB, in a sterile bottle mix 800 mL sterile distilled water, 100 mL of 10X PBS pH 7.4, and 100 mL of full-strength AMB. These volumes can be scaled up or down for analyzing a higher or lower number of samples, respectively. If sterile distilled water is not available, the buffer can be prepared with non-sterile water followed by filter sterilization using a 0.22-mm pore size filter.
11. Based on our previous work [14], PMAxx Dye can be substituted by PMA for the analysis of natural fire blight cankers. However, the use of PMA Enhancer or another bacterial cell membrane disruptor in low concentrations is necessary for a complete exclusion of dead-cell signals during dPCR analysis. We have successfully used PMAxx Dye combined with PMA Enhancer for fire blight canker analysis in apple, pear, and Asian pear. The analysis of cankers in other host species, or of other host tissues (shoots, leaves, flowers) might require additional optimization.
12. Because the protocol for PMAxx Dye treatment involves exposure to light, centrifugation, and removal of the supernatant, we recommend the use of 2-mL microcentrifuge tubes allowing the light to reach the samples. We used polypropylene 2-mL tubes of the brand NEST Scientific (Rahway, NJ, Ref. CGN-4310-020). We prefer to use 2-mL tubes instead of 1.5-mL tubes because of the rounded bottom, which makes it easier to decant the supernatant after centrifugation.
13. Handle this product using standard personal protective equipment and diligently follow the waste disposal regulations of your institution for this reagent.
14. Knowing the efficiency of the used extraction method is necessary to calculate the final bacterial cell concentration in your samples. Ideally, the DNA extraction method should be 100% efficient, meaning that the DNA of every single *E. amylovora* cell in the sample will be isolated. Because of the nature of plant

samples that contain lignified tissues and other compounds that interfere with column functions, the extraction efficiencies are rarely 100%, regardless of the extraction kit employed. In our experience, using the column-based DNeasy Plant Mini Kit (Qiagen) with canker samples, depending on the analyzed plant material (different cultivars of apple, pear, and Asian pear), the efficiency of DNA extractions can range between 18% to 75%. This variable efficiency can severely affect result interpretation, so DNA extraction efficiency must be included in calculations to determine the samples' final *E. amylovora* cell concentrations.

15. Samples must be cryopreserved as soon as possible after collection. Ideally, keep them in a cooler with ice until flash-frozen and stored at −80 °C.
16. Asymptomatic trees may still be infected with the pathogen. To ensure the absence of *E. amylovora* cells in negative control samples, we recommend performing classic isolations on general and selective media for *E. amylovora*, enrichments in liquid media, and molecular analysis of the samples by qPCR or regular PCR, as described elsewhere [2, 4].
17. Follow safety practices to handle liquid nitrogen. To avoid cross-contamination of samples, dip samples into the liquid nitrogen while inside individually labeled and closed bags. Because of the considerable sample sizes, we recommend first carefully filling the container with liquid nitrogen and then slowly immersing samples one by one to prevent splashing and spilling of liquid nitrogen outside the container.
18. Based on our results, there are considerable *E. amylovora* populations in the asymptomatic bark areas surrounding the visible margin of the canker. Therefore, we recommend processing a ring area surrounding the canker margin together with the canker itself. We standardized the analysis of a 4-mm ring area around the canker margin to make sure we were able to process cankers with different sizes, degrees of branch girdling, etc., similarly. To obtain more reproducible results, we strongly recommend processing samples consistently, using the same sampling criteria for all the analyzed cankers.
19. This ratio is recommended by the EPPO Standards PM 7/20 (2) [2]. Using the same proportion of sample/maceration buffer in every sample makes the analysis more reliable and reproducible because the concentration of plant tissue and the compounds released during plant processing are similar in all the samples. Hence, oxidation rates and the potential effect of secondary metabolites on *E. amylovora* cells become more homogeneous among samples.

20. Cutting cankers in small pieces helps with sample homogenization. To avoid breaking the bag while processing big pieces of lignified tissue, we recommend repeatedly hammering their edges and progressively moving to the sample center. When processed, the content in the bag is liquid, usually bright green, although depending on the sample, colors may vary from almost transparent to orange or dark brown. It may be necessary to process large samples (weighing more than 5 g) in multiple separate bags, combining their content afterward.
21. We usually process one sample, and while it is incubating on ice, we start homogenizing the next one. Depending on the canker sizes and how difficult they are to break into small pieces, one can process between 4 and 6 samples and proceed to PMAxx Dye treatments with the same group of samples simultaneously. We do not recommend incubating processed samples on ice for more than 30 min.
22. Because we usually work with high sample numbers, we have standardized the use of 0.4 mL sample macerate to analyze as many samples as possible per vial of PMAxx Dye without compromising DNA extractions. However, the volumes can be scaled up or down depending on your needs.
23. These conditions induce the covalent binding of PMAxx Dye to dead-cell DNA, which makes the treated DNA unsuitable for PCR amplification. The described treatment conditions in this protocol have been successfully used to analyze fire blight canker samples of different sizes, collected in different periods of the year from various cultivars of diverse host species (apple, pear, Asian pear). PMAxx Dye concentration, staining period, incubation temperature, and photo-activation times might need adjustments to analyze other types of samples and if buffer other than 0.1X AMB is used to homogenize plant material, the ratio sample/buffer is not 1/50 or the pathogen studied is not *E. amylovora*.
24. We start the protocol directly, without using a tissue lyser, by adding 400 μL buffer AP1 and 4 μL RNase A stock solution to the pellet of the sample already treated with PMAxx Dye. The next steps are performed according to the manufacturer's instructions, with all the centrifugation steps at 15,800 × *g* and elution of DNA samples in 200 μL EB buffer.
25. The recommended final concentration of target DNA in the final reaction mixture is between 200 and 2,000 copies/μL. Under our conditions, sample dilution is rarely required. However, if working with different pathosystems or with other sample processing protocols, a dilution of sample DNA might be necessary.

26. Both, loading the disposable blade and applying the Immersion Fluid to the chip surface are very sensitive steps and require some practice, precision, and manual dexterity. The presence of air bubbles in the blade compartment may lead to an improper sample distribution on the chip surface, with undesired effects on target DNA quantification. Applying the Immersion Fluid too fast, with too much pressure, or touching the chip surface with the syringe tip may have similar consequences. Additionally, undesired drops of Immersion Fluid on the chip edges may lead to improper sealing of the chips. It is essential to remove every drop of Immersion Fluid with a wipe moistened with isopropanol. We strongly recommend performing several chip loading training sessions before working with the experimental samples.
27. Drops of Immersion Fluid on the chip edges may hamper the sealing of the chip lid, leading to the leak of Immersion Fluid and sample evaporation during thermal cycling.

Acknowledgments

This material is based upon work supported by the National Institute of Food and Agriculture, U.S. Department of Agriculture, Hatch/Multistate Research Program projects NYG-625835 and NYG-625857 under 1009897 and 1014444, the NY State Farm Viability Institute grant number 81927/A001-FVI 17 006, and the NY State Specialty Crop Block Grant Program grant number SCG 82535/A001-SCG 17 005 to SGA, and by the Ministry of Education, Science and Technological Development of the Republic of Serbia, Contract No. 451-03-9/2021-14/200010 to KG.

References

1. Pirc M, Ravnikar M, Tomlinson J et al (2009) Improved fireblight diagnostics using quantitative real-time PCR detection of *Erwinia amylovora* chromosomal DNA. Plant Pathol 58: 872–881
2. EPPO (2013) Diagnostics PM 7/20 (2) *Erwinia amylovora*. EPPO Bull 43:21–45
3. Dreo T, Pirc M, Ramšak Ž et al (2014) Optimising droplet digital PCR analysis approaches for detection and quantification of bacteria: a case study of fire blight and potato brown rot. Anal Bioanal Chem 406:6513–6528
4. ISPM 27 (2016) Annex 13: *Erwinia amylovora*. In: Diagnostic protocols for regulated pests. International Plant Protection Convention, Food and Agriculture Organization of the United Nations. https://www.ippc.int/en/publications/83443/. Accessed 25 May 2021
5. Ordax M, Biosca EG, Wimalajeewa SC et al (2009) Survival of *Erwinia amylovora* in mature apple fruit calyces through the viable but nonculturable (VBNC) state. J Appl Microbiol 107:106–116
6. Santander RD, Català-Senent JF, Marco-Noales E et al (2012) In planta recovery of *Erwinia amylovora* viable but nonculturable cells. Trees 26:75–82
7. Wagner M, Amann R, Lemmer H et al (1993) Probing activated sludge with oligonucleotides specific for proteobacteria: inadequacy of culture-dependent methods for describing microbial community structure. Appl Environ Microbiol 59:1520–1525

8. Biosca EG, Flores R, Santander RD et al (2016) Innovative approaches using lichen enriched media to improve isolation and culturability of lichen associated bacteria. PLoS One 11:e0160328
9. Wang B, Liu W, Liu X et al (2017) Comparative analysis of microbial communities during enrichment and isolation of DDT-degrading bacteria by culture-dependent and -independent methods. Sci Total Environ 590–591: 297–303
10. Tewari S, Sharma S (2019) Molecular techniques for diagnosis of bacterial plant pathogens. In: Das S, Dash HR (eds) Microbial diversity in the genomic era, 1st edn. Academic Press, Chennai
11. Maes M, Garbeva P, Crepel C (1996) Identification and sensitive endophytic detection of the fire blight pathogen Erwinia amylovora with 23S ribosomal DNA sequences and the polymerase chain reaction. Plant Pathol 45:1139–1149
12. Evrenosoglu Y, Misirli A, Gülcan R (1999) Determination of phenolic compounds in pear cultivars resistant and susceptible to *Erwinia amylovora*. Acta Hortic 2:327–334
13. Spinelli F, Speakman J-B, Rademacher W et al (2005) Luteoforol, a flavan 4-ol, is induced in pome fruits by prohexadione-calcium and shows phytoalexin-like properties against *Erwinia amylovora* and other plant pathogens. Eur J Plant Pathol 112:133–142
14. Santander RD, Meredith CL, Aćimović SG (2019) Development of a viability digital PCR protocol for the selective detection and quantification of live *Erwinia amylovora* cells in cankers. Sci Rep 9:11530
15. Vogelstein B, Kinzler KW (1999) Digital PCR. Proc Natl Acad Sci U S A 96:9236–9241
16. Morley AA (2014) Digital PCR: a brief history. Biomol Detect Quantif 1:1–2
17. Laig M, Fekete C, Majumdar N (2020) Digital PCR and the QuantStudio^{TM} 3D digital PCR system. In: Biassoni R, Raso A (eds) Quantitative real-time PCR: methods and protocols. Methods in molecular biology, 2nd edn. Springer Nature, Humana, New York
18. Kharadi RR, Schachterle JK, Yuan X et al (2021) Genetic dissection of the *Erwinia amylovora* disease cycle. Annu Rev Phytopathol 59. https://doi.org/10.1146/annurev-phyto-020620-095540
19. Şantander RD, Català-Senent JF, Figàs-Segura À et al (2020) From the roots to the stem: unveiling pear root colonization and infection pathways by *Erwinia amylovora*. FEMS Microbiol Ecol 96:fiaa2010
20. Quan P, Sauzade M, Brouzes E (2018) dPCR: a technology review. Sensors 18:1271

Chapter 15

Innovative Detection of the Quarantine Plant Pathogen *Curtobacterium flaccumfaciens* pv. *flaccumfaciens*, Causal Agent of Bacterial Wilt of Leguminous Plants

Stefania Tegli, Dario Gaudioso, and Damiano Stefanucci

Abstract

Specific, sensitive, and rapid detection of quarantine and regulated plant pathogens is pivotal for the control of the diseases they cause. Here, we describe the PCR-based methods which have been developed for *Curtobacterium flaccumfaciens* pv. *flaccumfaciens* (*Cff*), quarantine plant pathogenic bacterium for EU and causal agent of bacterial wilt of Leguminous plants. These methods include an end-point and a real-time PCR test, and a LAMP assay. Their threshold analytical limits range from 100 to 10 fg of DNA template per reaction and are currently used worldwide for routine testing for *Cff* from laboratory to field scale.

Key words *Curtobacterium flaccumfaciens* pv. *flaccumfaciens*, Bacterial wilt of bean, rep-PCR, End-point PCR, qPCR, Visual LAMP

1 Introduction

Curtobacterium flaccumfaciens pv. *flaccumfaciens* (*Cff*) is a Gram-positive plant pathogenic bacterium, causing bacterial wilt of bean and bacterial tan spot of soybean. Together with cowpea and mungbean, bean and soybean are worldwide the major hosts for *Cff*, who also attacks other leguminous plants [1, 2]. Since its first report in South Dakota in 1922 [3], *Cff* spread and became established in the North and Midwest States of the USA, with recurrent resurgence episodes. Then, *Cff* was later recorded also in Canada, Mexico, Brazil, Southwest Australia, and Iran [2, 4]. As far as the European Union Member States are concerned, since 2019 *Cff* is included into the A1 Quarantine pest list, Annex II A [5, 6], because *Cff* is currently absent on the EU territories, after several sporadic and recent outbreaks in Spain and Germany, followed by the prompt *Cff* eradication [7, 8].

It is able to survive on several alternative host plants, including some *Solanaceae* and *Graminaceae*, as well as on crop debris and

Nicola Luchi (ed.), *Plant Pathology: Methods and Protocols*, Methods in Molecular Biology, vol. 2536,
https://doi.org/10.1007/978-1-0716-2517-0_15,

into the soil for variable periods according to the pedoclimatic conditions [9–11]. *Cff* colonizes its hosts systemically via the vascular system, and thus, infected seeds are also produced. In these seeds, mostly asymptomatic, *Cff* is able to survive for long time, both in controlled and in field conditions. The movement of *Cff*-infected seeds is the main mean for *Cff* spreading at short and long distances, as well as for its introduction into countries where it was previously absent [1, 12]. Up to now, no effective chemical treatments have been found against *Cff*, while the use of resistant cultivars is an important mitigation measure. However, as occurring with other seedborne plant pathogens, the use of certified healthy seeds is a pivotal step for disease control and to successfully prevent epidemic events. Surprisingly, the official testing methods for *Cff* are still based on the visual inspection of seeds, while highly specific and sensitive seed health tests are essential for an efficient phytosanitary surveillance and for the operative implementation of quarantine regulatory procedures.

Here, three PCR-based approaches based on end-point PCR, real-time PCR, and LAMP assay are described for the specific *Cff* detection. These methods are able to detect small amounts of *Cff* DNA pathogen showing a threshold analytical sensitivity ranging from 100 to 10 fg/reaction. These three procedures can be considered a sort of diagnostic toolkit for *Cff* that can differentially be applied accordingly to the instruments and equipment available, as well as to their transferability from lab to field or on-site testing [13, 14].

2 Materials

Solutions have to be prepared using ultrapure/double-distilled water, having a resistivity of 18.2 MΩ·cm at 25 °C. The reagents are molecular biology grade. Solutions and reagents have to be stored properly.

2.1 Bacterial Cultures

1. Luria–Bertani medium (LB) [15]: 10 g tryptone, 5 g yeast extract, 10 g NaCl, up to 1 L with distilled water. Agar (16 g/L) has to be added in case of solid culture medium. Autoclave for 20 min at 120 °C.
2. Nutrient broth yeast extract medium (NBY): 8 g nutrient broth, 2 g yeast extract, 2 g K_2HPO_4, 0.5 g KH_2PO_4, 0.25 g $MgSO_4{\cdot}7H_2O$, 10 g glucose, 1 L distilled water. Agar (16 g/L) has to be added in case of solid culture medium. Bring to volume up to 1 L. Autoclave for 20 min at 120 °C.
3. 80% glycerol solution, in double-distilled water and autoclaved.

2.2 DNA Template Extraction

1. "GRS Genomic DNA kit—BroadRange" (GRISP, Porto, Portugal).
2. Other commercial kits containing enzymatic steps with lysozyme and Proteinase K.
3. PCR grade sterile water.
4. 1 M Tris buffer pH 8.0: dissolve 121.14 g Tris base in 650 mL of double-distilled water, adjust to desired pH using concentrated HCl (i.e., about 49 mL at 25 °C), add double-distilled water up to 1 L final volume, and then autoclave for 20 min at 120 °C or filter sterilize (*see* **Note 1**).
5. 0.5 M EDTA pH 8.0: on a magnetic stirrer, dissolve 186.1 g disodium EDTA·$2H_2O$ in 700 mL of double-distilled water, by simultaneously adding NaOH and monitoring pH variations until pH 8.0 is reached (i.e., about 20 g NaOH pellets) (*see* **Note 2**). Add double-distilled water up to 1 L final volume, dispense into aliquots, and then sterilize by autoclaving at 120 °C for 20 min.
6. Tris–EDTA (TE) buffer 1X: 10 mM Tris–HCl pH 8.0 (i.e., 1 mL of 1 M Tris–Cl pH 8.0), 1 mM EDTA pH 8.0 (i.e., 0.2 mL of 0.5 M EDTA pH 8.0), add distilled water up 1 L, dispense into aliquots and then sterilize by autoclaving at 120 °C for 20 min (*see* **Note 3**).
7. Thermomixer or dry block/water bath incubator.
8. Vortex.

2.3 Agarose Gel Preparation

1. 50X TAE Buffer: 242 g Tris base, 57.1 mL glacial acetic acid, 100 mL 0.5 M EDTA pH 8.0, then add sterile double-distilled water up to 1 L. Dilute with sterile double-distilled water 50X TAE buffer to 1X before its use in gel electrophoresis (*see* **Note 4**).
2. Electrophoretic agarose gel: the concentration of agarose in a gel depends on the sizes of the DNA fragments that have to be resolved, and most used concentrations range from 0.5% to 2% w/v. Weigh the desired amount of agarose, put it into an Erlenmeyer flask, and add the corresponding volume of 1X TAE, which should not more than 1/3 of the Erlenmeyer flask volume. Swirl the flask to mix manually and then melt the mixture by heating into a microwave, until the agarose has dissolved (*see* **Note 5**). Cool the agarose solution down to 60 °C by incubation into a water bath. Then, add ethidium bromide to a final concentration of 0.5 μg/mL, or alternative and safer dyes. Pour the molten agarose into a gel tray, equipped with the most appropriate well comb, and let the agarose gel to set at room temperature. Keep away the gel from light to avoid ethidium bromide degradation.

3. 6X loading dye: 0.25% bromophenol blue, 0.25% xylene cyanol FF, 15% Ficoll 400 in water.
4. Molecular weight DNA Ladder (*see* **Note 6**).
5. Agarose.
6. A power supply.
7. Electrophoretic horizontal chamber, trays, and combs.

2.4 End-Point, Real-Time, and LAMP PCR

1. DreamTaq DNA polymerase (5 U/μL, Thermo Fisher Scientific) and its 10X buffer.
2. 2X SsoFast EvaGreen Supermix (Bio-Rad Laboratories).
3. Primers pairs specific for *Cff*, at 50 μM working solution in PCR grade sterile double-distilled water, unless otherwise specified. For end-point PCR, the primers pair is *CffFOR2/CffREV4* [13] (*CffFOR2*: 5′- GTTATGACTGAACTT CACTCC -3′; *CffREV4*: 5′-GATGTTCCCGGTGTTCAG-3′). For real-time PCR, the primers pair is *Cff*F3/*Cff*B3 [14] (*Cff*F3: 5′-CGTTAGTGAAGGCTGACGAA-3′; *Cff*B3: 5′-TT CCCGGTGTTCAGTTGAC-3′). In LAMP PCR assay, the primers *Cff*F3 and *Cff*B3 are used as outer primers, while the inner primers are *CffFIP* and *CffBIP* [14] (*CffFIP*: 5′-GTTT GCATCCGTACGGGGCGACTAGCACCGACGGAACC-3′ and *CffBIP*: 5′-TTCGGTCCTGCAGTTAGTCAGCGAAT AGTTCGCGGCGTGG-3′).
4. 10 mM dNTPs mix, in sterile deionized water at pH 7.0, DNase and RNase free.
5. Variable volumes pipettes.
6. Sterile disposable RNase/DNase-free tips with aerosol barriers.
7. PCR tubes or strips.
8. Optical tubes or 96-well microplate for real-time PCR.
9. Adhesive seals for PCR plates.
10. Thermocycler.
11. Thermocycler with real-time detection system.

3 Methods

3.1 Bacterial Culture Preparation

1. Bacterial cultures, used for extraction of DNA template for PCRs, are aseptically prepared under a laminar flow hood.
2. *Cff* strains are long term conserved at −80 °C, in a 40% glycerol aqueous sterile solution.
3. When needed, bacteria are streaked on Petri dishes containing LB or NBY agarized medium and let grown at 26 °C for 48–72 h, to obtain single *Cff* colonies.

4. Liquid *Cff* cultures are prepared by suspending two or three single *Cff* colonies, freshly obtained on LB or NBY agar plates, with a sterile loop in 3 mL of LB or NBY liquid medium. Then *Cff* cultures are grown overnight under shaking at 160 rpm, at 26 °C (*see* **Note 7**).

3.2 DNA Template Extraction

Cff is a Gram-positive bacterium, and thus, its bacterial cell wall is difficult to be lysed in order to release DNA molecules. Therefore, DNA yield strongly depends on the extraction protocol used. Here, two different approaches are reported, thermal lysis and DNA extraction by commercial kits.

3.2.1 Thermal Lysis

1. Preheat the incubator at 95 °C (a dry bath incubator, a thermomixer, or even a thermocycler could be used for this purpose).
2. Pour 50 μL of sterile double-distilled water in a 0.5-mL Eppendorf tube.
3. With a sterile loop (or a sterile wooden toothpick), pick up a bacterial single colony to be then transferred into the Eppendorf tube with sterile double-distilled water.
4. Transfer the tube at 95 °C and incubate for 20 min.
5. Immediately put into ice, for at least 5 min.
6. Centrifuge at 14,000 g for 2 min to let the cell debris to pellet. Keep in ice until its utilization. Use DNA immediately or store at −20 °C for few days for further use (*see* **Note 8**).

3.2.2 Commercial Kits

1. Several kits could be used to perform *Cff* genome extraction to obtain purified DNA. The commercial kits suggested to be used contain lysozyme and Proteinase K.
2. DNA is eluted in TE buffer and then properly diluted (i.e., 10 ng/μL concentration).
3. Extracted DNA can be used immediately or stored at −20 °C for a few months.

3.3 End-Point PCR for Cff

1. According to the number of samples to be tested, including at least a positive control (i.e., *Cff* pure genomic DNA as template) and a negative control (i.e., a no DNA template sample), prepare the PCR Master Mix.
2. The final reaction volume for each sample is 25 μL, with 2.5 μL of 10X PCR buffer, 100 μM dNTPs for each deoxynucleotide, 0.5 μM of each primer (*CffFOR2* and *CffREV4*), 1U Taq DNA polymerase, and with PCR grade sterile double-distilled water up to 25 μL.
3. Unless otherwise required, the volume of DNA template for each reaction is 2 μL.

4. The PCR master mix is the mixture of all the PCR reagents, that is prepared as a batch and then divided in aliquots corresponding to the total number of samples, including the controls.
5. The PCR components are added into the PCR master mix, that is prepared on ice, according to the following order: PCR grade water, PCR buffer, dNTPs, PCR primers, and Taq polymerase.
6. Mix thoroughly the PCR master mix, and immediately aliquot 23 μL for each sample into a PCR tube or in a well of a PCR plate.
7. For each tube or well, add a 2 μL solution of DNA template for unknown samples and positive controls, or PCR grade sterile double-distilled water for negative controls.
8. The PCR samples are mixed by a brief centrifugation and then placed into a thermocycler.
9. The *Cff*-specific end-point amplification protocol consists of an initial denaturation at 94 °C for 3 min, followed by 30 cycles of denaturation (at 94 °C for 1 min), primer annealing (at 62 °C for 45 s), and primer extension (at 72 °C for 30 s), and by a final extension at 72 °C for 5 min.
10. In positive samples, a 306-bp amplicon is obtained [13].

3.4 Real-Time PCR for Cff

1. Unknown samples are tested in triplicate, while control samples in duplicate.
2. A standard curve is generated using DNA template serially 10-fold diluted over the range 10 ng to 10 fg, with three replicates for each standard dilution point in each run (*see* **Note 9**) (Fig. 1a).
3. Prepare the Real Time PCR Master Mix. Gently mix 5 μL of SsoFast EvaGreen 2X (Bio-Rad Laboratories) with 0.5 μM for primer *Cff*F3 and primer *Cff*B3 [14] and 2.3 μL of sterile PCR grade double-distilled water (*see* **Note 10**).
4. Load 7.5 μL of Master Mix into a PCR tube or into a well of a 96-well PCR plate.
5. Load 2.5 μL of DNA template solution, to obtain a 10 μL final reaction volume.
6. Make sure there are no bubbles into the PCR mixture, by a brief spin (*see* **Note 11**).
7. Place the samples into a thermocycler with real-time detection system (i.e., CFX96, Bio-Rad Laboratories), to detect the fluorescence emitted by the fluorophore at each amplification cycle.
8. The amplification protocol consists of an initial incubation at 95 °C for 3 min, followed by 40 cycles at 95 °C for 5 s and at 60 °C for 1 min (optics on); melt curve analysis is carried out at

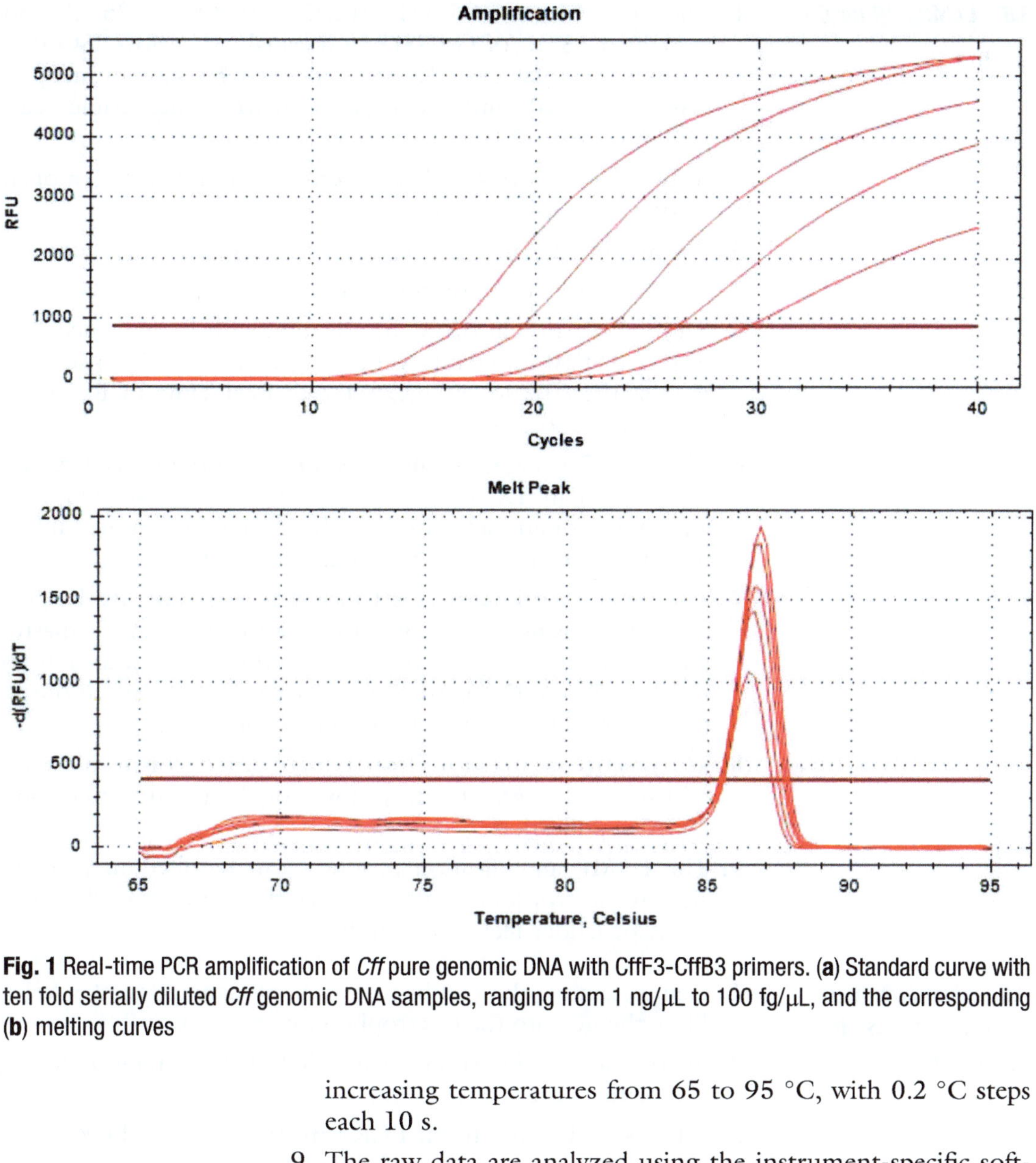

Fig. 1 Real-time PCR amplification of *Cff* pure genomic DNA with CffF3-CffB3 primers. (**a**) Standard curve with ten fold serially diluted *Cff* genomic DNA samples, ranging from 1 ng/μL to 100 fg/μL, and the corresponding (**b**) melting curves

increasing temperatures from 65 to 95 °C, with 0.2 °C steps each 10 s.

9. The raw data are analyzed using the instrument-specific software installed on the real-time PCR apparatus used (i.e., CFX Maestro, Bio-Rad Laboratories).
10. The analysis of melting curves confirms that the amplicons produced belong to a single discrete species and that no contaminating DNA is present, when a single distinct peak is obtained, having the specific melting temperature (Tm) that identifying the DNA target both in the standard curve and in positive controls (Fig. 1b).
11. Unknown samples are quantified by absolute quantification, that is by comparison with the standard curve.

3.5 LAMP PCR for *Cff*

1. The final volume of LAMP reaction mixture is 25 μL and contained 15 μL of GspSSD Isothermal MasterMix ISO-001 (Optigene, Horsham, UK), the *CffF3/CffB3* outer primer pair (200 nM each), and the *CffFIP/CffBIP* inner primer pair (800 nM each) [14].
2. As template, a 5 μL solution of purified genomic DNA or of an unknown sample is used.
3. Positive and negative controls are strongly recommended.
4. Reactions have to be set up on ice.
5. Run the reaction at 63 °C for 30 min, with fluorescence data recorded at 60 s time intervals, using a real-time thermocycler, such as the CFX96™ CFX96 Touch Real Time PCR System (Bio-Rad Laboratories).
6. The LAMP generated amplicons can be also analyzed on 2% agarose gel in 1X TAE (5 μL reaction/well), stained with ethidium bromide, and visualized by UV light using the Molecular Imager Gel Doc™ XR System (Bio-Rad).
7. The visual colorimetric LAMP assay for *Cff* can also be performed by using the WarmStart Colorimetric LAMP master mix (New England BioLabs, Ipswich, MA, USA), according to the manufacturer's instructions. The reaction is carried out at 63 °C for 30 min in a 25 μL final reaction.
8. The positive samples turn from pink to yellow, for the presence of a redox dye and the drop down of the reaction pH upon DNA amplification.
9. The LAMP amplification products can be also analyzed by agarose electrophoresis gel, to check the presence of the characteristic ladder-like DNA bands.

3.6 Electrophoresis of PCR Products on Agarose Gel

1. Gently remove the comb, to avoid any damage to the wells. Place the gel into the electrophoretic apparatus.
2. Fill the electrophoresis tank with 1X TAE as running buffer, to completely cover the gel.
3. Add 6X loading dye to the PCR samples, unless the PCR buffer used already contains a loading dye. Loading dye provides a visible tracking of DNA migration on the gel and increases the density of PCR samples to allow their setting to the bottom of gel well.
4. Carefully load the samples (i.e., 5 μL/well or accordingly to the well volume) and a DNA molecular weight ladder into the wells. Generally, the DNA size marker is placed on the first and the last lanes of the gel.
5. Set the power supply to the desired voltage (i.e., 1–5 V/cm between electrodes).

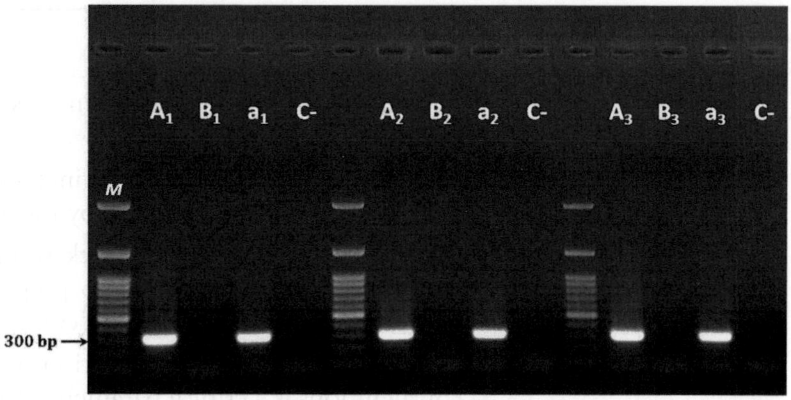

Fig. 2 Image under UV light of agarose gel electrophoresis of the PCR products obtained with *Cff*-specific primers CffFOR2 and CffREV4. DNA templates were obtained as following: A = by thermal lysis of a Cff single bacterial colony; B = by thermal lysis of a no-*Cff* single bacterial colony; a = by genome DNA extraction and purification from *Cff* bacterial cultures. C- = negative no-template control. 1, 2, and 3 represent three different *Cff* and no-*Cff* strains. M = Molecular DNA size marker (ExcelBand™ 100 bp+3K DNA Ladder, SMOBIO Techn. Inc.)

6. Place the on the electrophoretic apparatus, connect it to the power supply, and turn on the power (*see* **Note 12**).
7. Run the gel at 80–100 V and turn off the power when the dye frontline has approximately run down about 2/3 of the gel and the bands of the DNA markers are separated (*see* **Note 13**). The time required for the run depends on the gel concentration as well as on the voltage applied.
8. When the electrophoresis run is completed, turn off the power supply and remove the lid from the gel tank.
9. Carefully remove the agarose gel from the tank together with its tray, and absorb the running buffer in excess by using a small lab paper towel.
10. The agarose gel is then removed from the tray, and transfer on a UV transilluminator and exposed to UV light.
11. Under UV, DNA fragments and amplicons appear as fluorescent bands on a dark background (Fig. 2).
12. Pictures are taken of DNA agarose gel under UV light. Gel images are recorded and analyzed by a digital imaging detection system developed for electrophoretic applications.
13. At last, agarose gel and running buffer are discarded as hazardous waste.

4 Notes

1. Tris is a buffering agent and in Tris–HCl the Tris base pH is adjusted by adding HCl.
2. The EDTA disodium salt goes into solution when until the pH of the solution is around 8.0 by the addition of NaOH.
3. Use sterile 0.5 M EDTA stock solution, and sterile beakers, cylinders, and bottles used to prepare 50× TAE solution in order to avoid the presence of contaminant DNA.
4. For some downstream applications where the concentration of divalent ions is a critical parameter, EDTA concentration in TE is 0.1 mM instead of 1 mM, to avoid EDTA chelating Mg^{2+} as well as other divalent cations.
5. At 30 s intervals, remove the flask from the microwave to mix well. Do not bring the solution to boil to avoid evaporation of 1X TAE buffer and thus improper gel polymerization.
6. The molecular DNA Ladder to be used is selected according to the size of the target amplicons. A DNA Ladder ranging from 100 bp to 3 kb is generally recommended.
7. To promote *Cff* growth in liquid culture, increase shaking up to 200 rpm.
8. It is recommended to centrifuge thermal lysed samples before use in PCR, to remove any insoluble PCR inhibitors. It is also advisable to dilute samples from thermal lysis to dilute possible residual inhibitors.
9. To produce the standard curve, analyze several serial 10-fold dilutions of the target DNA, over 5–7 log. The coefficient of correlation obtained for the standard curve (R^2) has to be in the range in the range 0.98–1, while the reaction efficiency has to be between 90% and 105%.
10. The length of the target amplicon in real-time PCR is generally recommended to be not greater than 250 bp.
11. Air bubbles present in the reaction volume could partially inhibit amplification reaction.
12. Verify that power supply and electrophoresis apparatus are working properly, by the air bubbling occurring to the cathode when the electric circuit is on.
13. DNA is negatively charged and amplicons run toward the anode, that is the positive electrode. Generally, in the electrophoretic tanks the negative electrode (i.e., cathode) is black, while the anode is red.

Acknowledgments

This work was supported by Ministero della Difesa, SFINGE project, coordinated by S.T.

References

1. Jeger M, Bragard C, Caffier D et al (2018) Pest categorisation of *Curtobacterium flaccumfaciens* pv. *flaccumfaciens*. EFSA J 16:e05299
2. Osdaghi E, Young AJ, Harveson RM (2020) Bacterial wilt of dry beans caused by *Curtobacterium flaccumfaciens* pv. *flaccumfaciens*: a new threat from an old enemy. Mol Plant Pathol 21: 605–621
3. Hedges F (1922) A bacterial wilt of beans caused by *Bacterium flaccumfaciens* nov. sp. Science 55:433–434
4. Tegli S, Cerboneschi M, Vidaver AK (2017) Detection of *Curtobacterium flaccumfaciens* pv. *flaccumfaciens* in bean seeds and in seeds of other Leguminosae crops. In: Fatmi M, Walcott RR, Schaad NW (eds) Detection of plant-pathogenic bacteria in seed and other planting material, 2nd edn. APS Press, St Paul, pp 77–83
5. EPPO (2011) *Curtobacterium flaccumfaciens* pv. *flaccumfaciens*. Bull. EPPO OEPP 41: 320–328
6. The European Commission (2019) European Commission Implementing Regulation (EU) 2019/2072. Off J Eur Union. https://eur-lex.europa.eu/eli/reg_impl/2019/2072/oj
7. González AJ, Tello JC, Rodicio MR (2005) Bacterial wilt of beans (*Phaseolus vulgaris*) caused by *Curtobacterium flaccumfaciens* in Southeastern Spain. Plant Dis 89:1361
8. Sammer UF, Reither K (2012) *Curtobacterium flaccumfaciens* pv. *flaccumfaciens* on soybean in Germany – a threat for farming. J Phytopathol 160:314–316
9. Harveson RM, Schwartz HF, Urrea CA et al (2015) Bacterial wilt of dry-edible beans in the central high plains of the US: past, present, and future. Plant Dis 99:1665–1677
10. Gonçalves RM, Schipanski CA, Koguishi L et al (2017) Alternative hosts of *Curtobacterium flaccumfaciens* pv. *flaccumfaciens*, causal agent of bean bacterial wilt. Eur J Plant Pathol 148: 357–365
11. Osdaghi E, Taghavi SM, Hamzehzarghani H et al (2018) Epiphytic *Curtobacterium flaccumfaciens* strains isolated from symptomless solanaceous vegetables are pathogenic on leguminous but not on solanaceous plants. Plant Pathol 67:388–398
12. Camara RC, Vigo SC, Maringoni AC (2009) Plant to seed transmission of *Curtobacterium flaccumfaciens* pv. *flaccumfaciens* in a dry bean cultivar. J Plant Pathol 91:549–554
13. Tegli S, Sereni A, Surico G (2002) PCR-based assay for the detection of *Curtobacterium flaccumfaciens* pv. *flaccumfaciens* in bean seeds. Lett Appl Microbiol 35:331–337
14. Tegli S, Biancalani C, Ignatov AN et al (2020) A powerful LAMP weapon against the threat of the quarantine plant pathogen *Curtobacterium flaccumfaciens* pv. *flaccumfaciens*. Microorganisms 8:1705
15. Bertani G (1951) Studies on lysogenesis. I. The mode of phage liberation by lysogenic *Escherichia coli*. J Bacteriol 62:293–300

Chapter 16

Specific Detection of *Pseudomonas savastanoi* Pathovars: From End-Point PCR to Real-Time PCR and HRMA

Stefania Tegli and Chiara Pastacaldi

Abstract

Pseudomonas savastanoi is a phytopathogenic bacterium causing severe disease on olive, oleander, ash, and other *Oleaceae*. Three main pathovars belong to this species: *P. savastanoi* pv. *savastanoi*, pv. *nerii*, and pv. *fraxini*. Detection methods are mostly based on the visual inspection of the typical symptoms (i.e., knots and galls). However, this bacterium can survive on the host plant also as an epiphyte without giving any symptom. To avoid the spread of *P. savastanoi* to areas where it is absent, it is necessary to develop efficient and sensitive detection methods. Here, we reported three different PCR-based techniques, able to discriminate the three *P. savastanoi* pathovars attacking woody plants.

Key words *Pseudomonas savastanoi*, Detection, End-point PCR, Real-time PCR, HRM analysis

1 Introduction

Pseudomonas savastanoi is a phytopathogenic Gram negative bacterium that belongs to the *Pseudomonas syringae* complex. It causes diseases on woody plants, mainly belonging to the genus of *Oleaceae*, and to *Nerium oleander* [1, 2]. As demonstrated by several biochemical, physiological, and DNA–DNA hybridization analyses, three main different pathovars attacking woody plants have been included in this species: *P. savastanoi* pv. *savastanoi* (*Psv*), *P. savastanoi* pv. *nerii* (*Psn*), and *P. savastanoi* pv. *fraxini* (*Psf*), isolated from olive, oleander, and ash (*Fraxinus* spp.) plants, respectively [2, 3]. The typical symptoms caused by *P. savastanoi* are knots or galls on branches and stems of the host plant, and rarely on leaves and fruits [2]. Knot formation is induced by the production of indole-3-acetic acid and cytokinins by the pathogen [4–6]. *P. savastanoi* is not able to survive in soil, where its population decreases rapidly. However, it can persist on the host plant as an epiphyte, without inducing the formation of knots [2, 3]. In its epiphytic state, *P. savastanoi* is a potential source of inoculum for

Nicola Luchi (ed.), *Plant Pathology: Methods and Protocols*, Methods in Molecular Biology, vol. 2536,
https://doi.org/10.1007/978-1-0716-2517-0_16,

the development of new knots and it can be spread to healthy plants or to new areas by environmental factors or by managing procedures [7, 8]. Nowadays, strategies to control and manage this pathogen are limited and they are based mainly on pruning and on the use of copper-based compounds. However, the increasing development of *P. savastanoi* copper-resistant strains is making this approach less effective [3, 9]. The monitoring of *P. savastanoi* epiphytic populations is essential also to limit the spread of *P. savastanoi* on long distances and through asymptomatic materials. Traditional microbiological detection assays for *P. savastanoi* are based on the visual inspection for the typical knots, followed by bacterial isolation on semiselective media [10, 11], and also by pathogenicity and biochemical tests. However, these methods have low-sensitivity and specificity and are time-consuming. Thus, several PCR-based detection methods for *P. savastanoi* have been developed and whose primers anneal on the *iaaL* gene [8, 12–14]. Unfortunately, most pathovars of the *P. syringae* complex own this gene [15] and therefore these PCR-based assays are not able to properly discriminate the three pathovars *Psv*, *Psn*, and *Psf*. Conversely, it is essential to discriminate isolates belonging to these *P. savastanoi* pathovars: little is still known about their natural distribution as well as on the possibility of cross-infections in nature, although *Psn* was demonstrated to be able to infect olive and ash when artificially inoculated [16].

Here, we reported three different procedures developed for *P. savastanoi* detection, which are able to differentiate *Psv*, *Psn*, and *Psf* and that are based on end-point PCR, real-time PCR, and high-resolution melting analysis (HRMA) [9, 17]. Primers and probes for end-point and real-time PCR are pathovar-specific, while primers for the HRMA-based assay have been designed to detect single nucleotide polymorphisms discriminating the type three secretion system (TTSS) sequences of these three *P. savastanoi* pathovars.

2 Materials

Ultrapure/double-distilled water, having a resistivity of 18.2 MΩ·cm at 25 °C, has to be used to prepare solutions for molecular biology procedures. The reagents are molecular biology grade. Solutions and reagents have to be stored properly.

The following instruments and sterile disposable materials are needed for both microbiological and molecular biology procedures here reported, while instruments used for specific purposes are listed into the corresponding section.

1. Vortex.
2. Centrifuges and microcentrifuges.

3. Thermomixer or dry block/water bath incubator.
4. Orbital shaker.
5. Incubators.
6. Autoclave.
7. Micropipettes (2.5, 10, 20, 100, 200, and 1,000 μL).
8. Sterile inoculating loops.
9. Glass or sterile plastic spreaders.
10. Sterile disposable pipettes (1, 2, 5, 10, and 25 mL).
11. Sterile disposable micropipette tips (2, 10, 20, 100, 200, and 1,000 μL).
12. Sterile Eppendorf tubes (1.5 and 2 mL).
13. Sterile tubes (15 and 50 mL).

2.1 Bacterial Growth Media

1. King's B medium: 10 g tryptone, 10 g proteose peptone n°3, 1.5 g K_2HPO_4, 10 g glycerol, 1.5 g $MgSO_4{\cdot}7H_2O$, 16 g BactoAgar (only in case of solid medium), distilled water up to 1 L. Autoclave for 20 min at 121 °C.
2. 50 mg/mL nitrofurantoin: dissolve the antibiotic powder in dimethyl formamide (DMF). Store it at room temperature. Use the antibiotic with a final concentration of 50 μg/mL.
3. 80% glycerol solution, in double-distilled water and autoclaved.

2.2 DNA Extraction

1. "GRS Genomic DNA kit—BroadRange" (GRISP, Porto, Portugal).
2. Tris–EDTA (TE) Buffer: 1 mM EDTA pH 8.0, 10 mM Tris–HCl pH 8.0, distilled water up to final volume. Autoclave the solution at 121 °C for 20 min. Prepare in aliquots of 50–100 μL each and store at room temperature.
3. NanoDrop ND-1000 UV-Vis Spectrophotometer (Thermo Fisher Scientific).

2.3 Gel Electrophoresis Reagents

1. 50X TAE Buffer: 242 g Tris base, 57.1 mL glacial acetic acid, 100 mL 0.5 M EDTA pH 8.0. Add distilled water up to 1 L. Dilute the 50X TAE buffer to 1X before using it in gel electrophoresis (*see* **Note 1**).
2. 0.8–1% Agarose gel: add 0.8–1 g of agarose to 100 mL of 1X TAE. Heat the solution in a microwave until the agarose is completely dissolved. Add ethidium bromide to a final concentration of 0.5 μg/mL or other safer stain at the proper concentration. Pour the solution in a tray with the most suitable comb to allow its solidification (*see* **Note 2**).
3. 6X loading dye: 0.25% Bromophenol blue, 0.25% xylene cyanol FF, 15% Ficoll 400 in water.

4. Molecular weight DNA Ladder (*see* **Note 3**).
5. A voltage source.

2.4 End-Point PCR Materials

1. DreamTaq DNA polymerase (5 U/μL, Thermo Fisher Scientific) with its specific buffer (10X Green DreamTaq Buffer).
2. dNTPs mix, 10 mM for each deoxynucleotide.
3. Specific primers pair *PsvF/PsvR* for *Psv* (5′-GGCGATGTTCT CAGCGGATTTG -3′/5′- GATCAAGTGTCCAAGGAAGT GAAGG -3′), *PsnF/PsnR* for *Psn* (5′-ACCCCTCATTGTAACGGATG -3′/5′- TCCCCGGAATT CAACACTTA -3′) and *PsfF/PsfR* for *Psf* (5-′-CGCCTGCTGTACTCCTCGG-3′/5-′-TCGACCTGTCTAAGGCCC-3′).
4. PCR grade ultrapure water, DNase/RNase free.
5. PCR strips or 0.2-mL tubes.
6. Thermocycler.

2.5 Real-Time PCR Materials

1. iQ SYBR® Green Supermix (Bio-Rad, Hercules, CA, U.S.A) and iQ® Multiplex Powermix (Bio-Rad).
2. Specific real-time primers and probes (*see* **Note 4**). The primer pair and probe combinations *Psv*RT-F/*Psv*RT-R and *Psv*RT-P (5′-CGGATTTGGTTTGCGGGGTA-3′/5-′-AATGGGGTGACACTAAAATTGTGAA-3′, and 5-′-HEX-CTCGTGCGATCTAAACAGCCGTAGC-BHQ-1-3′), *Psn*RT-F/*Psn*RT-R and *Psn*RT-P (5-′-GCTCATTCGCTTGTTATCACTTCA-3′/5-′-TCCCCGGAATTCAACACTTA-3′, and 5′-FAM-TACGCCCGACGCCCGAGCCA-BHQ-1-3′), and *Psf*RT-F/*Psf*RT-R and *Psf*RT-P (5-′-CAGCTCATCCATTAATAGGGCAAG-3′/5-′-GGGCAGTGTCAGGGGATG-3′, and 5′-Texas Red-CTTGTACCGAAGCGTGCCGTCTGC-BHQ-2-3′), are used for *Psv*, *Psn*, and *Psf*, respectively.
3. PCR grade ultrapure water, DNase/RNase free.
4. 96-well PCR plate, low profile.
5. Adhesive seals for PCR plates.
6. Thermocycler with real-time detection system.

2.6 Multiplex-HRM Assay Materials

1. 2X SsoFast EvaGreen Supermix (Bio-Rad).
2. Primers pairs specific for the SNP markers present in the TTSS sequences and discriminating the *P. savastanoi* pathovars *Psv*, *Psn*, and *Psf* (*hrcC2_for/hrcC2_rev*, 5-′-CGCCAGCCACAGATATCG-3′/ 5-′-AACAGCACCAGCACAACT-3′; *hrpL1_for/hrpL1_rev*,

5′-GTCAACTGACGGCTGATC-3′/ 5-′-GCCTCCAGAAATACGCATT-3′; *hrcR_for/hrcR_rev*, 5-′-CACCTGCTGAACGCCAATT-3′/5-′-TGTTTCTCGGCTCGCTGTC-3′).

3. PCR grade ultrapure water, DNase/RNase free.
4. 96-well PCR plate, low profile.
5. Adhesive seals for PCR plates.
6. Thermocycler with real-time PCR detection system.

3 Methods

3.1 Bacterial Cultures and Growth Conditions

1. Prepare bacterial cultures under a sterile flow hood (i.e., laminar or vertical), to avoid airborne contaminations.
2. Streak *Psv*, *Psn*, and *Psf* on KB agar plate containing 50 μg/mL nitrofurantoin, to obtain single colonies. Incubate plates at 26 °C for 48 h (*see* **Note 5**).
3. Inoculate a single colony of *Psv*, *Psn*, or *Psf* into KB liquid medium (e.g., 5 mL in a 15-mL sterile tube).
4. Incubate at 26 °C overnight with shaking at 200 rpm.

3.2 Bacterial Genomic DNA Extraction

1. Prepare pre-cultures of *Psv*, *Psn*, and *Psf* under a sterile flow hood cabinet: inoculate a single colony into 3 mL of KB liquid medium. Incubate overnight at 26 °C, with shaking.
2. Extract genomic DNA using the "GRS Genomic DNA kit—BroadRange" (GRISP), following the manufacturer's protocol for Gram-negative bacteria. To elute the genomic DNA, use the buffer provided by the kit or TE buffer (*see* **Note 6**).
3. Check genomic DNA concentration and purity with NanoDrop spectrophotometer (*see* **Note 7**) or through gel electrophoresis (*see* **Note 8**).

3.3 End-Point PCR

Before preparing any PCR reaction (i.e., end-point, real-time, and multiplex-HRM assay), thaw reaction components on ice. Gently mix each one and spin down to collect all the content at the bottom of the tubes. Always set up each reaction on ice.

1. Prepare the end-point PCR reaction, in a 25 μL volume per sample and variable amounts of DNA template (*see* **Note 9**), with the following PCR reagents at their final concentration: 1X Green DreamTaq Buffer, 0.2 mM dNTPs, 0.5 μM each primer, 1U/reaction DreamTaq polymerase (*see* **Note 10**).
2. Load all the reagents into PCR strips or tubes, maintained on ice during sample preparation.
3. Briefly vortex and spin all the samples.

4. Set a thermocycler with the following PCR protocol: first denaturation step at 95 °C for 3 min, followed by 40 cycles of 95 °C for 30 s, 60 °C for 30 s, and 72 °C for 1 min; perform the final extension step at 72 °C for 10 min.
5. Insert the samples in the thermocycler and start the run.
6. In the meantime, prepare the agarose gel. After solidification, place the gel in an electrophoresis tank and completely cover it with 1X TAE buffer.
7. At the end of the run, load each sample into a well of the agarose gel to visualize PCR products (*see* **Note 11**). Load 5 μL of the molecular weight DNA ladder into the lanes on the left and right borders of the gel.
8. Run the electrophoresis for 40–60 min at 90–100 V.
9. Analyze the gel through a UV-Transilluminator (*see* **Note 12**).

3.4 Real-Time PCR

All the protocols here described are optimized for CFX96 thermocyclers (Bio-Rad), and it will be necessary to adjust them when using other platforms.

3.4.1 SYBR® Green Real-Time PCR

1. Set the SYBR® Green real-time PCR in a 25 μL volume per sample and variable amounts of DNA template (*see* **Note 9**), with the following reagents at their final concentrations: 1X iQ SYBR® Green Supermix (Bio-Rad), 0.3 μM for each primer (*see* **Note 10**).
2. Maintain the 96-well PCR plate on ice and load the reaction mixture.
3. Close the plate with the adhesive seal and spin it briefly.
4. Place the plate in the thermocycler and run the real-time PCR with the following protocol: initial denaturation at 95 °C for 3 min, followed by 40 cycles of 95 °C for 20 s and 60 °C for 25 s.
5. Raw data from SYBR® real-time PCR experiments are analyzed with the "CFX Manager" software (Bio-Rad) or any other comparable software (*see* **Notes 13** and **14**).

3.4.2 TaqMan® Real-Time PCR

1. Set the TaqMan® Real-Time PCR in a 25 μL volume per sample and variable amounts of DNA template (*see* **Note 12**), with the following reagents at their final concentrations: 1X iQ® Multiplex Powermix (Bio-Rad), 0.3 μM for each primer, 0.2 μM for the TaqMan® probe (*see* **Note 9**).
2. Load the reaction mix in each well of the 96-well PCR plate maintaining it on ice.
3. Cover the plate with the adhesive seal and spin briefly.

4. Set the thermocycler with the following protocol: first denaturation step at 95 °C for 3 min, followed by 40 cycles of 95 °C for 10 s, 60 °C for 30 s.
5. Raw data from TaqMan® Real-Time PCR experiments are analyzed with the "CFX Manager" software (Bio-Rad) or any other comparable software (*see* **Notes 13** and **14**).

3.5 Multiplex-HRM Assay

1. Set the Multiplex-HRM assay in a 10 μL volume per sample and DNA template (1 ng/μL), with the following reagents at their final concentrations: 1X SsoFast EvaGreen Supermix (Bio-Rad), and 0.5 μM for each primer (*see* **Notes 10** and **15**).
2. Load the mixture in each well of a 96-well PCR plate.
3. Cover the plate with the adhesive seal and spin briefly.
4. Set the thermocycler with the protocol reported in Fig. 1 and insert the plate inside the machine.
5. Start the run.
6. Analyze the raw data using both the "CFX Manager" and "Precision Melt Analysis" software (Bio-Rad).

4 Notes

1. To avoid DNA contamination, prepare the 0.5 M EDTA stock solution and autoclave it for 20 min at 121 °C. It is also better to autoclave beaker, cylinder, and bottle used to prepare the 50X TAE solution.
2. During the melting of the agarose with the microwave, do not boil the solution to avoid the evaporation of the 1X TAE buffer and thus an improper gel polymerization. Moreover, it is advisable to keep the gel away from light during solidification to prevent ethidium bromide degradation.
3. Choose the proper molecular weight DNA Ladder in relation to the size of the target amplicon. We recommend the use of DNA Ladder with a size ranging between 100 bp and 3 kb.
4. HEX, FAM, and Texas Red are the fluorophores used, while BHQ-1 and BHQ-2 are the quencher molecules used for TaqMan® probes.
5. For long-term storage of *Psn*, *Psv*, and *Psf*, use bacterial glycerol stocks. Prepare a culture inoculating a single colony of the target bacterium in 10 mL of KB medium without antibiotic. Incubate overnight at 26 °C, with shacking. The day after centrifuge the culture at 8,000 × *g* for 5–10 min, at 4 °C. Discard the supernatant and resuspend the pellet with 500 μL of fresh KB medium. Add 500 μL of 80% glycerol (previously

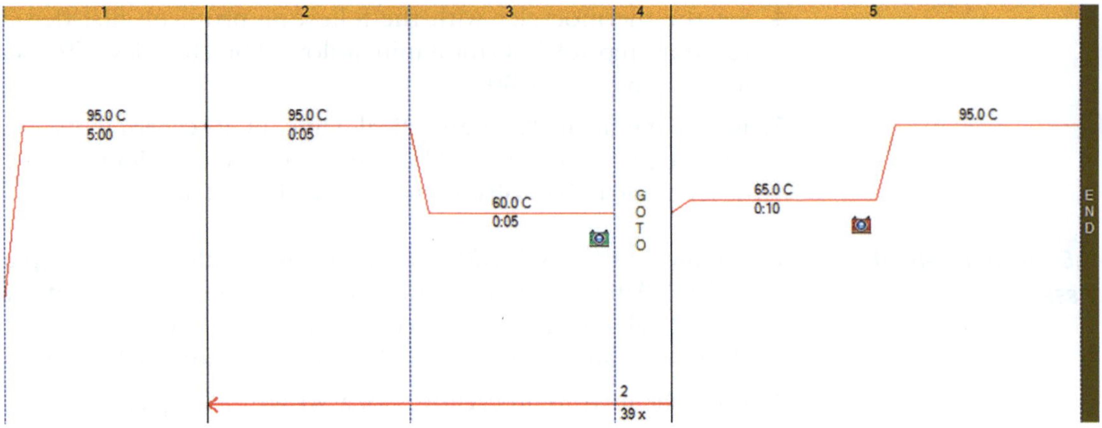

Fig. 1 Cycling parameters used for *P. savastanoi* multiplex HRMA assay

sterilized), in order to maintain a 1:1 ratio. Flash-frozen in liquid nitrogen and store at −80 °C.

6. To elute genomic DNA, use TE-low buffer or MilliQ H_2O to minimize the effects of salts in PCRs. The TE-low buffer is composed by 0.1 mM EDTA pH 8.0 and 10 mM Tris–HCl pH 8.0. However, in H_2O, the DNA is not stable for long time. For long-term storage, it is better to store DNA in TE buffer at −20 °C. Avoid thawing and freezing several times.
7. When using NanoDrop, pay attention to ratios between absorbance at 280 nm (A_{280}) and at 260 nm (A_{260}), and between the absorbance at 230 nm (A_{230}) and at 260 nm. The A_{280}/A_{260} ratio is used to determine protein contamination and it should have a value between 1.8 and 2.0. On the other hand, the A_{230}/A_{260} gives information about contamination with salts and aromatic compounds and its value should be around 2.2.
8. Remember to add 6X loading dye before loading the genomic DNA on the agarose gel. This dye has a double role: to increase the density of DNA samples in order to properly load them into gel wells and to allow the tracking of DNA fragments during the electrophoretic run. It is usually preferable to prepare a sample composed by 2 μL of genomic DNA, 3 μL of TE buffer, and 1 μL of 6X loading dye.
9. The amount of DNA that should be used as template is variable. However, it is known that these end-point and real-time detection methods can be successfully performed with DNA amounts ranging from 50 ng to 0.5 pg [9].
10. It is usually preferable to prepare a PCR Master Mix with all reagents except the target DNA. The volume of each component should be multiplied for the number of samples. Moreover, it is important to add a negative control (i.e., a sample without DNA template) to evaluate contaminations by foreign

DNA. It is also strongly recommended to add a positive control.

11. Add 6X loading dye to each sample, in case a loading dye is not already part of the PCR buffer.
12. In the *P. savastanoi*-specific end-point PCR assays, the positive controls and the positive samples for the primer pair *Psv*F/*Psv*R, *Psn*F/*Psn*F, or *Psf*F/*Psf*F, show an amplicon whose size is 388, 349, or 412 bp, respectively.
13. A different melting temperature is obtained for each of the three pathovars *Psv, Psn,* and *Psf.* A single peak is obtained at 85.5 °C for *Psv,* at 85.0 °C for *Psn,* and at 86.5 °C for *Psf.*
14. After SYBR® and TaqMan® Real-Time PCR experiments, it is also recommended to run the samples on agarose gel electrophoresis to further confirm the data (*see* **Note 11**). The amplicons produced by the primer pairs *Psv*RT-F/*Psv*RT-R, *Psn*RT-F/*Psn*RT-R, and *Psf*RT-F/*Psf*RT-R are 298 bp, 169 bp, and 227 bp, respectively.
15. To perform the Multiplex-HRM assay, remember to add to the reaction mixture all the primer pairs discriminating the SNPs of interest.

Acknowledgment

This work was supported by INSTM with a grant to C.P. on a project coordinated by S.T.

References

1. Gardan L, Bollet C, Ghorrah Abu M et al (1992) DNA relatedness among the pathovar strains of *Pseudomonas syringae* subsp. *savastanoi* Janse (1982) and proposal of *Pseudomonas savastanoi* sp. nov. Int J Syst Bacteriol 42: 606–612
2. Ramos C, Matas IM, Bardaji L et al (2012) *Pseudomonas savastanoi* pv. *savastanoi*: some like it knot. Mol Plant Pathol 13:998–1009
3. Young JM (2004) Olive knot and its pathogens. Australas Plant Pathol 33:33–39
4. Cerboneschi M, Decorosi F, Biancalani C et al (2016) Indole-3-acetic acid in plant–pathogen interactions: a key molecule for *in planta* bacterial virulence and fitness. Res Microbiol 167: 774–787
5. Tegli S, Bini L, Calamai S et al (2020) A MATE transporter is involved in pathogenicity and IAA homeostasis in the hyperplastic plant pathogen *Pseudomonas savastanoi* pv. *nerii*. Microorganisms 8:156
6. Añorga M, Pintado A, Ramos C et al (2020) Genes *ptz* and *idi*, coding for cytokinin biosynthesis enzymes, are essential for tumorigenesis and *in planta* growth by *P. syringae* pv. *savastanoi* NCPPB 3335. Front Plant Sci 11:1294
7. Quesada JM, Penyalver R, Pérez-Panadés J et al (2010) Dissemination of *Pseudomonas savastanoi* pv. *savastanoi* populations and subsequent appearance of olive knot disease. Plant Pathol 59:262–269
8. Quesada JM, García A (2007) Recovery of *Pseudomonas savastanoi* pv. *savastanoi* from symptomless shoots of naturally infected olive trees. Int Microbiol 10:77–84
9. Tegli S, Cerboneschi M, Libelli IM et al (2010) Development of a versatile tool for the simultaneous differential detection of *Pseudomonas*

savastanoi pathovars by End Point and Real-Time PCR. BMC Microbiol 10:156
10. Surico G, Lavermicocca P (1988) A semiselective medium for the isolation of *Pseudomonas syringae* pv. *savastanoi*. Phytopathology 79: 185–190
11. Azad Hamir R, Cooksey DA (1995) A semiselective medium for detecting epiphytic and systemic populations of *Pseudomonas savastanoi* from oleander. Phytopathology 85:740–745
12. Bella P, Licciardello G, Tessitori M et al (2008) A real-time PCR quantitative detection assay for *Pseudomonas savastanoi* pv. *nerii* in *Nerium oleander*. Phytopathol Mediterr 47:204–213
13. Penyalver R, García A, Ferrer A et al (2000) Detection of *Pseudomonas savastanoi* pv. *savastanoi* in olive plants by enrichment and PCR. Appl Environ Microbiol 66: 2673–2677
14. Bertolini E, Olmos A, López MM et al (2003) Multiplex nested reverse transcription-polymerase chain reaction in a single tube for sensitive and simultaneous detection of four RNA viruses and *Pseudomonas savastanoi* pv. *savastanoi* in olive trees. Phytopathology 93:286–292
15. Glickmann E, Gardan L, Jacquet S et al (1998) Auxin production is a common feature of most pathovars of *Pseudomonas syringae*. Mol Plant Microbe Interact 11:156–162
16. Iacobellis NS, Caponero A, Evidente A (1998) Characterization of *Pseudomonas syringae* ssp. *savastanoi* strains isolated from ash. Plant Pathol 47:73–83
17. Gori A, Cerboneschi M, Tegli S (2012) High-resolution melting analysis as a powerful tool to discriminate and genotype *Pseudomonas savastanoi* pathovars and strains. PLoS One 7: e30199

Part IV

High-Throughput Sequencing

Chapter 17

Computational Analysis of HTS Data and Its Application in Plant Pathology

Edoardo Piombo and Mukesh Dubey

Abstract

High-throughput sequencing is a basic tool of biological research, and it is extensively used in plant pathology projects. Here, we describe how to handle data coming from a variety of sequencing experiments, focusing on the analysis of Illumina reads. We describe how to perform genome assembly and annotation with DNA reads, correctly analyze RNA-seq data to discover differentially expressed genes, handle amplicon sequencing data from microbial communities, and utilize small RNA sequencing data to predict miRNA sequences and their putative targets.

Key words High-throughput sequencing, Genome assembly, Functional annotation, Transcriptomics, Differential expression, Metabarcoding, Small RNA, miRNA, Target prediction

1 Introduction

High-throughput sequencing has become a fundamental tool in all fields of biology, including plant pathology [1–4]. Illumina sequencing is the most widespread HTS technique today, due to the fact that it provides the best sequenced bases/cost ratio. Moreover, Illumina sequencing is very precise, reaching 99.9% accuracy, making it a favorite of the scientific community despite its short reads (150–300 bp). PacBio is another very popular choice: This third-generation method allows the production of 30–100 kb, which can greatly improve genome assemblies or enable the discovery of gene isoforms [5]. The Oxford Nanopore MinION sequencing allows for even longer reads while also being faster the Illumina or PacBio, at the cost of a reduced accuracy [6]. Furthermore, this sequencer is portable and can be operated by a simple laptop, making it ideal for situations in which sample transport is not possible because of physical or legal restrictions [7].

High-throughput sequencing provides an incredible amount of applications, but some of them are consistently used by the

Nicola Luchi (ed.), *Plant Pathology: Methods and Protocols*, Methods in Molecular Biology, vol. 2536,
https://doi.org/10.1007/978-1-0716-2517-0_17,

scientific community in a vast array of projects. In this chapter, we will provide guidelines for performing genome assembly and annotation, analyzing transcriptomic and metabarcoding data, and using small RNA (sRNAs) sequencing data for microRNAs (miRNAs) prediction.

Given the low genome size of most plant pathogens, genome assembly and annotation tend to be simple and straightforward procedures. A good assembly and good gene models will make many other bioinformatic analyses more informative, as well as giving important information about the characteristics of the organism of interest, be this a pathogen, a host or a biocontrol organism [8–10].

Transcriptome sequencing enables users to detect which genes are differentially expressed in a situation of interest. In plant pathology, this technique is commonly used to detect which plant genes respond to a treatment or an infection [11, 12], but it can also be used to observe the effect of the deletion of a gene [13], as well as for countless other situations, such as predicting genes involved in the biocontrol activity of mycoparasites [14].

Metabarcoding analyses allow scientists to study the microbial composition of a matrix. While most commonly used in ecology, these techniques can be used in plant pathology to detect pathogens [15, 16], study the microbial communities in disease suppressive soils [17], or compare the microbial communities living in association with susceptible or resistant plants [18].

The study of small RNAs is a recent application in plant pathology, but it has been proven that miRNAs regulate various cellular processes important for growth and development and mediate communications between interacting organisms [19, 20]. During plant–microbe interactions, miRNAs can move bidirectionally and modulate regulatory machinery of recipient organisms. sRNA sequencing and analysis help to expand the understanding of the molecular mechanisms underlying plant–pathogen interactions [21–23].

2 Materials

1. A computer with sufficient computational power. An i7 quad-core processor, 16 GB of RAM, and 1TB of storage should be sufficient for many analyses. However, the requirements may vary depending on the genome size of the organism of interest and the chosen software packages and operations.
2. Ubuntu operating system or windows subsystem for Linux (available on windows 10). However, some of the described programs can be used remotely and with most Web browsers through the Galaxy platform [24]. These are reported in Table 1.

Table 1
Software packages cited in the chapter and their application

Software	Can it be run remotely	Application
Bbduk	No	Quality trimming and adapter removal
Trimmomatic	Through Galaxy	
TrimGalore	No	
FastQC	Through Galaxy	Checking read quality
SPAdes	Through Galaxy	Genome assembly
Velvet	Through Galaxy	
ABySS	No	
QUAST	Through Galaxy	Checking assembly quality
ABySS-Sealer	No	Gap closing
BUSCO	Through Galaxy	Check genome completeness
CheckM	No	
RepeatModeler	No	Repeat masking
RepeatMasker	Through Galaxy	
Augustus	Yes	Gene prediction
Exonerate	Through Galaxy	
MAKER	Through Galaxy	
BRAKER	No	
SNAP	Through Galaxy	
GeneMark	No	
FGENESH	No	
BLAST2GO	Partially	Functional annotation
InterProScan	Yes	
DbCAN2	Yes	
antiSMASH	Yes	
EMBLmyGFF3	No	Prepare online submission
STAR	Through Galaxy	Splicing aware read mapping
Hisat2	Through Galaxy	
HTSeq	Through Galaxy	Read counting
featureCounts	Through Galaxy	
Trinity	Through Galaxy	De novo transcriptome assembly
DESeq2	Through Galaxy	Differential expression analysis
edgeR	Through Galaxy	
QIIME plugin itsxpress	No	Metabarcoding analysis
QIIME plugin DADA2	No	
QIIME plugin ANCOM	No	
QIIME plugin ALDEX2	No	
FUNGuild	No	
UNOISE2	No	
Deblur	No	
VSEARCH	Through Galaxy	
reformath.sh	No	Removing short and long reads
sortmerna	No	Structural RNA filtering
filterbyname.sh	No	

(continued)

Table 1
(continued)

Software	Can it be run remotely	Application
miRDeep2	No	miRNA prediction
RNAhybrid	No	
MIREAP	No	
MFOLD	No	
sRNA-tools-cli	No	
miRCheck	No	
MIRANDA	Through sRNA-toolbox	miRNA target prediction
TargetSpy	Through sRNA-toolbox	
PsRNATarget	Yes	
TAPIR	Through sRNA-toolbox	
TargetFinder	No	
PITA	Through sRNA-toolbox	
psRobot	Through sRNA-toolbox	
Cleaveland	No	Degradome-based miRNA target prediction
SeqTar	No	
PAREsnip2	No	

3 Methods

3.1 Genome Assembly and Annotation

The procedure is summarized in Fig. 1.

1. Adapter removal: Raw reads commonly include sequencing adapters that should be removed prior to any analysis. Bbduk [25], Trimmomatic [26], and TrimGalore [27] are all good software packages to perform this operation (*see* **Note 1**).
2. Quality trimming: This step consists in removing reads with low quality that could hamper successive analyses. The quality of each base is evaluated through an index called "Phred score." The Phred score usually decrease toward the right side of the reads, and it is common to trim the reads so that the average quality at the right side remains above 10. The programs used for adapter removal can usually also perform quality trimming at the same time. The necessary option to do this with bbduk is "trimq=10," but Trimmomatic and TrimGalore also have similar options.

   A brief example for adapter removal and quality trimming on paired-end data with bbduk is provided here:

   bbduk.sh -Xmx1g in1=reads_R1_raw.fq in2=reads_R2_raw.fq out1=reads_R1_clean.fq out2=reads_R2_clean.fq ref=adapters.fa ktrim=r k=23 mink=11 hdist=1 tpe tbo qtrim=r trimq=10
3. Quality checking: Before proceeding with further analysis, the quality of cleaned reads should be checked with software

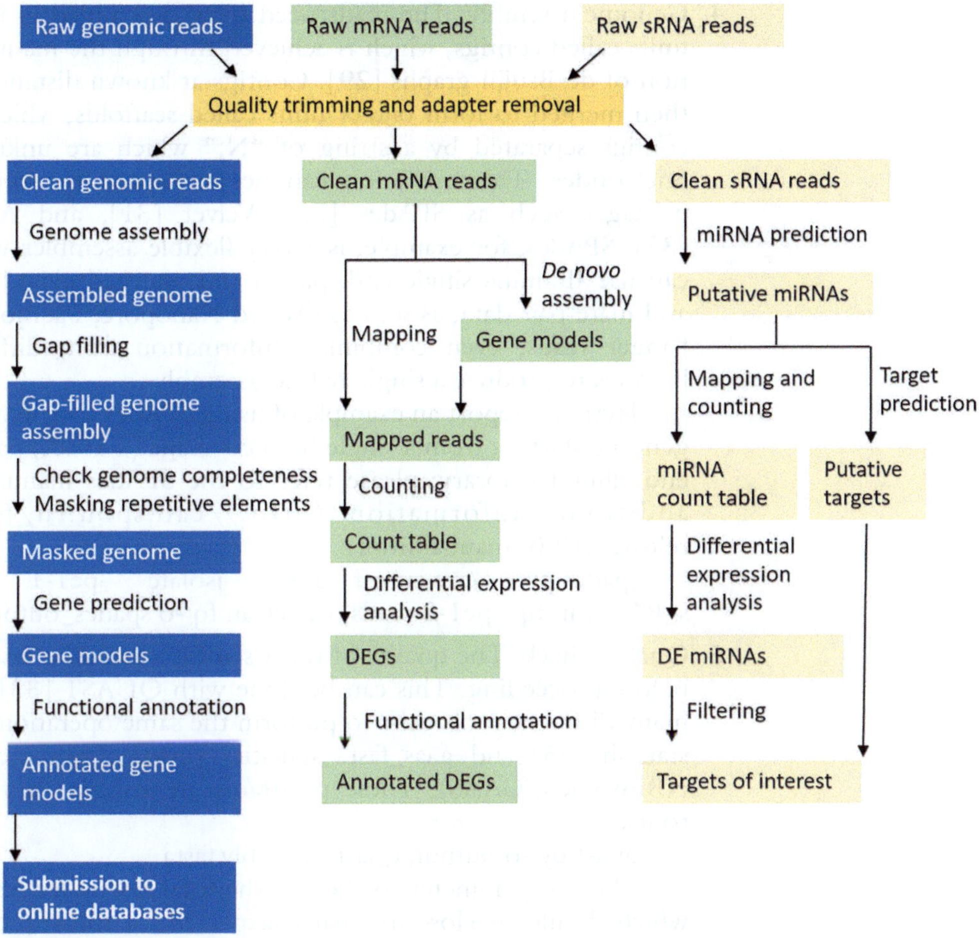

Fig. 1 Example workflow for genome assembly and annotation, transcriptome analysis, and miRNA prediction. Note that the starting raw reads will need to be genomic reads, mRNA reads, or sRNA reads, depending on the type of experiment

packages such as FastQC [28], using scripts like the following one:

```
fastqc reads_R1_clean.fq
```

This will give information regarding remaining adapters (which indicates that something went wrong with the adapter removal) and the average quality of reads. Among the others shown statistics, the "per sequence GC content" section should also be checked, since multiple peaks in this region can indicate contamination from a separate organism (*see* **Note 2**). The FastQC Web page offers examples of good (https://www.bioinformatics.babraham.ac.uk/projects/fastqc/good_sequence_short_fastqc.html) and bad (https://www.bioinformatics.babraham.ac.uk/projects/fastqc/bad_sequence_fastqc.html#M5) Illumina runs.

4. Genome assembly: The reads need to be assembled in larger units called contigs, which is achieved through the manipulation of de Bruijn graphs [29]. Contigs at known distance are then merged to form bigger units called scaffolds, which are contigs separated by a string of "N," which are unknown nucleotides. This operation can be performed by software packages such as SPAdes [30], Velvet [31], and ABySS [32]. SPAdes, for example, is a very flexible assembler which can use Illumina single-end, paired-end, merged paired-end, and mate pair data, as well as Oxford Nanopore, PacBio, and Sanger reads, even combining information from different libraries to produce a single hybrid assembly.

   Here, we report an example of using SPAdes to assemble a genome starting from a single high coverage (>100X) paired-end Illumina library; please refer to the SPAdes manual for additional information (http://cab.spbu.ru/files/release3.12.0/manual.html):

   spades.py --cov-cutoff auto --isolate --pe1-1 reads_R1_clean.fq --pe1-2 reads_R2_clean.fq -o spades_output

5. Quality check: The quality of the assembly should be assessed before proceeding. This can be done with QUAST [33], but many different tools exist to perform the same operation, like stats.sh [25] and gaas_fasta_statistics (https://github.com/NBISweden/GAAS). These programs are usually very easy to use:

   quast.py -o output_quast assembly.fasta

   The first parameter to check is the total assembled length, which should be close to what is expected for the sequenced organism. After this, the most important metrics to evaluate are the N50 and the number of scaffolds. N50 is the shortest scaffold length needed to cover 50% of the genome, which means that half of the genome sequence is covered by scaffolds larger than or equal to the N50 size (http://www.metagenomics.wiki/pdf/definition/assembly/n50). Generally, the higher the N50, the better the assembly, and this coincides with a lesser number of scaffolds, which in an ideal assembly is equal to the number of chromosomes of the organism. Some analyses, such as synteny analyses and full genome alignments, require higher N50s, while gene prediction and annotation can be performed even with lower values; therefore, the requirements in this step strongly depend on the objective of the study.

6. Gap closing: Low sequence coverage, repetitive elements, and short read length can cause the introduction of non-templated nucleotides (N) in the genome assembly. These values represent indeterminate nucleotides, which mean that their position is predicted, but their base (A, C, G or T) is not known.

Gap-closing algorithms use latent information in the reads to close as many gaps as possible, reducing the total number of Ns. This operation can be performed with software packages like ABySS-Sealer [34]. An example of Abyss-Sealer usage is the following:

abyss-sealer -b20G -k64 -k96 -k128 -o sealer_output -S assembly.fasta reads_R1_clean.fq reads_R2_clean.fq

7. Check genome completeness: Before proceeding with the next analyses, software packages such as BUSCO [35] should be used to evaluate the completeness of the assembly, an operation particularly important when there is no expected genome size, such as during the genome assembly of a novel species. This is done by detecting the presence of benchmarking universal single-copy orthologs (BUSCOs), sequences that are expected to be present only once in a genome of a precise taxon. The program will return information regarding the percentage of expected BUSCOs that are present in single copy, multiple copy, or no copy in the genome. As a rule of thumb, a genome assembled in single copy for at least 90% of its length is a good assembly (*see* **Note 3**).
8. Repeat masking: Repetitive sequences can be mistaken for genes by many predictors, causing the false prediction of new gene families. To avoid this, they should be "masked" through the generation of a repeat library. A good procedure to do this is the one suggested before running the gene predictor MAKER (http://weatherby.genetics.utah.edu/MAKER/wiki/index.php/Repeat_Library_Construction%2D%2DBasic), which utilizes the programs RepeatModeler and ProtExcluder [36]. The resulting repeat library can be used for gene prediction through MAKER (*see* following section), or it can be given as input to RepeatMasker [37] to mask the genome before gene prediction with other softwares.
9. Gene prediction: Gene prediction is a fundamental step for following analyses, and there are many software packages to do it. Some of these are de novo, like Augustus [38], while others are homology-based, such as exonerate (https://www.ebi.ac.uk/about/vertebrate-genomics/software/exonerate). Augustus, in particular, has the advantage of being possible to run entirely from the Web [39], a feature that makes it exceptionally user-friendly. Other popular tools are those that integrate both homology-based and de novo predictions in their pipelines, such as MAKER and the more recent BRAKER [40, 41] (*see* **Note 4**). These software packages can be tricky to use, requiring the setting of many parameters through control files (http://weatherby.genetics.utah.edu/MAKER/wiki/index.php/The_MAKER_control_files_explained) and the training of individual predictors. In the case of MAKER, the gene predictors SNAP (https://github.com/KorfLab/SNAP),

augustus [38], GeneMark [42], and FGENESH (http://www.softberry.com/berry.phtml?topic=fgenesh&group=help&subgroup=gfind) can be used as part of the pipeline and we recommend the use of at least two of them. These tools should be trained for predicting genes on the species of interest, by using cDNA and protein sequences from the species of interest, or, if these are unavailable, from close species. SNAP (*see* **Note 5**) can be easily trained for the genome of interest through the suggested pipeline (http://weatherby.genetics.utah.edu/MAKER/wiki/index.php/MAKER_Tutorial_for_GMOD_Online_Training_2014#Training_ab_initio_Gene_Predictors), while augustus can be trained online [39] and is also already optimized to work on several species (https://github.com/Gaius-Augustus/Augustus/tree/master/config/species) (*see* **Note 6**). Finally, when using MAKER on a fungal genome, we also recommend lowering the "pred_flank" parameter to 100 and activating the option "correct_est_fusion" to prevent the merging of gene models because of overlapping UTRs. Both options are activated by changing the respective parameters in the control files.

10. Functional annotation: Understanding the functions of the different genes is a core aspect of most genomics projects. An excellent software to perform functional annotation of protein sequences is BLAST2GO [43]. This software allows to do complete functional characterization and gene ontology attribution of genes in 4 steps. The first two steps are analyses with BLAST and InterProScan [44], which can be done online (very slow) or locally (*see* **Note 7**). The third step is mapping and usually takes the longest, but should finish within a few days given the small genome size of most plant pathogens. The last step is annotation, and it is very quick. After this, it will be possible to export the resulting annotation in a number of different file formats. These steps are easily run from the graphical interface (http://docs.blast2go.com/user-manual/quick-start/). Other software packages and services can be used to run more specific kinds of functional annotation regarding specific gene categories (*see* **Note 8**).

11. Submission to online databases: Most journals require new genomes to be uploaded to online databases such as GenBank and ENA. To do this, the fasta and gff files produced by gene prediction must be converted into a gbk or embl file, a procedure which is often difficult to perform for occasional users. For this reason, we suggest using the program EMBLmyGFF3 (https://github.com/NBISweden/EMBLmyGFF3), which makes the procedure quick and user-friendly. The simplest example of use is the following:

    EMBLmyGFF3 gene_coordinates.gff3 assembly.fasta -o result.embl

3.2 Transcriptome Analysis

The procedure is summarized in Fig. 1.

1. Adapter removal: The adapter removal procedure during transcriptome analysis is equivalent to what is performed for genome assembly (*see* Subheading 3.1, Step **1**).
2. Quality trimming: The quality trimming procedure during transcriptome analysis is equivalent to what is performed for genome assembly (*see* Subheading 3.1, Step **2**) (*see* **Note 9**).
3. Read mapping and counting: The clean reads must be mapped to the genome of the organism of interest, if available (*see* **Note 10**). Since transcripts are made up of multiple exons, which can be divided by introns in the genomic sequence, the gold standard for this is using splice-aware mappers such as STAR [45] or hisat2 [46]. It is first necessary to index the genome of interest; then, the mapping will produce output in bam format, which can be used together with gene coordinates files, such as gff files, to obtain a count table with the number of reads mapped to every gene for each sample (*see* **Note 11**). Good software packages to complete this operation are featureCounts [47] and HTSeq [48].

   A brief example of usage of STAR and featureCounts for 3 samples is reported here; we refer to the respective manuals for further explanations:

   Genome indexing:

   STAR --runMode genomeGenerate --genomeDir index/ --genomeFastaFiles genome.fasta --sjdbOverhang 100 --sjdbGTFtagExonParentTranscript Parent --sjdbGTFfile genes_coordinates.gff --genomeSAindexNbases 12

   Alignment:

   STAR --genomeDir index/ --outSAMtype BAM Unsorted --readFilesIn reads_sample1_R1_clean.fq reads_sample1_R2_clean.fq --outFileNamePrefix sample1_mapped

   STAR --genomeDir index/ --outSAMtype BAM Unsorted --readFilesIn reads_sample2_R1_clean.fq reads_sample2_R2_clean.fq --outFileNamePrefix sample2_mapped

   STAR --genomeDir index/ --outSAMtype BAM Unsorted --readFilesIn reads_sample3_R1_clean.fq reads_sample3_R2_clean.fq --outFileNamePrefix sample3_mapped

   Assemble count table:

   featureCounts -p -t exon -g Parent -a gene_coordinates.gff -o count_table.txt sample1_mapped.bam sample2_mapped.bam sample3_mapped.bam
4. Differential expression analysis: Count tables can be used to determine which genes are differentially expressed in some condition. Two very popular solutions to perform this operation are DESeq2 and edgeR [49, 50], two R packages easy to use through Galaxy [24] or RStudio [51]. Both of these

packages have excellent tutorials that can enable users to adapt them to any number of experimental setups (http://www.bioconductor.org/packages/release/bioc/vignettes/DESeq2/inst/doc/DESeq2.html; http://www.nathalievialaneix.eu/doc/html/solution_edgeR-tomato-withcode.html) (*see* **Note 12**). These packages do not return absolute expression values, but can determine which genes are more or less expressed in a situation rather than in another. This is done through a measure called fold change (FC), which is equivalent to how many times a gene is more expressed in an analyzed situation rather than in a reference one. Both packages will produce tables with log2(FC) and adjusted *p*-value (*see* **Note 13**) values for each gene. Usually, genes with log2(FC) > 1.5 or < -1.5 and with adjusted *p*-value < 0.05 are considered to be differentially expressed (*see* **Note 14**).

5. Analysis of differentially expressed genes (DEGs): Ordinarily, not all DEGs will be correlated to the object of the experiment. For this reason, it is necessary to filter them depending on their function. If all the genome has undergone functional annotation, it is possible to run Fisher tests to determine which gene ontology terms are enriched in the overexpressed or underexpressed DEGs. This can be done through the basic version of BLAST2GO [43] and allows to determine which functions are particularly present in genes differentially expressed in a situation. Instructions on how to perform this step from the graphical interface are available on the BLAST2GO Website (http://docs.blast2go.com/user-manual/quick-start/). Specific other group of genes might be interesting, depending on the aim of the experiment (*see* **Note 15**) (Fig. 2).

3.3 Analysis of Metabarcoding Data with QIIME2

The procedure is summarized in Fig. 3.

1. Install QIIME2: The installation can be done quickly through a conda environment (https://anaconda.org/qiime2/qiime2), by running the following command:

   conda install -c qiime2 qiime2

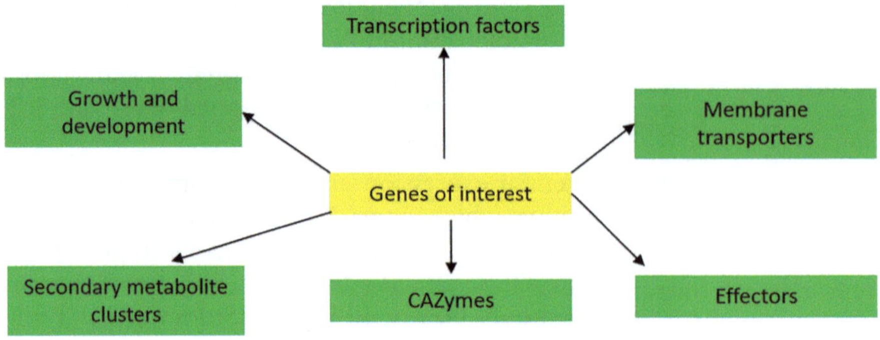

Fig. 2 Group of genes that are usually of interest in plant pathology projects

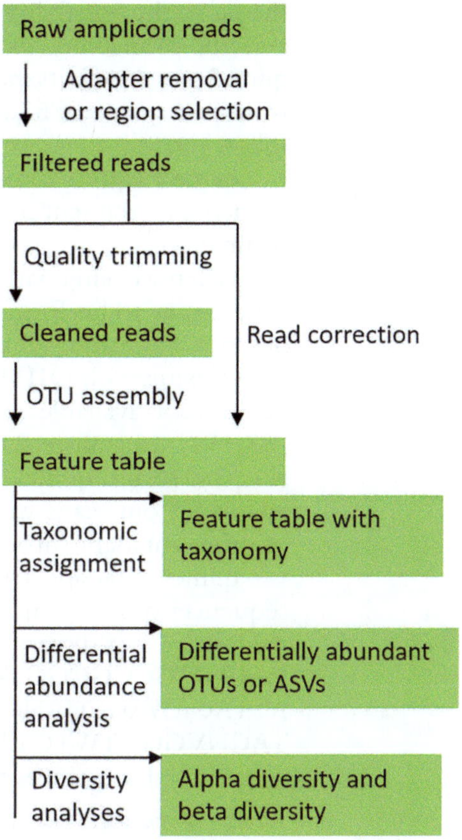

Fig. 3 Example workflow for metabarcoding data analysis

2. Import data into QIIME: In order to work in QIIME, all read files should be converted into a QIIME artifact file (qza file). This can be done through the command "qiime import" (*see* **Note 16**), and the following is an example of its use:

   qiime tools import --type SampleData[PairedEndSequencesWithQuality] --input-format PairedEndFastqManifestPhred33 --input-path manifest.txt --output-path sequences.qza

3. Create a metadata file: A metadata file is a tsv file listing on each row the information regarding a sample that will be used to divide biological replicates in groups for the analysis, such as the origin tissue, the time point, and the treatment (*see* **Note 17**).

4. Adapter removal or region selection: If the amplified region is the ITS1 or ITS2, the software ITSxpress [52] allows for the selection of the ITS region in every amplicon, automatically discarding sequence fragments that are not part of it, including adapters. This is particularly useful because the ITS length can vary between organisms, and the program is very user-friendly

thanks to the possibility of utilizing it through a QIIME plugin and the availability of an excellent tutorial (https://forum.qiime2.org/t/q2-itsxpress-a-tutorial-on-a-qiime-2-plugin-to-trim-its-sequences/5780). An example of its use is the following:

qiime itsxpress trim-pair-output-unmerged --i-per-sample-sequences sequences.qza --p-region ITS2 --o-trimmed trimmed.qza

When working with other regions, such as bacterial 16S, it is necessary to perform adapter removal, similarly to what was described in Subheading 3.1, Step **1**. However, since now we are working with QIIME artifacts (qza files), it is necessary to use a tool available as a QIIME plugin, such as cutadapt [53, 54] (*see* **Note 18**). A tutorial is available to do this (https://forum.qiime2.org/t/demultiplexing-and-trimming-adapters-from-reads-with-q2-cutadapt/2313) and an example of using cutadapt through QIIMEis given below:

qiime cutadapt trim-paired --i-demultiplexed-sequences sequences.qza --p-match-read-wildcards --p-match-adapter-wildcards --p-discard-untrimmed --p-adapter-f '^GTGY-CAGCMGCCGCGGTAA...
ATTAGAWACCCBNGTAGTCC' --p-adapter-r '^GGAC-TACNVGGGTWTCTAAT...TTACCGCGGCKGCTGRCAC' -o-trimmed-sequences trimmed_reads.qza

5. Read correction: Errors in Illumina reads can make it so that amplicons generated from the same sequence appear as several different sequences in the data, and therefore it is necessary to use programs that can correct the errors, such as DADA2 [55]. This program is available through a QIIME2 plugin, and it can be used on QIIME artifacts through the command "qiime dada2 denoise-paired" (*see* **Note 19**). Fundamental options for this command are "--p-trunc-len-r" and "--p-trunc-len-f," necessary to truncate low quality bases that can be present at the sides of reads. Each of the two options requires as input a number representing the position at which reads should be truncated, and deciding it is not trivial. Firstly, the command "qiime demux summarize" should be used to generate a qzv file with interactive quality plots similar to those in Fig. 4 (*see* **Note 20**). Then, the truncating positions should be chosen so that, ideally, the majority of bases retain a quality of at least 20. However, the total length of the reads should still be sufficient to allow them to overlap for at least 20 nucleotides, and therefore, lower qualities should be allowed if this is necessary to retain a length sufficient to make overlapping possible. An example of DADA2 use through QIIME is provided:

   qiime dada2 denoise-paired --i-demultiplexed-seqs trimmed.qza --p-trunc-len-r 235 --p-trunc-len-f 283 --output-dir dada2_output

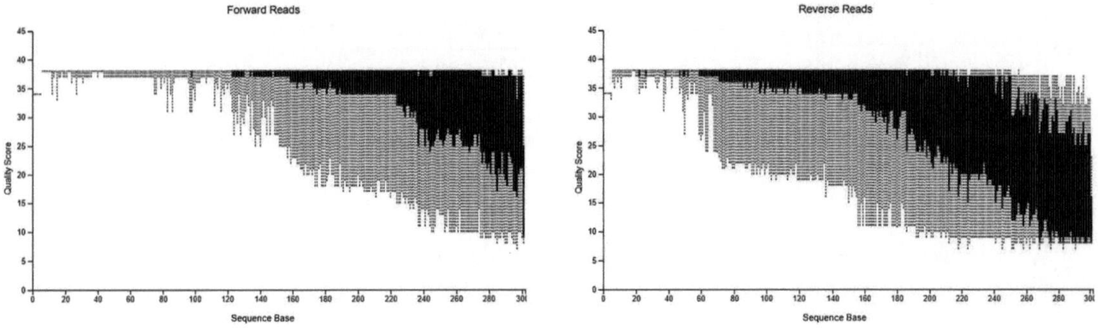

Fig. 4 Interactive quality plots generated with "qiime demux summarize" and visualized at https:/view.qiime2.org/

6. Choosing sampling depth for rarefaction: DADA2 will produce a feature table, which is a table listing, for each sample, the read count for each identified amplicon sequence, which are called amplicon sequence variants (ASVs). Before moving on to the following analyses, it is necessary to normalize the table, and the most common form of normalization is rarefaction (*see* **Note 21**), which is the discarding of reads up to a fixed value called sampling depth, so that the number of total reads assigned to ASVs in every sample is the same. This process will also remove from the analysis any sample with total sequencing depth inferior to the indicated sampling depth (*see* **Note 22**). To choose the correct value, we suggest using the command "qiime diversity alpha-rarefaction" using as input the feature table produced by DADA2. An example of this is provided here:

   qiime diversity alpha-rarefaction --i-table table.qza --p-max-depth 10,000 --m-metadata-file metadata.txt --p-steps 30 --output-dir alpha_rarefaction

   This command will produce rarefaction curves (Fig. 5) that enable the user to see at which value it is possible to rarefy without losing too much diversity. The correct value should be greater than 1000 and allow the conservation of the maximum possible diversity, while also leaving in the analyses at least 3 samples per group.

7. Calculating alpha (*see* **Note 23**) and beta diversities (*see* **Note 24**): Both of these operations can be performed as part of the command "qiime diversity core-metrics" (https://docs.qiime2.org/2018.4/plugins/available/diversity/core-metrics/), an example of which is provided here:

   qiime diversity core-metrics --i-table table.qza --m-metadata-file metadata.txt –p-sampling-depth 2000 --output-dir core_metrics

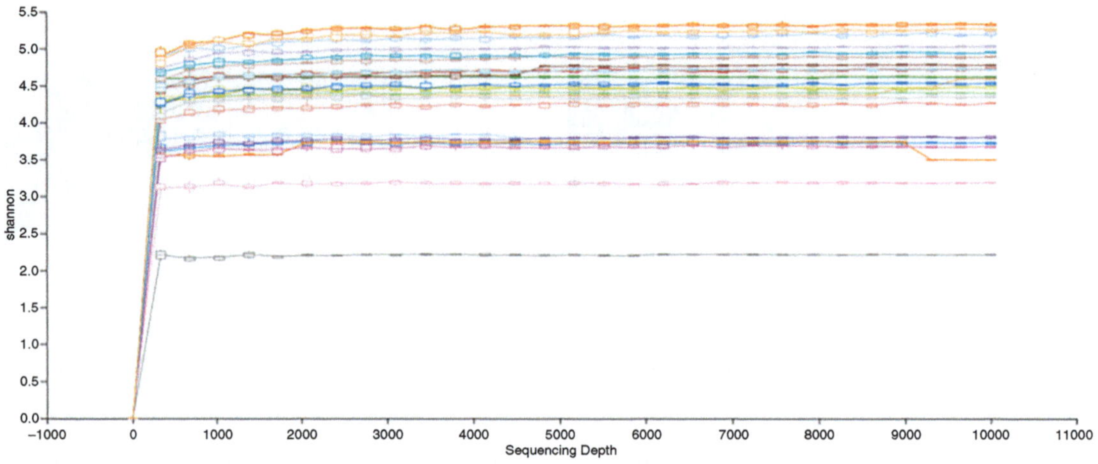

Fig. 5 Rarefaction curves generated with "qiime diversity alpha-rarefaction" and visualized at https:/view.qiime2.org/. In this example, most of the diversity of the samples can be retained with a sampling depth of 1000

This will produce a number of output files, including an emperor plot [56] based on the desired beta diversity index, alpha diversity vectors in qza format, and a rarefied feature table. The emperor plot is incredibly useful for the efficient visualization of relationships among samples. The alpha diversity vectors, on the other hand, can be used as input for the command "qiime diversity alpha-group-significance," which will determine which samples are significantly more or less diverse between each other, while also generating box plots useful for visualization. An example of this command is provided here:

qiime diversity alpha-group-significance --i-alpha-diversity shannon_vector.qza --m-metadata-file metadata.txt --o-visualization shannon_visualization.qzv

8. Taxonomic assignment: Each ASV represents a sequence, and it is possible to classify it taxonomically by comparing it to a database. For ITS sequences, the most common database is UNITE [57], while SILVA is popular for 16S sequences [58]. To perform the taxonomic assignment on the table obtained with DADA2, it is first necessary to train the chosen classifier on the database of interest. Therefore, it is necessary to download the latest QIIME release version of UNITE (https://unite.ut.ee/repository.php, *see* **Note 25**) or SILVA (https://docs.qiime2.org/2020.11/data-resources/), which should then be imported in QIIME through the command "qiime tools import" with the option "--type 'FeatureData [Sequence]'." The same command should then be used to import the taxonomy file of the chosen database version, with options "--type 'FeatureData[Taxonomy]'" and "--input-

format HeaderlessTSVTaxonomyFormat". The QIIME feature classifier can then be trained on the database with "qiime feature-classifier fit-classifier-naive-bayes," and the resulting "classifier.qza" file can be used to perform the taxonomic assignment on the DADA2 representative sequences output with the command "qiime feature-classifier fit-classifier-naive-bayes," which will produce a "taxonomy.qza" file (*see* **Note 26**). An interactive barplot figure with taxonomic information can be obtained with the command "qiime taxa barplot." An example of all the listed commands is provided here:

qiime tools import --type 'FeatureData[Sequence]' --input-path sh_refs_qiime_ver8.2_dynamic_04.02.2020.fasta --output-path unite_ver8.2.qza

qiime tools import --type 'FeatureData[Taxonomy]' --input-format HeaderlessTSVTaxonomyFormat --input-path sh_taxonomy_qiime_ver8.2_dynamic_04.02.2020.txt --output-path unite-taxonomy_ver8.2.qza

qiime feature-classifier fit-classifier-naive-bayes --i-reference-reads unite_ver8.2.qza --i-reference-taxonomy unite-taxonomy_ver8.2.qza --o-classifier classifier.qza

qiime feature-classifier classify-sklearn --i-classifier classifier.qza --i-reads representative_sequences.qza --o-classification taxonomy.qza

qiime taxa barplot --i-table table.qza --i-taxonomy taxonomy.qza --m-metadata-file metadata.txt --o-visualization taxa-bar-plots.qzv

9. Differential abundance analysis: This step (*see* **Note 27**) can be done within QIIME with a variety of tools. Two popular ones are ANCOM (*see* **Note 28**) and ALDEX2 [59, 60]. Both methods take into account the compositional nature of the data, and therefore, they can be used directly on the table produced by DADA2, without the need for rarefaction or other normalization techniques. The aldex2 plugin is particularly user-friendly, and a tutorial can be found here (https://library.qiime2.org/plugins/q2-aldex2/24/). The analysis can be completed with just two commands: "qiime aldex2 aldex2" to determine differential abundance and "qiime aldex2 extract-differences" to extract differentially abundant ASVs, which can later be exported in tsv format. Furthermore, "qiime aldex2 effect-plot" can be used to generate informative plots regarding the data (http://bioconductor.org/packages/release/bioc/vignettes/ALDEx2/inst/doc/ALDEx2_vignette.pdf). An example is provided here:

qiime aldex2 aldex2 --i-table table.qza --m-metadata-file metadata.txt --m-metadata-column Treatment --output-dir aldex2_output

qiime aldex2 effect-plot --i-table aldex2_output/differentials.qza --o-visualization aldex2_output/aldex2_plots.qzv qiime aldex2 extract-differences --i-table aldex2_output/differentials.qza --o-differentials aldex2_output/significant_-differentials --p-sig-threshold 0.05 --p-effect-threshold 0 --p-difference-threshold 0

qiime tools export --input-path aldex2_output/significant_differentials.qza --output-path differentially-abundant-features

10. Data interpretation: The significance of the results will, naturally, depend upon the role of the detected organisms in the studied ecological niche. Often, however, the high number of detected ASVs will make it impossible to manually analyze each of them. For this reason, it is possible to use the software FUNguild [61] to assign a putative functional role to the detected populations. The most abundant ASVs, however, should always be analyzed manually and according to the existing literature to avoid reporting imprecise results. The following is an example of FUNguild usage:

    python FUNGuild.py taxa -otu feature_table.txt -format tsv -column taxonomy -classifier unite

    python FUNGuild.py guild -taxa feature_table.taxa.txt

3.4 Analysis of sRNA Sequencing Data for miRNA Prediction

The procedure is summarized in Fig. 1.

1. Adapter removal: The adapter removal procedure during transcriptome analysis is equivalent to what is performed for genome assembly (*see* Subheading 3.1, Step **1**).
2. Quality trimming: The quality trimming procedure during transcriptome analysis is equivalent to what is performed for genome assembly (*see* Subheading 3.1, Step **2**).
3. Remove short and long reads: miRNAs are usually 22 bp long in animals, while in plants they tend to be 21bp long, with a small number of them being 23 or 24 bp long. In fungi, the situation is less clear, but it is still a good practice to remove all reads that are not between 18 bp and 32 bp in length. This can be done through the command "reformath.sh" of the BBTools software package [62], using the "minlength" and "maxlength" options. An example of this is provided:

    reformath.sh in=all_reads.fq out=18-32_reads.fq minlength=18 maxlength=32

4. Structural RNA filtering: Among the sRNA reads, there will inevitably be many sequences belonging to structural RNAs, such as rRNAs, snRNAs, snoRNAs, and tRNAs. A good tool to remove them is sortmerna [63], which can be used together with RNA sequence databases to identify structural RNAs in fastq files. SILVA is a good database for rRNA sequences [58], while snRNA, snoRNA, and tRNA can be downloaded from the NRDR database [64]. Structural RNAs detected with

sortmerna can then be removed with the command "filterbyname.sh" of BBTools [25], as shown in the following example:

sortmerna --ref structural_sRNAs.fasta --reads 18-32_reads.fq --blast 1 --aligned sortmerna_output/aligned_reads.txt --threads 20 --workdirwork_sortmerna

cut -f 1 sortmerna_output/aligned_reads.txt > list_structural_reads.txt

filterbyname.sh in=18-32_reads.fq out=filtered_18-32_reads.fq names=list_structural_reads.txt include=f

5. miRNAs prediction: Several programs exist to predict miRNAs in animal and plants [65], but many plant pathogens are fungi, and no specific software exist for this taxonomic group. However, some of the existing programs have been applied successfully to predict miRNA sequences in fungi [21, 66–70], and these include miRDeep2 [71], RNAhybrid [72], MIREAP (https://sourceforge.net/projects/mireap/), MFOLD [73], srna-tools-cli [74], and miRCheck [75]. An example of miRDeep2 usage is provided here. Before the actual miRNA prediction other two steps are reported: reads are collapsed in order to remove duplicates, and they are mapped to the reference genome in order to produce a mapping file in arf format (*see* **Note 29**). All the necessary steps can be performed with mirdeep2 commands, and we refer readers to the mirdeep2 user guide for further details (https://www.mdc-berlin.de/content/mirdeep2-documentation).:

   collapse_reads.pl reads.fasta > reads_collapsed.fasta

   mapper.pl reads_collapsed.fa -I -c -p genome -t reads_collapsed_vs_genome.arf

   miRDeep2.pl reads_collapsed.fa genome.fa reads_collapsed_vs_genome.arf known_miRNAs_from_this_species.fa known_miRNAs_from_related_species.fa precursors_from_this_species.fa 2>report.log

   The predicted miRNAs should be compared with those present in established databases such as miRBase [76] and Rfam [77] and those that do not have a match should be considered novel. Some organisms, such as wheat, also have dedicated sRNA databases [78].

6. Differential expression analysis: The methods suggested for the differential expression analysis of mRNA transcripts (Subheading 3.2, Step **4**) can be used also for miRNAs. When generating the count table for the differential expression analysis, it is suggested to allow for 1 mismatch when counting how many instances of each mature miRNA are present in each sample. One way to do this is counting the reads directly from the cleaned fastq files, using a software such as "agrep," which allows to look for matches while including a certain number of mismatches (*see* **Note 30**).

7. Target prediction: Several software packages can be used to predict targets for putative miRNAs, but they are mainly divided into two great categories: Animal-based tools look for 2–8 bp complementarity areas in the 3′ untranslated region (3′ UTR) [79, 80], which enable animal miRNAs to inhibit translation of mRNAs , while plant-based tools look for longer complementary areas along all the transcript [81], which enable plant miRNAs to induce the cleavage of target mRNAs [82–84] (*see* **Note 31**). Another common step of miRNA target prediction is a calculation of free energy, to make sure that the hypothetical bond between miRNA and mRNA would be stable, and an analysis of site accessibility to check that the hypothetical mRNA secondary structure does not prevent the binding. Among the most used animal-based tools we find MIRANDA [85], PITA [86] and TargetSpy [87], while on the plant side PsRNATarget [88], TAPIR [89], psRobot [90], and TargetFinder [91] are among the most popular. For fungi, however, there is no dedicated tool for predicting miRNA targets. For this reason, we suggest to try a combination of animals and plant-based methods, accepting all targets predicted by a combination of tools of the same type. The srna-toolbox [92] offer access as a Web service to the tools MIRANDA, PITA, TargetSpy, TAPIR, and psRobot.
8. Target filtering: The previous listed methods, especially the animal-based ones, will generate a large amount of putative miRNA targets, so an additional filtering step is needed before moving to biological validation. This additional step will require transcriptome and/or degradome sequencing data from the same samples used for the miRNA prediction.

   Validation through transcriptome: miRNAs tend to negatively regulate their targets; therefore, usually the targets of a miRNA will be overexpressed in a condition where the miRNA is underexpressed and vice versa [82, 93–95]. Screening for this "inverse expression" between miRNAs and transcripts will greatly reduce the number of putative targets.

   Validation through degradome: miRNA-mediated cleavage happens exactly between the 10th and 11th nucleotide of complementarity relative to the small RNA 5′-end. This divides the target transcript into an upstream unstable fragment, which is quickly degraded, and a stable downstream fragment [96]. Degradome sequencing is the sequencing of polyadenylated RNA sequences with uncapped 5′-ends, which will include mRNAs cleaved in a miRNA-dependent way [97–99]. Some programs can align sRNA reads and degradome reads to reference transcripts, searching the transcripts for spikes in degraded reads at locations compatible with miRNA-dependent cleavage. An user-friendly tool to do this is Cleaveland [100], but SeqTar and PAREsnip2 [101, 102]

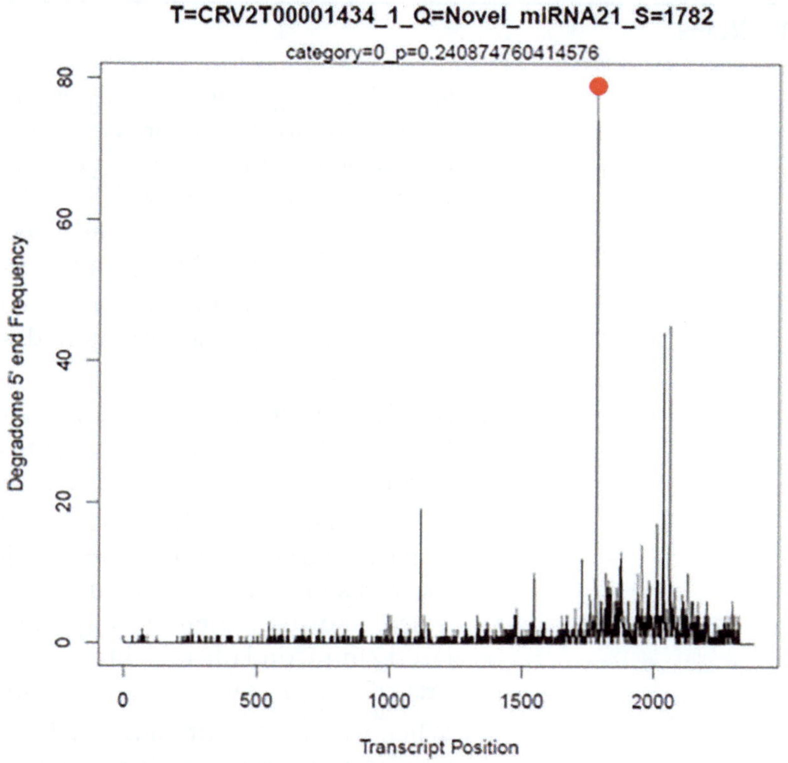

Fig. 6 Example of tplot generated with Cleaveland v. 4

can perform the same operation. In brief, Cleaveland will categorize possible mRNA-degraded transcripts into one of four categories: transcripts assigned a category of 2 or lower have an above-average number of degradome reads mapped to the putative cleavage site where there is alignment to one miRNA and are putative targets of miRNA-dependent degradation. The software can also generate automatic tplots that clearly show this phenomenon, an example of which is given in Fig. 6. We suggest users to read the Cleaveland manual for a more detailed explanation (https://github.com/MikeAxtell/CleaveLand4/blob/master/README), but an example is provided here:

CleaveLand4.pl -t -u miRNAs.fasta -n transcripts.fasta -e degradome_reads.fasta -o directory_for_tplots > results.txt

4 Notes

1. Sometimes, especially when working with data that was not sequenced directly by the user, it is possible to not know the used adapters. In these situations, the program FastQC [28] can be used to discover which adapters are contaminating the sample, since they will be listed under the "Overrepresented sequences" section of the program output.
2. In some cases, a double peak in GC content is expected, such as when sequencing a sample infected with a parasite or a mix of two symbionts.
3. Another tool to evaluate the completeness of an assembly is CheckM [103], which, however, can only be used for bacterial genomes.
4. MAKER and BRAKER can be difficult to install, so we suggest using a conda environment, which allows for the installation of the software packages and the necessary dependencies in a quick and user-friendly way (https://docs.conda.io/en/latest/miniconda.html; https://anaconda.org/bioconda/braker2; https://anaconda.org/bioconda/maker).
5. When working with fungi, SNAP can sometimes predict very small introns (less than 10 bp) which are not accepted by online repositories such as GenBank and EMBL. If such introns are being predicted, we recommend avoiding snap-based gene prediction.
6. Before doing your own augustus training, always check if the species of interest is phylogenetically close to those for which optimized parameters are already available. If this is the case, the training can be skipped and pre-set parameters can be used.
7. BLAST2GO can import pre-obtained blast results, or it can perform blast locally. In either case, it is necessary to provide the program with a local protein database. When studying fungal proteomes, we suggest the fungal part of the UniProt database [104]. Similarly, InterProScan results can be imported to avoid having to wait for the online tool. This program is easy to install and run following the instructions of the providers (https://interproscan-docs.readthedocs.io/en/latest/). A brief example of InterProScan use is provided here:

   ./interproscan.sh -i proteome.fasta -d output_directory -goterms -iprlookup

   The xml output file can be imported in BLAST2GO from the graphical interface, after which the command "Merge InterProScan GOs to annotation" should be used, always on the BLAST2GO graphical interface application, to add the new gene ontology terms to the annotation.

8. Depending on the organism of interest, it is often interesting to look at specific gene categories. CAZymes (genes involved in the metabolism of carbohydrates) often include genes involved in the degrading of plant cell walls by pathogens [105, 106]. Furthermore, glycoside hydrolases include chitinases, which are important for the mycoparasitic action of biocontrol organisms [107–109] and the resistance to plant pathogens [110, 111]. This type of genes can be easily predicted through the DbCAN2 online service [112], available at this URL: http://bcb.unl.edu/dbCAN2/blast.php. Another interesting category of genes to predict is secondary metabolite gene clusters, which can be detected through antiSMASH [113] and can be involved in the development of diseases [11, 114], the mycoparasitic action of biocontrol organisms [115], and the plant resistance to pathogens [116–120]. Finally, secreted proteins can also be of interest, because they often include effectors, which are proteins involved in the overcoming of the plant immune system by the pathogens [121, 122]. A good pipeline for the detection of secreted proteins has been developed by Levin et al. [123].
9. If the reads will be mapped to a pre-annotated genome, both adapter removal and quality trimming are redundant operations, and they can be skipped [124].
10. If the organism of interest is not sequenced and/or annotated, it is possible to run de novo transcriptome assembly, predicting transcript sequences that can be used as reference for read mapping. A user-friendly and popular program to do this is Trinity [125]. However, we suggest using a sequenced annotated genome to map reads whenever this is possible. If a genome is available but not annotated, transcript sequences produced with trinity can be used to train gene predictors to annotate it.
11. Most programs used for mapping and/or counting, including STAR and featureCounts, will generate summary files besides their normal output. Carefully check these files and particularly the percentage of reads successfully mapped to the target genome, before moving forward with the analysis. A large number of unmapped sequences can indicate that the sample is contaminated.
12. Both edgeR and DESeq2 require raw read counts, be sure to not normalize the data in any way before giving it as input to these two modules. They also require a metadata file listing information for each sample. This is a table listing for each sample the tissue of origin, the treatment, the time point, and/or any other information required to divide the samples into groups: be sure to have the samples in the same order in this metadata file and in the count table, as differences in the order of samples can create problems during the analysis. An example of simple metadata file is given in Table 2.

Table 2
Simple metadata file, describing an experiment in which three biological replicates of two cultivars were sequenced

ID	Cultivar
Sample1	Susceptible
Sample2	Susceptible
Sample3	Susceptible
Sample4	Resistant
Sample5	Resistant
Sample6	Resistant

13. There are many ways to adjust the *p*-value when estimating differential expression. The most used method is the false discovery rate (FDR), which is the default for both DESeq2 and edgeR. Be sure to use this value and not the non-adjusted *p*-value when estimating which genes are significantly differentially expressed.
14. These thresholds can be changed if the number of differentially expressed transcripts is particularly high or low. Log2 (FC) thresholds of 1 or 2 are widely used, and the maximum adjusted *p*-value can be lowered to 0.01 to increase stringency.
15. In plant pathology, RNA-seq can be used for a variety of reasons, from determining which genes are involved in plant resistance to a pathogen, to observing the mechanism of actions of pathogens on a susceptible host, to studying how a biocontrol organism can attack a pathogen. There is, therefore, no unique group of genes that will be of interest in every study. However, some of them can be recurrently interesting for many applications. CAZymes, secreted proteins, and secondary metabolite clusters (*see* **Note 8**) are interesting when studying pathogens, plant resistance genes, and biocontrol organisms, as are membrane transporters [14, 126, 127]. Genes involved in regulation, such as chromatin remodeling genes and transcription factors, are also usually of interest, since they can influence virulence and pathogenicity by regulating the expression of other genes [128–130]. This type of genes can be identified through the analysis of the gene ontology terms assigned to proteins during functional annotation. One more interesting resource when studying plant pathogens is the PHI-base database that contains the effect of the deletion or overexpression of several genes on the virulence and/or pathogenicity of many plant pathogens [131]. PHI-base can be mined through online blast (http://phi-blast.phi-base.org/) or downloaded locally (http://www.phi-base.org/downloadLink.htm).

When studying the plant side of the pathosystem, on the other hand, among the genes of interest we find PR (pathogen-related) genes [132, 133], peroxidases [134–136], germin-like proteins (GLPs) [137–140], and MAP kinases [141].

16. The "qiime tools import" command can take many options, and it can be counter-intuitive at first. When importing paired-end fastq data, which is the most common form of raw metabarcoding data, the option "--type SampleData[PairedEndSequencesWithQuality]" is necessary, as well as the option "--input-path manifest_file.csv," where the manifest file is a csv file listing in each raw the path to a file, the id of the sample to which that file refers and the read orientation. An example is given in Table 3.
17. An example of metadata file is given in Table 4.
18. It is also possible to perform adapter removal before importing the data into qiime (Subheading 3.3, Step **3**), using therefore one of the programs described in Subheading 3.1, Step **1**.
19. Many other programs are available to perform this operation. Some, such as UNOISE2 and Deblur [142, 143], correct error in the reads in a way similar to DADA2. Others cluster all similar sequences (usually with a similarity threshold of 97% or 99%) into units called operational taxonomic units (OTUs) [144–146]. Pauvert and colleagues [147] tested 360 software packages and parameter combinations on mock communities composed of a wide array of Ascomycota and Basidiomycota, identifying DADA2-based approaches as the ones most effective at capturing the composition of the fungal communities. However, the method of choice should consider the objectives of the study, since, for example, VSEARCH is the most sensitive approach [146, 147].
20. Qzv files can be visualized at the url: https://view.qiime2.org/.
21. There are other forms of normalization besides rarefaction, and, while rarefaction is commonly employed and easy to

Table 3
Example of manifest file necessary for importing data in qiime with the command "qiime tools import --type SampleData[PairedEndSequencesWithQuality]"

sample-id,absolute-filepath,direction
Sample1,/home/User/metabarcoding/raw_reads/ITS_1.fastq.gz,forward
Sample1,/home/User/metabarcoding/raw_reads/ITS_2.fastq.gz,reverse
Sample2,/home/User/metabarcoding/raw_reads/ITS_3.fastq.gz,forward
Sample2,/home/User/metabarcoding/raw_reads/ITS_4.fastq.gz,reverse

Table 4
Example of metadata file, describing an experiment in which a control group and a treated group were sequenced, each in five replicates, at a single timepoint

sample-id	Treatment	Timepoint
Sample1	Control	Timepoint1
Sample2	Control	Timepoint1
Sample3	Control	Timepoint1
Sample4	Control	Timepoint1
Sample5	Control	Timepoint1
Sample6	Treatment 1	Timepoint1
Sample7	Treatment 1	Timepoint1
Sample8	Treatment 1	Timepoint1
Sample9	Treatment 1	Timepoint1
Sample10	Treatment 1	Timepoint1

apply, it can reduce statistical power depending on how much data is removed and the discussion regarding its application for alpha and beta diversity analyses is ongoing [148, 149]. There are several other normalization methods proposed, and scaling with ranked subsampling (SRS), trimmed mean of M-values (TMM), relative log expression (RLE), and cumulative sum scaling (CSS) were all proven to be reliable [150–154]. However, none of these methods is currently supported in QIIME2, and to apply them it is therefore necessary to export the feature table in a tsv format, normalize the data with the chosen method, and then re-import the data in the qza format to continue the analyses with QIIME2.

22. The individual sequencing depth of each sample can be obtained through the command "qiime feature-table summarize," using as input the feature table produced by DADA2.

23. With "alpha diversity," we indicate the amount of diversity in an individual sample, and this measure is therefore a value that will describe how diverse are the populations identified within a single sample. For example, a sample composed of many different ASVs will have high alpha diversity, while one almost exclusively composed of a single ASV will have low alpha diversity. There are several indices to measure alpha diversity. Among the most popular ones, we find the number of features (number of ASVs if DADA2 was used to perform the analysis) and the Shannon index [155], which takes into account how dominant a feature is (that is, a sample will have low diversity if

the vast majority of reads are assigned to a single ASV, even if the total amount of detected ASVs is high).

24. With "beta diversity," we indicate the diversity among samples. This term therefore indicates how different two samples are between each other and can be represented as a dissimilarity. Beta diversity can be measured with several indices too, and one of the most popular is the Bray Curtis dissimilarity, used in several works [156–158].

25. There are many options for every version of the UNITE database. In particular, from the Website it is possible to download a "large option" with singletons clustered together with a 97% threshold in similarity and a "small option" without singletons unless they are representative sequences. Each of these two options will download 3 versions of the database: a version with features clustered together according to a similarity threshold of 97%, one at 99% and one called "dynamic," which uses different similarity thresholds for different species, based on manual curation by experts. Furthermore, for each of these versions, there is also a developer version, which contains untrimmed sequences, while the main versions contain only the ITS sequence. We suggest using the non-developer dynamic version of the "small option" database, which is also used in the ITSxpress qiime tutorial (https://forum.qiime2.org/t/q2-itsxpress-a-tutorial-on-a-qiime-2-plugin-to-trim-its-sequences/5780), but the developer version is suggested by the QIIME team, which found that there is no benefit in training classifiers on the trimmed version of the database rather than the developer one (https://docs.qiime2.org/2018.6/tutorials/feature-classifier/).

26. The taxonomic assignment will often result in predictions at the species level, but it must be stressed that no analysis can overcome the limits of the amplified marker region of choice, which in the case of ITS1 and ITS2 will often not be trustworthy beyond the genus level [159–161]. This is a key limitation in the application of this technique to plant pathogens [16, 162].

27. Depending on the study, it might be of interest to check for the differential abundance of genera or species, rather than individual ASVs. To obtain a genus-based or species-based feature table, it is necessary to use the command "qiime taxa collapse," which takes as input the feature table produced by DADA2 and the taxonomy file obtained after the taxonomical assignment. A specific taxonomic level needs to be indicated when running the command: 1 corresponds to phylum, 6 to genus, and 7 to species.

28. ANCOM is the tool used in the QIIME2 Moving Pictures tutorial, which can be found at this link: https://docs.qiime2.org/2020.11/tutorials/moving-pictures/.

29. All these steps are necessary for miRNA prediction using mirdeep2. They are performed with mirdeep2 commands, and they could be unnecessary when using other techniques for miRNA prediction.

30. Alternatively, it is also possible to simply align the reads to the genome as explained in Subheading 3.2, Step **3**, using the coordinates predicted for miRNAs in place of the gene coordinates. However, some programs, like mirdeep, return miRNA coordinates in bed format, while most mappers require files in gff or gtf formats. Be aware that coordinates in bed files are 0-indexed (that is, the number of the first base is zero), while gff and gtf files are 1-indexed. It is necessary to keep this in mind when converting from one file format to the other. The same mapping programs listed in Subheading 3.2, Step **3** can be used, but splicing awareness and soft clipping should be disabled, and the maximum number of mismatches should be set to 5% of the read length (that is, 1 bp every 20). Given the small length of the alignment, score-based filters should also be deactivated. Using the STAR aligner [45], the necessary options to enable all the listed changes are the following:

    STAR --genomeDir index_genome --readFilesIn reads.fq --outFileNamePrefix output_name --outFilterMismatchNoverLmax 0.05 --outFilterMatchNmin 16 --outFilterScoreMinOverLread 0 --outFilterMatchNminOverLread 0 --alignIntronMax 1 –alignEndsType EndToEnd

    The programs that assemble the count table from the mapping files should also receive options that only allow them to count alignments in which the aligned read and the target miRNAs have the same length, enabling one mismatch. Furthermore, multimapping reads should be included in the analysis, because it is possible for the same mature miRNAs to be generated by multiple precursors in the genome. In featureCounts [47] all of this can be achieved with the following options:

    featureCounts --fracOverlapFeature 0.95 --fracOverlap 0.95 -M --fraction -t miRNA -g ID -f -a miRNA_coordinates.gff -o output_count_table.txt sample1-bam sample2.bam ...

31. However, it must be noted that there is also proof of some miRNAs acting through target cleavage in animals [163, 164] and through translation inhibition in plants [95, 165, 166].

References

1. Tan K, Ipcho SVS, Trengove RD et al (2009) Assessing the impact of transcriptomics, proteomics and metabolomics on fungal phytopathology. Mol Plant Pathol 10:703–715
2. Adams IP, Fox A, Boonham N et al (2018) The impact of high throughput sequencing on plant health diagnostics. Eur J Plant Pathol 152(4):909–919. https://doi.org/10.1007/s10658-018-1570-0
3. Sarethy IP, Saharan A (2021) Genomics, proteomics and transcriptomics in the biological control of plant pathogens: a review. Indian Phytopathol 74(1):3–12
4. Villamor DE V, Ho T, Al Rwahnih M et al (2019) High throughput sequencing for plant virus detection and discovery. Phytopathology 109:716–725
5. Rhoads A, Au KF (2015) PacBio sequencing and its applications. Genomics Proteomics Bioinformatics 13:278–289
6. Lu H, Giordano F, Ning Z (2016) Oxford Nanopore MinION sequencing and genome assembly. Genomics Proteomics Bioinformatics 14:265–279
7. Menegon M, Cantaloni C, Rodriguez-Prieto A et al (2017) On site DNA barcoding by nanopore sequencing. PLoS One. https://doi.org/10.1371/journal.pone.0184741
8. Wiemann P, Sieber CMK, Von Bargen KW et al (2013) Deciphering the cryptic genome: genome-wide analyses of the rice pathogen Fusarium fujikuroi reveal complex regulation of secondary metabolism and novel metabolites. PLoS Pathog 9:e1003475
9. Shimizu KK, Copetti D, Okada M et al (2021) De novo genome assembly of the Japanese wheat cultivar Norin 61 highlights functional variation in flowering time and fusarium-resistant genes in east Asian genotypes. Plant Cell Physiol 62:8–27
10. Karlsson M, Durling MB, Choi J et al (2015) Insights on the evolution of mycoparasitism from the genome of clonostachys rosea. Genome Biol Evol 7(2):465–480. https://doi.org/10.1093/gbe/evu292
11. Matić S, Bagnaresi P, Biselli C et al (2016) Comparative transcriptome profiling of resistant and susceptible rice genotypes in response to the seedborne pathogen Fusarium fujikuroi. BMC Genomics 17:608. https://doi.org/10.1186/s12864-016-2925-6
12. Zhang ET, Zhang H, Tang W (2021) Transcriptomic analysis of wheat seedling responses to the systemic acquired resistance inducer n-hydroxypipecolic acid. Front Microbiol 12:94
13. Zhang W, Gui Y, Short DPG et al (2018) Verticillium dahliae transcription factor VdFTF1 regulates the expression of multiple secreted virulence factors and is required for full virulence in cotton. Mol Plant Pathol 19:841–857
14. Nygren K, Dubey M, Zapparata A et al (2018) The mycoparasitic fungus Clonostachys rosea responds with both common and specific gene expression during interspecific interactions with fungal prey. Evol Appl 11:931–949. https://doi.org/10.1111/eva.12609
15. Tedersoo L, Drenkhan R, Anslan S et al (2019) High-throughput identification and diagnostics of pathogens and pests: overview and practical recommendations. Mol Ecol Resour 19(1):47–76
16. Piombo E, Abdelfattah A, Droby S et al (2021) Metagenomics approaches for the detection and surveillance of emerging and recurrent plant pathogens. Microorganisms 9:188
17. Siegel-Hertz K, Edel-Hermann V, Chapelle E et al (2018) Comparative microbiome analysis of a Fusarium wilt suppressive soil and a Fusarium wilt conducive soil from the Châteaurenard region. Front Microbiol 9:568
18. Vergine M, Meyer JB, Cardinale M et al (2020) The Xylella fastidiosa-resistant olive cultivar "Leccino" has stable endophytic microbiota during the olive quick decline syndrome (OQDS). Pathogens 9:35
19. Song L, Fang Y, Chen L et al (2021) Role of non-coding RNAs in plant immunity. Plant Commun 2(3):100180
20. Yang X, Zhang L, Yang Y et al (2021) miRNA mediated regulation and interaction between plants and pathogens. Int J Mol Sci 22:2913
21. Chen R, Jiang N, Jiang Q et al (2014) Exploring microRNA-like small RNAs in the filamentous fungus Fusarium oxysporum. PLoS One 9:e104956. https://doi.org/10.1371/journal.pone.0104956
22. Son H, Park AR, Lim JY et al (2017) Genome-wide exonic small interference RNA-mediated gene silencing regulates sexual reproduction in the homothallic fungus Fusarium graminearum. PLoS Genet 13:e1006595. https://doi.org/10.1371/journal.pgen.1006595

23. Xia Z, Wang Z, Kav NNV et al (2020) Characterization of microRNA-like RNAs associated with sclerotial development in Sclerotinia sclerotiorum. Fungal Genet Biol 144:103471. https://doi.org/10.1016/j.fgb.2020.103471
24. Blankenberg D, Gordon A, Von Kuster G et al (2010) Manipulation of FASTQ data with galaxy. Bioinformatics 26(14):1783–1785. https://doi.org/10.1093/bioinformatics/btq281
25. Bushnell B (2019) BBTools: a suite of fast, multithreaded bioinformatics tools designed for analysis of DNA and RNA sequence data. Joint Genome Institute, Berkeley
26. Bolger AM, Lohse M, Usadel B (2014) Trimmomatic: a flexible trimmer for Illumina sequence data. Bioinformatics 30:2114–2120. https://doi.org/10.1093/bioinformatics/btu170
27. Krueger F, Ewels P, Schuster-Boeckler B Taking appropriate QC measures for RRBS-type or other -Seq applications with Trim Galore! https://github.com/FelixKrueger/TrimGalore/blob/master/Docs/Trim_Galore_User_Guide.md
28. Andrews S FastQC: a quality control tool for high throughput sequence data. https://www.bioinformatics.babraham.ac.uk/projects/fastqc/
29. Compeau PEC, Pevzner PA, Tesler G (2011) How to apply de Bruijn graphs to genome assembly. Nat Biotechnol 29:987–991
30. Bankevich A, Nurk S, Antipov D et al (2012) SPAdes: a new genome assembly algorithm and its applications to single-cell sequencing. J Comput Biol 19:455–477
31. Zerbino DR, Birney E (2008) Velvet: algorithms for de novo short read assembly using de Bruijn graphs. Genome Res 18:821–829
32. Simpson JT, Wong K, Jackman SD et al (2009) ABySS: a parallel assembler for short read sequence data. Genome Res 19:1117–1123
33. Gurevich A, Saveliev V, Vyahhi N, Tesler G (2013) QUAST: quality assessment tool for genome assemblies. Bioinformatics 29:1072–1075
34. Paulino D, Warren RL, Vandervalk BP et al (2015) Sealer: a scalable gap-closing application for finishing draft genomes. BMC Bioinformatics 16:1–8
35. Simão FA, Waterhouse RM, Ioannidis P et al (2015) BUSCO: assessing genome assembly and annotation completeness with single-copy orthologs. Bioinformatics 31(19):3210–3212. https://doi.org/10.1093/bioinformatics/btv351
36. Flynn JM, Hubley R, Goubert C et al (2020) RepeatModeler2 for automated genomic discovery of transposable element families. Proc Natl Acad Sci 117:9451–9457
37. Tempel S (2012) Using and understanding RepeatMasker. In: Mobile genetic elements. Springer, pp 29–51
38. Stanke M, Keller O, Gunduz I et al (2006) AUGUSTUS: ab initio prediction of alternative transcripts. Nucleic Acids Res 34:W435–W439
39. Stanke M, Morgenstern B (2005) AUGUSTUS: a web server for gene prediction in eukaryotes that allows user-defined constraints. Nucleic Acids Res 33:W465–W467
40. Campbell MS, Holt C, Moore B, Yandell M (2014) Genome annotation and curation using MAKER and MAKER-P. Curr Protoc Bioinforma 48:4–11
41. Hoff KJ, Lomsadze A, Borodovsky M, Stanke M (2019) Whole-genome annotation with BRAKER. In: Gene prediction. Springer, pp 65–95
42. Borodovsky M, Mills R, Besemer J, Lomsadze A (2003) Prokaryotic gene prediction using GeneMark and GeneMark. hmm. Curr Protoc Bioinforma 1:4–5
43. Conesa A, Götz S, García-Gómez JM et al (2005) Blast2GO: a universal tool for annotation, visualization and analysis in functional genomics research. Bioinformatics 21:3674–3676. https://doi.org/10.1093/bioinformatics/bti610
44. Jones P, Binns D, Chang HY et al (2014) InterProScan 5: genome-scale protein function classification. Bioinformatics 30:1236–1240. https://doi.org/10.1093/bioinformatics/btu031
45. Dobin A, Davis CA, Schlesinger F et al (2013) STAR: Ultrafast universal RNA-seq aligner. Bioinformatics 29:15–21. https://doi.org/10.1093/bioinformatics/bts635
46. Wen G (2017) A simple process of RNA-sequence analyses by Hisat2, Htseq and DESeq2. In: Proceedings of the 2017 International Conference on Biomedical Engineering and Bioinformatics, pp 11–15
47. Liao Y, Smyth GK, Shi W (2014) FeatureCounts: an efficient general purpose program for assigning sequence reads to genomic features. Bioinformatics 30:923–930. https://doi.org/10.1093/bioinformatics/btt656
48. Anders S, Pyl PT, Huber W (2015) HTSeq—a Python framework to work with high-

throughput sequencing data. Bioinformatics 31:166–169

49. Love MI, Anders S, Huber W (2014) Differential analysis of count data – the DESeq2 package. Genome Biol 15:10–1186
50. Robinson MD, McCarthy DJ, Smyth GK (2010) edgeR: a bioconductor package for differential expression analysis of digital gene expression data. Bioinformatics 26:139–140. https://doi.org/10.1093/bioinformatics/btp616
51. Allaire J (2012) RStudio: integrated development environment for R. Boston, MA 770: 394
52. Rivers AR, Weber KC, Gardner TG et al (2018) ITSxpress: software to rapidly trim internally transcribed spacer sequences with quality scores for marker gene analysis. F1000Res 7:1418
53. Estaki M, Jiang L, Bokulich NA et al (2020) QIIME 2 enables comprehensive end-to-end analysis of diverse microbiome data and comparative studies with publicly available data. Curr Protoc Bioinforma 70:e100
54. Martin M (2011) Cutadapt removes adapter sequences from high-throughput sequencing reads. EMBnet J 17(1):10–12. https://doi.org/10.14806/ej.17.1.200
55. Callahan BJ, McMurdie PJ, Rosen MJ et al (2016) DADA2: high-resolution sample inference from Illumina amplicon data. Nat Methods 13:581-583. https://doi.org/10.1038/nmeth.3869
56. Vázquez-Baeza Y, Pirrung M, Gonzalez A, Knight R (2013) EMPeror: a tool for visualizing high-throughput microbial community data. Gigascience 2:16. https://doi.org/10.1186/2047-217X-2-16
57. Abarenkov K, Nilsson RH, Larsson KH et al (2010) The UNITE database for molecular identification of fungi - recent updates and future perspectives. New Phytol 186: 281–285. https://doi.org/10.1111/j.1469-8137.2009.03160.x
58. Quast C, Pruesse E, Yilmaz P et al (2013) The SILVA ribosomal RNA gene database project: improved data processing and web-based tools. Nucleic Acids Res 41:590–596. https://doi.org/10.1093/nar/gks1219
59. Mandal S, Van Treuren W, White RA et al (2015) Analysis of composition of microbiomes: a novel method for studying microbial composition. Microb Ecol Health Dis 26: 27663
60. Fernandes AD, Reid JN, Macklaim JM et al (2014) Unifying the analysis of high-throughput sequencing datasets: characterizing RNA-seq, 16S rRNA gene sequencing and selective growth experiments by compositional data analysis. Microbiome 2:1–13
61. Nguyen NH, Song Z, Bates ST et al (2016) FUNGuild: an open annotation tool for parsing fungal community datasets by ecological guild. Fungal Ecol 20:241–248
62. Bushnell B, Rood J, Singer E (2017) BBTools software package. PLoS One
63. Kopylova E, Noé L, Touzet H (2012) SortMeRNA: fast and accurate filtering of ribosomal RNAs in metatranscriptomic data. Bioinformatics 28:3211–3217. https://doi.org/10.1093/bioinformatics/bts611
64. Paschoal AR, Maracaja-Coutinho V, Setubal JC et al (2012) Non-coding transcription characterization and annotation: a guide and web resource for non-coding RNA databases. RNA Biol 9:274–282. https://doi.org/10.4161/rna.19352
65. Zhao Y, Kuang Z, Wang Y et al (2021) MicroRNA annotation in plants: current status and challenges. Brief Bioinform
66. Devers EA, Branscheid A, May P, Krajinski F (2011) Stars and symbiosis: Microrna- and microrna*-mediated transcript cleavage involved in arbuscular mycorrhizal symbiosis. Plant Physiol 156:1990–2010. https://doi.org/10.1104/pp.111.172627
67. Zhou J, Fu Y, Xie J et al (2012) Identification of microRNA-like RNAs in a plant pathogenic fungus Sclerotinia sclerotiorum by high-throughput sequencing. Mol Gen Genomics 287:275–282. https://doi.org/10.1007/s00438-012-0678-8
68. Carreras-Villaseñor N, Esquivel-Naranjo EU, Villalobos-Escobedo JM et al (2013) The RNAi machinery regulates growth and development in the filamentous fungus Trichoderma atroviride. Mol Microbiol 89(1): 96–112. https://doi.org/10.1111/mmi.12261
69. Wang L, Xu X, Yang J et al (2018) Integrated microRNA and mRNA analysis in the pathogenic filamentous fungus Trichophyton rubrum. BMC Genomics 19:1–14. https://doi.org/10.1186/s12864-018-5316-3
70. Kang K, Zhong J, Jiang L et al (2013) Identification of microRNA-Like RNAs in the Filamentous Fungus Trichoderma reesei by Solexa Sequencing. PLoS One 8:e76288. https://doi.org/10.1371/journal.pone.0076288
71. Friedländer MR, MacKowiak SD, Li N et al (2012) MiRDeep2 accurately identifies known and hundreds of novel microRNA

genes in seven animal clades. Nucleic Acids Res 40(1):37–52. https://doi.org/10.1093/nar/gkr688
72. Rehmsmeier M, Steffen P, Höchsmann M, Giegerich R (2004) Fast and effective prediction of microRNA/target duplexes. RNA 10: 1507–1517
73. Zuker M (2003) Mfold web server for nucleic acid folding and hybridization prediction. Nucleic Acids Res 31:3406–3415
74. Moxon S, Schwach F, Dalmay T et al (2008) A toolkit for analysing large-scale plant small RNA datasets. Bioinformatics 24:2252–2253
75. Jones-Rhoades MW, Bartel DP (2004) Computational identification of plant microRNAs and their targets, including a stress-induced miRNA. Mol Cell 14:787–799
76. Kozomara A, Birgaoanu M, Griffiths-Jones S (2019) MiRBase: from microRNA sequences to function. Nucleic Acids Res 47:155–162. https://doi.org/10.1093/nar/gky1141
77. Griffiths-Jones S, Bateman A, Marshall M et al (2003) Rfam: an RNA family database. Nucleic Acids Res 31:439–441
78. Remita MA, Lord E, Agharbaoui Z et al (2016) A novel comprehensive wheat miRNA database, including related bioinformatics software. Curr Plant Biol 7:31–33
79. Peterson SM, Thompson JA, Ufkin ML et al (2014) Common features of microRNA target prediction tools. Front Genet 5:23
80. Riffo-Campos ÁL, Riquelme I, Brebi-Mieville P (2016) Tools for sequence-based miRNA target prediction: what to choose? Int J Mol Sci 17:1987
81. Mishra AK, Duraisamy GS, Matoušek J (2015) Discovering microRNAs and their targets in plants. CRC Crit Rev Plant Sci 34: 553–571
82. Mourrain P, Béclin C, Elmayan T et al (2000) Arabidopsis SGS2 and SGS3 genes are required for posttranscriptional gene silencing and natural virus resistance. Cell 101: 533–542
83. Tang G, Reinhart BJ, Bartel DP, Zamore PD (2003) A biochemical framework for RNA silencing in plants. Genes Dev 17:49–63
84. Liu S-R, Zhou J-J, Hu C-G et al (2017) MicroRNA-mediated gene silencing in plant defense and viral counter-defense. Front Microbiol 8:1801
85. Enright A, John B, Gaul U et al (2003) MicroRNA targets in Drosophila. Genome Biol 4:1–27
86. Kertesz M, Iovino N, Unnerstall U et al (2007) The role of site accessibility in microRNA target recognition. Nat Genet 39:1278–1284
87. Sturm M, Hackenberg M, Langenberger D, Frishman D (2010) TargetSpy: a supervised machine learning approach for microRNA target prediction. BMC Bioinformatics 11:1–17
88. Dai X, Zhuang Z, Zhao PX (2018) psRNATarget: a plant small RNA target analysis server (2017 release). Nucleic Acids Res 46: W49–W54
89. Bonnet E, He Y, Billiau K, Van de Peer Y (2010) TAPIR, a web server for the prediction of plant microRNA targets, including target mimics. Bioinformatics 26:1566–1568
90. Wu H-J, Ma Y-K, Chen T et al (2012) PsRobot: a web-based plant small RNA meta-analysis toolbox. Nucleic Acids Res 40: W22–W28
91. Fahlgren N, Howell MD, Kasschau KD et al (2007) High-throughput sequencing of Arabidopsis microRNAs: evidence for frequent birth and death of MIRNA genes. PLoS One 2:e219
92. Rueda A, Barturen G, Lebrón R et al (2015) SRNAtoolbox: an integrated collection of small RNA research tools. Nucleic Acids Res 43:473. https://doi.org/10.1093/nar/gkv555
93. Farh KK-H, Grimson A, Jan C et al (2005) The widespread impact of mammalian MicroRNAs on mRNA repression and evolution. Science 310:1817–1821
94. Guo H, Ingolia NT, Weissman JS, Bartel DP (2010) Mammalian microRNAs predominantly act to decrease target mRNA levels. Nature 466:835–840
95. Iwakawa H, Tomari Y (2013) Molecular insights into microRNA-mediated translational repression in plants. Mol Cell 52: 591–601
96. Llave C, Xie Z, Kasschau KD, Carrington JC (2002) Cleavage of Scarecrow-like mRNA targets directed by a class of Arabidopsis miRNA. Science 297:2053–2056
97. Addo-Quaye C, Eshoo TW, Bartel DP, Axtell MJ (2008) Endogenous siRNA and miRNA targets identified by sequencing of the Arabidopsis degradome. Curr Biol 18:758–762
98. German MA, Pillay M, Jeong D-H et al (2008) Global identification of microRNA--target RNA pairs by parallel analysis of RNA ends. Nat Biotechnol 26:941–946
99. Gregory BD, O'Malley RC, Lister R et al (2008) A link between RNA metabolism and silencing affecting Arabidopsis development. Dev Cell 14:854–866

100. Addo-Quaye C, Miller W, Axtell MJ (2009) CleaveLand: a pipeline for using degradome data to find cleaved small RNA targets. Bioinformatics 25:130–131

101. Zheng Y, Li Y-F, Sunkar R, Zhang W (2012) SeqTar: an effective method for identifying microRNA guided cleavage sites from degradome of polyadenylated transcripts in plants. Nucleic Acids Res 40:e28–e28

102. Thody J, Folkes L, Medina-Calzada Z et al (2018) PAREsnip2: a tool for high-throughput prediction of small RNA targets from degradome sequencing data using configurable targeting rules. Nucleic Acids Res 46(17):8730–8739. https://doi.org/10.1093/nar/gky609

103. Parks DH, Imelfort M, Skennerton CT et al (2015) CheckM: assessing the quality of microbial genomes recovered from isolates, single cells, and metagenomes. Genome Res 25(7):1043–1055. https://doi.org/10.1101/gr.186072.114

104. Bateman A, Martin MJ, O'Donovan C et al (2015) UniProt: a hub for protein information. Nucleic Acids Res 43(D1):D204–D212. https://doi.org/10.1093/nar/gku989

105. Lorrai R, Ferrari S (2021) Host cell wall damage during pathogen infection: mechanisms of perception and role in plant-pathogen interactions. Plan Theory 10:399

106. Bellincampi D, Cervone F, Lionetti V (2014) Plant cell wall dynamics and wall-related susceptibility in plant–pathogen interactions. Front Plant Sci 5:228

107. Piombo E, Sela N, Wisniewski M et al (2018) Genome sequence, assembly and characterization of two Metschnikowia fructicola strains used as biocontrol agents of postharvest diseases. Front Microbiol 9:593. https://doi.org/10.3389/fmicb.2018.00593

108. Mukhopadhyay R, Kumar D (2020) Trichoderma: a beneficial antifungal agent and insights into its mechanism of biocontrol potential. Egypt J Biol Pest Control 30(1): 1–8

109. Tzelepis G, Dubey M, Jensen DF, Karlsson M (2015) Identifying glycoside hydrolase family 18 genes in the mycoparasitic fungal species clonostachys rosea. Microbiology 161(7): 1407–1419. https://doi.org/10.1099/mic.0.000096

110. Datta SK, Muthukrishnan S (1999) Pathogenesis-related proteins in plants. CRC Press

111. Sharma N, Sharma KP, Gaur RK, Gupta VK (2011) Role of chitinase in plant defense. Asian J Biochem 6:29–37

112. Zhang H, Yohe T, Huang L et al (2018) DbCAN2: a meta server for automated carbohydrate-active enzyme annotation. Nucleic Acids Res 46:95–101. https://doi.org/10.1093/nar/gky418

113. Blin K, Shaw S, Steinke K et al (2019) Anti-SMASH 5.0: updates to the secondary metabolite genome mining pipeline. Nucleic Acids Res 47:81–87. https://doi.org/10.1093/nar/gkz310

114. Suga H, Arai M, Fukasawa E et al (2019) Genetic differentiation associated with fumonisin and gibberellin production in Japanese Fusarium fujikuroi. Appl Environ Microbiol 85(1):e02414–e02418

115. Fatema U, Broberg A, Jensen DF et al (2018) Functional analysis of polyketide synthase genes in the biocontrol fungus Clonostachys rosea. Sci Rep 8:1–17. https://doi.org/10.1038/s41598-018-33391-1

116. Vergine M, Nicolì F, Sabella E et al (2020) Secondary Metabolites in Xylella fastidiosa—Plant Interaction. Pathogens 9:675

117. Shah N, Li Q, Xu Q et al (2020) CRb and PbBa8. 1 synergically increases resistant genes expression upon infection of Plasmodiophora brassicae in Brassica napus. Genes (Basel) 11: 202

118. Li H, Zhou Z, Hua H, Ma W (2020) Comparative transcriptome analysis of defense response of rice to Nilaparvata lugens and Chilo suppressalis infestation. Int J Biol Macromol 163:2270–2285

119. Boba A, Kostyn K, Kozak B et al (2020) Fusarium oxysporum infection activates the plastidial branch of the terpenoid biosynthesis pathway in flax, leading to increased ABA synthesis. Planta 251:1–14

120. Ghosh S, Malukani KK, Chandan RK et al (2019) How plants respond to pathogen attack: interaction and communication. In: Sensory biology of plants. Springer, pp 537–568

121. Ceulemans E, Ibrahim HMM, De Coninck B, Goossens A (2021) Pathogen effectors: exploiting the promiscuity of plant signaling hubs. Trends Plant Sci 26(8):780–795

122. Martel A, Ruiz-Bedoya T, Breit-McNally C et al (2021) The ETS-ETI cycle: evolutionary processes and metapopulation dynamics driving the diversification of pathogen effectors and host immune factors. Curr Opin Plant Biol 62:102011

123. Levin E, Kishore A, Ballester AR et al (2019) Identification of pathogenicity-related genes and the role of a subtilisin-related peptidase S8 (PePRT) in autophagy and virulence of

Penicillium expansum on apples. Postharvest Biol Technol 149:209–220

124. Liao Y, Shi W (2019) Read trimming is not required for mapping and quantification of RNA-seq reads. NAR Genom Bioinform 2(3):lqaa068
125. Grabherr MG, Haas BJ, Yassour M et al (2011) Trinity: reconstructing a full-length transcriptome without a genome from RNA-Seq data. Nat Biotechnol 29:644
126. Tegli S, Bini L, Calamai S et al (2020) A MATE Transporter is Involved in Pathogenicity and IAA Homeostasis in the Hyperplastic Plant Pathogen Pseudomonas savastanoi pv. nerii. Microorganisms 8:156
127. Thakur K, Chawla V, Bhatti S et al (2013) De novo transcriptome sequencing and analysis for Venturia inaequalis, the devastating apple scab pathogen. PLoS One 8:e53937
128. Zhao C, Waalwijk C, De Wit PJGM et al (2011) EBR1, a novel Zn2Cys6 transcription factor, affects virulence and apical dominance of the hyphal tip in Fusarium graminearum. Mol Plant-Microbe Interact 24:1407–1418. https://doi.org/10.1094/MPMI-06-11-0158
129. Oh M, Son H, Choi GJ et al (2016) Transcription factor ART1 mediates starch hydrolysis and mycotoxin production in Fusarium graminearum and F.verticillioides. Mol Plant Pathol 17(5):755–768. https://doi.org/10.1111/mpp.12328
130. Brown DW, Lee SH, Kim LH et al (2015) Identification of a 12-gene fusaric acid biosynthetic gene cluster in Fusarium species through comparative and functional genomics. Mol Plant-Microbe Interact 28:319–332. https://doi.org/10.1094/MPMI-09-14-0264-R
131. Urban M, Cuzick A, Rutherford K et al (2017) PHI-base: a new interface and further additions for the multi-species pathogen-host interactions database. Nucleic Acids Res 45:604–610. https://doi.org/10.1093/nar/gkw1089
132. Hugot K, Riviere M-P, Moreilhon C et al (2004) Coordinated regulation of genes for secretion in tobacco at late developmental stages: association with resistance against oomycetes. Plant Physiol 134:858–870
133. Niderman T, Genetet I, Bruyere T et al (1995) Pathogenesis-related PR-1 proteins are antifungal (isolation and characterization of three 14-kilodalton proteins of tomato and of a basic PR-1 of tobacco with inhibitory activity against Phytophthora infestans). Plant Physiol 108:17–27
134. Chittoor JM, Leach JE, White FF (1997) Differential induction of a peroxidase gene family during infection of rice by Xanthomonas oryzae pv. oryzae. Mol Plant-Microbe Interact 10:861–871
135. Passardi F, Cosio C, Penel C, Dunand C (2005) Peroxidases have more functions than a Swiss army knife. Plant Cell Rep 24:255–265
136. Van Loon LC (1997) Induced resistance in plants and the role of pathogenesis-related proteins. Eur J Plant Pathol 103:753–765
137. Lane BG (2002) Oxalate, germins, and higher-plant pathogens. IUBMB Life 53:67–75
138. Lou Y, Baldwin IT (2006) Silencing of a germin-like gene in Nicotiana attenuata improves performance of native herbivores. Plant Physiol 140:1126–1136
139. Godfrey D, Able AJ, Dry IB (2007) Induction of a grapevine germin-like protein (VvGLP3) gene is closely linked to the site of Erysiphe necator infection: a possible role in defense? Mol Plant-Microbe Interact 20:1112–1125
140. Guevara-Olvera L, Ruíz-Nito ML, Rangel-Cano RM et al (2012) Expression of a germin-like protein gene (CchGLP) from a geminivirus-resistant pepper (Capsicum chinense Jacq.) enhances tolerance to geminivirus infection in transgenic tobacco. Physiol Mol Plant Pathol 78:45–50
141. Pedley KF, Martin GB (2005) Role of mitogen-activated protein kinases in plant immunity. Curr Opin Plant Biol 8:541–547
142. Edgar R (2016) UNOISE2: improved error-correction for Illumina 16S and ITS amplicon sequencing. bioRxiv. https://doi.org/10.1101/081257
143. Amir A, McDonald D, Navas-Molina JA et al (2017) Deblur rapidly resolves single-nucleotide community sequence patterns. mSystems 2(2):e00191–e00116. https://doi.org/10.1128/msystems.00191-16
144. Schloss PD, Westcott SL, Ryabin T et al (2009) Introducing mothur: open-source, platform-independent, community-supported software for describing and comparing microbial communities. Appl Environ Microbiol 75(23):7537–7541. https://doi.org/10.1128/AEM.01541-09
145. Durling MB, Clemmensen KE, Stenlid J, Lindahl B (2011) SCATA-An efficient bioinformatic pipeline for species identification and quantification after high-throughput sequencing of tagged amplicons

146. Rognes T, Flouri T, Nichols B et al (2016) VSEARCH: a versatile open source tool for metagenomics. PeerJ 4:e2584. https://doi.org/10.7717/peerj.2584
147. Pauvert C, Buée M, Laval V et al (2019) Bioinformatics matters: the accuracy of plant and soil fungal community data is highly dependent on the metabarcoding pipeline. Fungal Ecol 41:23–33. https://doi.org/10.1016/j.funeco.2019.03.005
148. Weiss S, Xu ZZ, Peddada S et al (2017) Normalization and microbial differential abundance strategies depend upon data characteristics. Microbiome 5:1–18
149. Willis AD (2019) Rarefaction, alpha diversity, and statistics. Front Microbiol 10:2407
150. Pereira MB, Wallroth M, Jonsson V, Kristiansson E (2018) Comparison of normalization methods for the analysis of metagenomic gene abundance data. BMC Genomics 19:1–17
151. Beule L, Karlovsky P (2020) Improved normalization of species count data in ecology by scaling with ranked subsampling (SRS): application to microbial communities. PeerJ 8:e9593
152. Robinson MD, Oshlack A (2010) A scaling normalization method for differential expression analysis of RNA-seq data. Genome Biol 11:1–9
153. Anders S, Huber W (2010) Differential expression analysis for sequence count data. Nat Preced 1
154. Paulson JN, Colin Stine O, Bravo HC, Pop M (2013) Differential abundance analysis for microbial marker-gene surveys. Nat Methods 10:1200–1202. https://doi.org/10.1038/nmeth.2658
155. Spellerberg IF, Fedor PJ (2003) A tribute to Claude Shannon (1916–2001) and a plea for more rigorous use of species richness, species diversity and the 'Shannon–Wiener' Index. Glob Ecol Biogeogr 12:177–179
156. Liu J, Abdelfattah A, Norelli J et al (2018) Apple endophytic microbiota of different rootstock/scion combinations suggests a genotype-specific influence. Microbiome 6:1–11
157. Abdelfattah A, Sanzani SM, Wisniewski M et al (2019) Revealing cues for fungal interplay in the plant–air interface in vineyards. Front Plant Sci 10:922. https://doi.org/10.3389/fpls.2019.00922
158. Piombo E, Abdelfattah A, Danino Y et al (2020) Characterizing the fungal microbiome in date (Phoenix dactylifera) fruit pulp and peel from early development to harvest. Microorganisms 8:641
159. Schoch CL, Seifert KA, Huhndorf S et al (2012) Nuclear ribosomal internal transcribed spacer (ITS) region as a universal DNA barcode marker for Fungi. Proc Natl Acad Sci U S A 109(16):6241–6246. https://doi.org/10.1073/pnas.1117018109
160. Stielow JB, Lévesque CA, Seifert KA et al (2015) One fungus, which genes? Development and assessment of universal primers for potential secondary fungal DNA barcodes. Persoonia 35:242–263. https://doi.org/10.3767/003158515X689135
161. Samson RA, Visagie CM, Houbraken J et al (2014) Phylogeny, identification and nomenclature of the genus Aspergillus. Stud Mycol 53(1):1–27. https://doi.org/10.1016/j.simyco.2014.07.004
162. Abdelfattah A, Malacrinò A, Wisniewski M et al (2018) Metabarcoding: a powerful tool to investigate microbial communities and shape future plant protection strategies. Biol Control 120:1–10. https://doi.org/10.1016/j.biocontrol.2017.07.009
163. Bagga S, Bracht J, Hunter S et al (2005) Regulation by let-7 and lin-4 miRNAs results in target mRNA degradation. Cell 122:553–563
164. Gu K, Mok L, Chong MMW (2018) Regulating gene expression in animals through RNA endonucleolytic cleavage. Heliyon 4:e00908
165. Brodersen P, Sakvarelidze-Achard L, Bruun-Rasmussen M et al (2008) Widespread translational inhibition by plant miRNAs and siRNAs. Science 320:1185–1190
166. Djuranovic S, Nahvi A, Green R (2012) miRNA-mediated gene silencing by translational repression followed by mRNA deadenylation and decay. Science 336:237–240

Chapter 18

Biomonitoring of Fungal and Oomycete Plant Pathogens by Using Metabarcoding

Émilie D. Tremblay and Guillaume J. Bilodeau

Abstract

Fungal and oomycete plant pathogens are responsible for the devastation of various ecosystems such as forest and crop species worldwide. In an effort to protect such natural resources for food, lumber, etc., early detection of non-indigenous phytopathogenic fungi in new areas is a key approach in managing threats at their source of introduction. A workflow was developed using high-throughput sequencing (HTS), more specifically metabarcoding, a method for rapid and higher throughput species screening near high-risk areas, and over larger geographical spaces. Biomonitoring of fungal and oomycete entities of plant pathogens (e.g., airborne spores) regained from environmental samples and their processing by metabarcoding is thoroughly described here. The amplicon-based approach goes from DNA and sequencing library preparation using custom-designed polymerase chain reaction (PCR) fusion primers that target the internal transcribed spacer 1 (ITS1) from fungi and oomycetes and extends to multiplex HTS with the Ion Torrent platform. In addition, a brief and simplified overview of the bioinformatics analysis pipeline and other available tools required to process amplicon sequences is also included. The raw data obtained and processed enable users to select a bioinformatics pipeline in order to directly perform biodiversity, presence/absence, geographical distribution, and abundance analyses through the tools suggested, which allows for accelerated identification of phytopathogens of interest.

Key words Oomycete, Fungi, Metagenomics, Amplicon sequencing, Ion Torrent S5, Metabarcoding, Bioinformatics

1 Introduction

Plant pathogenic fungal spores spread by wind, rain, international trading of goods, and even through direct or incidental vectors (e.g., insects), a phenomenon that can cause the introduction of exotic diseases sometimes responsible for the devastation of various plant species. New technologies such as metagenomics represent a high-throughput option for quicker phytopathogen screening by avoiding time-consuming and lower-throughput traditional methods such as culturing. Compared to insect surveillance, surveying pathogenic fungi is much harder logistically (size, type of material,

Nicola Luchi (ed.), *Plant Pathology: Methods and Protocols*, Methods in Molecular Biology, vol. 2536,
https://doi.org/10.1007/978-1-0716-2517-0_18,

quantities, etc.) and because many plants remain asymptomatic for a long time, when signs are noticeable it can already be too late for efficient containment. Diversity and incidence information obtained through metagenomics done on environmental samples such as airborne spores or insect traps can detect the presence of unwanted organisms, even in very low concentrations. The approach described can also identify hotspot areas and thereafter provide guidance for pest management, growers, and regulatory organizations. In the workflow presented in this chapter, genomic DNA (gDNA) was extracted from filtrates originating from spore traps and preservation liquids from baited beetle traps (stacked funnels). DNA amplicons were processed on the Ion Torrent S5 sequencer using custom fusion primers designed to multiplex hundreds of samples and two genic regions (350–400 bp amplicons from the ribosomal DNA region ITS1 (fungi and oomycetes)) [1] simultaneously.

However, the ITS1 region does not always allow species resolution (n.b. some groups of fungi/oomycetes can only be accurately identified at the genus level using the ITS1, and other genic region(s) must be used for proper resolution), and this proof of concept led to the finding of several closely related fungal or oomycete species, which were confirmed by species-specific quantitative PCR (qPCR) assays. That is because although bioinformatics analysis is an important step to process the millions of sequences generated by HTS from which the alignment results are used for the identification of targets, they should then be compared with more traditional/validated methods—such as species-specific qPCR assays—for confirmation done on the samples initially narrowed down by HTS.

In this chapter, we will describe the method and technology we used and developed to allow biomonitoring of fungal and oomycete plant pathogens by using a metabarcoding protocol, from DNA extraction, amplicon and library preparation, sequencing, and data analysis. This technology gives users access to multiple outputs including sample abundance, biodiversity, and phylogeny, in order to rapidly identify the current risks/areas associated with exotic plant pathogens.

Considerations to keep in mind when using this technology for the purpose of biomonitoring the diversity of oomycetes and fungi will be discussed, followed by an in-depth description of the protocols necessary to realize this work.

1.1 Databases (Public Versus Curated), Genic Regions, Their Strength and Weaknesses, and Their Importance

The National Center for Biotechnology Information (NCBI, www.ncbi.nlm.nih.gov/) advances science and health by providing open access to biomedical and genomic information and databases. The ITS sequences in NCBI nucleotide database (nt) are among the most commonly used one for the identification of organisms. This general and open-source database has the dual advantage of getting regular updates and providing wide-ranging sources of species sequence information. However, being an open-source resource, the data deposited in the NCBI are not curated prior to be logged in so their proper identification can be wrongly inputted and depend on the identification of the material deposited into the repository system, their authors' expertise, the identification methods and tools used, etc. Therefore, users of open-source databases shall treat their sequence alignments cautiously and use more robust datasets for validation, when possible. One main limitation of metabarcoding is its dependence on databases, as analyses are directly affected by their contents [2, 3]. As a matter of fact, even curated databases, which are highly reliable, are less comprehensive and inclusive given the time, expertise, and resources necessary to verify each sequence and therefore may not include any, or not as many species representatives compared to open-source databases. For example, fewer oomycetes are described compared to plants and fungi, limiting accessibility to realistically and accurately represented diversity [4–6]. A combination of open-source and curated databases is a valuable solution to cover each type of databases' weaknesses. For instance, to screen for our fungal species (ITS1), in addition to the NCBI nt, we also used UNITE [7], a curated database centered on the eukaryotic nuclear ribosomal ITS region of fungi. Since there is no equivalent publicly available curated database for oomycetes' species inference, an oomycete database, and taxonomic map, "OmDB" [8, 9], consisting of an extraction of all oomycete ITS sequences within the NCBI nt, was generated and used as the sole database.

The genic region ITS also has its pros and cons for species identification in metagenomics. Pros include multiple copy numbers, which allows for high detection sensitivity, sufficient variability for genus and sometimes species resolution, gene marker universality among all eukaryotes, and availability of a large variety of reference sequences [10–12]. One con is that not all fungi or oomycetes can be identified down to the species level due to insufficient variability within the region. In addition, any given bioinformatics pipeline used will affect the final taxonomic assignments and downstream analyses, which is why it is critical to select an approach that will be the best fit for the results aimed for and the type of samples to be tested [13].

In this chapter, we used curated and open-source databases to identify our amplicons clustered in operating taxonomic units (OTUs). OTUs are a largely used concept that groups sequence reads into clusters based on their level (%) of dissimilarity. For fungi and oomycetes, reads that differ $\geq$3% will generally be placed in different clusters, and those above that threshold will be gathered together. This concept allows for less computer resource-demanding analyses as it reduces the amount of sequences to analyze once grouped into one representative sequence [14, 15]. The OTUs obtained through bioinformatics analyses were used to infer genus and/or species identity, but there is a more recent concept, Amplicon Sequence variants (ASVs, also known as ESVs; Exact Sequence Variants), that has been gaining in popularity and some scientists even believe they shall replace the OTUs [16]. Without going deep into the details, the main advantages of ASVs over OTUs are that they allow for comparison of data from different sequencing runs, and they do not rely on an predefined arbitrary threshold value and can potentially allow for better resolution down to the species level (up to one nucleotide difference could be caught) [16–18].

1.2 Material for Biomonitoring and DNA Extraction

For biomonitoring and mostly when looking for rare species such as potential invasive species, it is important to get environmental material that will simultaneously provide accurate information on the source of the organism, while also achieving great detection sensitivity. Biomonitoring is not necessarily about looking at overall diversity or ecological niches, as it can alternatively target-specific organisms present in very low abundance. Although not touched on in this chapter, it is critical that the extraction method chosen to access gDNA is best suited (i.e., optimized for recovery and yield) for the type of material that samples originate from. Depending on the source of the material (airborne spores, insect trap, plants, soil, etc.), often consisting of an heterogeneous mix of organisms and organic matter that contains PCR inhibitors, the extraction and purification treatments will impact the quantity and the quality of the gDNA recovered and the ability to retrieve specimens present in minute concentration. For those reasons, although commercially available extraction/purification kits may be faster and more user-friendly (large-scale kits often avoid the use of less hazardous chemicals such as organic solvents), they may not allow for the optimal level of DNA recovery needed for metabarcoding from environmental samples. While commercial DNA extraction kits are generally adequately suited for extracting nucleic acids, the extraction procedures often need to be optimized to ensure a robust extraction for specific material or pathogens with survival structures (e.g., thick-walled spores) that can be difficult to disrupt. PCR inhibitors include but are not limited to humic acids, phenolic compounds, heavy metals, clay particles that bind to DNA, components of bacterial cells, and other contaminants from the extraction workflow [19–22]. A commonly used approach to reduce the impact of PCR inhibitors is to dilute the extracted DNA to minimize their concentration. Another approach, presented in this chapter, is to use magnetic particle purification that binds to DNA [23]. However, both techniques also lessen the sensitivity of the assay and can prevent detection when the pathogen is present at low levels as the DNA is also diluted/reduced in concentration. This is a reason why an improved method for purification of nucleic acids and tools to evaluate PCR amplification efficiency and independently of the pathogen quantification assay would need to be developed.

1.3 Popular Sequencing Platforms

The most frequently used sequencing platform, Illumina, uses fluorescence-based nucleotide sequence detection. Another popular platform, the Ion Torrent that uses pH measurements to read nucleotide sequences, is the technology opted for and described in this book chapter. Usually, Illumina sequence reads that are generated during a single experiment have the same length and the sequence reads from both ends of a fragment ("paired-end" reads, options of 150 × 2 bp or 300 × 2 bp), whereas those generated through the Ion Torrent sequencer vary in length, are single-ended, and must be selected for a specific size of up to 400–500 base pairs [24–27]. There are also other popular sequencing platforms that are better suited for long fragments including the Oxford Nanopore and PacBio, which could represent an option if users were targeting longer reads for their specific purposes.

1.4 Validation of Findings

Validation using species-specific assays or traditional methods is a highly recommended practice to confirm both the presence and identity of a target species putatively detected via metagenomics. Tremblay et al. [1] used qPCRs, targeting different and/or multiple genic regions, to validate findings on the few samples that were suspected of containing target species after HTS and bioinformatics processing. Another confirmation approach consists of sequencing specific genic region(s) (using highly specific primers, if available) by traditional Sanger sequencing. The same method can be completed for other gene regions as well. Please note that difference in sensitivity in environmental sample can vary from a PCR and a qPCR amplification, in part due to the difference in copy numbers for the targeted gene.

2 Materials

2.1 Instruments Needed

1. Ion S5 System (Thermo Fisher Scientific).
2. Ion Chef Instrument (Thermo Fisher Scientific).
3. ViiA 7 Real-Time PCR System (Thermo Fisher Scientific).
4. Veriti Thermal Cycler (Thermo Fisher Scientific).
5. Qubit Fluorometer (Thermo Fisher Scientific).
6. FastGene MagnaStand 96-well plate (Bulldog Bio Inc.).
7. MagneSphere(R) Magnetic Separation Stand 12-Hole, 1.5 mL Vial (Promega).
8. A benchtop centrifuge for microtubes.
9. A vortex for microtubes.

2.2 Reagents for Ion Torrent and Sequencing Library Preparation

1. QuickPick XL Plant DNA Magnetic Particles (Bio-Nobile).
2. QuickPick XL Plant DNA Binding Buffer (Bio-Nobile).
3. QuickPick XL Plant DNA Wash Buffer (Bio-Nobile).
4. QuickPick XL Plant DNA Elution Buffer (Bio-Nobile).
5. Agencourt AMPure XP Reagent 5 mL (Beckman Coulter).
6. Ion 510 & 520 & 530 Kit—Chef (Thermo Fisher Scientific).
7. Ion Universal Library Quantitation Kit (Thermo Fisher Scientific).
8. Qubit 1× dsDNA HS, 500 assays kit (Thermo Fisher Scientific).
9. Platinum Taq DNA Polymerase (Thermo Fisher Scientific).
10. DNA Gel Loading Dye (6×) (Thermo Fisher Scientific).
11. Low Molecular Weight DNA Ladder (New England Biolabs).

2.3 Other Solution or Reagents

1. TBE Buffer (Tris-borate-EDTA) (10×) (Thermo Fisher Scientific).
2. UltraPure agarose, 500 g (Thermo Fisher Scientific).
3. UltraPure 1 M Tris-HCI Buffer, pH 7.5 (Thermo Fisher Scientific).
4. TE Buffer, pH 8.0, RNase-free (Thermo Fisher Scientific).

3 Methods

3.1 gDNA Purification Using QuickPick

1. Depending on the volume of gDNA available, you will need to adjust the protocol for this purification. You can use as a little as 10 μL of DNA and adjust the volume of reagents used proportionally. In this example, 12.5 μL will be used.
2. In a clear or light-colored 1.5-mL Eppendorf tube for each sample (labeled accordingly and uniquely), pipet 312.5 μL of QuickPick XL Plant DNA Binding Buffer (*see* **Note 1**).
3. Add 12.5 μL of QuickPick XL Plant DNA Magnetic Particles [23] into each sample tube (*see* **Note 2**).
4. Add 25 μL of gDNA solution in separate respective sample tubes.
5. Gently and continuously mix the suspension with a tube rotator (or manually inverting tubes) at room temperature for 10 min. DNA will pair with magnetic particles, allowing for physical liquid separation from this step until Step **10**.
6. Using a benchtop centrifuge designed for microtubes, spin down tubes quickly for 3–5 s, at a low speed, to pull solutions at the bottom of the tubes.
7. Place sample tubes into the slots of a magnetic bar or rack for 2 min. Initially brown and homogenous, the solution will

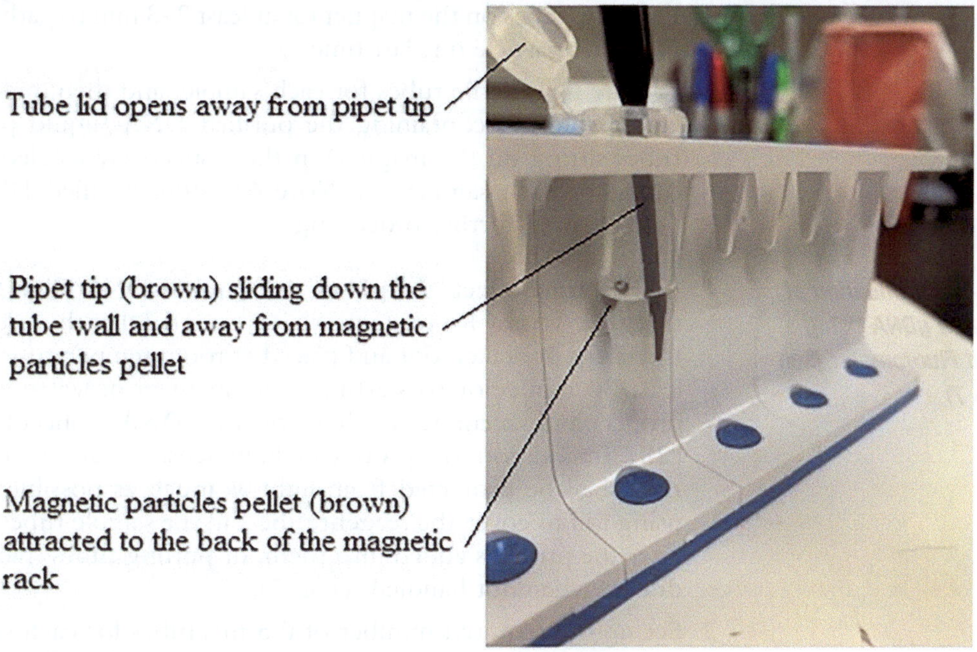

Fig. 1 Magnetic bead captured on magnet

steadily clarify and concentrate at the magnet contact point. When placing the tubes in, ensure their opening is accessible (Fig. 1) given that next, you will need to remove the liquid part, while leaving the tubes in the rack and being careful not to dislodge or touch the clump of DNA and magnetic particles attracted to the bar or rack (*see* **Note 3**). Discard.

8. Remove tubes from the magnet device. Dispense 500 μL of QuickPick XL Plant DNA Wash Buffer. Gently mix with a few quick finger flicks, and then, repeat Step **5**.

9. Continue with the first wash by repeating what was done in Step **6**, but after the 2 min on the magnet stage, set your pipet at a 500 μL volume (e.g., 500 for a P1000) for liquid collection. Discard the wash solution (*see* **Note 4**).

10. Perform a second wash by repeating Steps 7 and **8**; i.e., add 500 μL of QuickPick XL Plant DNA Wash Buffer, place on magnet for 2 min, and then pipet out and discard the total volume, being careful not to dislodge the pellet.

11. Remove tubes from magnet and add 25 μL of QuickPick XL Plant DNA Elution Buffer (*see* **Note 5**).

12. Gently and continuously mix the suspension with a tube rotator (or manually inverting tubes) at room temperature for 10 min.

13. Spin down the tubes as described in Step **6**.

14. Place the tubes on the magnet for at least 2–3 min to gather the magnetic particle one last time.
15. Prepare new sterile tubes for each sample, and then, pipet the eluate solution containing the purified DNA (liquid part in tubes sitting on the magnet) in the respectively labeled fresh tubes for each sample (*see* **Note 6**). Store purified DNA at −20 °C until further processing.

3.2 Quantification of Purified gDNA Using a Qubit Fluorometer (See Note 7)

1. Temperature affects the pH of the Qubit dsDNA High Sensitivity (HS) Assay Kit reagents, so it is essential that those kept in the fridge are taken out and placed at room temperature for at least 30 min prior to starting experiment in order to ensure proper and accurate reading of the DNA concentration [28]. In addition, the probes are light-sensitive, and therefore, they shall be protected from light as much as possible (e.g., using foil to cover the reagent tube and the sample tubes/rack after the probe is added into them, or putting tubes/rack in a drawer when not handled) (Fig. 2).
2. Set up the required number of 0.5 mL tubes for each sample and add two extra ones for the standards (*see* **Note 8**).
3. Prepare the Qubit working solution following this calculation, in μL:

   [number of samples + 2 standards + 2 extras (to avoid volume shortages from pipetting)] × 199 (each reaction's final volume is 200 μL) = required volume (in μL) of Qubit dsDNA HS Buffer to add in one large tube (*see* **Note 9**).

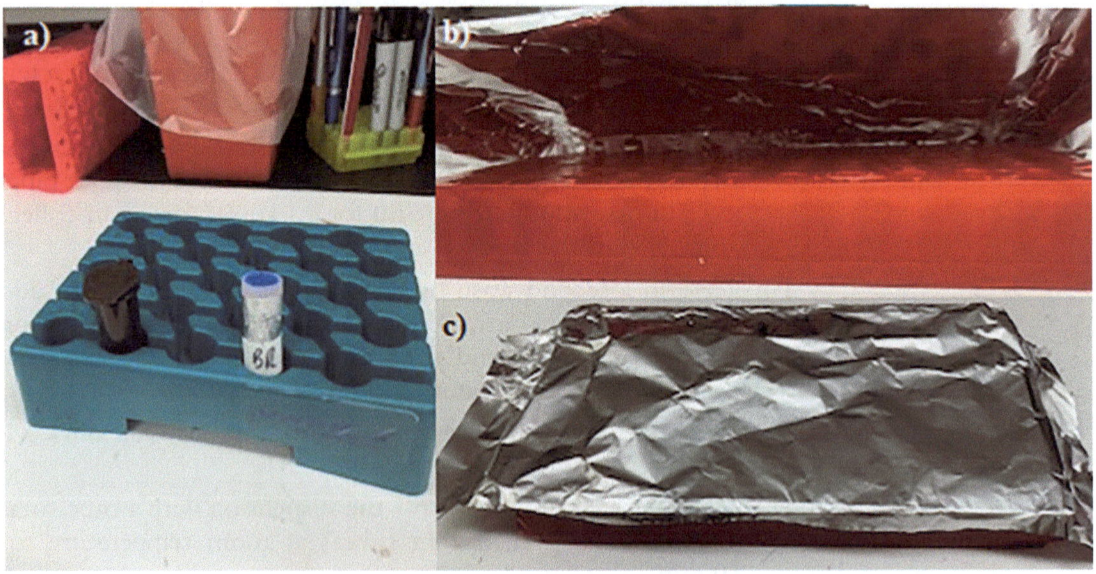

Fig. 2 Representation of the Qubit dsDNA High Sensitivity (HS) Assay probe preparation and protection from light, (**a**) foil to cover the reagent tube and (**b**–**c**) the sample tubes and rack before and after the probe is added into them

4. Add Qubit dsDNA HS Reagent (light-sensitive) calculated volume to your larger tube containing the buffer, following this formula: [number of samples + 2 standards + 2 extras] (*see* **Note 10**). Flick, gently vortex, or pipet up and down to mix, then use a benchtop centrifuge designed for microtubes and proper-sized tube adaptor, and spin down tubes quickly for 3–5 s, at a low speed, to pull solutions at the bottom of the tubes (*see* **Note 11**).
5. Add 190 μL of working solution to each of the 2 tubes that will be used for standards.
6. Add 10 μL of Qubit standard #1 to the appropriately labeled tube, which contains 190 μL of the working solution. Similarly, add 10 μL of Qubit standard #2 to the appropriately labeled tube, which also contains 190 μL of the working solution. As described in Step **4**, flick, gently vortex, or pipet up and down to mix, and then spin down tubes quickly for 3–5 s (*see* **Note 12**).
7. Add 199 μL of working solution to individual assay tubes labeled for your samples. You will likely have a little bit of extra volume left of the working solution as there were two extra "assays" included in Step **3**.
8. Add 1 μL of each sample (or of their dilution, if applicable) to the assay tubes containing 199 μL of working solution. As described in Step **4**, flick, gently vortex, or pipet up and down to mix, and then spin down tubes quickly for 3–5 s (*see* **Note 13**). Protect tubes from light until reading.
9. Incubate all standard and sample tubes (protected from light) at room temperature for 2 min.
10. On the "Home touch screen" of the Qubit Fluorometer instrument, press "DNA" and then select "dsDNA High Sensitivity" as the assay type. The "Read standards" screen is displayed. Ensure that your instrument is set for a 200 μL assay and a 1 μL volume of DNA (*see* **Note 14**).
11. When the reading is complete (~3 s), remove Standard #1. Repeat the same process for Standard #2.
12. After both standards are read, press Read samples, insert your first sample tube into the chamber, close the lid, and then hit "Read tube." When the reading is complete (~3 s), remove the sample tube, note the raw data value, and if you wish, scroll on the screen to get the calculated concentration (samples are diluted 1:200). Alternatively, the raw data values could be converted by multiplying them by a 0.2 factor. Read next tubes and continue noting the values until you are finished. If the instrument cannot detect DNA in your tube, it will output a $\leq$0.5 ng/μL value (*see* **Note 15**).
13. Discard assay tubes after you are done.

3.3 Target Region PCR Amplification with Fusion Primers Technology

1. It is required to order the desired custom-made, bidirectional, and barcoded fusion primers for each region targeted. Those primers include different sections such as sequencing adaptors, a key, a barcode, and general PCR primers, and the sequence order is critical [1, 29] (*see* **Note 16**). Each sample requires two primers sets (one set for forward and the other one for reverse sequencing) for each genic region amplified. So, if one sample is to be screened for both fungi and oomycete ITS1 region, then eight different primers will be needed. In order to keep track of each sample, a unique barcode (10–12 bases long) will be assigned to it and remain the same sequence used for the primer sets used. Barcode sequences are predetermined by the Ion Torrent technology, and a list of 96 of them can be found in Table 1. Each environmental sample shall have their own barcode assigned (*see* **Note 17**). Table 2 provides further detail on how to construct the sequences of the primers to be ordered (*see* **Note 18**).
2. Resuspend and dilute the primer stock received at a 10 μM concentration for running primers. Typically, manufacturers indicate the volume of DNA-free water to add for resuspending stock primers at a 100 μM concentration. A simple 1:10 dilution can then be done to reach the desired 10 mM concentration (*see* **Note 19**).
3. If needed, dilute aliquots of your purified DNA at a concentration of ~0.1 ng/μL to normalize your starting material (*see* **Note 20**).
4. Prepare your PCR master mixes for the different directions and regions targeted. Each environmental sample requires 2 PCR (bidirectional) per genomic region so, if both the ITS1 for fungi and oomycetes are planned, 4 different PCR master mixes per environmental sample will be needed (*see* **Note 21**). The amplicons generated will be ~350–400 bp. Keep solutions on a cold block at all times. Each 25 μL PCR needs a) a list of reagents commonly used in each PCR (Table 3, List A) and b) reagents uniquely added to each PCR well (Table 3, List B). As shown in Table 3, List A comprises 1× PCR buffer, 1 mM $MgCl_2$, 0.25 mM dNTP, 0.50 mM of the P1-appended primer (e.g., ITS2-P1 in Fig. 5a, or Omlo5.8S47-P1 in Fig. 5b, or ITS1F-P1 in Fig. 5c, or Omup8S67-P1 in Fig. 5d), 0.04U of Platinum Taq polymerase, and DNA-free water to complete the reaction volume. A large batch master mix can be prepared for those reagents. However, the rest of the components for the 25 μL PCR (Table 3, List B) must be added one at a time, given that the barcodes ***and*** the PCR/sequencing directions vary for each PCR well. Those are the A1-appended barcoded primers (0.50 mM) and 1 μL of purified genomic DNA ($\leq$0.1 ng).

Table 1
List of Ion Xpress barcodes 1–96 sequences compatible with the Ion Torrent platform [29]

Index	Identifier	Sequence	Number of bases
1	Ion Xpress 1	CTAAGGTAAC	10
2	Ion Xpress 2	TAAGGAGAAC	10
3	Ion Xpress 3	AAGAGGATTC	10
4	Ion Xpress 4	TACCAAGATC	10
5	Ion Xpress 5	CAGAAGGAAC	10
6	Ion Xpress 6	CTGCAAGTTC	10
7	Ion Xpress 7	TTCGTGATTC	10
8	Ion Xpress 8	TTCCGATAAC	10
9	Ion Xpress 9	TGAGCGGAAC	10
10	Ion Xpress 10	CTGACCGAAC	10
11	Ion Xpress 11	TCCTCGAATC	10
12	Ion Xpress 12	TAGGTGGTTC	10
13	Ion Xpress 13	TCTAACGGAC	10
14	Ion Xpress 14	TTGGAGTGTC	10
15	Ion Xpress 15	TCTAGAGGTC	10
16	Ion Xpress 16	TCTGGATGAC	10
17	Ion Xpress 17	TCTATTCGTC	10
18	Ion Xpress 18	AGGCAATTGC	10
19	Ion Xpress 19	TTAGTCGGAC	10
20	Ion Xpress 20	CAGATCCATC	10
21	Ion Xpress 21	TCGCAATTAC	10
22	Ion Xpress 22	TTCGAGACGC	10
23	Ion Xpress 23	TGCCACGAAC	10
24	Ion Xpress 24	AACCTCATTC	10
25	Ion Xpress 25	CCTGAGATAC	10
26	Ion Xpress 26	TTACAACCTC	10
27	Ion Xpress 27	AACCATCCGC	10
28	Ion Xpress 28	ATCCGGAATC	10
29	Ion Xpress 29	TCGACCACTC	10
30	Ion Xpress 30	CGAGGTTATC	10
31	Ion Xpress 31	TCCAAGCTGC	10
32	Ion Xpress 32	TCTTACACAC	10

(continued)

Table 1
(continued)

Index	Identifier	Sequence	Number of bases
33	Ion Xpress 33	TTCTCATTGAAC	12
34	Ion Xpress 34	TCGCATCGTTC	11
35	Ion Xpress 35	TAAGCCATTGTC	12
36	Ion Xpress 36	AAGGAATCGTC	11
37	Ion Xpress 37	CTTGAGAATGTC	12
38	Ion Xpress 38	TGGAGGACGGAC	12
39	Ion Xpress 39	TAACAATCGGC	11
40	Ion Xpress 40	CTGACATAATC	11
41	Ion Xpress 41	TTCCACTTCGC	11
42	Ion Xpress 42	AGCACGAATC	10
43	Ion Xpress 43	CTTGACACCGC	11
44	Ion Xpress 44	TTGGAGGCCAGC	12
45	Ion Xpress 45	TGGAGCTTCCTC	12
46	Ion Xpress 46	TCAGTCCGAAC	11
47	Ion Xpress 47	TAAGGCAACCAC	12
48	Ion Xpress 48	TTCTAAGAGAC	11
49	Ion Xpress 49	TCCTAACATAAC	12
50	Ion Xpress 50	CGGACAATGGC	11
51	Ion Xpress 51	TTGAGCCTATTC	12
52	Ion Xpress 52	CCGCATGGAAC	11
53	Ion Xpress 53	CTGGCAATCCTC	12
54	Ion Xpress 54	CCGGAGAATCGC	12
55	Ion Xpress 55	TCCACCTCCTC	11
56	Ion Xpress 56	CAGCATTAATTC	12
57	Ion Xpress 57	TCTGGCAACGGC	12
58	Ion Xpress 58	TCCTAGAACAC	11
59	Ion Xpress 59	TCCTTGATGTTC	12
60	Ion Xpress 60	TCTAGCTCTTC	11
61	Ion Xpress 61	TCACTCGGATC	11
62	Ion Xpress 62	TTCCTGCTTCAC	12
63	Ion Xpress 63	CCTTAGAGTTC	11
64	Ion Xpress 64	CTGAGTTCCGAC	12
65	Ion Xpress 65	TCCTGGCACATC	12

(continued)

Table 1
(continued)

Index	Identifier	Sequence	Number of bases
66	Ion Xpress 66	CCGCAATCATC	11
67	Ion Xpress 67	TTCCTACCAGTC	12
68	Ion Xpress 68	TCAAGAAGTTC	11
69	Ion Xpress 69	TTCAATTGGC	10
70	Ion Xpress 70	CCTACTGGTC	10
71	Ion Xpress 71	TGAGGCTCCGAC	12
72	Ion Xpress 72	CGAAGGCCACAC	12
73	Ion Xpress 73	TCTGCCTGTC	10
74	Ion Xpress 74	CGATCGGTTC	10
75	Ion Xpress 75	TCAGGAATAC	10
76	Ion Xpress 76	CGGAAGAACCTC	12
77	Ion Xpress 77	CGAAGCGATTC	11
78	Ion Xpress 78	CAGCCAATTCTC	12
79	Ion Xpress 79	CCTGGTTGTC	10
80	Ion Xpress 80	TCGAAGGCAGGC	12
81	Ion Xpress 81	CCTGCCATTCGC	12
82	Ion Xpress 82	TTGGCATCTC	10
83	Ion Xpress 83	CTAGGACATTC	11
84	Ion Xpress 84	CTTCCATAAC	10
85	Ion Xpress 85	CCAGCCTCAAC	11
86	Ion Xpress 86	CTTGGTTATTC	11
87	Ion Xpress 87	TTGGCTGGAC	10
88	Ion Xpress 88	CCGAACACTTC	11
89	Ion Xpress 89	TCCTGAATCTC	11
90	Ion Xpress 90	CTAACCACGGC	11
91	Ion Xpress 91	CGGAAGGATGC	11
92	Ion Xpress 92	CTAGGAACCGC	11
93	Ion Xpress 93	CTTGTCCAATC	11
94	Ion Xpress 94	TCCGACAAGC	10
95	Ion Xpress 95	CGGACAGATC	10
96	Ion Xpress 96	TTAAGCGGTC	10

Table 2

Barcoded fusion primers designed for fungi or oomycetes multiplexing and their associated partner primer

PCR primer sense	Target organism	Barcode identifier	PCR primer partner	Custom primer sequence: Seq.[a] adaptor	Custom primer sequence: key	Custom primer sequence: Barcode (IonXpress 1-3)	Custom primer sequence: Spacer[b] and general PCR primer	General primer reference
Forward	Fungi	ITS[c]1FAB[d]-1	ITS2-P1[e]	CCATCTCATCCCT GCGTGTCTCCGAC	TCAG	CTAAGGTAAC	GATCTTGGTCATT TAGAGGAAGTAA	[48]
		ITS1FAB-2				TAAGGAGAAC		
		ITS1FAB-3				AAGAGGATTC		
		ITS2AB-1	ITS1F-P1			CTAAGGTAAC	GATGCTGCGTTCT TCATCGATGC	[49]
		ITS2AB-2				TAAGGAGAAC		
		ITS2AB-3				AAGAGGATTC		
		Omup18S67AB-1	Omlo5.8S47-P1			CTAAGGTAAC	GATCTCGCCATTT AGAGGAAGGT	C.A. Levesque, personal communication
		Omup18S67AB-2				TAAGGAGAAC		
		Omup18S67AB-3				AAGAGGATTC		
	Oomycete	Omlo58S47AB-1	Omup18S67-P1			CTAAGGTAAC	GATATTACGTATC GCAGTTCGCAG	[50, 51]
		Omlo58S47AB-2				TAAGGAGAAC		
		Omlo58S47AB-3				AAGAGGATTC		
Reverse	Fungi	ITS1F-P1	ITS2AB-1, ITS2AB-2, ITS2AB-3.	CCACTACGCCTCCGCTTTC CTCTCTATGGGCAGTCGGT	None	None	GATCTTGGTCATT TAGAGGAAGTAA	[48]
		ITS2-P1	ITS1FAB-1, ITS1FAB-2, ITS1FAB-3.				GATGCTGCGTTCT TCATCGATGC	[49]
		Omup18S67-P1	Omlo58S47AB-1, Omlo58S47AB-2, Omlo58S47AB-3.				GATCTCGCCATTT AGAGGAAGGT	C.A. Levesque, Personal communication
	Oomycete	Omlo5.8S47-P1	Omup18S67AB-1, Omup18S67AB-2, Omup18S67AB-3.				GATATTACGTATC GCAGTTCGCAG	[50, 51]

Adapted from Tremblay et al. [1]

[a]Sequencing

[b]The spacer (nucleotides G-A-T) is inserted between the barcode and the specific primer to help the sequencer's software discriminate properly the barcode sequences which are all by a "C." This helps properly sorting reads

[c]Internal Transcribed Spacer

[d]Sequencing adaptor A1

[e]Sequencing adaptor P1

Table 3
Reagent volumes and concentrations required for each 25 μL PCR, listed by commonly shared components (List A) and unique to each PCR well ones (List B)

Reagent	Stock concentration	Vol.[a] (μL) per 25 μL reaction	Final concentration
List A			
PCR Buffer	10×	2.50	1×
$MgCl_2$	50 mM	0.50	1 mM
dNTP[b] mix	10 mM	0.63	0.25 mM
P1-appended primer	10 μM	1.25	1.25 μM
Platinum Taq polymerase	5 U/μL	0.20	0.04 U/μL
DNA-free water	Not applicable	17.67	Not applicable
List B			
Barcoded primer	10 μM	1.25	1.25 μM
Purified genomic DNA	≤0.1 ng/μL	1.00	≤0.004 ng/μL
Final vol.		25.00	

[a]Volume
[b]Deoxynucleotide mixture consisting of an equimolar solution of the four DNA bases: dATP, dCTP, dGTP, and dTTP

5. Tightly seal tubes or plates (*see* **Note 22**). Mix by inverting 5 times, and then using a benchtop centrifuge designed for microtubes, gently spin down tubes or plate for ~1 min, at a low speed, to pull solutions at the bottom of the tubes, and dislodge trapped air bubbles.
6. Run your PCR on a thermocycler according to the following cycling conditions. One cycle consisting of 3 min at 95 °C; followed by 30 cycles for 30 s at 95 °C, 30 s at 52 °C, and 1 min at 72 °C for each cycle; and 1 cycle lasting 10 min at 72 °C. Optionally, users could also set up an idling cycle at 4–10 °C (∞) at the end (*see* **Note 23**). This step is designed to better preserve the PCR products and could be applied overnight if needed.
7. Store your PCR products at −20 °C until further processing.

3.4 Electrophoresis on PCR Products (i.e., Indexed Amplicons)

1. Prepare a 1.5% agarose gel to confirm proper DNA amplification of your amplicons [30] (*see* **Note 24**). Each sample will require one well, 1–2 and extra wells for the molecular markers will have to be included (minimum of one per gel row). Heat the solution until crystals have disappeared using a microwave (in 30 s intervals) while being careful with the heated solution which can boil and spill easily.
2. Let the solution cool down to about 50 °C, but not too much as the gel would solidify, and then add 11 μL of GelRed stain. Carefully swirl to mix, and avoid introducing air bubbles.

Table 4
Gel electrophoresis recipes for different dimensions [47]

Tray/gel dimension (cm)	0.5× TBE (mL)	Agarose (g)	GelRed (μL)	Running time (min)	Voltage (V)
7	40	0.6	4	~45–50	~100
10	90	1.35	9		
15	110	1.65	11	~80–90	~120–130
19	150	2.25	15		
25	225	3.38	22.5		

3. Steadily pour gel into the assembled gel mold while avoiding the introduction of air bubbles. Ensure the volume is evenly distributed and add combs.
4. Wait until the gel has solidified at room temperature and then place the gel into the electrophoresis cell/apparatus filled with running buffer (i.e., 0.5× TBE). Ensure that your running buffer volume is sufficient to just cover the wells.
5. Gently remove the combs pulling up straight.
6. After thawing your DNA (samples and controls), add 1 μL of 6× loading dye to 2–4 μL of DNA aliquots for each sample and mix both solutions on a mixing plate or a piece of Parafilm. Keep the mixture on cold block as the dye evaporates fast at room temperature.
7. Load your gel wells with the DNA and loading dye mixture. For each gel row, add 4 μL of a low molecular weight DNA ladder/marker (e.g., 100 bp DNA ladder, which covers 25 bp to 766 bp fragments) that covers the 300–500 bp size range in one or 2 well(s) and allows to space out bands at every ~100 bp near the desired 400 bp band.
8. Close the apparatus, and connect the leads. The black one (cathode) shall be the one closer to the wells, whereas the red one (anode) shall be the one at the other end.
9. Program the apparatus to run for 90 min at 120 V and start the instrument (*see* **Note 25**).
10. Once the program is finished, capture a photograph of your wells/amplification bands using a UV light system designed for that purpose.
11. Your DNA bands shall be located around the 400 bp marker. It is possible that you do not get bands for some samples as the fungal or oomycete portion may be in very low concentration. This does not automatically mean that you do not have usable material, so do not skip further processing of those samples yet. You will likely notice another faint band at the 50–100 bp mark; this is expected and represents the primer–dimers formed, which need to be cleaned up (*see* next step).

3.5 Sequentially Purify PCR Products at Two Different Concentrations

This step is to remove excess primers and primer–dimers. Ratios are determined for a specific size clean-up. The 0.7× purification will remove fragments ≥500 bp, whereas the 1.4× ratio will remove everything ≤100 bp. The combination of those successive cleaning will therefore keep only fragments of the desired length (~150–400 bp) [31].

1. Resuspend Agencourt AMPure XP reagent by mixing the bottle up and down, and let it reach room temperature before use.
2. Select tubes, tube strips, or a 96-well plate that will be compatible with the magnet type/format you will be using for this step. Uniquely label tubes or use a system to keep track of the sample location in the plate/strip used (*see* **Note 26**). Most 96-well plates have suitable 8-strip caps.
3. Add 0.7× volume of Agencourt AMPure XP reagent to an aliquot of your amplified genomic DNA volume (i.e., a 0.7:1 beads:DNA ratio). For example, to 15 μL of gDNA, 10.5 μL of the reagent is required (*see* **Note 27**).
4. Mix well the suspension by pipetting up and down several times. This will ensure contact between the particles and your DNA so that they pair together.
5. Incubate your tubes at room temperature for 5 min.
6. Place tubes on magnet plate for 2 min. Similar to what was explained in *Step 7* in Subheading 3.1, sit your tubes or plate and covers strategically, depending on where the magnetic part of your device is located.
7. Once solution is clear and the particles have pelleted, remove and discard supernatants without disturbing pellet while keeping the samples on the magnetic device.
8. While keeping samples on magnet still, perform the washing step by adding 200 μL of *freshly prepared* 70% ethanol (*see* **Note 28**).
9. Incubate the solution at room temperature for 30–45 s.
10. Remove and discard ethanol supernatant without disturbing pellet and repeat the wash (*i.e. Steps 8 and 9*) with another 200 μL of *freshly prepared* 70% ethanol.
11. After the second wash, remove residual ethanol by aspirating it with a 20 μL pipet, avoiding pellet and on magnet still.
12. On the magnet, allow pellets to air-dry (i.e., lids opened) for ~5 min at room temperature (*see* **Note 29**).
13. Remove samples from the magnet and then add 15 μL of TE buffer pH 8,0 (*see* **Note 30**).
14. Place tubes on magnet for ~1 min or until the solution clarifies and particles pellet neatly in a clump.

15. The TE buffer has now separated your DNA from the particles, so while avoiding the pellets, collect your DNA samples—the supernatant—by pipetting the volume into a new collection tube.
16. Store product at −20 °C or proceed with the next purification process described next.
17. Repeat *Steps 1* and *2*.
18. From the purified products (at 0.7×) obtained in *Step 16*, add a 1.4:1 beads:DNA ratio (1.4×) volume of Agencourt AMPure XP reagent to an aliquot of your purified (0.7×) genomic DNA volume (*see* **Note 31**).
19. Repeat *Steps 4–15* with the same volumes and concentrations of ethanol and TE buffer.
20. Store purified product at −20 °C until further processing.

3.6 Quantify Bidirectional Barcoded Sample Library by qPCR, Dilute and Pool Your Library

1. Prior to being pooled for the sequencing, your samples need to be adjusted at an equimolar concentration. The Ion Universal Library Quantitation qPCR Kit used on serial dilutions of your samples will allow you to determine the concentration of properly tagged DNA fragments within each sample tube using a standard curve [32, 33]. A standardized library control (*E. coli* DH10B library control) with a known concentration of 68 pM will be used to quantify your samples. Following the volumes detailed in Table 5, prepare 6 serial dilutions as follows.
2. While ensuring that you keep the stock and your dilutions on a cold block/ice at all time, slowly thaw the *E. coli* DH10B library control (*see* **Note 32**).
3. Using a benchtop centrifuge designed for microtubes, spin down tubes quickly for 3–5 s, at a low speed, to pull solutions at the bottom of the tubes.

Table 5
Quantification with serial dilutions using the Ion Universal Library Quantitation qPCR Kit

Standard	Volume (μL)	Water volume (μL)	Concentration (pM)
STD 1	5 from stock *E. coli* DH10B library control (68 pM)	45	6.80E+00
STD 2	5 from STD 1	45	6.80E−01
STD 3	5 from STD 2	45	6.80E−03
STD 4	5 from STD 3	45	6.80E−04
STD 5	5 from STD 4	45	6.80E−05
STD 6	5 from STD 5	45	6.80E−06

4. Prepare the STD 1 (Table 5) by adding 45 μL of molecular grade water to a sterile tube and then add 5 μL of the stock *E. coli* DH10B library control (68 pM). Gently mix by inverting the tube or pipetting up and down and then spin down the tube quickly for 3–5 s, at a low speed.
5. Using a new pipet tip, prepare the STD 2 (Table 5) by adding 45 μL of molecular grade water to a sterile tube and then add 5 μL of the STD 1 solution. Gently mix by inverting the tube or pipetting up and down and then spin down the tube quickly for 3–5 s, at a low speed.
6. Using a new pipet tip, prepare the STD 3 (Table 5) by adding 45 μL of molecular grade water to a sterile tube and then add 5 μL of the STD 2 solution. Gently mix by inverting the tube or pipetting up and down and then spin down the tube quickly for 3–5 s, at a low speed.
7. Continue to prepare the rest of the serial dilutions until you reach STD 6.
8. In a similar manner as for the standards, you need to prepare serial dilutions of your samples to try "hitting" the concentration range of your control library. To do so, you need to prepare 2 dilutions, namely 1 in 20 and then 1 in 2000 as follows.
9. For each sample, and direction, and genic region (fungi, oomycete), prepare the 1 in 20 dilution by adding 38 μL of molecular grade water to a sterile tube/plate and then add 2 μL of the stock purified DNA from *Step 20* of Subheading 3.5. Gently mix by inverting the tube or pipetting up and down and then spin down the tube quickly for 3–5 s, at a low speed.
10. For each sample, direction, and genic region (fungi, oomycete), using a new pipet tip, prepare an intermediate dilution of 1 in 200 by adding 18 μL of molecular grade water to a sterile tube/plate and then add 2 μL of the dilution (1 in 20) prepared in *Step 9*. Doing this intermediate dilution step will, avoid adding too little volume of DNA into a large volume of water at once such as a direct "1 in 200 dilution". Gently mix by inverting the tube or pipetting up and down and then spin down the tube quickly for 3–5 s, at a low speed.
11. For each sample, direction, and genic region (fungi, oomycete), prepare the 1 in 2000 dilution by adding 18 μL of molecular grade water to a sterile tube/plate and then add 2 μL of the dilution (1 in 200) prepared in *Step 10*.
12. Prepare your qPCR master mix for the different dilutions of control and samples. Each environmental sample requires 2 PCR (bidirectional) per genomic region (fungi, oomycete) and two dilutions (1 in 20 and 1 in 2000) of them. Keep

solutions on a cold block at all time. Each 10 μL qPCR needs the reagents, concentrations, and volumes as follows: 1× (or 5 μL/10 μL reaction) of TaqMan Fast Universal PCR Master Mix, 1× (or 0.5 μL/10 μL reaction) of Ion Library TaqMan Quantitation Assay, and 2.5 μL of DNA or standard and DNA-free water to complete the reaction volume. A large batch of master mix can be prepared to accommodate duplicates or, ideally triplicates, of each standard and dilutions/sample/direction/genic region.

13. Tightly seal tubes or plates (*see* **Note 33**). Mix by inverting 5 times, and then using a benchtop centrifuge designed for microtubes, gently spin down tubes or plate for ~1 min, at a low speed, to pull solutions at the bottom of the tubes and dislodge trapped air bubbles.
14. Before you run your qPCR on a thermocycler according to the cycling conditions described here, it is essential that you program the instrument software for a standard curve run. Most instruments will allow for the input of the concentrations of the standard at the predefined locations in your tubes/plate loaded. Cycling conditions are 1 cycle (hold) of 2 min at 95 °C, followed by 40 cycles for 15 s at 95 °C, and 1 min at 60 °C (*see* **Note 34**).
15. Once the qPCR is completed, you will need to analyze your results for each sample by confirming that each of them falls within the standard curve quantification range (*see* **Note 35**).
16. Once all samples are within the standard curve range, it is important to determine the concentration, in pM, while considering the dilution factor used (*see* **Note 36**).
17. The sample library to be pooled needs to be at a 35 pM concentration, but it is safer to pool all samples at 100 pM and then dilute the pooled library at 35 pM (*see* **Note 37**). This means that each sample shall be diluted at a 100 pM (if possible, otherwise some concentration higher than 35 pM) concentration prior to being pooled into the four respective combined library tubes. Based on the quantification performed with the qPCR, handlers can use either their non-diluted sample tube or those already diluted to reach the 100 pM concentration. Always mix and spin down the tubes between dilution process, as explained in the earlier steps.
18. Once all samples, directions, and regions are at 100 pM, you can pool them into four different tubes, one per region and direction (i.e., barcoded ITS1F, ITS2, Omup8S67, and Omlo5.8S47) by adding 2 μL of each sample into the designated pooled library tube (*see* **Note 38**).
19. Then, using the $C_1V_1 = C_2V_2$ formula presented in Subheading 3.3, dilute your pooled library from a 100 pM to 35 pM, ensuring that you generate at least 30 μL of the 35 pM library.

20. Make sure to proceed with the sequencing of your libraries within 48 h of their dilution.

3.7 Plan Run and Prepare Template (Ion Chef)

1. You need to plan a run by setting up all your parameters (including flows, kits/chip/reagents versions used, samples' barcodes, and their metadata), into the Torrent Server to be able to use the Ion Chef and Ion S5 sequencer [34–36]. Login into the Torrent Server (using a unique IP address provided by your Thermo Fisher Scientific sales representative). If it is a plan you will use quite frequently, you can save the run as a template to keep all parameters recorded for future use.
2. To define your barcodes, click on the gear-shaped icon at the top right corner of the Torrent Server Web page, then select "References Sequences." On the list of options displayed on the left, click Barcode Sets.
3. After the DNA barcodes page has opened, click on "Add new DNA Barcode Sets." Assign a name for your barcode set, then click on "Download the example file" below (*see* **Note 39**). Fill out the downloaded Comma-Separated Variable (.csv) template file with your sample names, primer, adaptor, and barcode sequences using Microsoft Excel or the equivalent, and then, save it (ensure it is still saved as a .csv file).
4. To allow the software to sort your samples by their respective barcodes and genic region assigned during the data analysis, upload your barcode .csv file for the first chip (e.g., ITS1F) into the Torrent browser as follows: Repeat *Step 2*, and save it with a unique name. There is an Upload button at the bottom of the form once a Name is set and you have selected the .csv file. Perform this process for all four barcoded primers, and upload them all into the browser.
5. The Chef System is designed to hold two chips, and each one requires a respective planned run. To do so, click on "Plan New Run" on the right upper tab (Fig. 8). On the server, under the Plan tab is where the templates are found and in the upper right corner is a "Plan a New Run" selection.
6. On the list of generic templates provided on the right, select DNA and Fusions. The planning wizard will open.
7. In the "Create Plan" tab, select the Ion S5 System and your Chip Version (e.g., Ion 530). Leave the following fields empty: Sample Preparation kit, Library Kit Type, and Control Sequence. Tick the "Ion Chef" template kit, and select the Ion 520 & Ion 530 Kit-Chef right below. Tick a read length of 400, and select the "Ion S5 Sequencing Kit." Ensure that the Base Calibration Mode is set at "Default Calibration."
8. Click Next to reach for the "Ion Reporter" tab of the wizard.

9. Select the following plugins: FileExporter and md5sum. Then, select (or create first) a project under which you will be storing your sequencing data to be associated with. Example: the laboratory and project names.
10. Click Next. You will be brought into the "SavePlan" tab of the wizard.
11. Define a name for the Sequencing Run.
12. Tick the "Custom" option for the Analysis Parameters and then in the "Alignment Args." section; add this parameter to enable soft clipping: -g 0 (Fig. 7).
13. In the "Default Reference & BED Files, leave the "Reference Library," "Target Region," and "Hotspot Regions" to "None." Tick the "Use same reference and BED files for barcodes." Tick the "Use same reference & BED files for all chips" box.
14. Enter the number of different barcode/primer combinations (i.e., the number of rows in your .csv file) in the "number of Barcode" box. In the "Sample Tube Label" field, enter a name for your first run/chip and enter/scan the barcode of the first tube that you will use to load your first sequencing library in. The tube is provided in the kit and has already a label on it. Those Ion Chef library sample tubes have unique barcode numbers labeled on them, used to track which library will be loaded in which of the two chips. Enter those respective barcode numbers into the "Sample Tube Labels" fields, and keep track of which one goes where.
15. Then, tick the check box for the barcodes to be loaded into the table (barcode kit saved in *Steps 2–4*) (*see* **Note 40**). After you upload the table, save the run at the bottom of the screen.
16. Return to *Step 5,* and repeat the procedure from the beginning to create a Planned Run for a second chip. You can copy the first Planned Run in the Planned Run page and edit appropriately.
17. Once the first sequencing run is set, plan a second sequencing run for the third and fourth chips to be prepared.

3.8 Ion Chef Protocol

1. Confirm that the instrument has been cleaned before starting the process; otherwise, start with the cleaning procedure immediately.
2. Take the Ion 510, 520 & Ion 530 Chef Reagents cartridge out of the freezer, and place it at room temperature ~45 min prior to using it (*see* **Note 41**).
3. Check that the server is connected to your instrument by clicking on "Settings" and "Torrent Server" on the Touchscreen.
4. Following the four respective barcodes assigned to the four respective tubes (*see* Subheading 3.7, *Step 14*), pipet 25 μL of

each diluted (35 pM) library into their previously assigned tube. A sequencing run can hold two libraries, so store the 2 others at −20 °C.

5. Add 4 μL of Ion S5 Calibration Standard to each diluted library tube.
6. Close the tubes, flick, gently vortex, or pipet up and down to mix; then, spin down tubes quickly for 3–5 s. Keep library tubes on a cold block/ice until you load them.
7. Unpack all consumables consisting of:
   (a) 2 chip adapters.
   (b) 1 enrichment cartridge.
   (c) 1 tip cartridge.
   (d) 1 seal for the PCR plate and frame.
   (e) 2 disposable lids for the recovery station.
   (f) 12 recovery tubes.
   (g) 1 Chef reagents cartridge (from *Step 2*).
   (h) 1 Ion S5 Chef Solution cartridge.
8. Gently tap cartridges (b, g, and h) on the bench to pull down their reagents.
9. For the same reason, spin down the NaOH tube.
10. Using the touchscreen button, open the instrument's door. After you hear a latch noise, you can gently pull the door up.
11. Select the "Step by Step" procedure on the touch screen options (Fig. 9a). Otherwise, you will have to refer to the user manual from the Thermo Fisher Scientific Website (*see* **Note 42**).
12. Select the "Prepare Chip" option (Fig. 9b). Follow the instructions highlighted to load the instrument properly.
13. Once loading is completed, close the door and hit "Start Check" (Fig. 9c) on the touchscreen. This will take about 5 min, after which a message will appear: Deck Scan complete.
14. Press NEXT.
15. You will be asked about data management, and get a list generated from the Ion Torrent Browser to allow you to select the following: chipID; sampleID and your program.
16. Press NEXT. Two run options will appear: TIME or PAUSE. Select "TIME" and plan the run strategically as this will take 12 h, 45 min after which you must immediately recover your samples and proceed with the sequencing of the first chip. The second one must be stored at 4 °C for a most of 6–8 h. It is recommended to start the Ion Chef run at the end of the day (e.g., after 3 pm) in order to be ready to start with the

sequencing the very next morning for the first chip and the same afternoon for the second chip.

17. Remember that the sequencer must be initialized 40 min prior to loading a chip. The steps are detailed in the next section.
18. Once your Ion Chef run is complete, unload the instrument as instructed on the touchscreen and proceed to sequencing immediately. When you unload the centrifuge buckets, be careful not to throw away the metal adaptor: those are not single-use and shall be put back into their respective slot in the centrifuge.
19. As per indicated on the instrument, proceed with the cleaning and UV sterilizing of it by following the steps indicated on the screen.
20. Repeat this whole section for the two next sequencing libraries prepared.

3.9 Ion S5 Initialization and Sequencing Protocol

1. On the sequencer's touchscreen, hit "Initialize." Proceed with this step at least 40 min prior to the loading of your first chip. In addition, the sequencing reagent cartridge must be left at room temperature for 45 min before the initialization. You also need to invert the wash solution bottle 5 times, and swirl it at an angle to ensure proper mixing of the contents (*see* **Note 43**).
2. Remove the red caps from the wash and cleaning solution bottles prior to beginning the initialization.
3. Since you will be performing runs of 850 flows, you will have to perform the manual cleaning before you proceed with your sequencing (*see* **Note 44**), and repeat so before you load your second chip as well.
4. Follow the steps indicated on the screen to load the sequencer for the initialization while keeping the sequencing chip from the previous run into the instrument for that process.
5. Once fully loaded, close the door and hit "Next." The instrument will be running for about 40 min, after which you need to hit "Home" for the sequencer to be ready for sequencing your first chip (*see* **Note 45**).
6. Press "Run" on the touchscreen, and then, follow the steps indicated (i.e., unload the used chip, and replace it with your first chip to sequence).
7. Ensure/review that the planned run displayed corresponds to the chip you are about the sequence; then, press "Start Run" (*see* **Note 46**).
8. After the first sequencing run is completed (~4.5 h), proceed with the manual cleaning of the instrument, following the instructions indicated on the screen.

9. If you are sequencing a second chip, proceed with the initialization step as previously described. Then, proceed with the sequencing for your second chip (stored in the fridge for 6–8 h).
10. After the sequencing is done, the data analysis will still need to go for several hours before you can see your results in the Ion Torrent Browser. However, data analysis does not prevent users from starting another run on the sequencer. The analysis will be queued and resume after the sequencing is done.
11. You can see your results in the Data Tab of the Browser where you will select your project and be able to generate a PDF report, and export your raw data (fastq files) for further bioinformatics analysis.
12. Repeat this whole section for the 2 other chips prepared.

3.10 Downstream Data Analysis Using Bioinformatics

Given that the intention for this book chapter is to focus on the sample processing, this last section is only an overview of some tools that could be used for further analysis. The pipeline and tools used will greatly impact the output results and shall be picked strategically, with the best suited method for the type of samples treated. There are many approaches already developed and publicly available for bioinformatics analysis and even specifically for metabarcoding. Each step shall be customized for the type of data sequenced, the genomic region targeted, the organisms screened for, and even depending on the sequencer used. More importantly, the database(s) (e.g., curated vs non-curated) and the clustering method (e.g., OTUs vs ASVs) opted for will profoundly influence the results as well, as previously mentioned in the introduction.

For the specific methodology described in this chapter, the HTS datasets consisting of fungi and oomycete amplicons were processed as described by Tremblay et al. [1], which combined functions from multiple tools into a whole customized workflow/ automated script. Briefly, sequences were trimmed for length and quality with the open-source software Mothur [37] (version 1.37.2). Then, with the ITSx open-source software [38] (version 1.0.11), the ITS1 sequences were extracted. A series of scripts from the QIIME [39] (version 1.9.1) pipeline were then used to further process the data, cluster reads into OTUs (97% cutoff), and generate OTU tables. Next, the OTU tables were imported in the R environment and analyzed with the RAM package [40] for diversity, abundances, and community profiling analyses. It is important to understand that some of those tools (including QIIME) and concepts (OTUs) used at the time of publication, have since evolved, or are now considered outdated by some people. Newer or updated tools are now available. For instance, a newer version of QIIME—QIIME2—[41] has been released and is now commonly used for similar purposes. This new version of the software also allows for clustering ASVs rather than OTUs.

The workflow presented above is only one option among many, and this section will not elaborate further on the bioinformatics analyses, but below is a list of a few popular tools/pipelines/software/package, etc., that users shall consider for their own analyses:

1. DADA2 [17]
2. USEARCH [42, 43]
3. Phyloseq [44]
4. Snakemake [45]
5. FUNGuild [46]
6. Mothur [37]
7. QIIME2 [41]

4 Notes

1. It is recommended to close caps between each step to avoid cross-contamination (aerosols, splashing, handler breathing, etc.). For the same reason, it is essential that sterile tips (with filters) are changed and discarded between each reagent and for each sample.
2. The particles are heavy and tend to settle at the bottom of the tube, so users shall invert tubes, then pipet up and down 3–5 times between each aliquot, and immediately take the volume needed.
3. Sliding the pipet tips along the opposite side of the pellet is a good practice, as well as slowly pulling volume in one shot, at a steady rhythm. Opt for the smallest pipet that can hold your total volume, and set it at a volume slightly higher to avoid sucking up too much air bubbles. Discard, unless you accidentally suck up the dark pellet or some of it, then dispense all liquid back into the tube, and remove it from the magnet and repeat Steps **4–6** (inclusively).
4. Never set a pipet at a higher volume than its limit; this will damage your pipet's components.
5. This will separate your DNA from the magnetic particle and release it in the liquid phase. Do not discard your liquid from now on as it contains your purified DNA.
6. Given that there will be multiple processes performed on the samples; it is recommended that the unique identifier assigned describes what was done. For example, one could label a tube "Sample001-QP" (for QuickPick) or "Sample001-pur." (indicating it was purified). Organization is a key in metabarcoding; therefore, handlers shall establish a system for storage (e.g., assign a box for genomic DNA, another one for purified DNA,

and so on) and store samples in a logical manner such as numerical order and perhaps using color coding or other visual cues. Information on samples, such as collection year and type of sampler may be a good idea to include in the short names assigned.

7. It is important to select the right Qubit assay as they are designed for different DNA concentration ranges. The following protocol uses the Qubit dsDNA High Sensitivity (HS) Assay Kit, which is designed for the lowest throughput DNA concentration and covers 10 pg/μL–100 ng/μL. If users do not know their starting concentration, they can try the HS assay and see if the concentration is out of the maximum range, in which case they could either use the Qubit dsDNA Broad Range (BR) Assay Kit assay instead, or simply dilute their DNA (e.g., 1:10, or 1:100) for quantification. Dilutions shall always be done serially, in 1:10 factors rather than combining a very low volume of sample into a large volume of buffer or water.
8. Label tubes' caps rather than sides because the fluorescence is read through the tubes walls. For the same reason, do not handle the tubes without gloves.
9. Example for 16 samples: $[16 + 2 + 2] \times 199 = 3980$ μL of buffer would be needed.
10. Example for 16 samples: $[16 + 2 + 2] \times 1$ μL $= 20$ μL of the reagent would be needed.
11. Ensure that there are no air bubbles trapped in the solution. This completes your working solution.
12. Ensure that there are no air bubbles trapped in the solution. This completes your 2 standards solution, which final volumes should be of 200 μL. Protect tubes from light until reading.
13. Ensure that there are no air bubbles trapped in the solution. This completes your sample solution, which final volumes should be of 200 μL.
14. After inserting your prepared Standard #1 in the tube chamber and closing the instrument flap, press "Read Standards" to proceed. Always read new standards (hit yes when prompted).
15. A ≤ 0.5 ng/μL value does not necessarily translate in the absence of DNA, so users are encouraged to proceed with those very low-throughput samples still. Plant pathogens can be present in very low concentration yet still be detectable by HTS.
16. Those for fungi and oomycetes were previously designed by Tremblay et al. [1] and will allow users to uniquely label each samples, along with amplifying the ITSI region from fungi and from oomycetes.

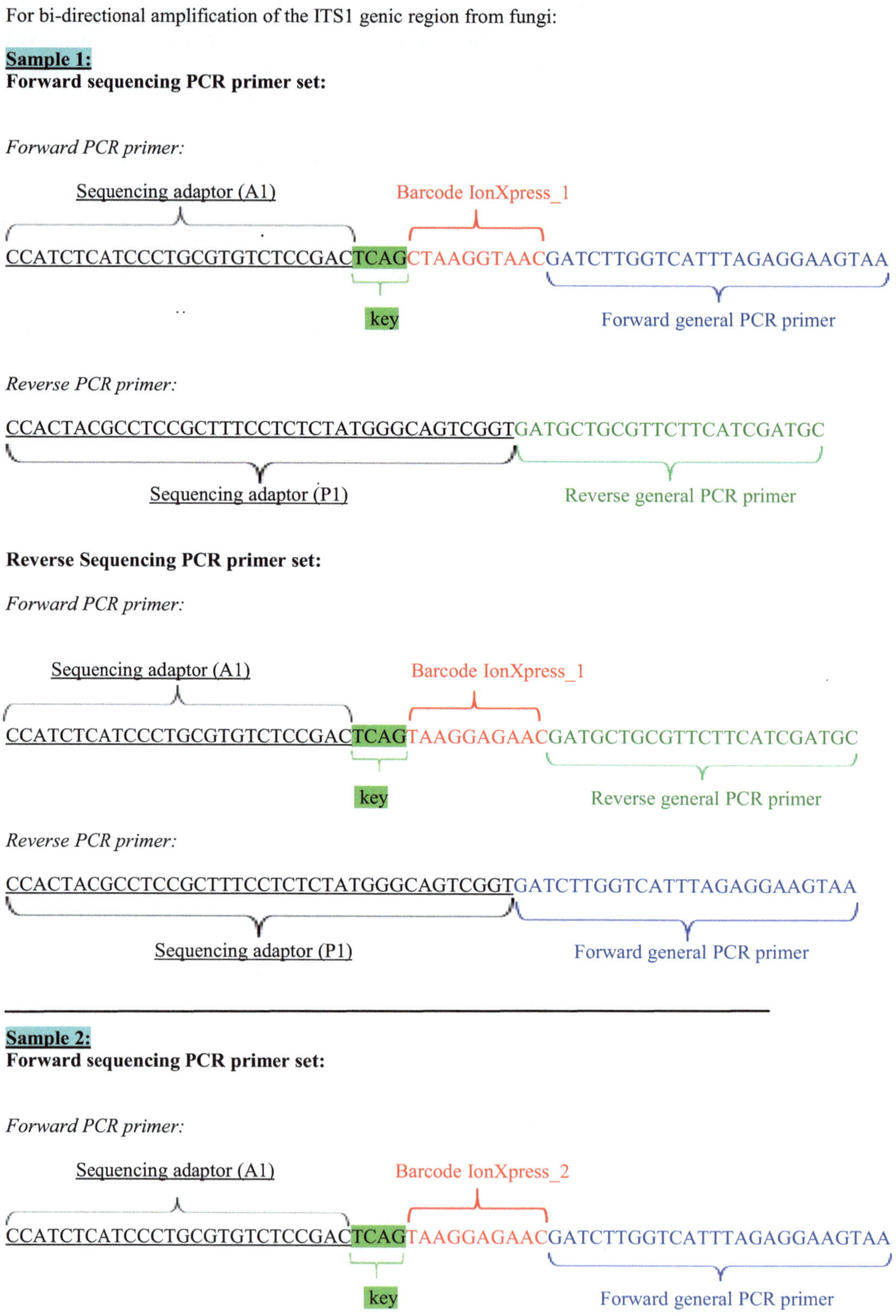

Fig. 3 For fungi, examples for "Environmental Sample 1" and "Environmental Sample 2" custom fusion primers sequences, assigned with the Ion Xpress barcodes 1 and 2, respectively

Reverse PCR primer:

CCACTACGCCTCCGCTTTCCTCTCTATGGGCAGTCGGTGATGCTGCGTTCTTCATCGATGC

Sequencing adaptor (P1) | Reverse general PCR primer

Reverse Sequencing PCR primer set:
Forward PCR primer:

Sequencing adaptor (A1) | Barcode IonXpress_2

CCATCTCATCCCTGCGTGTCTCCGACTCAGCTAAGGTAACGATGCTGCGTTCTTCATCGATGC

key | Reverse general PCR primer

Reverse PCR primer:

CCACTACGCCTCCGCTTTCCTCTCTATGGGCAGTCGGTGATCTTGGTCATTTAGAGGAAGTAA

Sequencing adaptor (P1) | Forward general PCR primer

Fig. 3 (continued)

17. Users shall order sufficient number of barcoded primers for each environmental sample to be processed. The custom fusion primers can be ordered from many companies such as LGC Biosearch Technologies (www.biosearchtech.com) or Integrated DNA Technologies (www.idtdna.com).
18. Examples for three environmental samples are included in Table 3. In addition, Figs. 3 (fungi) and 4 (oomycetes) illustrate two examples of the order of the different parts and primers required per sample. The Ion Torrent sequencing requires adapters A (forward) and P1 (reverse) for the instrument to read the sequences [4].
19. Example:

$$C_1 \times V_1 = C_2 \times V_2$$

$$100 \times V_1 = 10 \times 80$$

$$V_1 = 8\ \mu l$$

C_1 = initial "stock" concentration (i.e., 100 μM)
V_1 = volume of C_1 needed (unknown)
C_2 = final concentration desired (i.e., 10 μM)
V_2 = final volume of 10 mM solution desired (e.g., 80 μL)

For bi-directional amplification of the ITS1 genic region from oomycetes:

Sample 1:
Forward sequencing PCR primer set:

Forward PCR primer:

Sequencing adaptor (A1) | Barcode IonXpress_1

CCATCTCATCCCTGCGTGTCTCCGACTCAGCTAAGGTAACGATCTCGCCATTTAGAGGAAGGT

key | Forward general PCR primer

Reverse PCR primer:

CCACTACGCCTCCGCTTTCCTCTCTATGGGCAGTCGGTGATATTACGTATCGCAGTTCGCAG

Sequencing adaptor (P1) | Reverse general PCR primer

Reverse Sequencing PCR primer set:

Forward PCR primer:

Sequencing adaptor (A1) | Barcode IonXpress_1

CCATCTCATCCCTGCGTGTCTCCGACTCAGTAAGGAGAACGATATTACGTATCGCAGTTCGCAG

key | Reverse general PCR primer

Reverse PCR primer:

CCACTACGCCTCCGCTTTCCTCTCTATGGGCAGTCGGTGATCTCGCCATTTAGAGGAAGGT

Sequencing adaptor (P1) | Forward general PCR primer

Sample 2:
Forward sequencing PCR primer set:

Forward PCR primer:

Sequencing adaptor (A1) | Barcode IonXpress_2

CCATCTCATCCCTGCGTGTCTCCGACTCAGTAAGGAGAACGATCTCGCCATTTAGAGGAAGGT

key | Forward general PCR primer

Fig. 4 For oomycetes, examples for "Environmental Sample 1" and "Environmental Sample 2" custom fusion primers sequences, assigned with the Ion Xpress barcodes 1 and 2, respectively

Reverse PCR primer:

CCACTACGCCTCCGCTTTCCTCTCTATGGGCAGTCGGTGATATTACGTATCGCAGTTCGCAG

Sequencing adaptor (P1) | Reverse general PCR primer

Reverse Sequencing PCR primer set:

Forward PCR primer:

Sequencing adaptor (A1) | Barcode IonXpress_2

CCATCTCATCCCTGCGTGTCTCCGACTCAGCTAAGGTAACGATATTACGTATCGCAGTTCGCAG

key | Reverse general PCR primer

Reverse PCR primer:

CCACTACGCCTCCGCTTTCCTCTCTATGGGCAGTCGGTGATCTCGCCATTTAGAGGAAGGT

Sequencing adaptor (P1) | Forward general PCR primer

Fig. 4 (continued)

So in this example, 8 μL of stock primer solution (at 100 μM) shall be added to 72 μL (80 – 8) of DNA-free water.

20. Samples that had a concentration below the Qubit detection limit should also be processed and detected, though not diluted, given that they may still have sufficient DNA to be processed and detected with the metagenomics process.
21. Please note that sequencing with Ion Torrent is not paired-end and that sequencing amplicon length cannot exceed 500 bp. Figure 5 illustrates examples of PCRs required to generate amplicons from 3 different samples indexed with unique barcodes/identifiers and the primer setup.
22. It is recommended to avoid using films or foil for sealing plates as they can cause cross-contamination through aerosols further in the process because handlers will need to re-open the tubes after the PCR. Most 96-well plates have suitable 8-strip caps that can be gently opened and reduce spraying.
23. While the last step (idle) is optional, it is highly recommended to include it if user would not take their tubes/plate out of the thermocycler immediately after it is completed.

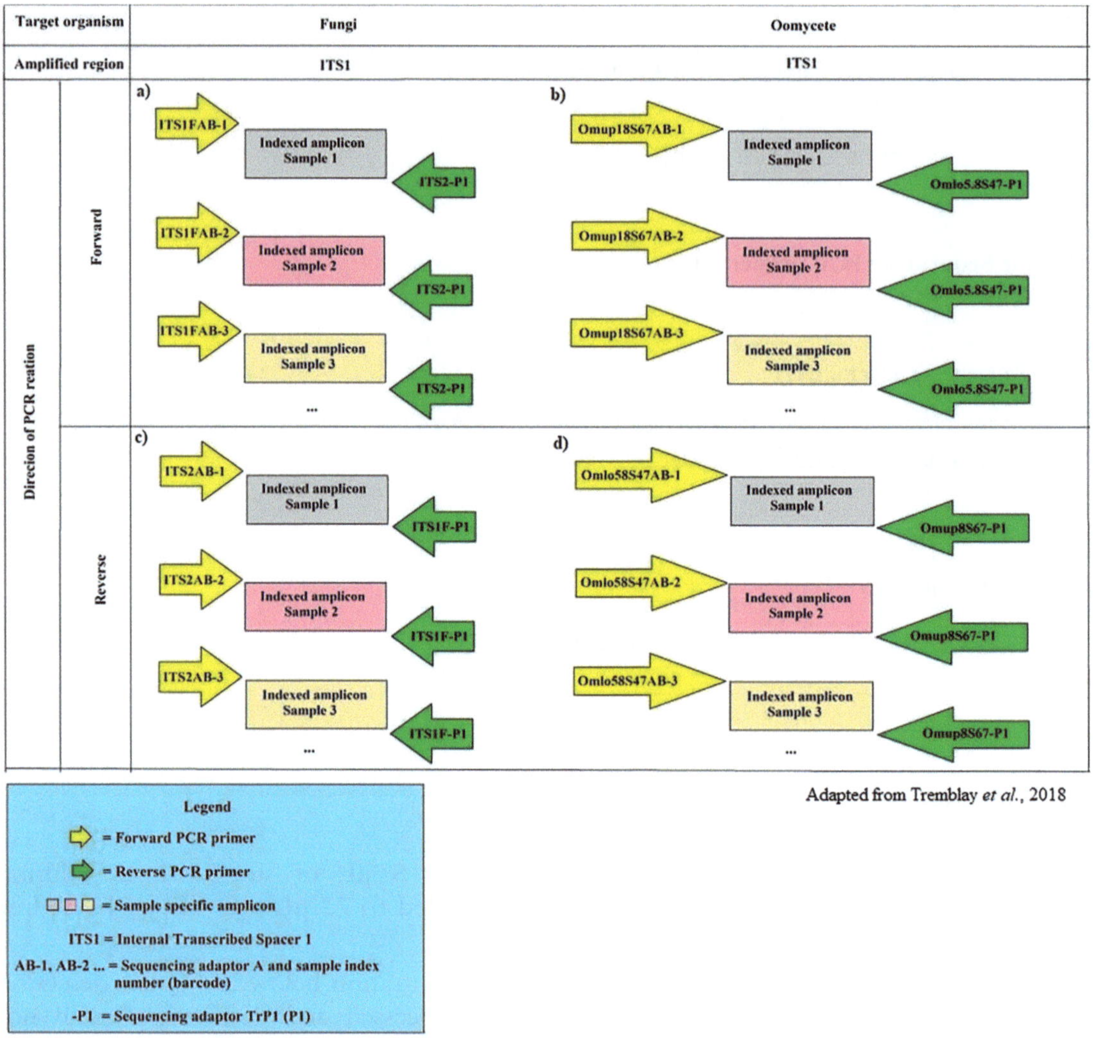

Fig. 5 Example of three (B1, B2, and B3) barcoded PCR primers setup used to, respectively, amplify bidirectionally the ITS1 intergenic region (amplicons) from fungi and oomycete. The Ion Torrent sequencing adaptors A and TrP1 (P1) are also appended to the generated amplicons during this process. (Adapted from Tremblay et al. [1])

24. An example recipe and setup are provided here, but users shall adjust their volumes and quantities according to the size of their gels and sample number to be prepared, as well as the cell system dimensions. Table 4 presents 4 gel recipes examples for different device/gel dimensions, including the following one: For a 15 cm gel, mix 110 mL of 0.5× TBE (Tris-borate-EDTA) buffer with 1.65 g of agarose. Do not forget to dilute it from the original 10× solution.

25. Refer to Table 4 for those settings, depending on the size of your gel and of your device. Watch the stain migration a few times to ravoid losing your DNA at the far end of your gel.

26. Like before, it is recommended to close caps between each step to avoid cross-contamination (aerosols, splashing, handler breathing, etc.). For the same reason, it is essential that sterile tips (with filters) are changed and discarded between each reagent and for each sample. Users should avoid using films or foil for plates as they can cause cross-contamination through aerosols further in the process.
27. The particles are heavy and tend to settle at the bottom of the tube, so users shall invert tube, then pipet up and down 3–5 times between each aliquot, and immediately take the volume needed.
28. Do not mix or pipet to mix; let the pellet be as much as possible.
29. Do not leave to dry for too long as it could reduce the quality of the purified DNA.
30. Make sure samples are well mixed by pipetting up and down several times while adding the buffer.
31. For example, to 15 μL of purified gDNA, 21 μL of the reagent is required. Users shall invert the reagent tube, then pipet up and down 3–5 times between each aliquot, and immediately take the volume needed.
32. Once liquid, gently mix by inverting the tube or pipetting up and down. Do not vortex the stock or dilutions.
33. It is recommended to avoid using films or foil for sealing plates as they can cause cross-contamination through aerosols further in the process. Most 96-well plates have suitable 8-strip caps.
34. The STD 1 has a concentration of 6.8 pM, STD 2 of 0.68 pM, STD 3 of 0.068 pM, etc.; this protocol is described for using with a ViiA7 instrument, but depending on which one the users will opt for, there are differences in the cycling profiles. Those technicalities can be found on page 21 of the manufacturer's user guide (www.thermofisher.com/document-connect/document-connect.html?url=https%3A%2F%2Fassets.thermofisher.com%2FTFS-Assets%2FLSG%2Fmanuals%2FMAN0010799_Ion_16S_Metagenomics_UG.pdf&title=VXNlciBHdWlkZTogSW9uIDE2UyBNZXRhZ2Vub21pY3MgS2l0).
35. Depending on the instrument you use, the parameters and setting for baseline and cycle threshold vary. Figure 6a shows an example of the amplification curves for each standard (STD 1-STD 6) dilution, which are between ~14–33 cycle threshold (Ct) values, and Fig. 6b shows the associated standard curve (log) for those controls. Figure 6c, d is an example of samples for which the dilution picked (1 in 2000) all fell within the standard curve range (i.e., ~14–33 Ct values). However, it is not always the case that all, or even any, samples are within such

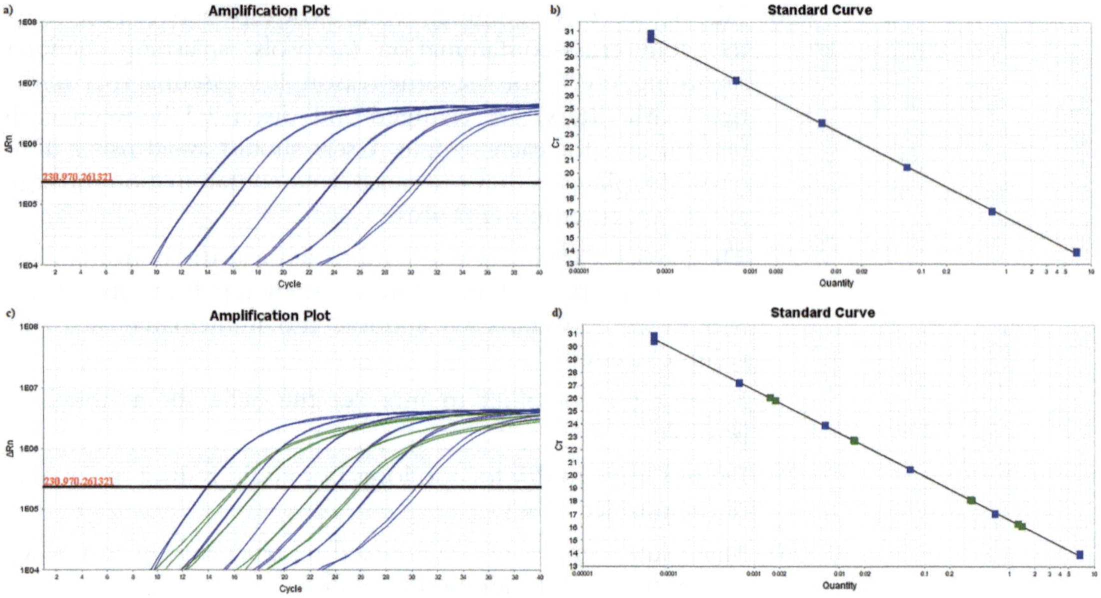

Fig. 6 Standard curve for qPCR quantification. (**a**) Amplification curves for each standard dilution (1–6) and (**b**) associated standard curve (log) for those controls. (**c**, **d**) Samples for which the dilution picked (1 in 2000)

range because they are either too concentrated (e.g., Ct value ≤ 14) or too diluted (e.g., Ct value ≥ 33). In such cases, while the instrument's software may still output a value, it would not be accurate enough and should never be used to quantify samples. Rather, users shall check for a less/more concentrated dilution of their samples and may have to do additional dilutions and qPCR to fall within the desired range.

36. To determine the "starting" sample concentration, dilution calculations must be taken into account. For example, if the concentration determined by the qPCR assay of a sample diluted 1 in 2000 was equal to 3.16 pM, then the concentration of your non-diluted samples would be $3.16 \times 2000 = 6320$ pM. Users shall determine the concentration of their tubes for each sample, direction, and genic region.

37. Depending on the target fragment and the sequencing chip version used, the optimal concentration of the pooled library can differ. Based on testing performed with mixed regions and non-mixed regions (Bilodeau, unpublished), It is also recommended to perform dilutions using at least 2 μL of the stock/diluted sample to avoid pipetting fluctuation.

38. It is recommended to load all ITS1F libraries together, and separately from those for ITS2, and so on for the oomycete barcoded primers. The sequencing process can support 2 different chips and will therefore accommodate for two libraries

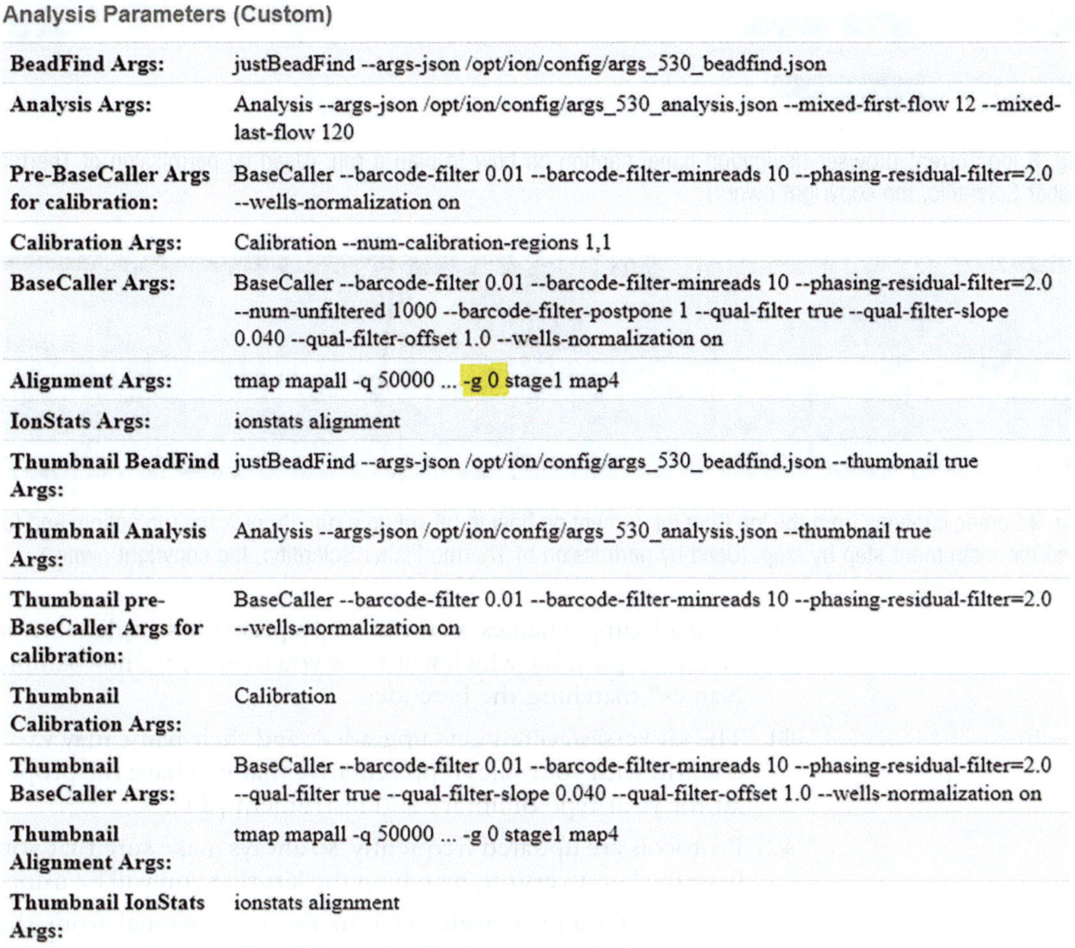

Analysis Parameters (Custom)	
BeadFind Args:	justBeadFind --args-json /opt/ion/config/args_530_beadfind.json
Analysis Args:	Analysis --args-json /opt/ion/config/args_530_analysis.json --mixed-first-flow 12 --mixed-last-flow 120
Pre-BaseCaller Args for calibration:	BaseCaller --barcode-filter 0.01 --barcode-filter-minreads 10 --phasing-residual-filter=2.0 --wells-normalization on
Calibration Args:	Calibration --num-calibration-regions 1,1
BaseCaller Args:	BaseCaller --barcode-filter 0.01 --barcode-filter-minreads 10 --phasing-residual-filter=2.0 --num-unfiltered 1000 --barcode-filter-postpone 1 --qual-filter true --qual-filter-slope 0.040 --qual-filter-offset 1.0 --wells-normalization on
Alignment Args:	tmap mapall -q 50000 ... -g 0 stage1 map4
IonStats Args:	ionstats alignment
Thumbnail BeadFind Args:	justBeadFind --args-json /opt/ion/config/args_530_beadfind.json --thumbnail true
Thumbnail Analysis Args:	Analysis --args-json /opt/ion/config/args_530_analysis.json --thumbnail true
Thumbnail pre-BaseCaller Args for calibration:	BaseCaller --barcode-filter 0.01 --barcode-filter-minreads 10 --phasing-residual-filter=2.0 --wells-normalization on
Thumbnail Calibration Args:	Calibration
Thumbnail BaseCaller Args:	BaseCaller --barcode-filter 0.01 --barcode-filter-minreads 10 --phasing-residual-filter=2.0 --qual-filter true --qual-filter-slope 0.040 --qual-filter-offset 1.0 --wells-normalization on
Thumbnail Alignment Args:	tmap mapall -q 50000 ... -g 0 stage1 map4
Thumbnail IonStats Args:	ionstats alignment

Fig. 7 Ion Torrent Browser navigation panel caption on how to select customized analysis parameters in the alignment argument line. (Used by permission of Thermo Fisher Scientific, the copyright owner)

per sequencing run. If you are processing the four regions, you will need to plan two sequencing runs consisting of four pooled libraries and four chips (Fig. 7).

39. You may want to keep a copy the example file as a reference for preparing additional runs, so it is recommended to save it, and name your filled-out .csv file under a different new name when saving that one. It is also recommended to separate the barcoded forward and reverse primers and organisms (fungi vs oomycete) onto four different chips (i.e., two sequencing runs) to be used. For instance, the barcoded ITS1F primers would get loaded onto the first chip, and those for the ITS2 barcoded primers would get loaded onto the second chip and that would complete the first sequencing run. Similarly, a second sequencing run would include the barcoded Omup18S67 primers, loaded on a third chip, and the barcoded Omlo58S47 primer library, loaded on a fourth chip.

Fig. 8 Ion Torrent Browser navigation panel caption on how to plan a run. (Used by permission of Thermo Fisher Scientific, the copyright owner)

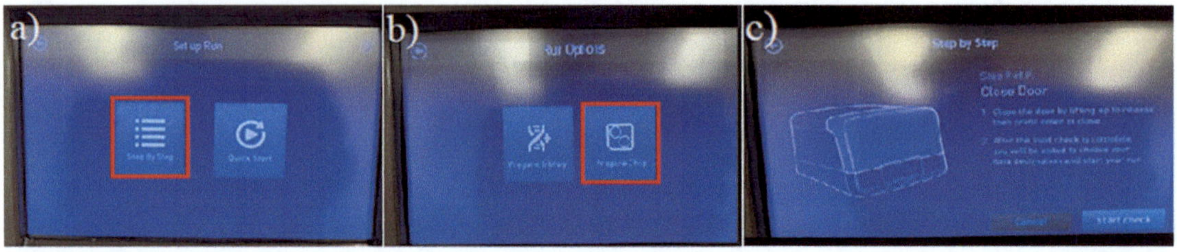

Fig. 9 Screen captions from the Ion Chef instrument on how to (**a**) set up a run, (**b**) pick the run option, and (**c**) load the instrument step by step. (Used by permission of Thermo Fisher Scientific, the copyright owner)

40. To load sample names, you can load a premade sample table or save a sample table which will allow you to enter in the "Sample Names" matching the barcodes.
41. The kit version often gets upgraded, and their name may vary; confirm with your sales representative that you have the proper kit for your type of library and instrument [34].
42. Protocols are updated frequently, so always make sure that you have the latest version, matching the kits that you will be using.
43. Users could alternatively refer to the user manual from the Thermo Fisher Scientific Website [34]. Again, the manual is often updated so users must ensure to have the latest version of it, matching the kits that they intend to use.
44. If it was not already done by the previous user, the instrument will prompt you to select your choice (clean or initialize) on its screen (Fig. 9).
45. Once initialized, the instrument can wait for up to 24 h before your sequencing gets started.
46. The sequencer's door must make a "click" noise to be properly closed.

Acknowledgements

The authors would like to thank Debbie Shearlaw, Benoit Goulet, Adam Colville, and Mathieu Larivière for offering their support with editing and technical inputs. Their valuable support and suggestions significantly

References

1. Tremblay É, Duceppe M, Bérubé J et al (2018) Screening for exotic forest pathogens to increase survey capacity using metagenomics. Phytopathology 108(12):1509–1521. https://doi.org/10.1094/PHYTO-02-18-0028-R
2. Nilsson RH, Hyde KD, Pawłowska J et al (2014) Improving ITS sequence data for identification of plant pathogenic fungi. Fungal Divers 67(1):11–19
3. Nilsson RH, Ryberg M, Kristiansson E et al (2006) Taxonomic reliability of DNA sequences in public sequence databases: a fungal perspective. PLoS One 1:1–4. https://doi.org/10.1371/journal.pone.0000059
4. Blackwell M (2011) The Fungi: 1, 2, 3... 5.1 million species? Am J Bot 98(3):426–438
5. Hawksworth DL, Rossman AY (1997) Where are all the undescribed fungi? Phytopathology 87(9):888–891
6. Rossman A, Palm M (2006) Why are *Phytophthora* and other Oomycota not true Fungi? The American Phytopathological Society. https://www.apsnet.org/edcenter/intropp/pathogengroups/pages/oomycetes.aspx. Accessed 10 Feb 2019
7. Kõljalg U, Larsson K-H, Abarenkov K et al (2005) UNITE: a database providing web-based methods for the molecular identification of ectomycorrhizal fungi. New Phytol 166(3):1063–1068. https://doi.org/10.1111/j.1469-8137.2005.01376.x
8. Duceppe M-O, Tremblay ED (2019) OmDB. figshare. https://figshare.com/articles/OmDB_fasta_gz/7615931/1. Accessed 26 Feb 2019
9. Duceppe M-O, Tremblay ED (2019) OmDB taxonomic map. figshare. https://figshare.com/articles/OmDB_tax_txt/7615886/2. Accessed 26 Feb 2019
10. Edger PP, Tang M, Bird KA et al (2014) Secondary structure analyses of the nuclear rRNA internal transcribed spacers and assessment of its phylogenetic utility across the Brassicaceae (mustards). PLoS One 9(7):e101341
11. Lamarche J, Potvin A, Pelletier G et al (2015) Molecular detection of 10 of the most unwanted alien forest pathogens in Canada using real-time PCR. PLoS One 10(8):1–37
12. Lamarche J, Stewart D, Pelletier G et al (2014) Real-time PCR detection and discrimination of the *Ceratocystis coerulescens* complex and of the fungal species from the *Ceratocystis polonica* complex validated on pure cultures and bark beetle vectors. Can J For Res 44(9):1103–1111
13. Majaneva M, Hyytiäinen K, Varvio SL et al (2015) Bioinformatic amplicon read processing strategies strongly affect eukaryotic diversity and the taxonomic composition of communities. PLoS One 10(6):1–18
14. Westcott S, Schloss P (2015) De novo clustering methods outperform reference-based methods for assigning 16S rRNA gene sequences to operational taxonomic units. PeerJ 3:e1487
15. Kopylova E, Navas-Molina J, Céline M et al (2016) Open-source sequence clustering methods improve the state of the art. MSystems 1(1):e00003–e00015
16. Callahan B, McMurdie P, Holmes S (2017) Exact sequence variants should replace operational taxonomic units in marker-gene data analysis. ISME J 11(12):2639–2643
17. Callahan BJ, McMurdie PJ, Rosen MJ et al (2016) DADA2: high-resolution sample inference from Illumina amplicon data. Nat Methods 13(7):581
18. Tikhonov M, Leach R, Wingreen N (2015) Interpreting 16S metagenomic data without clustering to achieve sub-OTU resolution. ISME J 9(1):68–80
19. Haudenshield J, Hartman G (2011) Exogenous controls increase negative call veracity in multiplexed, quantitative PCR assays for *Phakopsora pachyrhizi*. Plant Dis 95(3):343–352
20. Wilson I (1997) Inhibition and facilitation of nucleic acid amplification. Appl Environ Microbiol 63(10):3741
21. Frostegård Å, Courtois S, Ramisse V et al (1999) Quantification of bias related to the extraction of DNA directly from soils. Appl Environ Microbiol 65(12):5409–5420
22. Schneegurt M, Dore S (2003) Direct extraction of DNA from soils for studies in microbial ecology. Curr Issues Mol Biol 5(1):1–8
23. Anonymous (2015) QuickPick SML Plant DNA – instruction for use. Bio-Nobile, Online
24. Lahens N, Ricciotti E, Smirnova O et al (2017) A comparison of Illumina and Ion Torrent sequencing platforms in the context of differential gene expression. BMC Genomics 18(1):1–13
25. Quail M, Smith M, Coupland P et al (2012) A tale of three next generation sequencing platforms: comparison of Ion Torrent, Pacific Biosciences and Illumina MiSeq sequencers. BMC Genomics 13:341. https://doi.org/10.1186/1471-2164-13-341
26. Salipante S, Kawashima T, Rosenthal C et al (2014) Performance comparison of Illumina

and ion torrent next-generation sequencing platforms for 16S rRNA-based bacterial community profiling. Appl Environ Microbiol 80(24):7583–7591

27. Hertke S (2015) S5 & S5XL (Ion Torrent)... closing the gap on Illumina. https://www.linkedin.com/pulse/s5-s5xl-ion-torrent-closing-gap-illumina-scott-herke/. Accessed 17 Oct 2018
28. Anonymous (2018) Qubit 4 fluorometer. ThermoFisher Scientific, Online
29. Anonymous (2012) Ion amplicon library preparation (fusion method). ThermoFisher Scientific, Online
30. Lee P, Costumbrado J, Hsu C-Y et al (2012) Agarose gel electrophoresis for the separation of DNA fragments. J Vis Exp 62:e3923. https://doi.org/10.3791/3923
31. Edwards D (2012) PCR purification: AMPure and simple. www.keatslab.org/blog/pcrpurificationampureandsimple. Accessed 5 Jul 2017
32. Anonymous (2015) Ion 16S metagenomics kit – user guide, vol 2021. ThermoFisher Scientific, Online
33. Anonymous (2014) Ion library quantitation kit – user guide. ThermoFisher Scientific, Online
34. Anonymous (2015) Ion 520 & ion 530 kit – chef – user guide. ThermoFisher Scientific, Online
35. Anonymous (2017) Torrent suite software 5.6 – Help. ThermoFisher Scientific, Online
36. Anonymous (2021) Torrent suite software – user guide. ThermoFisher Scientific, Online
37. Schloss PD, Westcott SL, Ryabin T et al (2009) Introducing mothur: open-source, platform-independent, community-supported software for describing and comparing microbial communities. Appl Environ Microbiol 75(23):7537–7541
38. Bengtsson-Palme J, Ryberg M, Hartmann M et al (2013) Improved software detection and extraction of ITS1 and ITS2 from ribosomal ITS sequences of fungi and other eukaryotes for analysis of environmental sequencing data. Methods Ecol Evol 4(10):914–919
39. Caporaso JG, Kuczynski J, Stombaugh J et al (2010) QIIME allows analysis of high-throughput community sequencing data. Nat Methods 7(5):335–336
40. Chen W, Simpson J, Levesque C (2016) RAM: R for amplicon-sequencing-based microbial-ecology R package, version 1.2.1.3. Vienna, Austria
41. Bolyen E, Rideout J, Dillon M et al (2019) Reproducible, interactive, scalable and extensible microbiome data science using QIIME 2. Nat Biotechnol 37(8):852–857
42. Edgar RC (2013) UPARSE: highly accurate OTU sequences from microbial amplicon reads. Nat Methods 10(10):996–998
43. Edgar RC (2010) Search and clustering orders of magnitude faster than BLAST. Bioinformatics 26(19):2460–2461
44. McMurdie P, Holmes S (2013) phyloseq: an R package for reproducible interactive analysis and graphics of microbiome census data. PLoS One 8(4):e61217
45. Mölder F, Jablonski K, Letcher B et al (2021) Sustainable data analysis with Snakemake. F1000Res 10:33
46. Nguyen N, Song Z, Bates S et al (2016) FUNGuild: an open annotation tool for parsing fungal community datasets by ecological guild. Fungal Ecol 20:241–248
47. Balletbó A, Kuiper S (2019) Agarose gel electrophoresis V.3. https://doi.org/10.17504/protocols.io.7ukhnuw
48. White TJ, Bruns T, Lee S et al (1990) Amplification and direct sequencing of fungal ribosomal RNA genes for phylogenetics. In: PCR protocols: a guide to methods and applications. Academic Press, Cambridge
49. Gardes M, Bruns TD (1993) ITS primers with enhanced specificity for basidiomycetes – application to the identification of mycorrhizae and rusts. Mol Ecol 2:113–118
50. Decock C (2012) Fungal Planet description sheets: 128–153. Persoonia 29:146
51. Bilodeau G, Lévesque C, De Cock A et al (2007) Molecular detection of *Phytophthora ramorum* by real-time polymerase chain reaction using TaqMan, SYBR Green, and molecular beacons. Phytopathology 97(5):632–642

Chapter 19

Phytobiome Metabarcoding: A Tool to Help Identify Prokaryotic and Eukaryotic Causal Agents of Undiagnosed Tree Diseases

Carrie J. Fearer, Antonino Malacrinò, Cristina Rosa, and Pierluigi Bonello

Abstract

Recent advancements in high-throughput sequencing have provided scientists with vastly enhanced tools to diagnose unknown tree diseases. One of these techniques is referred to as metabarcoding, which uses phylogenetically informative reference genes to taxonomically classify short DNA sequences amplified from environmental samples. Using metabarcoding, we are able to compare the microbiota of symptomatic and asymptomatic (including presumably naïve) samples and identify microbe(s) that are only present in symptomatic samples and could therefore be responsible for the undiagnosed disease. Metabarcoding involves two main steps: library preparation and bioinformatic processing. For library preparation, the appropriate reference gene for the organism of interest (i.e., bacteria, phytoplasma, fungi, or other eukaryotes, such as nematodes) is amplified from the DNA extracted from the environmental samples using PCR and prepared for sequencing. The bioinformatic processing includes four major steps: (1) quality check and cleanup on raw reads; (2) classification of the sequences into taxonomically informative groups (ASVs or OTUs); (3) taxonomy assignments based on the reference database; and (4) differential abundance and diversity analyses to identify microbes that are significantly associated with just symptomatic samples and that point toward the putative causal agent of the disease.

Key words Sequencing, Bioinformatics, Microbiome, Forest pathology

1 Introduction

High-throughput sequencing (HTS) techniques have transformed the methods that forest pathologists use to detect and diagnose tree diseases [1]. Conventional diagnostic methods mostly rely on the isolation and in vitro cultivation of the putative agent of disease or on the a priori knowledge of specific markers that can be used to detect its presence. However, microbial agents of disease cannot always be isolated in vitro, and this prevents the identification of markers that can help the diagnosis. Metagenomic techniques represent a helpful tool to overcome such limitations. The direct

Nicola Luchi (ed.), *Plant Pathology: Methods and Protocols*, Methods in Molecular Biology, vol. 2536,
https://doi.org/10.1007/978-1-0716-2517-0_19,

analysis of the collective genome of microorganisms associated with environmental samples avoids the in vitro cultivation step and allows for a quicker identification of potential disease agents. Indeed, compared to conventional diagnostic processes, metagenomics provides an efficient method for rapid detection of known and unknown microbial pathogens [2], allows for a more accurate assessment of microbial diversity, and increases the resolution of taxonomic identification from mixed community samples [3].

While the process of assembling genomes from metagenomics data can be costly and computationally demanding, forest pathology can benefit from focusing on small areas of the genome which nucleotide composition can be used for taxonomical identification of microorganisms. This process extends the concept of molecular barcoding to a high-throughput workflow known as metabarcoding. In metabarcoding, the DNA obtained from an environmental sample is used as template to amplify short DNA sequences targeting a specific gene, which is selected to target anywhere from a broad range of organisms (e.g., bacteria, fungi) to a narrow range of taxa (e.g., haplotypes of a pathogen). These amplicons are then sequenced using HTS technology, generating millions (or billions) of short reads that provide a deep taxonomic analysis of the microbial diversity of environmental samples. Many metabarcoding studies rely on the use of reference genes coding for the small-subunit (SSU) ribosomal RNA (rRNA) to serve as references, since these regions are highly conserved within species but variable between species while being present in virtually all organisms within the group of interest, such as fungi and bacteria [4]. Specifically, the 16S rRNA SSU is mostly used to study communities of prokaryotic organisms [5], while the 18S rRNA SSU is mostly used for eukaryotic organisms [6]. However, in regard to fungi, the 18S region does not evolve rapidly enough to encompass full fungal diversity [7]. Thus, it has been widely accepted that the internal transcribed spacer (ITS) region between the small (18S) and large (28S) subunits is more useful, as this region is highly conserved within most fungal species but varies between species, increasing the sensitivity and selectivity for species identification [7].

In this chapter, we present a comprehensive pipeline (Fig. 1) based on a recent case study aimed at identifying the causal agent of beech leaf disease [8], describing how metabarcoding techniques using the three aforementioned reference genes can be used to determine the causal agent of an unknown tree disease. Briefly, metabarcoding analyses can be divided into two major steps: library preparation and bioinformatic analysis. The library preparation procedure uses

DNA extracted from environmental samples as template to amplify the target gene chosen on the basis of the taxonomical group of interest. This process is done using PCR and basically enriches the library with the target gene from all target

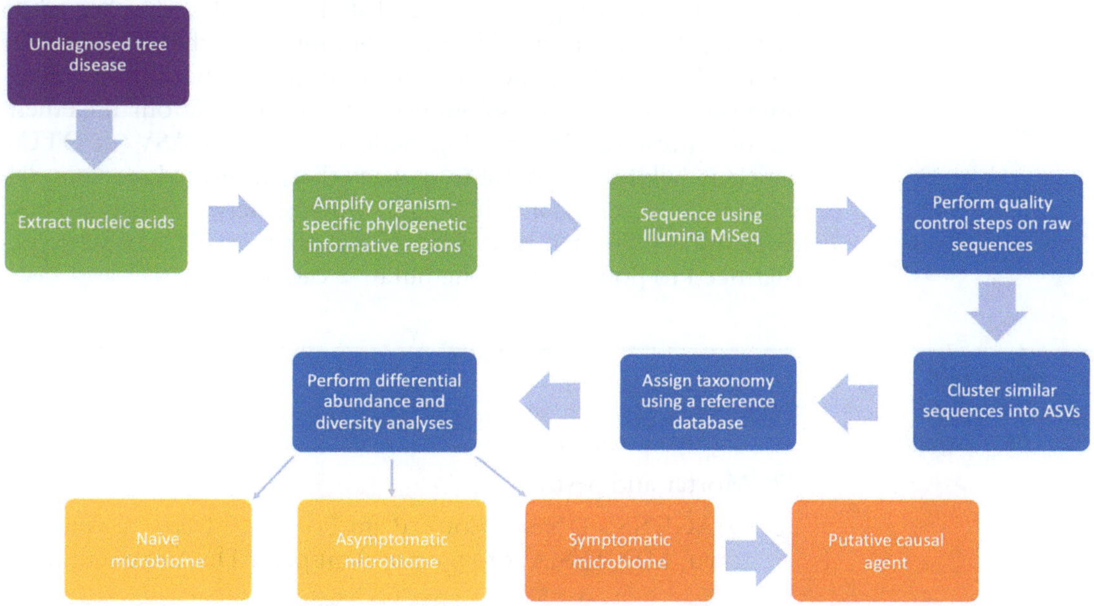

Fig. 1 Pipeline for completing a metabarcoding study of an unknown tree disease. Green boxes represent the processes for library preparation. Blue boxes represent the bioinformatic processes. The yellow and orange boxes represent the comparison of microbial communities to determine the suspected causal agent

microorganisms. This library is then processed to add the adaptors (i.e., short nucleotide sequences) that allow the sequencing machine to recognize and process the amplicons. The sequencing machine returns a collection of short reads that need to be processed using bioinformatics tools. There are several ways to perform the bioinformatics steps, which might change according to the specific research question or intrinsic nature of the experiment. Previously published manuscripts provide more details for the laboratory and bioinformatics aspects [9–12], to which readers can refer for greater details. Generally, the bioinformatics analysis involves four major steps. First, raw data is subjected to a quality control to remove low-quality information that might lead to misleading results. Second, the raw data is clustered into bins according to the similarity between sequences (similar sequences are grouped within the same bin), and this bin can represent an OTU (operational taxonomic unit) or an ASV (amplicon sequence variant), according to the algorithm we used to generate it. Third, a sequence from each bin is used to assign a taxonomical identification through the comparison toward a reference database (e.g., SILVA, UNITE). Fourth, we generate the three basic pieces of information to run the downstream analysis: (i) the ASV (or OTU) table, which is a matrix having each ASV (or OTU) as a row, each sample as a column, and the read counts for all combinations of ASVs and samples as the cell values; (ii) the metadata table, with each sample as a row and all the variables we collected on that sample as columns; (iii) the taxonomy table, that has a row for each ASV (or OTU) and separate columns for the different taxonomical levels (from kingdom down to species). A fourth element can be represented by a phylogenetic tree of all the ASV (or OTU), which is helpful to compute metrics that need phylogenetic distance (e.g., UniFrac metric, Faith's phylogenetic diversity). These elements can be then processed with a diverse variety of strategies that need to point toward the putative causal agent.

2 Materials

2.1 DNA Extraction

1. Liquid nitrogen.
2. Mortar and pestle.
3. QIAGENS's DNeasy PowerPlant Pro kit (QIAGEN, Germantown, MD, U.S.A) or an equivalent plant DNA extraction kit.

2.2 PCR Amplification

1. 10X PCR buffer with KCl (Thermo Fisher Scientific, Waltham, MA, U.S.A.).
2. 25 mM $MgCl_2$.
3. 5 U/μL *Taq* polymerase (Thermo Fisher Scientific).
4. 2 mM dNTPs.
5. 16S V5-V7 primers [13]:

   799F – 5'-AACMGGATTAGATACCCKG-3'

   1193R – 5'-ACGTCATCCCCACCTTCC-3'
6. 16S phytoplasma primers [14, 15]

   R16F2n – 5'-GAAACGACTGCTAAGACTGG-3'

   R16R2 – 5'-TGACGGGCGGTGTGTACAAACCCCG-3'

   M1(785F) – 5'-GTCTTTACTGACGCTGAGGC-3'

   M2(123R) - 5'-CTTCAGCTACCCTTTGTAAC-3'
7. ITS primers [16]:

   ITS1F – 5'-CTTGGTCATTTAGAGGAAGTAA-3'

   ITS2 – 5'-GCTGCGTTCTTCATCGATGC-3'
8. 18S primers [17]:

   NemF – 5'-GGGGAAGTATGGTTGCAAA-3'

   NF1 – 5'-GGTGGTGCATGGCCGTTCTTAGTT-3'

   18Sr2b - 5'-TACAAAGGGCAGGGACGTAAT-3'
9. Thermocycler

2.3 Gel Electrophoresis

1. Agarose.
2. 1 X Tris-acetate-EDTA (TAE) buffer.
3. 1 X GelRed (Biotium, Fremont, CA, U.S.A) with 6X loading dye (Thermo Fisher Scientific).
4. 1 kb DNA ladder (Thermo Fisher Scientific).

2.4 Illumina Library Preparation and Sequencing

1. Nextera XT Index Kit (Illumina, San Diego, CA, U.S.A.).
2. Agilent Bioanalyzer or Agilent TapeStation (Agilent, Santa Clara, CA, U.S.A.).
3. Illumina MiSeq System and reagents (Illumina).

2.5 Bioinformatic Processing

1. QIIME2 [18].
2. 16S and 18S SILVA classifier available at: https://www.arb-silva.de/download/arb-files/.
3. ITS UNITE99 classifier available at: https://unite.ut.ee/repository.php.
4. R version 4.0.3 [19].
5. R packages: "phyloseq" [20], "ape" [21], "agricolae" [22], "metacoder" [23], "DESeq2" [24], "picante" [25], "vegan" [26], "emmeans" [27], "lme4" [28].

3 Methods

Prior to starting, control (i.e., naïve, which is defined by samples collected well outside the zone of infestation that can be considered disease-free samples), asymptomatic, and symptomatic samples, such as foliage or roots, must be collected for analysis and flash frozen in liquid nitrogen. Store samples at −80 °C. All remaining procedures are completed at room temperature unless otherwise specified.

3.1 DNA Extraction

1. Beginning with naïve samples followed by asymptomatic samples and then symptomatic samples to avoid potential cross-contamination, grind frozen samples into a find powder using a mortar and pestle that has been frozen in liquid nitrogen. Ensure samples stay frozen by grinding the sample in liquid nitrogen.
2. Using the ground material, follow extraction instructions provided by QIAGEN's DNeasy PowerPlant Pro Kit.
3. Measure the quality and amount of extracted DNA using a spectrophotometer (e.g., NanoDrop, Thermo Fisher Scientific) or a fluorometer (e.g., Qubit, Thermo Fisher Scientific). Each sample must contain at least 5 ng/μL of DNA with 260/280 and 260/230 values between 1.8–2.0.

3.2 Bacterial 16S Amplification

1. Perform all PCR reactions in a volume of 25 μL containing 1X PCR buffer with KCl, 2 mM $MgCl_2$, 0.2 mM dNTP mix, 0.4 μM of 799F primer, 0.4 μM of 1199R primer, 2 U *Taq* polymerase, and 10 ng of template DNA [29].
2. Amplify the samples using a thermocycler and setting the following conditions: initial denaturation at 95 °C for 3 min, followed by 35 cycles of denaturation at 95 °C for 45 s, annealing at 54 °C for 30 s, and extension at 72 °C for 1 min, with a final extension performed at 72 °C for 10 min. Note that some steps might vary on the basis of the polymerase, and we suggest to always refer the producer's protocol.
3. Prepare a 1.5% agarose gel. Add 1 μL of the GelRed/loading dye mixture to 5 μL of PCR product and load into the gel wells. Prepare 1 kb DNA ladder according to product specifications and add to gel well. Submerge gel in 1 X TAE buffer, and allow to run in the electrophoretic chamber at 100–120 V for ~45 min. The loading dye can be used to track samples during the run.
4. Upon completion of gel run, two bands will be produced at approximately 1000 bp (chloroplast DNA) and 500 bp (bacterial DNA). Perform a gel extraction of the 500 bp band by excising the band from the gel using a sterile, aerosol resistant 1000 μl pipette tip, extruding the mini gel plug into 100 μL of Milli-Q water, and heating to 90 °C in a thermocycler for five minutes.
5. Perform a second PCR using the gel purified product as template using the same reagent concentrations as in **step 1** and reaction conditions as in **step 2**.

3.3 Phytoplasma 16S Nested PCR

1. Perform the first set of PCR reactions in a volume of 25 μL containing 1X PCR buffer with KCl, 3.5 mM $MgCl_2$, 0.2 mM dNTP mix, 0.4 μM of R16F2n primer, 0.4 μM of R16R2 primer, 1.25 U *Taq* polymerase, and 50 ng of template DNA [14].
2. Amplify the samples in a thermocycler using the following conditions: initial denaturation at 94 °C for 3 min, followed by 35 cycles of denaturation at 94 °C for 1 min, annealing at 55 °C for 2 min, and 72 °C for 3 min, with a final extension at 72 °C for 10 min.
3. Dilute the PCR product 1:30 and use 1 μL as template for the second PCR reaction using the same reagent concentrations as in step 1 using the M1 and M2 primers.
4. Amplify the samples in a thermocycler using the following conditions: 5 cycles of 94 °C for 3 min, 50 °C for 1 min, and 72 °C for 1 min followed by 30 cycles of 94 °C for 30 s, 50 °C for 30 s, and 72 °C for 30 s, with a final extension of 72 °C for 90 s [15]. The expected product is approximately 550 bp.

3.4 Fungal ITS PCR

1. Carry out PCR reactions in a volume of 25 μL containing 1X PCR buffer with KCl, 3.5 mM $MgCl_2$, 0.2 mM dNTP mix, 0.4 μM of ITS1F primer, 0.4 μM of ITS2 primer, 1.25 U *Taq* polymerase, and 20 ng of template DNA [30].
2. Amplify samples in a thermocycler at the following conditions: initial denaturation at 95 °C for 3 min, followed by 35 cycles of denaturation at 95 °C for 30 s, annealing at 55 °C for 30 s, and extension at 68 °C for 2 min, with a final extension performed at 68 °C for 10 min [16]. The expected product size is approximately 250 bp.

3.5 Nematode 18S Semi-Nested PCR

1. Carry out the first PCR reactions in 25 μL volumes containing 1 X PCR buffer with KCl, 1.5 mM MgCl2, 0.2 mM dNTPs, 1 μM of NemF primer, and 1 μM of 18Sr2b primer, 1 U *Taq* polymerase, and 1 ng of template DNA [17].
2. Amplify the samples in a thermocycler at the following conditions: initial denaturation at 94 °C for 5 min, followed by 20 cycles of denaturation at 94 °C for 30 s, annealing at 53 °C for 30 s, and extension at 72 °C for 1 min, with a final extension performed at 72 °C for 10 min [17].
3. Make a 1:10 dilution of the PCR product and use as the template for the second PCR using the same reagent concentrations as in step 1 using the NemF primer instead of the NF1 primer.
4. Amplify the samples using the same conditions as in step 2 using 58 °C for the annealing temperature instead of 53 °C. The expected product size is 420 bp.

3.6 Illumina Library Preparation and Sequencing

1. Follow instructions provided by the Nextera XT Index Kit for library preparation and multiplexing.
2. Quantify each library using a Qubit fluorometer.
3. Pool libraries to an equimolar ratio. For more information, please visit: https://support.illumina.com/help/pooling-calculator/pooling-calculator.htm.
4. Perform quality control using an Agilent Bioanalyzer or TapeStation to ensure the amplicon library has the correct size and fragment distribution.
5. Sequence samples on Illumina MiSeq platform using 300 bp paired-end chemistry.

3.7 Quality Control and Taxonomic Assignments in QIIME2 [18]

1. Download and install QIIME2 according to directions at: https://docs.qiime2.org/2021.2/install/ (*see* **Note 1**).
2. Create a QIIME2 manifest .txt file of Illumina forward and reverse .fastq files following the guide in Fig. 2.

ManifestExample.txt

```
SampleID    forward-absolute-filepath            reverse-absolute-filepath            symptom-type    Additional-info-if-necessary
Sample1     $PWD/Sample1forwardread.fastq        $PWD/Sample1reverseread.fastq        Symptomatic     Location
Sample2     $PWD/Sample2forwardread.fastq        $PWD/Sample2reverseread.fastq        Asymptomatic    Location
etc.
```

Fig. 2 An example of a manifest file accepted by QIIME2. An explanation of each column is as follows: "sampleID" is the name of the sample, "forward-absolute-filepath" is the location of the sample's forward read file, "reverse-absolute-filepath" is the location of the sample's reverse read file, "symptom-type" is the symptom state of each sample, and "additional-info-if-necessary" is a column for any other variables or sample information you would like to include, such as location (you may add as many additional information columns as is necessary for your analysis)

3. Open QIIME2 and import manifest text file to QIIME2. Process .fastq files into QIIME2-generated artifact files using the following command (*see* **Note 2**):

   qiime tools import \

   --type 'SampleData[PairedEndSequencesWithQuality]' \

   --input-path manifest.txt \

   --output-path manifest.qza \

   --input-format PairedEndFastqManifestPhred33V2

4. Generate and view demultiplexed reads (Fig. 3).

   qiime demux summarize \

   --i-data manifest.qza \

   --o-visualization demux.qzv

   qiime tools view demux.qzv

5. Perform quality control and clustering using the DADA2 pipeline. Trim any base pairs with a quality score less than 20 (determined based on the visualization from the previous step). Use the "trim-left" command to trim bases at the start of each sequence and "trunc-len" to trim bases from the end of each sequence. Change "000" from both of these commands to reflect the base pairs where you would like the trimming to start.

   qiime dada2 denoise-single \

   --i-demultiplexed-seqs manifest.qza \

   --p-trim-left 000 \

   --p-trunc-len 000 \

   --o-representative-sequences rep-seqs-dada2.qza \

   --o-table table-dada2.qza \

   --o-denoising-stats stats-dada2.qza

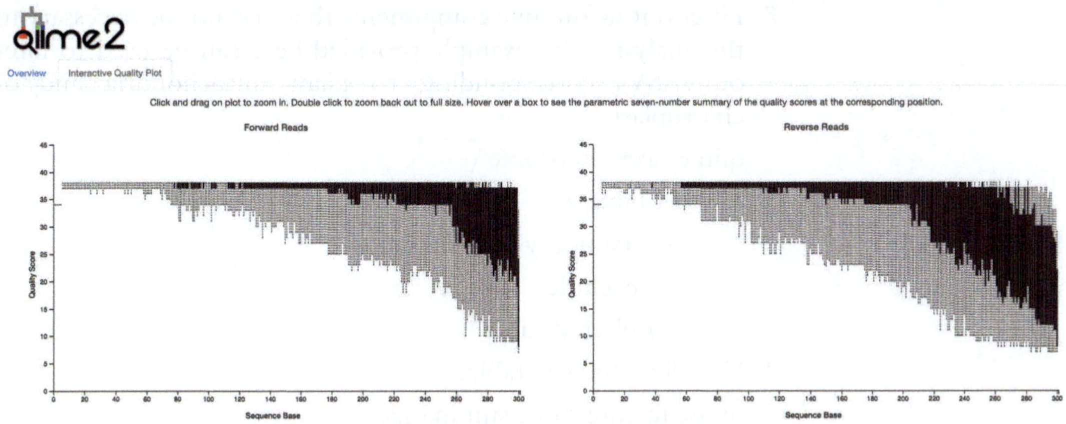

The plot at position 260 was generated using a random sampling of 9399 out of 3790953 sequences without replacement. This position (260) is greater than the minimum sequence length observed during subsampling (35 bases). As a result, the plot at this position is not based on data from all of the sequences, so it should be interpreted with caution when compared to plots for other positions. Outlier quality scores are not shown in box plots for clarity.

Parametric seven-number summary for **position 260**

Box plot feature	Percentile	Quality score
(Not shown in box plot)	2nd	8
Lower Whisker	9th	17
Bottom of Box	25th	27
Middle of Box	50th (Median)	37
Top of Box	75th	37
Upper Whisker	91st	38
(Not shown in box plot)	98th	38

Download forward parametric seven-number summaries as CSV

The plot at position 149 was generated using a random sampling of 9998 out of 3790953 sequences without replacement. This position (149) is greater than the minimum sequence length observed during subsampling (35 bases). As a result, the plot at this position is not based on data from all of the sequences, so it should be interpreted with caution when compared to plots for other positions. Outlier quality scores are not shown in box plots for clarity.

Parametric seven-number summary for **position 149**

Box plot feature	Percentile	Quality score
(Not shown in box plot)	2nd	10
Lower Whisker	9th	23
Bottom of Box	25th	35
Middle of Box	50th (Median)	38
Top of Box	75th	38
Upper Whisker	91st	38
(Not shown in box plot)	98th	38

Download reverse parametric seven-number summaries as CSV

Demultiplexed sequence length summary

Forward Reads

Total Sequences Sampled	10000
2%	255 nts
9%	260 nts
25%	269 nts
50% (Median)	299 nts
75%	300 nts
91%	300 nts
98%	300 nts

Reverse Reads

Total Sequences Sampled	10000
2%	255 nts
9%	260 nts
25%	273 nts
50% (Median)	299 nts
75%	300 nts
91%	300 nts
98%	300 nts

Fig. 3 An example of the quality plot generated using QIIME

```
qiime metadata tabulate \
    --m-input-file stats-dada2.qza \
    --o-visualization stats-dada2.qzv
```

6. Assign taxonomy using the taxonomic classifier specific to the gene of interest obtained from the UNITE or SILVA databases.

```
qiime feature-classifier classify-sklearn \
    --i-classifier classifier.fasta
    --i-reads rep-seqs-dada2.qza
    --o-classification taxonomy.qza
```

7. Filter out taxonomic components that may not be necessary to the analysis. The example provided here can be used to filter out ASVs corresponding to plant mitochondria and/or chloroplast.

```
qiime taxa filter-table \
    --i-table table-dada2.qza \
    --i-taxonomy taxonomy.qza \
    --p-exclude chloroplast,mitochondria \
    --o-filtered-table table-no-chloroplast-mitochondria.qza
```

8. Visualize the new table.

```
qiime feature-table summarize \
    --i-table table-no-chloroplast-mitochondria.qza\
    --o-visualization table-no-chloroplast-mitochondria.qzv
```

9. Rarefy the samples in order to prevent variation in sampling depth among samples [31]. The appropriate sampling depth will be approximately around the median number of reads that eliminates the fewest samples. This can be determined by viewing the Frequency per sample section in the "table-no-chloroplast-mitochondria.qzv." This number should replace the "000" as the "max-depth" and "sampling-depth" (*see* **Note 3**).

```
qiime diversity alpha-rarefaction \
    --i-table table-no-chloroplast-mitochondria.qza \
    --p-max-depth 000 \
    --m-metadata-file manifest.txt \
    --o-visualization rarefaction-plots.qzv
qiime feature-table rarefy \
    --i-table table-no-chloroplast-mitochondria.qza \
    --p-sampling-depth 000 \
    --o-rarefied-table rarefied-table.qza
qiime feature-table filter-seqs \
    --i-data rep-seqs-dada2.qza \
    --i-table rarefied-table.qza \
    --o-filtered-data rarefied-rep-seqs.qza
qiime feature-table summarize \
    --i-table rarefied-table.qza \
    --o-visualization rarefied-table.qzv \
    --m-sample-metadata-file manifest.txt
qiime feature-table tabulate-seqs \
    --i-data rarefied-rep-seqs.qza \
    --o-visualization rarefied-rep-seqs.qzv
```

10. Generate a phylogenetic tree.

```
qiime phylogeny align-to-tree-mafft-fasttree \
    --i-sequences rarefied-rep-seqs.qza \
    --o-alignment aligned-rarefied-rep-seqs.qza \
    --o-masked-alignment
        masked-aligned-rarefied-rep-seqs.qza \
    --o-tree unrooted-tree.qza \
    --o-rooted-tree rooted-tree.qza
```

11. Complete taxonomic classification with rarefied sequences.

```
qiime feature-classifier classify-sklearn \
    --i-classifier rRNA-specific-classifier \
    --i-reads rarefied-rep-seqs.qza \
    --o-classification rarefied-taxonomy.qza
qiime metadata tabulate \
    --m-input-file rarefied-taxonomy.qza \
    --o-visualization rarefied-taxonomy.qzv
qiime taxa barplot \
    --i-table rarefied-table.qza \
    --i-taxonomy rarefied-taxonony.qza \
    --m-metadata-file manifest.txt \
    --o-visualization taxa-bar-plots.qzv
```

3.8 Conduct Diversity Analyses Using R

1. Download and install all necessary R packages.
2. Import metadata from QIIME2 into RStudio (*see* **Note 4**).
   (a) Import ASV table (feature-table.txt), manifest table, and taxonomy table with headers and row names.

```
data<- read.table("feature-table.txt", header = T, sep = "\t", row.names = 1)
metadata<- read.table ("manifest.txt", header = T, sep = "\t", row.names = 1)
taxa<- read.table("taxonomy.tsv", header = T, sep = "\t", row.names = 1)
```

   (b) Import rooted phylogenetic tree.

```
tree<- read.tree("tree.nwk")
```

   (c) Reroot the phylogenetic tree based on the node below the tip of the new outgroup.

```
tree <- root (tree, 1, resolve.root = T)
```

   (d) Check if the column names from the data table match with the row names from the metadata.

```
colnames(data) %in% rownames(metadata)
```

(e) Split the taxa data into separate columns.

```
taxa$Taxon <- cSplit(data.frame(taxa), 'Taxon', ";")
```

(f) Format the taxa table into an R data frame.

```
taxa<- as.data.frame(taxa)
```

(g) Change the column names of the taxa table to reflect the seven taxonomic ranks.

```
names(taxa) <- c("Kingdom", "Phylum", "Class",
"Order", "Family", "Genus",

"Species")
```

3. Create phyloseq object by combining the ASV table (data), metadata, and taxonomic tree [20].

```
PO <- merge_phyloseq(otu_table(data, taxa_are_rows = T),

        sample_data(metadata),

        tree,

        tax_table(as.matrix(taxa))

)
```

4. Subset phyloseq object into the symptom types of the samples.

```
PO <- subset_samples(PO, symptom_type == "symptom1" |

      symptom_type == "symptom2")
```

5. Filter out samples with less than 1000 reads.

```
PO <- prune_samples(sample_sums(PO)>=1000, PO)
```

6. Remove singletons.

```
PO <- filter_taxa(PO, function (x) {sum(x > 0) > 1}, prune = TRUE)
```

7. Perform a permutational multivariate analysis of variance (PERMANOVA) with 999 permutations using the UniFrac distance matrix [26].

   (a) Create a separate data frame using the sample data.

   ```
   sampledf <- data.frame(sample_data(PO))
   ```

   (b) Specify the phylogenetic distance that will be used to compare communities. In this example, we will use the unweighted UniFrac distance metric.

   ```
   dist.mat <- phyloseq::distance(PO, method = "unifrac")
   ```

   (c) Specify the number of permutations to run for the PERMANOVA.

   ```
   perm <- how(nperm = 999)
   ```

   (d) Conduct a PERMANOVA comparing your main effects of interest and their interactions (*see* **Note 5**).set.seed (100)

   ```
   pmv <- adonis2(dist.mat ~ symptom_type* variable1, data = sampledf,
       permutations = perm)
   ```

8. Create a non-metric multidimensional scaling (NMDS) ordination to identify clustering based on sample type [26].

   (a) Specify the type of ordination to use.

   ```
   nmds_ord <- ordinate(physeq = PO, method = "NMDS", distance = "unifrac",
       formula = ~ symptom_type)
   ```

   (b) Plot NMDS chart with ellipses to identify clustering at 95% confidence intervals. Different colors will represent the different symptom types and will be indicated in a legend.

```
nmds_plot_source <- plot_ordination(physeq = PO, ordination =
nmds_ord, axes
    = c(1,2), color = "symptom_type") + stat_ellipse(mapping =
    aes(linetype = symptom_type),
    level = 0.95,
    type = "norm",
   show.legend = T) +
   theme(legend.background = element_rect(color = "black", size
   0.3, linetype = "solid"),
   legend.text = element_text(size = 12),
   axis.text.x = element_text(color = "black"),
   axis.text.y = element_text(color = "black"),
   panel.grid = element_blank())
```

9. Calculate the Phylogenetic, Shannon, and Simpson diversity indices and test for differences in microbial diversity between symptom types (Fig. 4) [25, 26]. To plot each diversity index, repeat step h. changing y = PD to y = Shannon or y = Simpson.
   (a) Create a separate data frame of the OTU table.

   ```
   otus <- as.data.frame(otu_table(PO))
   ```

   (b) Transpose OTU data frame.

   ```
   otus <- t(otus)
   ```

   (c) Calculate Faith's phylogenetic diversity index.

   ```
   comm.pd <- pd(otus, tree_Q, include.root = TRUE)
   ```

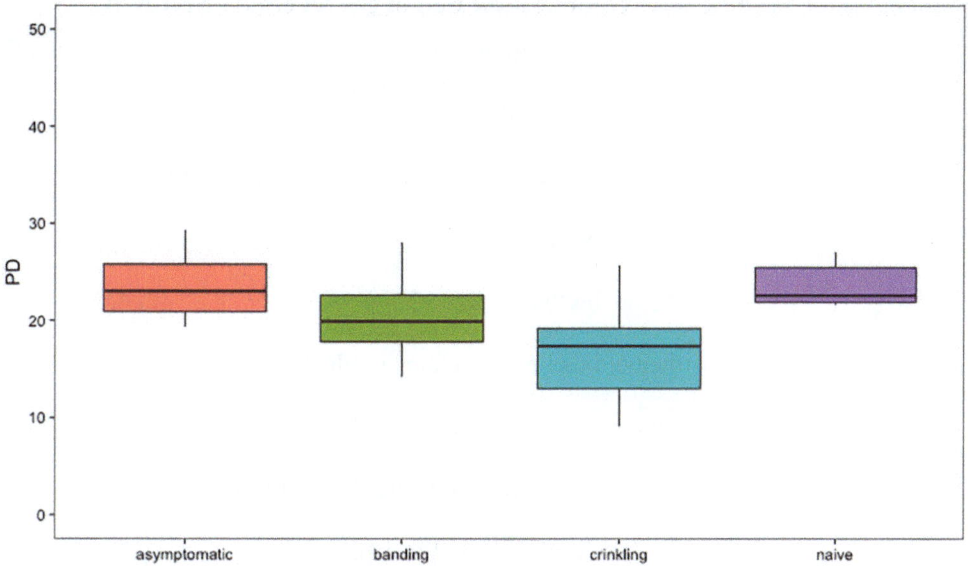

Fig. 4 An example of microbial diversity analysis. This example compares the microbiome using the phylogenetic diversity from four symptom types: banding and crinkling (two types of symptoms), asymptomatic, and naïve

(d) Calculate observed taxa index and Shannon's and Simpson's diversity indices.

```
diversity <- estimate_richness(PO, measures = c("Observed","Shannon",
    "Simpson"))
```

(e) Merge sample data and diversity data frames.

```
div2 <- cbind(sample_data(PO), diversity)
```

(f) Merge data frame from step e, with the calculations from Faith's phylogenetic diversity index.

```
div2 <- cbind(div2, comm.pd)
```

(g) Calculate the community evenness and combine with data frame from step f.

```
div2$Evenness <- div2$Shannon/log(div2$Observed)
```

(h) Use the package ggplot to plot the microbial diversity analysis with each symptom type represented by a different color as indicated in the legend (*see* **Note 6**).

```
phylogenetic_plot <- ggplot(div2, aes(x = symptom_type, y = PD, fill =
        symptom_type)) +
        geome_boxplot(outlier.colour = "black", notch = F, outlier.shape =
            NA) +
        theme_bw(base_size = 12) +
        theme(axis.title.x = element_blank(),
        axis.text.x = element_text(color =  "black"),
        axis.text.y = element_text(color = "black"),
        panel.grid = element_blank(),
        legend.position = "none",
        legend.background = element_rect(color = "black", size = 0.3,
        linetype =
            "solid"),
        legend.text = element_text(size = 12),
        panel.background = element rect(size = 0.5, linetype = "solid"),
        strip.background = element_rect(fill = "white")) +
ylim(0,10)
```

10. Fit a linear mixed-effects model [27]. Replace PD with Shannon or Simpson to conduct all diversity analyses.

```
model <- lmer(PD ~ symptom_type * (1|variable1), data = div2)
```

11. Generate a results table.

```
Anova(model)
```

12. Compute estimated marginal means, list the pairwise contrasts, and adjust the *p* value for multiple comparisons using Tukey's multiple comparison test.

```
emmeans(model, list(pairwise ~ symptom_type), adjust =
    "tukey")
```

3.9 Differential Abundance Analysis in RStudio

1. Convert the data from a phyloseq object into a DESeq object [24].

```
ds <- phyloseq_to_deseq2(PO, ~ symptom_type)
```

2. Perform the differential expression analysis.
   (a) Fit a DESeq model.

```
gm_mean <- function(x, na.rm=TRUE){exp(sum(log(x[x > 0]),
    na.rm=na.rm) / length(x))}

geoMeans <- apply(counts(ds), 1, gm_mean)

ds$group <- factor(paste0(ds$symptom_type))

design(ds) <- ~ group

diagds <- estimateSizeFactors(ds, geoMeans = geoMeans)

diagds <- DESeq(diagds, test = "Wald", fitType = "parametric")
```

   (b) Generate the results table by contrasting symptomatic/asymptomatic samples.

```
res <- results(diagds, cooksCutoff = FALSE, contrast = c("group",
    "symptomatic", "asymptomatic"))
```

3. Filter results that give only a significant adjusted *p* value with a positive log 2-fold change. A positive log 2-fold change indicates that those samples are significantly more abundant in symptomatic samples vs. control samples.

```
sigtab <- res[which(res$padj < 0.01 & res$log2FoldChange > 0), ]

sigtab <- as(sigtab, "data.frame")
```

4. Combine the differential ASV table and taxa data table to determine the ASVs that are more significantly abundant in symptomatic samples, which point toward a putative causal agent.

```
sigtab <- setDT(sigtab, keep.rownames = TRUE) []

taxa <- setDT(taxa, keep.rownames = TRUE) []

diff.ASVs <- merge(sigtab, taxa, by = "rn")
```

4 Notes

1. A more descriptive tutorial of the QIIME2 steps can be found at https://docs.qiime2.org/2021.2/tutorials/moving-pictures/.
2. All code written in green represents an input file while all code written in blue is an output file. All code written in orange indicates an item that must be changed to reflect your personal data.
3. Choosing the correct number for rarefaction can be difficult. It is recommended that you try this code with several different read numbers and view the rarefaction plots. When you see the curves in the top plot leveling off (Fig. 5), you can assume the maximum sampling depth has been reached and can use this number for rarefaction as long as it does not eliminate too many of your samples. The bottom plot illustrates the number of samples that remain in each sampling group after rarefaction to ensure that you have retained an appropriate number of samples in each category to conduct diversity metrics.
4. All QIIME2 .qza files can be opened using a text editing software. The file for the feature table that will be used for RStudio will be formatted as a .biom table. To convert to a useable format, use the following code in command-line:

   ```
   biom convert \
       -i feature-table.biom \
       -o feature-table.tsv \
       --to-tsv
   ```
5. Variable1 indicates a random variable in your analysis, such as site location. This allows you to test for both the main effects and the interaction of the two main effects to determine if another factor of your analysis is explaining the results of your data.
6. The parameters for the function "ylim" will need to be adjusted for each diversity index as they vary in ranges.

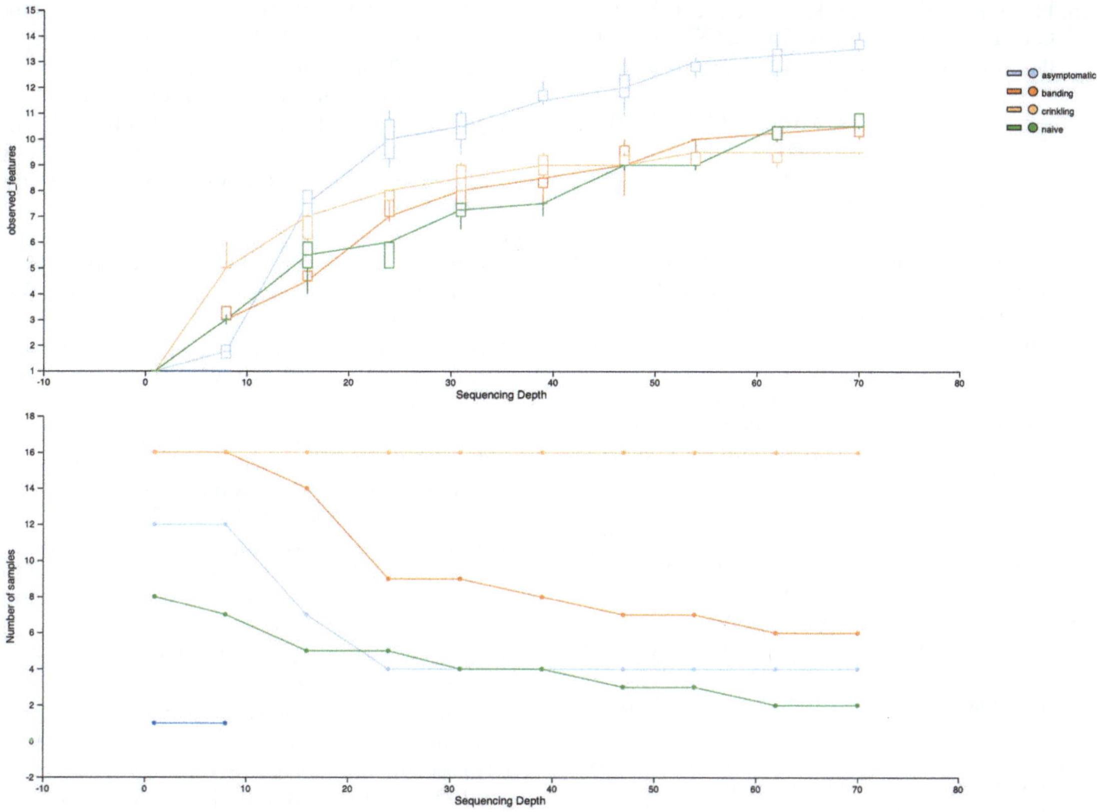

Fig. 5 Alpha rarefaction plots showing an appropriate sequencing depth. The top plot shows the curves leveling off, indicating that additional sampling will most likely not lead to an increase in diversity. The bottom plot shows the number of samples that remain in each group specified by the metadata after rarefaction

Acknowledgements

The authors thank Afaq M. Mohamed Niyas and Caterina Villari from University of Georgia for providing the modified 16S phytoplasma PCR protocol.

References

1. Abdelfattah A, Malacrinò A, Wisniewski M et al (2018) Metabarcoding: a powerful tool to investigate microbial communities and shape future plant protection strategies. Biol Control 120:1–10
2. Roman-Reyna V, Dupas E, Cesbron S et al (2021) Metagenomic sequencing for rapid identification of *Xylella fastidiosa* from leaf samples. bioRxiv 6:e0059121
3. Deiner K, Bik HM, Elvira M et al (2017) Environmental DNA metabarcoding: transforming how we survey animal and plant communities. Mol Ecol 26(21):5872–5895
4. O'Brien HE, Parrent JL, Jackson JA et al (2005) Fungal community analysis by large-scale sequencing of environmental samples. Appl Environ Microbiol 71(9):5544–5550
5. Janda JM, Abbott SL (2007) 16S rRNA gene sequencing for bacterial identification in the diagnostic laboratory: pluses, perils, and pitfalls. J Clin Microbiol 45(9):2761

6. Hadziavdic K, Lekang K, Lanzen A et al (2014) Characterization of the 18S rRNA gene for designing universal eukaryote specific primers. PLoS One 9(2):e87624
7. Prewitt ML, Diehl SV, Mcelroy TC et al (2008) Comparison of general fungal and basidiomycete-specific ITS primers for identification of wood decay fungi. For Prod J 58(4): 66–71
8. Ewing CJ, Slot J, Benitez Ponce M-S et al (2021) The foliar microbiome suggests fungal and bacterial agents may be involved in the beech leaf disease pathosystem. Phytobiomes J. (in press)
9. Creer S, Deiner K, Frey S et al (2016) The ecologist's field guide to sequence-based identification of biodiversity. Methods Ecol Evol 7(9):1008–1018
10. Nilsson RH, Anslan S, Bahram M et al (2019) Mycobiome diversity: high-throughput sequencing and identification of fungi. Nat Rev Microbiol 17(2):95–109
11. Pollock J, Glendinning L, Wisedchanwet T et al (2018) The madness of microbiome: attempting to find consensus "best practice" for 16S microbiome studies. Appl Environ Microbiol 84(7):e02627
12. Zinger L, Bonin A, Alsos IG et al (2019) DNA metabarcoding—need for robust experimental designs to draw sound ecological conclusions. Mol Ecol 28(8):1857–1862
13. Horton MW, Bodenhausen N, Beilsmith K et al (2015) Genome-wide association study of *Arabidopsis thaliana*'s leaf microbial community. Nat Commun 5:5320
14. Gundersen DE, Lee I-M (1996) Ultrasensitive detection of phytoplasmas by nested-PCR assays using two universal primer pairs. Phytopathol Mediterr 35(3):144–151
15. Gibb KS, Padovan AC, Mogen BD (1995) Studies on sweet potato little-leaf phytoplasma detected in sweet potato and other plant species growing in northern Australia. Phytopathology 85(2):169–174
16. Usyk M, Zolnik CP, Patel H et al (2017) Novel ITS1 fungal primers for characterization of the mycobiome. mSphere 2(6):e00488
17. Sapkota R, Nicolaisen M (2015) High-throughput sequencing of nematode communities from total soil DNA extractions. BioMed Cent Ecol 15:3
18. Bolyen E, Rideout JR, Dillon MR et al (2019) Reproducible, interactive, scalable and extensible microbiome data science using QIIME 2. Nat Biotechnol 37:852–857
19. R Core Team (2017) R: a language and environment for statistical computing, Vienna
20. McMurdie PJ, Holmes S (2013) phyloseq: an R package for reproducible interactive analysis and graphics of microbiome census data. PLoS One 8(4):e61217
21. Paradis E, Schliep K (2018) Ape 5.0: an environment for modern phylogenetics and evolutionary analyses in R. Bioinformatics 35: 526–528
22. de Mendiburu F (2020) agricolae: statistical procedures for agricultural research. R package version 1.3-3
23. Foster Z, Sharpton T, Grünwald N (2017) Metacoder: an R package for visualization and manipulation of community taxonomic diversity data. PLoS Comput Biol 13(2):1–15
24. Love MI, Huber W, Anders S (2014) Moderation estimation of fold change and dispersion for RNA-seq data with DESeq2. Genome Biol 15:550
25. Kembel SW, Cowan PD, Helmus MR et al (2010) Picante: R tools for integrating phylogenies and ecology. Bioinformatics 26: 1463–1464
26. Oksanen J, Blanchet FG, Friendly M et al (2020) vegan: Community ecology package. R package version 2.5-7
27. Lenth R (2021) emmeans: Estimated marginal means, aka least-squares means. R package version 1.5.5-1
28. Bates D, Maechler M, Bolker B et al (2015) Fitting linear mixed-effects models using lme4. J Stat Softw 67(1):1–48
29. Beckers B, Op BM, De TS et al (2016) Performance of 16s rDNA primer pairs in the study of rhizosphere and endosphere bacterial microbiomes in metabarcoding studies. Front Microbiol 7:650
30. White TJ, Bruns T, Lee SB et al (1990) Amplification and direct sequencing for fungal ribosomal RNA genes for phylogenetics. In: Innis MA, Gelfand DH, Sninsky JJ, White TJ (eds) PCR protocols: a guide to methods and amplifications. Academic Press, Inc, San Diego, pp 315–322
31. Ramakodi MP (2021) Effect of amplicon sequencing depth in environmental microbiome research. Curr Microbiol 78(3): 1026–1033

Part V

Plant-Pathogen Interactions

Chapter 20

Plant–Fungal Interactions: Laser Microdissection as a Tool to Study Cell Specificity

Raffaella Balestrini and Fabiano Sillo

Abstract

In the past 20 years, laser microdissection (LMD) technology has been widely applied to plant tissues, allowing to obtain new information on the role of different cell-type populations during plant development and interactions, including plant–pathogen interactions. The application of a LMD approach allowed verifying the response of plant and pathogen during the progression of the infection in different cell types, focusing both on gene expression in host plants and pathogens. Here, a protocol to apply the LMD approach to study plant and fungal transcript profiles in different cell-type populations is described in detail, from the biological material preparation to RNA extraction and gene expression analyses.

Key words Cell specificity, Fungal pathogen, Gene expression, Laser microdissection, Plant–microbe interaction

1 Introduction

Laser microdissection (LMD) is a technology that allows rapid collection of selected cell-type populations from a section of a sample composed of various cells with diverse functions [1], allowing the extraction DNA, RNA, proteins, and even metabolites from homogeneous cell-type populations. Plant–pathogen interaction studies are surely an area where LMD have an impact since these experiments provide information about the spatial expression of the genes involved in interaction processes. The responses in these interactions can in fact be localized in specific cell types or microbial structures. LMD methodology has been widely used to study cell specificity in plant-fungal symbiotic interactions [2–6]. Additionally, LMD has been used to dissect the complexity in several pathogenic systems, i.e., during the interaction with phytoplasmas [7, 8], fungal pathogens [9–12], and parasitic plants [13]. These studies were mainly focused on target gene expression analyses or transcriptomics, giving the possibility to analyze the contemporaneous

Nicola Luchi (ed.), *Plant Pathology: Methods and Protocols*, Methods in Molecular Biology, vol. 2536, https://doi.org/10.1007/978-1-0716-2517-0_20,

expression of several genes in homogenous cell-type populations and allowing showing differences in transcriptional regulations of infected cells and cells distant from the infection sites [14]. Although the application of this technology in microbiological research is still unusual [15], also due to the cost of the LMD system, different studies have been applied LMD technology to characterize the transcriptional changes of plant tissues infected by fungal pathogens, focusing both on plant and fungal genes. During the interaction with pathogens in fact only a small fraction of cells can be in contact with the pathogen infection structures, mainly at early infection stages, and the large portion of noninfected cells might mask, or dilute, transcriptional changes localized in specific infected cells [11]. Additionally, fungi (both symbiotic and pathogenic fungi) can form different colonization/infection structures during the infection process progression [6, 9, 10]. Thus, it is fundamental to study local plant response to fungal pathogens as well as the changes in gene expression in diverse infection fungal structures (e.g., the haustorium in biotrophic interactions) or at different pathogen developmental stages [9]. In some cases, before to work with infected tissues, protocols for the preparation of tissues subjected to fungal infection to be used for LMD application have been optimized [16, 17]. Interestingly, Falter et al. [18] provided a novel method to isolate the plant's adaxial epidermal cell layer containing localized, pathogen-induced cell wall structures, with the aim to perform quantitative proteome and carbohydrate analyses. This may be of great importance in plant–microbe interactions where pathogen-induced cell wall modifications are essential for plant defense [18].

This methodology has been already used for dissecting the complexity in infected tissues in diverse plant/fungal pathogen combinations, focusing on the gene expression in the two partners. In a first paper, an LMD approach was applied to the fungal pathogenic system maize and *Colletotrichum graminicola*. This work was aimed to collect maize stalk cells associated with *C. graminicola* hyphae at early infection stage. Fungal microarray analysis has been performed, leading to the identification of several transcripts highly accumulated and potentially involved in pathogenesis [9]. From a methodological point of view, the authors have used a paraffin embedding protocol useful to maintain the GFP signal, which allows following the fungal infection in plant tissues. Although the use of markers is not essential for the LMD approach, this could be very useful for the identification of target cells. The LMD approach was also used to dissect the complexity in soybean infected by the rust fungus *Phakopsora pachyrhizi*. In detail, RNA has been isolated from uredinia collected by an LMD system, allowing cDNA library construction and the identification of rust unisequences [19, 20]. Microdissection of infection sites from *Arabidopsis* leaves challenged by the powdery mildew

Golovinomyces orontii has been also performed, helping in identifying elements of the plant defense response not previously found through whole-leaf transcriptome analysis [11]. The same authors presented in detail an optimized protocol of tissue preparation, laser microdissection (LMD), and RNA isolation of homogeneous cell types at the site of *Golovinomyces orontii* infection of mature *Arabidopsis* leaves [21]. Looking at the fungal side, LMD was used to dissect three distinct zones from poplar leaves infected by *Melampsora larici-populina*, which belongs to Basidiomycota, at the uredinial stage, with the aim to separate fungal sporulation and fungal infection tissues [10]. RNA was extracted and amplified from these diverse zones to verify rust transcript profiles through transcriptome analysis using *M. larici-populina* whole-genome exon oligoarrays and reverse-transcription quantitative polymerase chain reaction (RT-qPCR). It is worth noting that in addition to the impact of fixation methods used for tissue preparation on RNA quality [22], loss of RNA integrity during slide mounting was previously reported and resolved using a tape transfer system [23]. However, other factors can negatively impact RNA quality during tissue preparation, such as sample infiltration and slide mounting. To overcome these problems, low-melting paraffin coupled with a direct transfer of paraffin ribbons onto microscope slides without water and slide drying led to a significant increase of RNA integrity [10]. More recently, an LMD approach has been used to investigate the processes involved in grapevine–*Plasmopora viticola* interaction, where only a few cells are in contact with the pathogen, leading to highly localized transcriptional changes in infected cells that might be masked by the large portion of noninfected cells [24]. In this direction, transcriptional changes in stomata cells and surrounding areas at early stages of *P. viticola* infection showed that the expression of seven *P. viticola*-responsive genes was higher in LMD-collected cells with respect to whole leaves, showing a site-specific transcriptional regulation of the host. Moreover, modulation of genes that were not regulated in the whole leaf analysis was also identified, confirming the LMD as a very powerful tool to study cell-specific response during plant–pathogen interaction. LMD methodology has been also used to verify the genes differentially expressed in the two-cell type populations (epidermis *vs.* hypodermis) forming the grape berry skin, suggesting a different function for the two-cell type populations [25]. Particularly, some genes putatively correlated to defense responses appeared in fact to be more or exclusively expressed in the external cell-type population. The aim of this chapter is to detail a method to apply the LMD technology to study gene expression in specific cell-type populations in infected plant tissues infected by pathogen fungi, starting from the preparation of the biological material to the gene expression analysis, also highlighting the variability in the already used protocols (*see* the Notes). It is worth noting that protocols already present in literature represent only a starting point and might need modifications to obtain the best results with different samples. Changes in tissue handling, such as different fixation times, may lead to different laser cutting parameters (e.g., longer fixation can change the hardness of the tissues).

2 Materials

The use of sterile materials (tweezers, scalpel, tubes, petri dishes, etc.) is required during all fixation and paraffin embedding steps to avoid RNA degradation as well as during molecular analyses (RT-PCR and RT-qPCR). All solutions must be prepared using ultrapure or sterilized water.

2.1 Major Equipment

1. Chemical hood (for fixation stage and embedding).
2. Binder oven.
3. Rotary microtome to obtain paraffin section using A35 Durable Low-profile Microtome Blades – Disposable – Stainless Steel, Tissue-Tek (Electron Microscopy Sciences, #63069).
4. Slide Warming Plate.
5. Laser Micro Dissection system.
6. PCR Thermal Cycler.
7. CFX Connect Real-Time PCR Detection System (Bio-Rad), Rotor-Gene Q (Qiagen), or an equivalent Real-Time PCR instrument equipped with a channel for detecting SYBR® Green dye.

2.2 LMD Supplies

1. Sterile falcon tubes, 50 mL.
2. Ethanol/Acetic Acid 3:1.
3. Neo-Clear or Xylene.
4. Paraplast Plus Paraffin. Paraffin/wax with a lower melting point (37–42 °C), that should avoid RNA degradation, may be used in alternative to Paraplast Plus to maintain RNA integrity [10, 26] (*see* **Note 1**).
5. Membrane Slides (PEN-Membrane 2.0 μm) (*see* **Note 2**).
6. Thin wall 0.5-ml PCR tubes (required for the Leica LMD systems; Greiner bio-one, #682201) (*see* **Note 2**).
7. PicoPure® RNA Isolation Kit (Termo Fisher Scientific, # KIT0204) or an equivalent RNA extraction kit for low amount of starting material.
8. DNase reagents, such as TURBO DNA- free™ Kit (Thermofisher).

9. Primers targeting plant and fungal housekeeping genes (at least two genes could be optimal) and specific candidate genes (such as genes involved in pathogenesis, etc. . .). Primer sequences depend on the studied organisms.
10. One-Step RT-PCR kit.
11. One-Step RT-qPCR kit.

3 Methods

The followed protocol is based on that published in Balestrini & Fiorilli [6], i.e., a protocol used for to study gene expression on arbuscular mycorrhizal interaction, with some modifications and notes on the base of the literature on plant–pathogen interactions. The several steps required by an experiment of LMD are reported in Fig. 1.

3.1 Fixation and Embedding in Paraffin

1. Segments from infected samples (and eventually noninfected ones as control) are cut and collected separately in RNase-free 50-mL tubes containing cold freshly prepared Farmer's fixative (absolute ethanol/glacial acetic acid, 3:1 v/v; 20 mL in each tube). After 20 min at room temperature (RT) under vacuum,

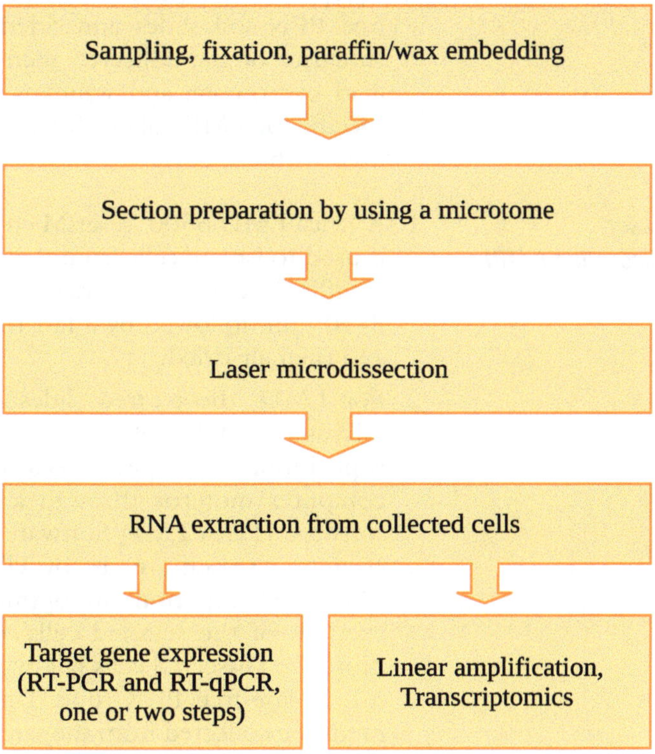

Fig. 1 Flowchart representing the major steps in an LMD experiment

to facilitate the fixative penetration, fixative solution is changed, and samples are incubated overnight at 4 °C (*see* **Notes 3** and **4**).

2. Samples are then dehydrated in a graded ethanol series (70%, 90% in ultrapure water, and 100% twice), followed by two steps in Neo-Clear® (*see* **Note 5**), 30 min on ice for each step.
3. A few chips of paraffin (approximately 10–15 pieces) are added to 20 mL of Neo-Clear to allow the gradual substitution with paraffin, and the samples are maintained at RT for about 2 h and then at 58 °C until the chips dissolved (for about 3 h).
4. This mixture was substituted with pure paraffin, previously melted at 58 °C, and tubes were then incubated overnight at 58 °C. In the first step with pure paraffin, tubes are kept open to permit evaporation of Neo-Clear traces. The following day, medium is replaced 1–2 times at approximately 6-h intervals and paraffin containing embedded root pieces is finally poured in Petri dishes (60 mm) and solidified by cooling. Protocols using low melting point paraffin or Steadman' wax are described in Hacquard et al. [10] and Gomez & Harrison [26], respectively.
5. Sections of about 10–12 μm of thickness are cut using a rotatory microtome. The section ribbons are placed on sterile ultrapure water (filtered with a 0.2 μm filter) on the RNase-free PEN foil slides and dried on a 40 °C warming plate. Integrity of the collected sections is checked using a bright field microscope and standard slides before to prepare those specific for LMD. Slides for LMD are then stored at 4 °C and used within 2 days.

3.2 Laser Microdissection (LMD)

1. A Leica LMD 6500 Laser Microdissection system (*see* **Note 6**) is used to isolate cells from the tissue sections. Just before use, paraffin sections are deparaffinized with Neo-Clear for 8–10 min, followed by a fast rinse in 100% ethanol for 1 min and then air-dried.
2. For LMD, the treated slides are placed face down on the microscope slide holder and a 0.5-mL RNase-free PCR tube is positioned in a special collector. Tissues are visualized on a computer monitor through a video camera and, thanks to software (Leica LMD Software), target cells encircled on the computer screen are cut by UV laser (337-nm wavelength). The LMD conditions are optimized to obtain a clean, narrow excision of the selected cells, depending on the sample (*see* **Note 7**). Dissected cells fall by gravity into the cap of the 0.5-mL RNase-free PCR tube. Different cell types (up to four) might be collected from the same slide using different collector positions.

3. After collection, 5–15 μL (depending on the number of the collected cells) of RNA extraction buffer is added in the tube cap containing the cell sections, and the tube is closed. Samples are centrifuged at 800 × *g* for 2 min, and then stored at −80 °C.

3.3 RNA Extraction, DNase Treatment, and RT-PCR/RT-qPCR

1. As a first step, samples containing the same dissected cell-type population in extraction buffer are pooled in a single tube (about 1000–1500 in total for each cell-type population for a target gene expression experiment, while more RNA is needed for large-scale transcriptomic analyses and RNA extracted from more than 3000–5000 cell sections is generally used) with a final volume of 50 μL. RNA is then extracted following the PicoPure™ RNA Isolation Kit with slight modifications (protocol IV in the manual) (*see* **Note 8**). Samples are first incubated at 42 °C for 30 min. Before to start, a preconditioning of the RNA Purification Column is needed. To this aim, 250 μL of the Conditioning Buffer (CB) is added onto the purification column filter membrane, inserted in the provided collection tube, and the tubes are incubated for 5 min at room temperature and then centrifuged at 16,000 x *g* for 1 min. After the incubation at 42 °C, 50 μL of 70% ethanol is added to each pooled sample and mixed well by pipetting up and down. Total volume (100 μL) bound to the preconditioned column is washed and eluted following the manufacture's protocol. RNA is eluted in 20–25 μL of Elution Buffer (EB) and collected into a new 0.5 mL microcentrifuge tube. The isolated RNA is ready for DNase treatment. Generally, 21 μL is used to allow the performing of the DNase treatment in 25 μL.
2. To avoid the risk of DNA contamination, RNA is subjected to a DNase treatment. The following protocol is referred to the use of the Turbo DNA-free™ Kit (*see* **Note 9**). In detail, 1.5 μL of enzyme and 2.5 μL of 10X buffer are added to the 21 μL of eluted RNA; samples are then mixed gently and incubated at 37 °C for 30 min. Turbo DNase incubation is stopped with 3 μL of DNase inactivation reagent, kept for 5 min at room temperature with occasional mixing. After a centrifugation at 10,000 × *g* for 1.5 min, to remove the inactivation buffer, the solution containing RNA is transferred to a fresh tube. The treated RNA is either used immediately for reverse transcription and amplification (RT-PCR) to verify the DNA contamination or stored at −80 °C until use.
3. RT-PCR reactions are carried out in a final volume of 10 μL containing 2 μL of 5x buffer, 0.5 μL of 10 mM dNTPs, 0.25 μL of each primer (diluted 1:10 from the 100 mM stock), 0.25 μL of One-Step RT-PCR enzyme mix, and 0.5 μL of total RNA (generally diluted 1:5). For the RT-PCR analyses, samples are

incubated for 30 min at 50 °C, followed by 15 min of incubation at 95 °C. Amplification reactions are run for 40 cycles at 94 °C for 30 s, 60 °C for 30 s, and 72 °C for 40 s using plant and fungal specific primers (*see* **Note 10**). Primers designed to the plant and fungal housekeeping genes are used to assess RNA amplification and possible contamination of RNA samples with genomic DNA. The absence of an amplified product in the RT-minus reactions (that are maintained in ice during the RT step at 50 °C, see above) excluded genomic DNA contamination. RT-PCR experiments are performed on at least two biological replicates, and PCR products are separated on 1.7–1.9% agarose gel. The best dilution of RNA to be used with specific primers (generally pure, 1:5, 1:10, depending on the samples and the target organism) is chosen on the basis of the first RT-PCR reactions with housekeeping gene(s).

4. Quantitative RT-PCR amplification reactions, performed with an iCycler apparatus, are carried out in a total volume of 25 μl, containing 2 μl RNA, 12.5 μl 2X SYBR Green RT-PCR Reaction Mix, 0.5 μl of each primer (10 μM), and 0.5 μl of iScript Reverse Transcriptase for One-Step RT-PCR. The following PCR program was used: 50 °C for 10 min, 95 °C for 5 min, 50 cycles of 95 °C for 10 s, and 60 °C for 30 s. A melting curve (80 steps with a heating rate of 0.5 °C per 10 s and a continuous fluorescence measurement) is recorded at the end of each run to exclude that the primers had generated nonspecific PCR products. Amplification reactions are successfully performed using specific primers for both fungal and plant genes. Baseline range and Ct values were automatically calculated using the iCycler software. Transcript levels were normalized to the Ct value of plant or fungal housekeeping gene(s). All the reactions are performed on at least two biological and two technical replicates with a good correlation between the replicates (*see* **Notes 11**, **12,** and **13**).

4 Notes

1. Use of cryosectioning (i.e., frozen sections) has been reported to yield high-quality RNA from plant cells [17, 27, 28]. As an example, in order to avoid degradation resulting from the fixation procedure, cryosectioning of frozen and embedded root material of *Medicago truncatula* plants inoculated with an arbuscular mycorrhizal fungus was performed to ensure sufficient RNA quality [27]. However, not always sections obtained by a cryostat have a sufficient quality to identify target cells. For this reason, generally an embedding protocol in paraffin is preferred although RNA quality cannot be optimal.

2. The products cited at points 5 and 6 (LMD supplies) are specifically requested for the Leica LMD apparatus. Other LMD systems could require different products.
3. Alternative fixatives may be used, e.g., 100% ice-cold acetone or a mixture of 60% absolute methanol, 30% chloroform and 10% glacial acetic acid (v/v/v) (methacarn fixative). Fixatives containing formaldehyde/formalin should be avoided since formaldehyde breaks nucleic acids and is therefore not suitable for subsequent RNA and DNA analyses.
4. Rapid microwave paraffin preparation method has been developed to avoid RNA degradation during this process [9, 16].
5. Neo-Clear® is a mixture of aliphatic hydrocarbons that can be used for the same applications as xylene (i.e., paraffin embedding and deparaffinization), without changes in protocols when Neo-Clear® is used instead of xylene. Isopropanol can be also used instead of xylene or Neoclear during paraffin-embedding process [16].
6. The cited system is only one among the several LMD systems present on the market. Several instruments have been in fact developed, by diverse companies, to isolate individual cells or groups of cells from intact tissues [29]. The procedure to prepare the biological materials can be the same for different LMD systems, although the maximum limit in section thickness may be different depending on the LMD system.
7. LMD parameters (e.g., objective to select and cut the targeted cells, laser power) must be optimized depending on the biological material, the section thickness and the used LMD system. Section thickness is depending on the samples (i.e., from the cell wall characteristics of the considered tissues) and should be optimized with the aim to improve the cutting. For very difficult material to be cut, instead of the slide with membrane it is possible to use "frame slides" (i.e., metal frame with membrane and without the glass, Fig. 2).
8. Other RNA extraction kits to be used with low quantity of starting material are also commercially available.
9. DNAse treatment directly performed in column, as described in PicoPure manufacturer's instructions, can be used for expression analyses depending on the target organism.
10. Primers that amplify small fragments are generally preferred [2]. The more common annealing temperature used is 60 °C. However, both for RT-PCR and quantitative PCR, this temperature is depending on the used primers, which have also to be checked before in standard PCR also to avoid cross-reactions between the plant and the fungus.

Fig. 2 A slide with membrane (above) and a metal frame (below) used for the LMD

11. For RT-qPCR, due to a difference in RNA integrity and quality, the comparison between RNAs extracted from microdissected samples and RNA obtained fresh material should be avoided.
12. RT-qPCR can be performed using a two-step or a one-step protocol on nonamplified [5, 8, 27, 31] or on aRNA (antisense RNA obtained by one/two rounds of amplification) [10, 30]. Generally, one or two rounds of RNA amplification may also be performed when LMD is associated to large-scale analysis. Using specific commercial kits to maintain the relative representation of RNA species are generally required to obtain the microgram amounts of mRNA needed for global profiling.
13. Diverse RT-qPCR kits can require different reagent quantities [5].

References

1. Emmert-Buck MR, Bonner RF, Smith PD et al (1996) Laser capture microdissection. Science 274:998–1001
2. Balestrini R, Gomez-Ariza J, Lanfranco L et al (2007) Laser microdissection reveals that transcripts for five plant and one fungal phosphate transporter genes are contemporaneously present in arbusculated cells. Mol Plant-Microbe Interact 20:1055–106210
3. Hacquard S, Tisserant E, Brun A et al (2013) Laser microdissection and microarray analysis of *Tuber melanosporum* ectomycorrhizas reveal functional heterogeneity between mantle and Hartig net compartments. Environ Microbial 15:1853–1869
4. Fochi V, Chitarra W, Kohler A et al (2017) Fungal and plant gene expression in the *Tulasnella calospora–Serapias vomeracea* symbiosis provides clues about nitrogen pathways in orchid mycorrhizas. New Phytol 213:365–379
5. Fochi V, Falla N, Girlanda M et al (2017) Cell-specific expression of plant nutrient transporter genes in orchid mycorrhizae. Plant Sci 263:39–45
6. Balestrini R, Fiorilli V (2020) Laser microdissection as a useful tool to study gene expression

in plant and fungal partners in AM symbiosis. In: Ferrol N, Lanfranco L (eds) Arbuscular mycorrhizal fungi: methods and protocols, series methods in molecular biology, protocol lab. Springer, Heidelberg

7. Santi S, Grisan S, Pierasco A et al (2013) Laser microdissection of grapevine leaf phloem infected by stolbur reveals site-specific gene responses associated to sucrose transport and metabolism. Plant Cell Environ 36:343–355
8. Rossi M, Pesando M, Vallino M et al (2018) Application of laser microdissection to study phytoplasma site-specific gene expression in the model plant *Arabidopsis thaliana*. Microbiol Res 217:60–68
9. Tang W, Coughlan S, Crane E et al (2006) The application of laser microdissection to in planta gene expression profiling of the maize anthracnose stalk rot fungus *Colletotrichum graminicola*. Mol Plant-Microbe Interact 19:1240–1250
10. Hacquard S, Delaruelle C, Legué V et al (2010) Laser capture microdissection of Uredinia formed by *Melampsora larici-populina* revealed a transcriptional switch between biotrophy and sporulation. Mol Plant-Microbe Interact 23: 1275–1286
11. Chandran D, Inada N, Hather G et al (2010) Laser microdissection of Arabidopsis cells at the powdery mildew infection site reveals site-specific processes and regulators. PNAS 107: 460–465
12. Klug K, Hogekamp C, Specht A et al (2015) Spatial gene expression analysis in tomato hypocotyls suggests cysteine as key precursor of vascular sulfur accumulation implicated in *Verticillium dahliae* defense. Physiol Plant 153:253–268
13. Honaas LA, Wafula EK, Yang Z et al (2013) Functional genomics of a generalist parasitic plant: laser microdissection of host-parasite interface reveals host-specific patterns of parasite gene expression. BMC Plant Biol 13:9
14. Coker TL, Cevik V, Beynon JL et al (2015) Spatial dissection of the *Arabidopsis thaliana* transcriptional response to downy mildew using fluorescence activated cell sorting. Front Plant Sci 6:527
15. Podgorny OV, Lazarev VN (2017) Laser microdissection: a promising tool for exploring microorganisms and their interactions with hosts. J Microbiol Methods 138:82–92
16. Inada N, Wildermuth MC (2005) Novel tissue preparation method and cell-specific marker for laser microdissection of *Arabidopsis* mature leaf. Planta 221:9–16
17. Luchi N, Capretti P, Fossdal CG et al (2012) Laser microdissection on Norway spruce bark tissue: a suitable protocol for subsequent real-time reverse transcription–polymerase chain reaction (RT-PCR) analysis. Plant Biosyst 146:92–98
18. Falter C, Ellinger D, von Hülsen B et al (2015) Simple preparation of plant epidermal tissue for laser microdissection and downstream quantitative proteome and carbohydrate analysis. Front Plant Sci 6:194
19. Tremblay A, Li S, Scheffler BE et al (2008) Laser capture microdissection and expressed sequence tag analysis of *Phakopsora pachyrhizi* uredinia. Physiol Mol Plant Pathol 73:163–174
20. Tremblay A, Hosseini P, Li S et al (2012) Identification of genes expressed by *Phakopsora pachyrhizi*, the pathogen causing soybean rust, at a late stage of infection of susceptible soybean leaves. Plant Pathol 61:773–786
21. Chandran D, Hather G, Wildermuth MC (2011) Global expression profiling of RNA from laser microdissected cells at fungal–plant interaction sites. In: McDowell J (ed) Plant immunity, methods in molecular biology (methods and protocols), vol 712. Humana Press
22. Kerk NM, Ceserani T, Tausta SL et al (2003) Laser capture microdissection of cells from plant tissues. Plant Physiol 132:27–35
23. Cai S, Lashbrook CC (2006) Laser capture microdissection of plant cells from tape transferred paraffin sections promotes recovery of structurally intact RNA for global gene profiling. Plant J 48:628–637
24. Lenzi L, Caruso C, Bianchedi PL et al (2016) Laser microdissection of grapevine leaves reveals site-specific regulation of transcriptional response to *Plasmopara viticola*. Plant Cell Physiol 57:69–81
25. Balestrini R, Ghignone S, Quiroga G et al (2020) Long-term impact of chemical and alternative fungicides applied to Grapevine *cv* Nebbiolo on Berry Transcriptome. Int J Mol Sci 21:6067
26. Gomez SK, Harrison MJ (2009) Laser microdissection and its application to analyze gene expression in arbuscular mycorrhizal symbiosis. Pest Man Sci 65:504–511
27. Gaude N, Bortfeld S, Duensing N et al (2012) Arbuscule-containing and non-colonized cortical cells of mycorrhizal roots undergo extensive and specific reprogramming during arbuscular mycorrhizal development. Plant J 69:510–528

28. Nakazono M, Qiu F, Borsuk LA et al (2003) Laser capture microdissection, a tool for the global analysis of gene expression in specific plant cell types: identification of genes expressed differentially in epidermal cells or vascular tissue of maize. Plant Cell 15:583–596
29. Balestrini R, Gómez-Ariza J, Klink VP et al (2009) Application of laser microdissection to plant pathogenic and symbiotic interactions. J Plant Interact 4:81–92
30. Hogekamp C, Arndt D, Pareira PA et al (2011) Laser microdissection unravels cell-type-specific transcription in arbuscular mycorrhizal roots, including CAAT-box transcription factor gene expression correlating with fungal contact and spread. Plant Physiol 157:2023–2043
31. Giovannetti M, Balestrini R, Volpe V et al (2012) Two putative-aquaporin genes are differentially expressed during arbuscular mycorrhizal symbiosis in *Lotus japonicus*. BMC Plant Biol 12:186

Chapter 21

Somatic Embryogenesis as a Tool for Studying Grapevine–Virus Interaction

Giorgio Gambino and Irene Perrone

Abstract

More than 80 viral species, many of which are not associated with a clear disease or symptomatology, can infect grapevine. The study of grapevine–virus interactions in recent years is playing an increasingly important role and these studies have shown that the molecular and physiological responses to a virus greatly vary depending on the viral strains, the presence of multiple viral infections, the grapevine genotype, and the environment. Moreover, due to the characteristics of the grapevine cultivation and its vegetative propagation, it is very difficult to find healthy plants in vineyards to use them as control in the experiments. Starting from these considerations, in order to investigate the plant–virus interaction in an unbiased way, it is important to set up an experimental system able to control as much of these variables as possible. The protocol here proposed provides the overcome some of these factors by: (i) the production of healthy plants by somatic embryogenesis; (ii) the virus transmission using in vitro micrografting, and (iii) the transfer of in vitro plants to ex-vitro conditions for the analysis of interest.

Key words somatic embryogenesis, healthy grapevines, micrografting, virus, diagnosis, plant–virus interaction

1 Introduction

Grapevine (*Vitis vinifera* L.), one of the most important and widespread fruit trees in the world, can be affected by several pathogens, such as fungi, oomycota, phytoplasma, viruses, and viroids, which can cause serious damage to the plants. Viruses and viroids represent very widespread classes of pathogens in grapevine and thanks to deep sequencing technologies more than 80 viral species have so far been identified in grapevine [1]. However, not all viruses are associated with "major diseases" in terms of symptomatology and economic relevance [2], for most of them, no clear data on the symptoms induced in grapevine are available. In recent years, in addition to the traditional works in which the molecular characteristics of the viruses and diagnostic approaches were studied, more attention is being paid to studies on the grapevine–virus

Nicola Luchi (ed.), *Plant Pathology: Methods and Protocols*, Methods in Molecular Biology, vol. 2536,
https://doi.org/10.1007/978-1-0716-2517-0_21,

interactions [3–8]. In general, these relationships are not limited to a simple plant–pathogen interaction, but can shift from antagonist to mutualistic or beneficial in response to environment [9, 10]. Also in grapevine, in some conditions and in specific grapevine genotype-virus strain combinations, some viruses are not always harmful to the plant [11–14]. These studies highlight how these interactions are often complex and suggest that the plant should be considered as a holobiont in which micro- and macro-biomes interact with the environment as a single organism [15].

In grapevine, different variables (e.g., viral strain, mixed infections with different viruses, grapevine genotype) can influence the plant–virus interactions and should be carefully considered before starting an experiment. The first critical point in these studies is the availability of virus and viroid-free plant material for healthy controls. It is very difficult to find healthy plants (completely free of viruses and viroids after check with high-throughput sequencing analyses) due to the high number of viral pathogens potentially infecting this species that can be transmitted by vegetative multiplication [1]. For example, from 2012 to 2016, our laboratory conducted a survey in more than 1000 accessions cultivated in the vineyards of northwestern Italy using three RT-PCR protocols [16–18] for the amplification of commonly monitored viruses in the grapevine germplasm of this area. Some virus- or viroid-free accessions were identified, but no accessions free of both viruses and viroids at the same time were detected. Consequently, to obtain healthy controls, an opportunity is the sanitation of infected plants through some approaches (i.e., meristem tip culture, in vivo and in vitro thermotherapy) among which the somatic embryogenesis has proven to be widely effective in eliminating viruses and viroids from grapevine [19–21].

Another critical point for these studies is the strong influence of the grapevine genotype. For this reason, it is advisable (i) to analyze different cultivars infected by the same virus [14] and (ii) to guarantee the genetic uniformity of healthy and infected samples. This last point can be obtained through the sanitation by somatic embryogenesis using a single plant regenerated by a somatic embryo multiplied in vitro (healthy control) and infected with the selected virus (infected samples). The regeneration of a single plant followed by in vitro multiplication is a strategy to avoid potential somaclonal variations arising from somatic embryogenesis with the risk to produce regenerated plants that are not genetically identical [22]. Once a healthy grapevine line has been produced, it is necessary to adopt an effective method for virus transmission. The mechanical inoculation of viruses in grapevine is poor or not efficient at all and other techniques such as the productions of infectious viral cDNA clone [23, 24] have been used with results not always reproducible in all laboratories. The graft can represent an effective strategy, as long as the viral status of the material used for the viral transmission is well characterized.

We report the setup of an experimental system that can be exploited to investigate if and what kind of interactions exist between viruses and grapevine.

2 Materials

2.1 Tissue Culture

1. Sterilization Solution: sodium hypochlorite in ultrapure water (1.5% available chlorine) containing several drops of Tween 20 as wetting agent.
2. Callus Induction Medium (CI): dissolve in ultrapure water NN mineral salts [25] and MS vitamins [26] according to the quantity indicated by the manufacturers (*see* **Note 1**), 60 g/L sucrose, 4.5 μM 2,4 dichlorophenoxyacetic acid (2,4-D), 8.9 μM 6-benzylaminopurine (BAP). Mix and adjust the pH at 5.8 with KOH, add 3 g/L of gerlite (or phytagel) and autoclave at 120 °C for 10 min. Distribute the medium under a laminar flow hood in sterile 60- or 90-mm Petri dishes. The CI can be used immediately or stored at 4 °C.
3. Embryo Differentiation Medium (ED): NN mineral salts and MS vitamins, 60 g/L sucrose, 10 μM naphthoxiacetic acid (NOA), and 1 μM BAP. Add ultrapure water to a volume of 980 mL, mix and adjust pH at 5.8 with KOH. Add 10 g/L of Agar and 2.5 g/L of activated charcoal and autoclave at 120 °C for 10 min. Let the ED cool down to about 40 °C, add 20 μM of filter sterilized indole-3-acetic acid (IAA, 20 mL/L) and distribute the medium in sterile 60- or 90-mm Petri dishes. The ED can be used immediately or stored at 4 °C.
4. Embryo Germination Medium (EG): MS mineral salts and vitamins, 30 g/L sucrose, 10 μM BAP dissolved in ultrapure water. Adjust pH at 5.7 with KOH, add 9 g/L of Agar, and autoclave at 120 °C for 10 min. Distribute the EG in sterile 60- or 90-mm Petri dishes (*see* **Note 2**), the medium can be used immediately or stored at 4 °C.
5. Root Medium (RO): Half-strength MS mineral salts, MS vitamins, 20 g/L sucrose, 2.5 μM indole-butyric acid (IBA), and 2.5 μM α-naphthalene-acetic acid (NAA) dissolved in ultrapure water. Mix and adjust pH at 5.7 with KOH, add 8 g/L of Agar, and autoclave at 120 °C for 10 min. Distribute the RO in sterile Magenta™ vessels, the medium can be used immediately or stored at 4 °C.
6. Maintenance Medium (MA): Half-strength MS mineral salts, MS vitamins, and 20 g/L sucrose dissolved in ultrapure water. Mix and adjust pH at 5.7 with KOH, add 8 g/L of Agar, and

autoclave at 120 °C for 10 min. Distribute the MA in sterile Magenta™ vessels, Microbox containers (*see* **Note 2**) or glass vessels. The medium can be used immediately or stored at room temperature.

2.2 RNA Extraction

1. Pestles, mortars, and all glassware are kept overnight at 180 °C, plasticware is autoclaved before use.
2. Diethyl pyrocarbonate (DEPC)-treated water, 10 M LiCl stored at 4 °C, 3 M sodium acetate pH 5.2.
3. Extraction Buffer: 100 mM Tris–HCl (pH 8.0), 25 mM EDTA (pH 8.0), 2% w/v cetyltrimethylammonium bromide (CTAB), 2% w/v PVP (MW 40,000), 2.5 M NaCl, autoclaved at 120 °C for 20 min. Store at room temperature.
4. SSTE Buffer: 10 mM Tris–HCl (pH 8.0), 1 mM EDTA (pH 8.0), 1% SDS, 1 M NaCl. Do not autoclave, filter sterilize. Store at room temperature.
5. Chloroform–isoamyl alcohol (24:1 v/v).
6. Isopropanol. Store at −20 °C.
7. 70% ethanol prepared with DEPC-treated water. Store at −20 °C.

2.3 RT-PCR

1. DNase I (Thermo fisher scientific).
2. High-Capacity cDNA Reverse Transcription Kit (Thermo fisher scientific).
3. Primers specific for viruses and/or viroids (Table 1).
4. Taq polymerase.
5. Agarose.
6. 0.5 x TBE (90 mM Tris–borate, 2 mM EDTA).
7. 100-bp ladder.

3 Methods

The protocol here proposed is organized in four steps: (i) production of healthy plants by somatic embryogenesis; (ii) diagnosis of the sanitary status of the plants by RT-PCR; (iii) virus transmission using micrograft, and (iv) transfer of in vitro plants to ex-vitro conditions for the analyses.

Table 1
Primer sequences and RT-PCR product size for the detection of the most common viruses and viroids infecting grapevine

Virus/Viroids	Primer	Primer sequences 5′-3′	Product size (bp)	References
Grapevine leafroll-associated virus 2	Forward	GGTGATAACCGACGCCTCTA	543	[16]
	Reverse	CCTAGCTGACGCAGATTGCT		
Grapevine virus B	Forward	GTGCTAAGAACGTCTTCACAGC	460	[16]
	Reverse	ATCAGCAAACACGCTTGAACCG		
Arabis mosaic virus	Forward	ATGCTGATGTTGGAGTTGCTCT	416	[16]
	Reverse	AGTGGCACTAAACCACAAAGC		
Grapevine leafroll-associated virus 3	Forward	TACGTTAAGGACGGGACACAGG	336	[16]
	Reverse	TGCGGCATTAATCTTCATTG		
Grapevine virus A	Forward	GAGGTAGATATAGTAGGACCTA	272	[16]
	Reverse	TCGAACATAACCTGTGGCTC		
Grapevine leafroll-associated virus 1	Forward	TCTTTACCAACCCCGAGATGAA	232	[16]
	Reverse	GTGTCTGGTGACGTGCTAAACG		
Grapevine fleck virus	Forward	TGACCAGCCTGCTGTCTCTA	179	[16]
	Reverse	TGGACAGGGAGGTGTAGGAG		
Grapevine rupestris stem pitting-associated virus	Forward	GGGTGGGATGTAGTAACTTTTGA	155	[16]
	Reverse	GCAAGTGAAATGAAAGCATCACT		
Grapevine fanleaf virus	Forward	ATGCTGGATATCGTGACCCTGT	118	[16]
	Reverse	GAAGGTATGCCTGCTTCAGTGG		
Grapevine pinot gris virus	Forward	ATTGCGGAGTTGCCTTCAAG	303	[18]
	Reverse	CTGAGAAGCATTGTCCCATC		

(continued)

Table 1
(continued)

Virus/Viroids	Primer	Primer sequences 5′-3′	Product size (bp)	References
Australian grapevine viroid	Forward Reverse	GTCC TCAGGAGAGCACCGG GAAAACTGG TTGGGACCGCTG	195	[17]
Citrus exocortis viroid	Forward Reverse	GTGTCCTTCCTTTCGC TGCTG TGGCCCGGAGAACAG TGAAG	143	[17]
Grapevine yellow speckle viroid 1	Forward Reverse	CAAAGCCCTTTTTC TTTCAACTGAG CCCAGAGCAGCGTGG TCC	249	[17]
Grapevine yellow speckle viroid 2	Forward Reverse	ACTAGTACTTTCTTCTATC TCCGAAGC ACTAGTCCGAGGACC TTTTCTAGCGCTC	375	[17]
Hop stunt viroid	Forward Reverse	AACCCGGGGCAACTCTTC TC AACCCGGGGCTCCTTTC TCA	298	[17]

3.1 Plant Regeneration From Somatic Embryogenesis

All the procedures related to tissue cultures must be conducted under a laminar flow hood using a glass bead sterilizer to sterilize all the tools such as forceps and scalpels.

1. Collect immature inflorescences of grapevine in vineyard (*see* **Note 3**) indicatively before the meiosis in the pollen mother cells (*see* **Notes 4** and **5**).
2. Sterilize the inflorescences the day of collection with the Sterilization solution for 15 min at room temperature, rinse several times with sterile distilled water, and store them at 4 °C for 2 days (*see* **Note 6**).
3. Sterilize the inflorescences for a second time with the Sterilization solution for 15 min at room temperature and rinse several times with sterile distilled water (*see* **Note 7**). Then, excise aseptically anthers and ovaries from flowers (Fig. 1a) under a stereomicroscope using forceps and scalpel. Transfer the anthers with the attached filaments on a plate with CI medium, with the adaxial side in contact with the medium (*see* **Note 8**), and transfer the ovaries (*see* **Note 9**). Maintain the cultures at 26 °C in the dark.

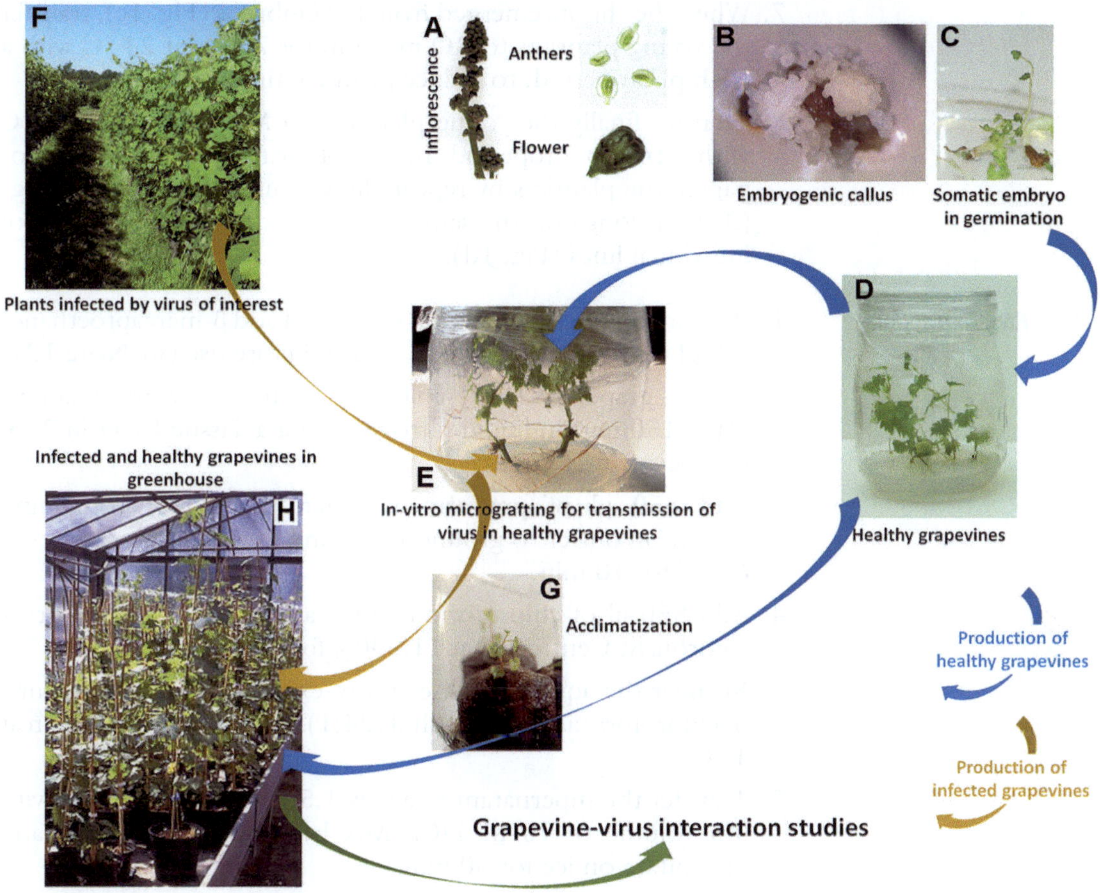

Fig. 1 Graphic description of the experimental system. Anthers and ovaries collected from immature inflorescences (**a**) are cultured on Callus Induction medium to obtain somatic embryogenesis (**b**). Shoots emerged from somatic embryos (**c**) are micropropagated by repeatedly subculturing apical cuttings until the rise of individual healthy plants (**d**). Grapevines infected with the virus of interest identified in the vineyard (**f**) are grafted in vitro (**e**) with healthy plants (**d**). Healthy (**d**) and infected grapevines after grafting (**e**) are acclimated to the greenhouse (**g**, **h**) for further analyses

4. Approximately after three months of culture, check the explants under a stereomicroscope and transfer pre-embryogenic masses or somatic embryos (Fig. 1b) to ED medium at 26 °C in the dark (*see* **Note 10**).
5. Keep embryogenic masses on ED medium for several months and transfer them every 4 weeks onto fresh medium until the formation and growth of somatic embryos (*see* **Note 11**).
6. Collect fully developed somatic embryos and after cutting the radical pole transfer them in EG medium at 24 °C with a 16-h photoperiod and a total energy of 50 μmol s^{-1} m^{-2} provided by white LED lamps.

7. When the shoots emerged from the embryos (Fig. 1c), transfer the young plantlets to RO medium for 7 days at 24 °C with a 16-h photoperiod, to induce plant rooting.
8. Transfer finally the young plantlets on MA medium at 24 °C with a 16-h photoperiod. In the following months, micropropagate the plantlets by repeatedly subculturing apical cuttings (3–4 cm long) on the same culture medium, until the rise of individual lines (Fig. 1d).

3.2 RNA Extraction

1. Prewarm extraction buffer at 65 °C and add β-mercaptoethanol (20 μl for 1 ml of extraction buffer) before use (*see* **Note 12**).
2. Grind grapevine leaf collected from in vitro plantlets (100–150 mg) in liquid nitrogen using a Tissue Lyser in 2-ml microcentrifuge tubes.
3. Add 850 μl of prewarmed β-mercaptoethanol containing extraction buffer to ground tissue and incubate the samples at 65 °C for 10 min.
4. Add 850 μl of chloroform:isoamyl alcohol (24:1) and vortex vigorously. Centrifuge at 11,000 *g* for 10 min at 4 °C.
5. Recover the aqueous phase and re-extract adding one volume of chloroform:isoamyl alcohol (24:1), at 11,000 *g* for 10 min at 4 °C.
7. Transfer the supernatant to a new 1.5-ml microcentrifuge with 0.5 volume of 10 M LiCl. Mix by inverting the tube and precipitate on ice for 30 min.
6. Centrifuge at 21,000 *g* for 30 min at 4 °C.
7. Remove the supernatant and dissolve the pellet in 500 μl of SSTE prewarmed at 65 °C by vortexing.
8. Add 500 μl of chloroform:isoamyl alcohol (24:1), vortex vigorously, and centrifuge at 11,000 *g* for 10 min at 4 °C.
8. Recover the aqueous phase and add 0.7 volume of isopropanol, 0.1 volume of 3M sodium acetate (pH 5.2), and 1 μl of glycogen. Mix by inverting the tube and precipitate at −20 °C for 30 min or overnight.
9. Centrifuge at 21,000 *g* for 20 min at 4 °C.
10. Wash the pellets with 70% ethanol and centrifuge at 21,000 *g* for 5 min at 4 °C.
11. Dissolve the pellet in 20 μl of DEPC-treated water and store at −80 °C.
12. Assess the RNA purity and concentration using nanodrop or spectrophotometer.

3.3 Detection of Viruses and Viroids

1. Verify the absence of viruses and viroids from plants regenerated from somatic embryos (*see* **Note 13**) by RT-PCR (*see* **Note 14**).
2. To treat the extracted RNA with DNase in order to remove DNA contaminations, dissolve 500 ng of total RNA in 8.5 μl of H_2O-DEPC.
3. Add 1 μl of 10X reaction buffer and 0.5 μl of DNase I.
4. Incubate at room temperature for 15 min and add 1 μl of EDTA solution.
5. Incubate at 65 °C for 10 min and transfer on ice or at −80 °C for long-term storage.
6. To obtain cDNA, add to 10 μl of total RNA treated with DNase, a mix containing: 2 μl of 10X RT Buffer, 2 μl of 10X RT Random Primers, 0.8 μl of 25X dNTP Mix (100 mM), 1 μl MultiScribe Reverse Transcriptase (50 U/μL), and 1 μl of RNase Inhibitor.
7. Incubate at 25 °C for 10 min, 37 °C for 120 min, and 85 °C for 5 min. Transfer on ice or at −20 °C for long-term storage.
8. Perform a PCR reaction: 2 μl of 10X Reaction Buffer, 0.8 μl of MgCl (50 mM), 0.4 μl of dNTP mix, 1 μl of each primer, 0.1 μl of Taq polymerase, and 1 μl of cDNA (*see* **Note 15**).
9. PCR cycling conditions: initial denaturation at 95 °C for 2 min, followed by 35 cycles at 95 °C for 30 s, annealing for 30 s at temperature requested by specific primers, and 72 ° C for 1 min. Final extension at 72 °C for 10 min.
10. Analyze amplicons by electrophoresis on 2.5% agarose gels buffered in 0.5X TBE and visualized by UV light after staining with ethidium bromide (*see* **Note 16**).

3.4 Virus Inoculation in Healthy Grapevine Plants

A grafting technique is adopted to transmit viruses from infected plants with the desired virus to healthy grapevine plants regenerated from somatic embryos (*see* **Note 17**).

1. Collect during winter woody cuttings from plants infected with the desired virus and keep them at 4 °C in a nylon bag to avoid dehydration.
2. Force woody cuttings to sprout in water at room temperature.
3. Collect the herbaceous green branches (about 5 cm) emerging from the buds, surface-sterilize them for 15 min with Sterilization solution, and rinse several times with sterile distilled water.
4. Graft apical cuttings (3–4 cm long) from one selected healthy plant, obtained in Subheading 3.1, **step 8**, on virus-infected herbaceous green branches and keep them onto MA medium in in vitro conditions (Fig. 1e).

5. After one month, verify the virus transmission by RT-PCR as described in the Subheading 3.3.
6. Micropropagate the single apical cutting infected with the desired virus by repetitively subculturing on MA medium (*see* **Note 18**).
7. Transfer healthy and infected plants to ex-vitro conditions by growing apical cutting in autoclaved peat pellets (Jiffy7®) for 2–4 weeks at 24 °C (Fig. 1g).
8. Transfer the plants in pots with soil to greenhouse conditions (Fig. 1h).
9. After 4–5 months of greenhouse cultivation, the plantlets can be used for physiological/molecular analyses (*see* **Notes 19 and 20**).

4 Notes

1. The composition of NN and MS mineral salts or vitamin is well known and reported in the original works [25, 26]; consequently it is possible to prepare the single stock solutions of all the components. However, it is more practical to buy ready-to-use salt and vitamin mixtures produced by several companies.
2. Microbox containers (https://saco2.com/microbox-containers-2/) can be used as alternative to Petri dishes. Microbox is a container of different sizes with covers containing filters for gas exchanges. In particular, for the embryo germination, Microbox of 60 mm in height and 80 mm in diameter is useful for the development of young plantlets.
3. The collecting period of immature inflorescences in the vineyard is limited to a few weeks in early spring. To have inflorescences available all year round, winter pruning wood can be stored at 4 °C in a nylon bag to avoid dehydration. During the season, the wood can be put simply in water or in pots with moist soil, and within few weeks, the buds will sprout with the emission of the inflorescences.
4. The optimal developmental stage for collecting inflorescences was identified for *V. vinifera* in the early stages of the microsporogenesis before the meiosis. However, this optimal stage can vary in the different genotypes; in particular in rootstocks, it is possible to collect the inflorescences even during or just after the meiosis.
5. The identification of the developmental stage of the grapevine inflorescences can be carried out according to the color of the anthers (green and translucent in the premeiosis phases) (Fig. 1a) or more accurately by observing under the microscope, the anthers squashed in acetocarmine [27].

6. After the collection and the first sterilization, keep the inflorescences at 4 °C for at least 2 days before the explants excision and use them up to 5 or 6 days from the collection time. After this period, the inflorescences tend to wither.
7. The second sterilization before the anthers and ovaries excision is not essential, but it is very useful to reduce the contamination of bacteria and fungi from the inflorescences collected in the field. Double sterilization does not damage the explants inside the flowers [28].
8. The CI medium [27, 29] over time has proved to be one of the most effective for inducing embryogenesis in grapevine. However, the grapevine aptitude for embryogenesis is closely linked to the genotype [30] and different culture substrates could perform better than CI for some cultivars [31]. Therefore, if the aptitude for embryogenesis of the cultivar of interest is not known, it is advisable to start from the CI; and in case of negative results, other cultural substrates can be tested.
9. The anthers (Fig. 1a) are the explants of choice to induce embryogenesis in grapevine; however, even the ovaries have proved to be very effective as well as the whole flowers [28].
10. For some genotypes and explants (e.g., ovaries), three months may not be sufficient to develop embryogenic masses, therefore undifferentiated calli can be transferred to ED; and after two months of culture, other embryogenic calli can be differentiated.
11. The protocol reported is specific for the rapid growth and germination of plantlets from somatic embryos. If the aim of the research is the maintenance of embryogenic calli for a long time for genetic transformation, after the CI medium, pre-embryogenic masses should be maintained on a medium containing 4.5 μM 2,4-D and 1 μM BAP [32].
12. The rapid CTAB protocol reported [33] gives high-quality RNA in few hours at low cost. However, any protocol can be used for RNA extraction, including commercial extraction kits. For grapevine tissues, the Spectrum™ Plant Total RNA Kit (Merck) is recommended following the manufacturer's instructions.
13. The somatic embryogenesis is a technique very effective to eliminate viruses and viroids from grapevine tissues [19, 20, 34]; however, it is always necessary to check the virological state of the regenerated plants [21, 35].
14. The diagnosis by RT-PCR of regenerated plantlets is effective if the virological status of mother plants from which the inflorescences were collected was already known. Otherwise, it is advisable to use high-throughput sequencing techniques [36] to ensure the absence of any type of viral and viroidal pathogens.

15. The PCR protocol reported represents only a general indication for viral diagnosis. Since the number of viruses and viroids infecting grapevine are very high, it will be necessary to use specific primers (Table 1) and protocols based on the viral status of the starting mother plant. For the major viruses and viroids infecting grapevine, very effective protocols are available, including rapid multiplex PCR approaches [16–18].
16. Ethidium bromide can be replaced with safer staining molecules.
17. The transmission of viruses to grapevine by mechanical inoculation is poor or not efficient at all; the grafting technique is the most effective technique to transmit viruses from infected to virus-free plants. However, it is important to know the viral status of the infected material used in the graft so that only the desired virus is transmitted to healthy plants. Therefore, high-throughput sequencing analysis of the infected material is always recommended.
18. It is essential to ensure the genetic uniformity of healthy and infected plants. As for healthy control, a plant regenerated from a single somatic embryo was used to ensure genetic uniformity and was multiplied in vitro; therefore also for infected plants, it is recommended to start from a single infected explant multiplied in vitro.
19. Young plants acclimatized to in vitro conditions are generally herbaceous and it is advisable to wait up to a year before carrying out the analyses of interest. Grapevine plants can survive many years in pots and in greenhouses, but it is essential to ensure a period of dormancy during the winter season and a pruning before spring budding. Continuous growth in conditioned greenhouse without dormancy is not recommended.
20. At this stage, healthy and infected grapevines with the same genotype are obtained. Once in the greenhouse, plants can be analyzed to understand the effects of virus in plants using physiological, metabolomic, and molecular approaches or to study the virus interactions with other pathogens.

References

1. Martelli GP (2018) Where grapevine virology is heading to. In: Proceedings of the 19th congress of International Council for the study of viruses and virus-lile diseases of the Grapevine, Santiago, Chile, 9–12 April 2018, pp 10–15
2. Martelli GP (2017) An overview on grapevine viruses, viroids, and the diseases they cause. In: Meng B, Martelli G, Golino D, Fuchs M (eds) Grapevine viruses: molecular biology, diagnostics and management. Springer, Cham, pp 31–46
3. Vega A, Gutiérrez RA, Peña-Neira A et al (2011) Compatible GLRaV-3 viral infections affect berry ripening decreasing sugar

accumulation and anthocyanin biosynthesis in *Vitis vinifera*. Plant Mol Biol 77:261–274
4. Gambino G, Cuozzo D, Fasoli M et al (2012) Co-evolution between Grapevine rupestris stem pitting-associated virus and *Vitis vinifera* L. leads to decreased defence responses and increased transcription of genes related to photosynthesis. J Exp Bot 63:5919–5933
5. Endeshaw ST, Sabbatini P, Romanazzi G et al (2014) Effects of grapevine leafroll-associated virus 3 infection on growth, leaf gas exchange, yield and basic chemistry of *Vitis vinifera* L. cv. Cabernet Franc. Sci Hortic 170:228–236
6. Montero R, Pérez-Bueno ML, Barón M et al (2016) Alterations in primary and secondary metabolism in *Vitis vinifera* 'Malvasia de Banyalbufar' upon infection with Grapevine Leafroll-associated Virus 3 (GLRaV-3). Physiol Plant 157:442–452
7. Chitarra W, Cuozzo D, Ferrandino A et al (2018) Dissecting interplays between *Vitis vinifera* L. and grapevine virus B (GVB) under field conditions. Mol Plant Pathol 19: 2651–2666
8. Tobar M, Fiore N, Pérez-Donoso AG et al (2020) Divergent molecular and growth responses of young 'Cabernet Sauvignon' (*Vitis vinifera*) plants to simple and mixed infections with Grapevine rupestris stem pitting-associated virus. Hortic Res 7:2
9. Roossinck MJ (2011) The good viruses: viral mutualistic symbioses. Nat Rev Microbiol 9: 99–108
10. Hily JM, Poulicard N, Mora MA et al (2016) Environment and host genotype determine the outcome of a plant–virus interaction: from antagonism to mutualism. New Phytol 209: 812–822
11. Perrone I, Chitarra W, Boccacci P, Gambino G (2017) Grapevine–virus–environment interactions: an intriguing puzzle to solve. New Phytol 213:983–987
12. Pantaleo V, Vitali M, Boccacci P et al (2016) Novel functional microRNAs from virus-free and infected *Vitis vinifera* plants under water stress. Sci Rep 6:20167
13. Repetto O, Bertazzon N, De Rosso M et al (2012) Low susceptibility of grapevine infected by GLRaV-3 to late *Plasmopara viticola* infections: towards understanding the phenomenon. Physiol Mol Plant Pathol 79:55–63
14. Gilardi G, Chitarra W, Moine A et al (2020) Biological and molecular interplay between two viruses and powdery and downy mildews in two grapevine cultivars. Hortic Res 7:188
15. Dicke M (2016) Plant phenotypic plasticity in the phytobiome: a volatile issue. Curr Opin Plant Biol 32:17–23
16. Gambino G (2015) Multiplex RT-PCR method for the simultaneous detection of nine grapevine viruses. In: Uyeda I, Masuta C (eds) Plant virology protocols: new approaches to detect viruses and host responses, vol 1236, 3rd edn. Humana Press Inc, pp 39–47
17. Hajizadeh M, Navarro B, Bashir NS et al (2012) Development and validation of a multiplex RT-PCR method for the simultaneous detection of five grapevine viroids. J Virol Methods 179:62–69
18. Glasa M, Predajňa L, Komínek P et al (2014) Molecular characterization of divergent Grapevine Pinot gris virus isolates and their detection in Slovak and Czech grapevines. Arch Virol 159:2103–2107
19. Gambino G, Bondaz J, Gribaudo I (2006) Detection and elimination of viruses in callus, somatic embryos and regenerated plantlets of grapevine. Eur J Plant Pathol 114:397–404
20. Gambino G, Navarro B, Vallania R et al (2011) Somatic embryogenesis efficiently eliminates viroid infections from grapevines. Eur J Plant Pathol 13:511–519
21. Turcsan M, Demian E, Varga T et al (2020) HTS-based monitoring of the efficiency of somatic embryogenesis and meristem cultures used for virus elimination in grapevine. Plan Theory 9:1782
22. Dalla Costa L, Malnoy M, Gribaudo I (2017) Breeding next generation tree fruits: technical and legal challenges. Hortic Res 4:7067
23. Meng B, Venkataraman S, Li C et al (2013) Construction and biological activities of the first infectious cDNA clones of the genus Foveavirus. Virology 435:453–462
24. Lovato A, Faoro F, Gambino G et al (2014) Construction of a synthetic infectious cDNA clone of Grapevine Algerian latent virus (GALV-Nf) and its biological activity in *Nicotiana benthamiana* and grapevine plants. Virol J 11:186
25. Nitsch JP, Nitsch C (1969) Haploid plants from pollen grains. Science 163:85–87
26. Murashige T, Skoog F (1962) A revised medium for rapid growth and bio assays with tobacco tissue cultures. Physiol Plant 15:473–497
27. Gribaudo I, Gambino G, Vallania R (2004) Somatic embryogenesis from grapevine anthers: identification of the optimal developmental stage for collecting explants. Am J Enol Vitic 55:427–430
28. Gambino G, Ruffa P, Vallania R, Gribaudo I (2007) Somatic embryogenesis from whole flowers, anthers and ovaries of grapevine (*Vitis* spp.). Plant Cell Tissue Organ Cult 90:79–83

29. Franks T, He DG, Thomas M (1998) Regeneration of transgenic *Vitis vinifera* L. Sultana plants: a genotypic and phenotypic analysis. Mol Breed 4:321–333
30. Gribaudo I, Gambino G, Boccacci P et al (2017) A multi-year study on the regenerative potential of several *Vitis* genotypes. Acta Hortic 1155:45–50
31. Martinelli L, Gribaudo I (2009) Strategies for effective somatic embryogenesis in grapevine: an appraisal. In: Roubelakis-Angelakis KA (ed) Grapevine Molecular Physiology & Biotechnology. Springer, Dordrecht
32. Torres-Viñals M, Sabaté-Casaseca S, Aktouche N et al (2004) Large-scale production of somatic embryos as a source of hypocotyl explants for Vitis vinifera micrografting. Vitis 43:163–168
33. Gambino G, Perrone I, Gribaudo I (2008) A rapid and effective method for RNA extraction from different tissues of grapevine and other woody plants. Phytochem analysis 19:520–525
34. Goussard P, Wiid J, Kasdorf G (1991) The effectiveness of in vitro somatic embryogenesis in eliminating fanleaf virus and leafroll- associated viruses from grapevines. S Afr J Enol Viticult 12:77–81
35. Gambino G, Di Matteo D, Gribaudo I (2009) Elimination of Grapevine fanleaf virus from three Vitis vinifera cultivars by somatic embryogenesis. Eur J Plant Pathol 123:57–60
36. Villamor DEV, Ho T, Al Rwahnih M et al (2019) High-throughput sequencing for plant virus detection and discovery. Phytopathology 109:716–725

Chapter 22

A Quantitative Luminol-Based Assay for ROS Burst Detection in Potato Leaves in Response to Biotic Stimuli

Muhammad Awais Zahid, Ramesh R. Vetukuri, and Erik Andreasson

Abstract

Reactive oxygen species (ROS) are important signaling agents in plants and animals. They are involved in diverse processes, including activation of immune responses to pathogen infection. Biphasic detection of ROS in response to pathogen perception is becoming more popular even in important crops like potato as means of screening different germ plasms and mutants generated by for example CRISPR-Cas9 as well as identifying signaling pathways. Here we describe a detailed protocol for quantifying ROS bursts induced in potato leaf discs in response to a bacterial elicitor and *Phytophthora infestans*.

Key words ROS, Reactive oxygen species, Potato, Immunity, Luminol, Oxidative burst, flg22

1 Introduction

Transient accumulation of reactive oxygen species (ROS) is an important signaling mechanism that is involved in regulation of diverse processes in plants and animals [1]. It is often referred to as oxidative burst and is believed to provide first-line plant defense (typically biphasic) against pathogens, triggered by associated elicitors in the apoplast called pattern association molecular patterns (PAMPs) and their receptors [2]. For example, in *Arabidopsis thaliana*, the leucine-rich repeat- receptor-like protein kinase FLS2 can recognize a 22-amino acid epitope of bacterial flagellin called "flg22" [3]. Other PAMPs include bacterial elongation factor (Ef-Tu) and lipopolysaccharides. This type of defense is called pattern-triggered immunity (PTI). However, virulent pathogens, such as *Pseudomonas syringae* pv. *tomato* DC3000 (Pto DC3000), secrete effectors into plant cells that can suppress PTI [4]. Biphasic detection of ROS in response to pathogen perception is becoming more prevalent even in important crops like potato as means of screening different germ plasms and mutants.

Nicola Luchi (ed.), *Plant Pathology: Methods and Protocols*, Methods in Molecular Biology, vol. 2536,
https://doi.org/10.1007/978-1-0716-2517-0_22,

Fast accumulation of ROS in the apoplast in response to PAMPs is strongly dependent on NADPH oxidase respiratory burst homolog D [5]. ROS may putatively have direct antimicrobial activity, but in this context, they mainly act as signaling molecules that activate further immune responses. The underlying mechanisms are still elusive [6], but recent data indicate that PTI and the other main pathogen recognition mechanism (effector-triggered immunity, ETI) are not exclusive and synergistically protect plants during pathogen infection [7, 8]. It has also been shown that treatment of potato (*Solanum tuberosum*) tubers with *Phytophthora infestans* causes biphasic ROS production [9], but many questions remain about the initiation and origin of the second ROS burst.

Detection of biphasic ROS induction in response to pathogens or associated elicitors is increasingly applied for screening varieties and mutants (generated by CRISPR-Cas9 editing, for example) for resistance to pathogens and identifying signaling mechanisms involved in ROS bursts and pathogen defense. Thus, it can play important roles in both crop development programs and functional genomics studies in the field [10]. To facilitate its exploitation, here we describe a protocol for a luminol-based assay for monitoring ROS bursts in potato leaves triggered by flg22 and *P. infestans*.

2 Materials

2.1 Sample Preparation and Plant Material

1. Distilled water.
2. Cork borer (*see* **Note 1**).
3. Potato plant leaf discs (*see* **Notes 2** and **3**).

2.2 ROS Assay

1. White, flat-bottomed 96-well plates.
2. Horseradish peroxidase, HRP (Sigma).
3. Luminol (Sigma).
4. Multichannel pipet (12 channels).
5. Dimethyl sulfoxide (DMSO, Sigma).
6. Elicitors (e.g., synthetic flg22 peptide—QRLSTGSRINSAKD DAAGLQIA) or *P. infestans* zoospores (*see* Subheading 3.3 for preparation).
7. Luminescence plate reader (e.g., Promega GloMax (R) Navigator).

3 Methods

3.1 Preparation of Stock Solutions

1. Prepare a 500× stock solution of luminol by dissolving 17 mg in 1 mL of water (or DMSO if the luminol used does not dissolve in water) and store aliquots (for up to 3 months) at −20 °C. Luminol is light-sensitive, so we recommend wrapping tubes containing the portions in aluminum foil or using dark tubes.
2. Prepare a 500× Horseradish Peroxidase (HRP) stock solution by dissolving 10 mg/mL in sterile water, make aliquots, and store (for up to 3 months) at −20 °C.
3. Prepare a 1-mM stock solution of flg22, by dissolving 2.27 mg in 1 mL of sterile water and storing at −20 °C.

3.2 Collection of Leaf Discs from Potato Plants

1. Leaf discs must be collected from potato plants carefully, by punching leaves with a cock borer (40 mm diameter) without injuring the leaves unnecessarily (*see* **Notes 1–3**).
2. We recommend taking 3–4 discs from each of 3–4 fully expanded leaves, well before the flowering stage (*see* **Note 4**), avoiding major leaf veins, the edges, and tips (Fig. 1) and consistently using the same area of all the leaves.
3. Place the leaf discs in a 12- or 24-well plate containing distilled water and wash them at least twice with water (*see* **Note 5**).
4. Place 200 μL of distilled water into wells of the 96-well plate, and carefully place washed leaf discs in them, adaxial side up. We use a flat spatula and a non-edge/sharp pipet tip to transfer leaf discs into 96-well plate (*see* **Notes 6** and **7**).
5. After placing the discs in the 96-well plate, cover the plate with a lid if provided by the supplier; otherwise wrap it in plastic film (*see* **Note 8**).
6. Incubate the 96-well plate containing the samples for 12–16 h to reduce confounding ROS signals associated with the wound response induced by punching the leaf discs (*see* **Note 9**).

3.3 ROS Measurement

1. After incubating the samples (usually overnight) and before starting the experiment, prepare fresh elicitation solution containing elicitor, luminol, and HRP (for example, for 10 mL of 100 nm of flg22 elicitation solution, add 20 μL of both the 500× HRP and luminol stock solutions to 9.96 mL of 104 nM flg22 solution (*see* **Note 10**).
2. As alternative agents for quantification of ROS burst in potato plants, we have used sporangia of *P. infestans* strain 88069. *P. infestans* was grown on rye agar medium plates [11]. To release sporangia, the plates are soaked with water and scraped with an L-shaped spreader, and then they are counted using a

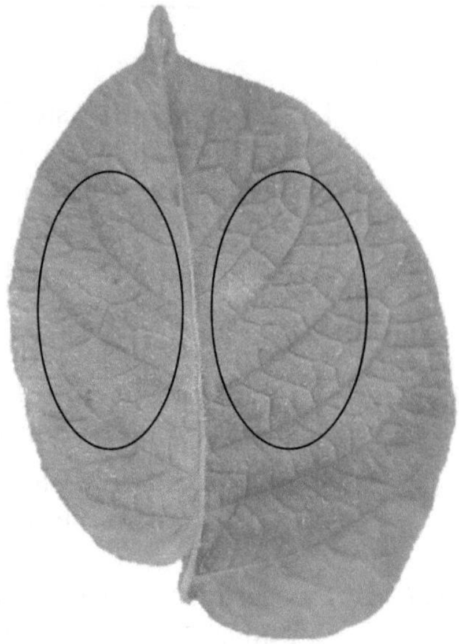

Fig. 1 Representative zone of potato leaflet disc collection for ROS assay. The preferred areas for collecting discs for the assay are inside the black marked ovals

hemocytometer after filtration through a 40-μm nylon cell strainer. The density is then adjusted to 1000 sporangia per mL. For example, if a 10 mL elicitation solution with 1000 sporangia per mL is wanted, add 20 μL of both 500× HRP and luminol stock solutions to 9.96 mL of a suspension with 1040 spores per mL (*see* **Note 11**).

3. Turn on the Promega GloMax(R) Navigator (or similar instrument) well before use and select an appropriate program. We usually use a program setting with 1-s integration per well, giving a high number of alterations. In our experiment, we set the interval for 3–5-h continuous measurement (*see* **Note 12**).
4. Carefully remove the water from the 96-well plate with a pipette. Many protocols suggest using a multichannel pipette for this, but it may injure leaf discs in the wells, so we prefer to use a single-channel pipette.
5. Add elicitation solution containing HRP and luminol to the 96-well plate with a multichannel pipette (*see* **Note 13**).
6. Immediately place the plate in the luminescence reader and run the program.
7. After completion of the program, export the data into Excel (or other data processing program) for further analysis.

3.4 Representation of Data

ROS burst is quantified by measuring the amount of light released due to the oxidation of luminol by HRP and expressed in relative light units (RLU). The data generated in a ROS assay can be examined and represented in various ways that should convey as much information as possible. Often this is done by displaying RLU generated per minute during the monitoring period, and/or total RLU generated during the period (*see* Figs. 2 and 3). Values (in RFU) collected at each point during the experiment are exported into an Excel file, then mean values for all the samples and standard errors are calculated.

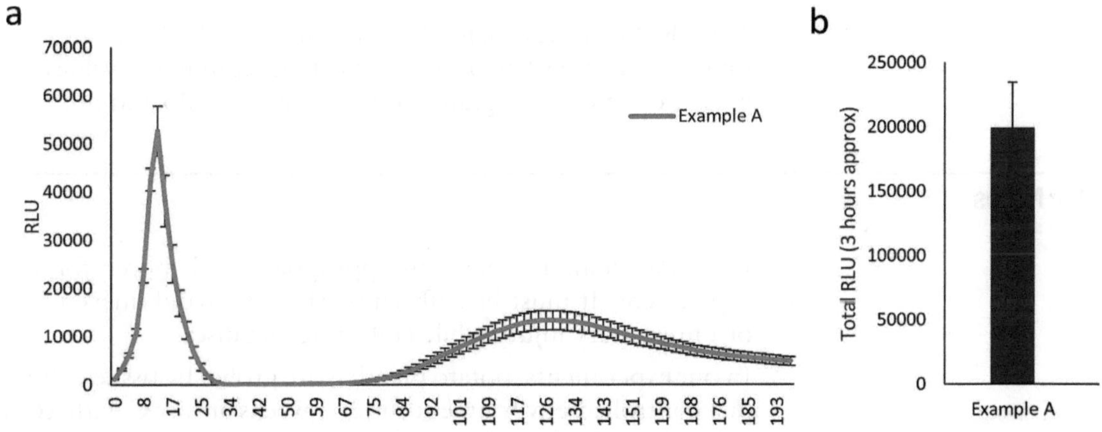

Fig. 2 Illustrative results of a typical assay of flagellin (flg22)-triggered ROS induction in potato plant leaf discs. (**a**) Bi-phasic ROS burst monitored for 3 h after treatment with 10 μm flg22. (**b**) Total ROS production during 3 h of continuous measurement

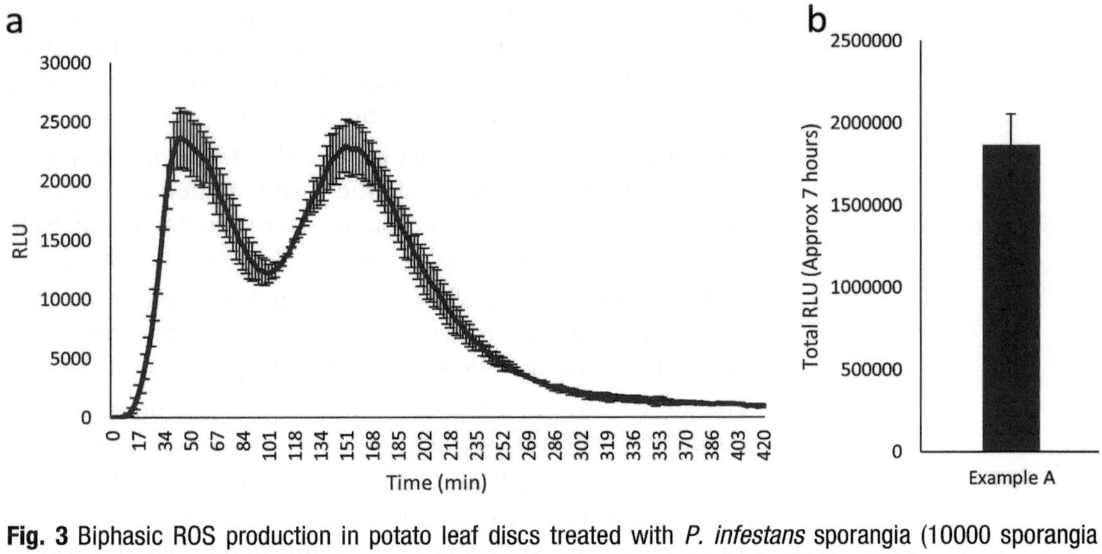

Fig. 3 Biphasic ROS production in potato leaf discs treated with *P. infestans* sporangia (10000 sporangia per/mL). Long-term ROS production was monitored for 7 h at 2-min intervals. (**b**) Total ROS production during 7 h of continuous measurement

3.4.1 ROS Production in Response to Different Elicitors

1. Measurements of ROS induced by flagellin (flg22) over 3 h, as described above, result in a biphasic curve. In the typical cases illustrated in Fig. 2a, the first ROS burst started instantly, peaked after 7 min, then declined after around 25 min. The second started after 75 min, had a longer peak and started to decline after 3 h.
2. A typical case of ROS production in response to *P. infestans* in potato leaf discs over 7 h (recorded at 3-min intervals) is shown in Fig. 3a. The first ROS burst started 17 min after elicitor treatment, peaked after ca. 40 min, and declined after ca. 90 min. The second ROS burst started after ca. 100 min, peaked after ca. 150 min, and returned to basal level after 5 h (*see* **Note 14**).
3. Total ROS was calculated by the sum of all RLU values over time, and standard error was calculated; represented values are presented in column graphs as shown in Figs. 2b and 3b.

4 Notes

1. It is important to select an appropriate cork borer for the experiment. It must be sufficiently sharp to avoid uneven cuts or unnecessary injury while collecting leaf discs.
2. In our experiments, potato plantlets are grown by tissue culture in a controlled environment for 2–3 weeks at 22 °C with 16-h light (140 μE)/8-h dark cycles. They are then transferred to 2 L plastic pots and grown in a Biotron controlled climate chamber providing 20 °C, 16-h (160 μmol/m2/s) light/8-h dark cycles, and 65% relative humidity, for 4–5 weeks.
3. Probably the most delicate part of the ROS assay protocol is handling the leaf discs. Unnecessary injury while punching them may significantly affect the outcome and thus must be minimized. Add the leaf discs directly to the water-containing wells so the wounded parts do not dry out.
4. Important factors to consider include the developmental stage and physiological status of the plant material, e.g., for immunity-related ROS studies, we recommend use of well-watered plants with consistent ages well before flowering.
5. Leaf discs must be carefully washed, and possibly more than twice if dirt particles are seen because they may carry microbes that cause unusual ROS bursts due to the presence of PAMPs.
6. The adaxial side of a leaf disc should be facing up when it is placed in a well, as it seems to result in less background noise.
7. Leaf discs should be floating on the water surface, if a leaf disc is injured, it will be submerged in the water, so it should not be used and discarded during the washing step to avoid inconsistent results.

8. It is important to cover the 96-well plates to prevent dehydration of the leaf discs and reduce the risk of contamination.
9. At least 12-h incubation is needed to reduce background noise caused by injury. However, the optimal incubation conditions will depend on the experimental conditions as well as the taxonomic, developmental, and physiological status of the material used.
10. The detection of ROS with a luminol-based method is sensitive, especially when using synthetic peptides, signals could be detected even with very low concentrations. Always use positive and negative controls where possible, such as water controls with substrate. If you are quantifying ROS using multiple treatments with the proposed Promega machine, we recommend designing a plate so that your samples will be placed column-wise. This is because the machine takes readings row-wise, and replicates in the column will increase the experimental confidence.
11. We have used spores from *P. infestans* and optimized conditions accordingly. However, other cultural filtrates, hyphal cell walls, or other elicitors can be used in the assay, with appropriate adjustment of the protocol if necessary.
12. It is important to turn on the mentioned instrument, select a program, and activate the photomultiplier tube before this step, because in some instruments it will take 5 min to warm up. Thus, part of the first burst induced by flg22 treatment (for example) will be missed if this is not done. Some other machines or newer versions may not need 5 min to warm up, but turning on the machine well before selecting a program is still recommended.
13. We recommend performing this step close to the plate reader as it must be quick to avoid missing some initial ROS measurements.
14. If measuring long-term ROS production (depending on experimental conditions), it may be advisable to increase the volume of the substrate from 200 to 300 μl to avoid samples drying out.

Acknowledgments

The Nilsson Ehle foundation, Formas (2020-01211) (2019-01316), Stiftelsen Lantbruksforskning R-19-25-282, and the Swedish Research Council (2019-04270) are acknowledged for financial support.

References

1. Wrzaczek M, Brosché M, Kangasjärvi J (2013) ROS signaling loops – production, perception, regulation. Curr Opin Plant Biol 16(5): 575–582
2. Qi J, Wang J, Gong Z et al (2017) Apoplastic ROS signaling in plant immunity. Curr Opin Plant Biol 38:92–100
3. Chinchilla D, Bauer Z, Regenass M et al (2005) The Arabidopsis receptor kinase FLS2 binds flg22 and determines the specificity of flagellin perception. Plant Cell 18(2):465–476
4. Jones JDG, Dangl JL (2006) The plant immune system. Nature 444(7117):323–329
5. Miller G, Schlauch K, Tam R et al (2009) The plant NADPH oxidase RBOHD mediates rapid systemic signaling in response to diverse stimuli. Sci Signal 2(84):ra45
6. Castro B, Citterico M, Kimura S et al (2021) Stress-induced reactive oxygen species compartmentalization, perception and signalling. Nat Plants 7(4):403–412
7. Ngou BPM, Ahn H-K, Ding P et al (2021) Mutual potentiation of plant immunity by cell-surface and intracellular receptors. Nature 592(7852):110–115
8. Resjö S, Zahid MA, Burra DD et al (2019) Proteomics of PTI and two ETI immune reactions in potato leaves. Int J Mol 20(19):4726
9. Kobayashi M, Ohura I, Kawakita K et al (2007) Calcium-dependent protein kinases regulate the production of reactive oxygen species by potato NADPH oxidase. Plant Cell 19(3): 1065–1080
10. Kieu NP, Lenman M, Wang ES et al (2021) Mutations introduced in susceptibility genes through CRISPR/Cas9 genome editing confer increased late blight resistance in potatoes. Sci Rep 11(1):4487
11. McLellan H, Boevink PC, Armstrong MR et al (2013) An RxLR effector from Phytophthora infestans prevents re-localisation of two plant NAC transcription factors from the endoplasmic reticulum to the nucleus. PLoS Pathog 9(10):e1003670

Part VI

Population Genomics in Plant Pathology

Chapter 23

Genetic Analysis of Plant Pathogens Natural Populations

Fabiano Sillo

Abstract

Population genetics allow to address fundamental questions about the biology of plant pathogens. By testing specific hypotheses, population genetics provide insights into the population genetic variability of pathogens across different geographical areas, time, and associated plant hosts, as well as on the structure and differentiation of populations, and on the possibility that a population is introduced and from where it has originated. In this chapter, basic concepts of population genetics are introduced, as well as the five evolutionary factors affecting populations, that is, mutations, recombination, variation in population size, gene flow, and natural selection. A step-by-step workflow, from sampling to data analysis, on how to perform a genetic analysis of natural populations of plant pathogens is discussed. Increased knowledge of the population biology of pathogens is pivotal to improve management strategies of diseases in agricultural and forest ecosystems.

Key words Population genetics, Genetic diversity, Evolution, Gene flow, Mutations, Natural selection, Hybridization

1 Introduction to Genetic Analysis of Natural Populations

1.1 Population Genetics of Plant Pathogens

The study of patterns of genetic variation, and genetic structure, within and between groups of individuals of the same species, underlies the field of biology defined as population genetics [1]. This field is represented by a theoretical component, using mathematical methods related to the theory of probability and statistics, and an empirical component, aimed at describing the genetic variability existing in populations as well as providing the estimation of the parameters that summarize their essential aspects. In particular, population genetics is focused on the origin of genetic variance, the transmission of variants from parents to offspring in the different generations, the changes occurring over time in a population due to evolutionary forces, and the existing genetic diversity in relation to the spatial distribution of individuals [1]. The study of population biology, including both population genetics and epidemiology (i.e., the study of distribution,

Nicola Luchi (ed.), *Plant Pathology: Methods and Protocols*, Methods in Molecular Biology, vol. 2536,
https://doi.org/10.1007/978-1-0716-2517-0_23,

frequency, and way of spread of disease agents), maintains a great importance in the context of the plant pathology. This is due to the fact that a significant outbreak, causing crop losses, usually consists of thousands if not millions of infection events, involving whole populations of the pathogen. In fact, while a single lesion, for example on a leaf, caused by a single event, has little or no significant ecological and economic impact, a progressing epidemic, affecting populations of host plants, causes significant damage to the whole productive supply chain. In order to understand the importance of population genetics in plant pathology, it is necessary to highlight a simple concept: pathogens evolve over time. Populations of pathogens, like all other organisms, continuously adapt to various environmental changes. Organism responses to these external pressures tend to be the result of the synergistic action of five major evolutionary factors: mutations, recombination, variation in population size, gene flow, and natural selection [2–4]. On these premises, it is possible to understand that the primary purpose of the population genetics applied to plant pathology is to identify which factors play a role in the pathogens evolution and how these evolutionary forces interact to determine the genetic background of pathogens, and their potential to be infectious [4]. Finally, the study of the population biology is needed to understand the reproductive potential (e.g., sexual, asexual) of the pathogens, the timing, and the way of spread, a fundamental requirement to implement effective control and management strategies [2, 5].

1.2 Basic Concepts in Population Genetics

One of the main concepts of population genetics is based is the concept of "population" itself. Although there are both a statistical and a biological definition of "population", and that a single definition is not applicable, it can be stated that a population is a group of organisms of the same species living in a specific geographical area in which individuals freely interact and mate [6]. Nevertheless, a population is in constant change, both in space and in time; members of a population, from an ecological point of view, share the same functional role and react in a similar way to environmental factors. Hence, it is clear that the population is the evolutionary unit, able to evolve over time through gene frequency changes [6]. In order to understand the forces driving the evolution of natural populations, it is necessary to compare natural populations with a theoretical "control" one, not subjected to evolution. In 1908, Hardy and Weinberg defined "population in equilibrium," a population in which neither the allele frequencies nor the distribution of the genotypes undergo variations through different generations. In the absence of allele frequency variations, equilibrium is maintained and the population does not evolve. This situation occurs if the following restrictive conditions are respected:

- No mutations occur.
- No gene flow between populations occur.
- Population is infinitely large in size.
- Selection does not occur, and all genotypes have the same reproductive and adaptive abilities.

If these conditions are met, allele frequencies within the population will remain constant for an indefinite period. It is evident that this condition does not occur in nature. However, the Hardy–Weinberg principle remains at the basis of population genetics: consequences of violation of the requirements of this model, such has, variation in allele frequencies, help to understand how recombination, mutations, migration, genetic drift, and natural selection influence the evolution of populations. In fact, upon deviated conditions compared to Hardy–Weinberg equilibrium, alleles of different loci can be transmitted in association with higher frequency than expected. This non-random association is called linkage disequilibrium (LD). Two loci are considered in linkage disequilibrium when the frequency by which they are inherited together is greater than the expected frequency, considering the principle of Mendelian independent assortment of alleles [7]. By using the central concept of Hardy–Weinberg, the Wright–Fisher model [8] has been developed to quantify the effects of the different evolutionary forces, such as genetic drift, mutations, selection, variation in population size, gene flow, on allele frequencies over time [8]. This model posed the basis for the so-called hierarchical F-statistics, that allow to quantify the degree of genetic differentiation between sub-populations.

1.3 Evolutionary Forces Affecting Plant Pathogen Populations

1.3.1 Mutations

Mutations represent the maximum source of genetic variation, causing direct changes in the DNA sequences of individuals and thus leading to emergence of new alleles in populations. This evolutionary force is particularly important for pathogens such as viruses and bacteria, able to rapidly reproduce and to invade hosts with extremely large populations. The rate of mutation in gene loci, although varying from locus to locus, is generally relatively low, but mutants with high fitness values can spread and infect before the mutation is lost due to genetic drift [2]. A plethora of different mutations have been identified and reported so far, including insertions and deletions of nucleotides (InDels) and DNA fragments, substitutions, inversions, and translocations. Small mutations affecting one or a few base pairs are considered Single Nucleotide Polymorphisms (SNPs). Mutations not influencing the reading frame are synonymous mutations; by contrast, non-synonymous mutations are often deleterious, but in several cases, they have significant effect on the product of a gene, also leading to changes in fitness (the survival and reproduction potential) of the organism [9]. Mutations might result in the creation of new virulent strains of

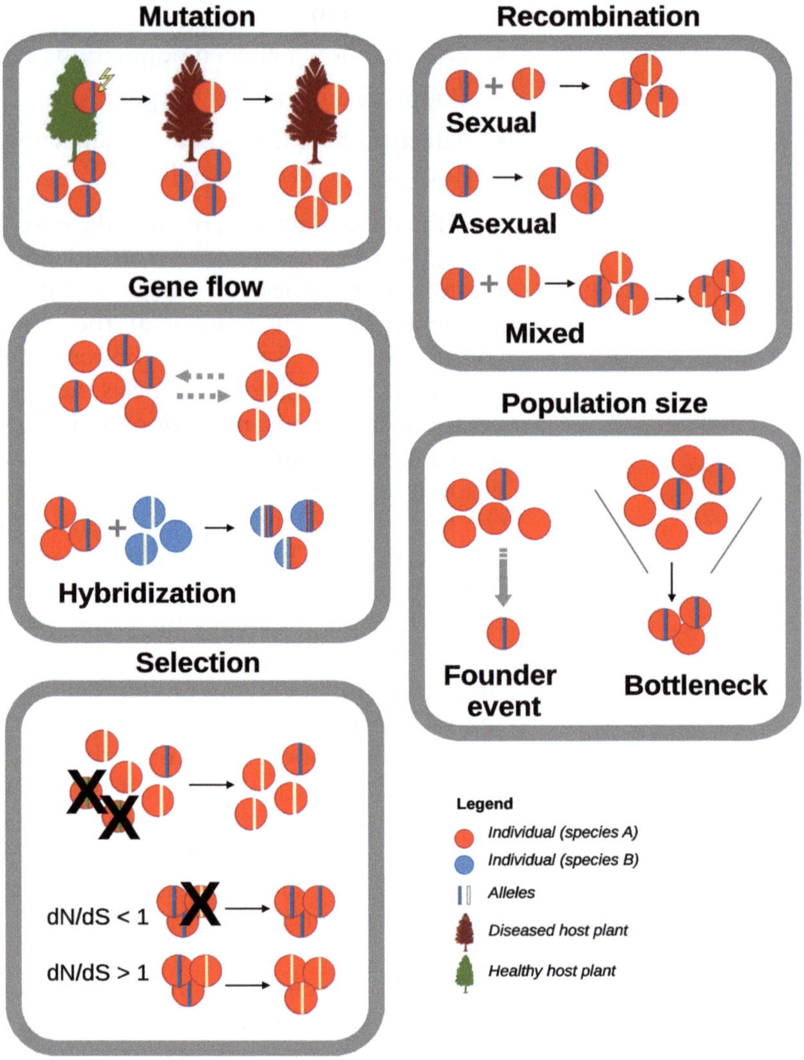

Fig. 1 Schematic representation of the five evolutionary forces acting on plant pathogen populations, and examples of their effects. From top left to bottom right: mutation affecting *avr* loci can generate virulent pathogen populations on resistant plants; recombination through a mixed reproduction mode may result in the generation of a large number of different genotypes, followed by a selection on the best fitting ones; gene flow during hybridization can result to introgression of adaptive alleles in hybrids; population size variation, as a result of founder event or a bottleneck process, can lead to dramatic changes in allele frequencies; purifying (ratio $dN/dS < 1$) and positive ($dN/dS > 1$) selection on genes/alleles can also affect the genotypic diversity of populations

a pathogen, for example, if avirulent alleles are affected (Fig. 1). The most known genetic interaction during host–pathogen interplay is the gene-for-gene (GFG) system, in which a gene product in the plant host recognizes a pathogen antigenic locus. In these GFG relationships, host plants resulted resistants if resistance (RES) gene(s) recognize a specific pathogen avirulence (AVR) protein [10]. By contrast, if the pathogen has a virulence allele (*avr*), or the plant harbors a susceptibility allele (*res*), resistance is not induced [10]. Selection and fixation of *AVR* and *RES* alleles in pathogens and plants, respectively, constitute the so-called "*arms race*" in evolutionary terms [11].

Several plant pathogens have been reported to possess *AVR* genes interacting with plant resistance *RES* genes. The list includes important pathogens such as *Blumeria graminis* [12], *Melampsora lini* [13], *Magnaporthe oryzae* [14], and *Fusarium oxysporum* [15]. Taking into account the above-mentioned system, it is easy to understand that mutation affecting these loci might play a crucial role in the overcoming of plant resistance. Different pathogen species have shown different mutation rate, and even at species level, this rate may vary depending on the chromosomal location of the considered locus [16, 17]. Genomes of important plant pathogens shared some common features, such as large size and abundance of transposable elements, leading to high level of genomic plasticity [18]. Plasticity of these genomes often results in their fast evolution, and consequently, to accumulation of mutations that may increase the fitness of the pathogens [18].

1.3.2 Recombination and Reproductive Strategies

Reproductive strategies, especially in fungi, dramatically influence the distribution of genetic variability within a population [17]. In fungi, reproduction can occur asexually, sexually, or both ("mixed" reproduction) [2, 3]. Furthermore, in the case of sexual reproduction, fungi exhibit different type of sexuality and mating systems, for example, homotallism or heterotallism. Pathogens that commonly undergo through sexual reproduction show a higher level of recombination compared to those that reproduce mostly asexually [2]. Genetic recombination is intrinsically linked to gamia, determining the emergence of novel combinations of alleles in the new generations. Pathogens with sexual reproduction phase are considered potentially more harmful (high recombination rate), rather than those that undergo clonal and self-reproduction (low recombination). However, the most dangerous pathogens are considered those showing both reproductive strategies, sexual and asexual [2, 3]. These organisms exploit the sexual cycle to assemble a large number of genotypes, which are then subjected to strong selective pressures by the environment. The genotypes suitable for survival will be asexually multiplied, and fast spreading of the pathogens is ensured [2] (Fig. 1). Clonal lineages are common in plant pathogens (e.g., some *Fusarium oxysporum* formae spaecialis), since several species are known to reproduce asexually in the agroecosystem. On the other hand, other pathogens need a sexual mating process to form infectious dykarion phase (e.g., *Ustilago* spp.) [19]. Reproductive structures in several cases are also used to withstand environmental stresses and overcoming seasons [3].

1.3.3 Population Size and Genetic Drift

Population size refer to the total number of individuals in the population. The number of mating individuals is referred to as the effective population size (*Ne*). As expected, *Ne* influences gene diversity by random genetic drift. This stochastic process affecting whole genomes of pathogens occurs when a population experiences a *bottleneck effect*, that is, a reduction in size (e.g., after climatic extremes, or as result of agronomic practices affecting plant pathogens), or a *founder event*, when a subset of genotypes colonize a new area or a new host (e.g., pathogens accidentally introduced in non-native areas) [2, 3] (Fig. 1). For example, depending on the type of between-season transmission, some plant pathogen populations experience severe demographic bottlenecks during the off-season [20].

It is clear that large populations generally harbor greater genetic diversity than small populations, because they have larger likelihood of developing mutations and maintaining favorable ones to increase their evolutionary potential. Small populations, on the other hand, are subjected to the deleterious effect of genetic drift [17]. For this reason, agronomic practices aimed at decreasing the size of populations such as crop rotation or alternate application of different fungicides, are deemed to play a crucial role in plant disease management programs [2]. The *Ne* is also influenced by the dispersal potential and the migration rate of pathogens (e.g., airborne vs. soilborne pathogens) [21].

1.3.4 Gene Flow

Gene flow is a process by which one or more alleles are exchanged between geographically separated populations, mainly due to migration of genotypes and mating. In the case of strictly asexual organisms, for which recombination is precluded, only genotypes are exchanged (genotypic flow). A high gene/genotype flow is a major advantage for pathogens because it allows to increase the genetic diversity within populations, by introducing alleles or genotypes that could favor a greater virulence [2, 3, 17]. Hence, the quantification of the existing gene flow between populations of a pathogen is useful for programming management strategies aimed at limiting the spread of the disease [22].

Recently, the exchange of genetic material between different but interfertile species through hybridization, defined as introgression, has started to be recognized as a significant catalyst of evolution of plant pathogens [23]. Thanks to advances in genetic characterization and to the democratization of technologies necessary for genomic research, molecular evidence of hybridization has exponentially risen in recent years, allowing to demonstrate that this phenomenon is common in plant pathogens [23, 24]. As stated before, the transfer of genetic material can be a stochastic process not necessarily involving the transfer of adaptive traits. During biological invasion, for instance, demographic expansion of the introduced species may result in massive allele introgression from native species into the invasive one, not necessarily related to gene functions [25]. However, introgressive events have often been documented to be driven by natural selection. Introgression events may shape the recipient genome by absorbing a novel gene or by generating favorable allele combinations, further accelerating adaptive processes, or by increasing hybrid fitness during competition with parental species [26]. As transient entities, hybrids can simply act as "*genetic bridges*" between native and invasive species, facilitating interspecific gene introgression through the repeated backcrossing with the parental species [27] (Fig. 1). Geographic areas where hybridization occurs, called "*hybrid zones*", can be considered as target for the study of ongoing evolution and adaptive selection of pathogens in nature [23].

1.3.5 Selection

Genetic diversity and population differentiation are strongly related to natural selection at phenotype level. Selection is a powerful force in evolution: if a population showed a degree of heritable phenotypic variation, and different phenotypes have different levels of fitness, selection occurs [17]. The environment where the pathogen and the plant host interact is influenced by the constant presence of anthropogenic activities. In this environment, selective pressures (e.g., use of resistant plant cultivars, use of fungicides, artificial irrigation, crop rotation) can be manipulated. For example, fungicide resistance is under strong selection in pathogen populations and may jeopardize the efficacy of disease management in several pathosystems. The development of fungicide resistance comprises the emergence of a resistant phenotype, followed by selection by fungicide application [17]. At gene level, purifying selection (negative selection) acts to remove deleterious mutations from populations, while positive selection promotes the spread of beneficial allele frequencies and may lead to selective sweeps. In the case of most neutral mutations, balancing selection maintains polymorphisms inside the populations [28]. The most direct method for showing the presence of positive selection is to demonstrate that the number of nonsynonymous substitutions per nonsynonymous site (dN) is significantly larger than the number of synonymous substitutions per synonymous site (dS) (Fig. 1). Selection can be inferred from the ratio of nonsynonymous to synonymous nucleotide substitutions in protein-coding regions (dN/dS ratios) among closely related species. Ratio of $dN/dS > 1$ is interpreted as a sign of positive selection, that is, natural selection promotes changes in the protein sequence, whereas $dN/dS < 1$ is generally accepted as a sign of purifying selection, where some replacement substitutions have been purified by natural selection, presumably because of their deleterious effects [28].

1.4 From Population Genetics to Population Genomics: The Rising of NGS

The rising use of next-generation sequencing (NGS) due to the decreasing of costs and the advance in technology boosted the generation of an unprecedented amount of sequence data. Genotyping moved from analysis of limited set of genetic markers (3–20) to thousands of loci. In contrast to traditional Sanger sequencing, NGS allows high throughput parallel sequencing of millions of DNA nucleotides at reduced cost and time [29]. Recently, second- and third-generation sequencing technology (e.g., Illumina and PacBio/MinION technologies, respectively) significantly improved the progress of genome sequencing. International efforts in sequencing activities are aimed at increasing the dimension of genome sequence databases For example, the ongoing "1000-fungal genomes" project of the U.S. Department of Energy (DOE) Joint Genome Institute (JGI) [30] is aimed at covering all fungal taxonomic units, including plant pathogens, through genome sequencing of representative strains. Thanks to data on whole genomes, role of natural selection on genes and functionally characterized loci can be deciphered. These information can be useful to understand genomic regions under selection within and among populations, the genetic traits linked to phenotypic ones, the correlation between adaptation and host specificity to genomic profiles [31]. The generation of low-cost and high-density genotype-by-sequencing (GBS) data is now easier than in the past, and they can replace the use of anonymous DNA-based markers, promoting the shift from "population genetics" to "population genomics" [31]. The development and assessment of new DNA-based markers is a laborious process that involve some time-consuming steps, including a pre-screening on a subset of samples to verify level of polymorphisms before assessment on large populations [31]. The use of NGS overcome these issues and a popular GBS method applied to population studies is represented by Restriction site-associated DNA sequencing (RAD-Seq). RAD-Seq combines the use of restriction enzymes to cut DNA into fragments, and the use of specific molecular identifiers to associate sequence reads to specific individual raw sequences. These two techniques are applied to an Illumina sequencing platform to obtain millions of sequences ("*reads*") [32]. Data from GBS provides perfect input for detection and characterization of SNP-based markers, through SNP-calling process.

2 Genetic Analysis Workflow

2.1 Sampling

A correct sampling is the first critical step in population genetic analyses. It requires a clear definition of the studied target population, also considering pathogen life cycle to be sampled [17]. Random sampling is essential to ensure independence of genotyped individuals. Although there is no universal rule for a reasonable number of genotypes to be sampled, since it is dependent on the pathogen species and on the tested hypothesis, it has been reported that a good size sampling should include at least 30–40 to 100–200 samples [2, 3]. It is worth noting that some analyses, such as those serving to detect linkage disequilibrium, require larger sample sizes [17]. Currently, softwares such as *SPOTG*, aimed at optimizing sampling strategy for common population genetic study, are available [33]. After sampling, isolation in vitro of the pathogens is a crucial step to both obtain enough biological material for DNA extraction and to ensure that DNA is free of contamination from other environmental-living organisms, even if, occasionally, pathogen DNA could be directly isolated from symptomatic plant tissue or other fungal structures (e.g., fruiting bodies, sclerotia). Nucleic acid extraction from pathogen materials can be carried out by using several commercial kit, for example, NucleoSpin Plant II kit (Macherey Nagel, Germany), EZNA Fungal DNA Mini Kit (Omega Bio-Tek, Darmstadt, Germany), DNeasy Plant Mini Kits (Qiagen, Valencia, CA, USA), for fungi or E.Z.N.A.® Bacterial DNA Kit (Macherey Nagel, Germany) for bacteria, or by traditional manual protocols. High quality and yield of extracted DNA guarantee high resolution in the genotyping step.

2.2 Genotyping

In the past, analyses aimed at studying population genetics of plant pathogens were mostly based on the use of morphological markers, linked to phenotype of individuals and populations studied. The advent of molecular biology techniques in the 1980s allowed the use of DNA-based markers. This has promoted in-depth analyses of populations of many plant pathogens. An ideal DNA-based molecular marker must meet some specific requirements: it should be polymorphic, to allow the detection of the variation between different samples, multiallelic, with at least two different alleles for the same locus, stable and non-epistatic, that is, not being influenced by the environment or by epigenetics or pleiotropic processes. In addition, it has been suggested that DNA-based markers should be (when possible) neutrals and codominants, to differentiate heterozygous condition from homozygotes [34]. Moreover, there are also some technical conditions that should be respected. For example, the marker must be simple and relatively cheap to develop and to detect, and it should ensure the reproducibility of all steps of the experiment [34].

Molecular markers can be divided into two main classes: those based on hybridization techniques and those that require the reaction of PCR (Polymerase Chain Reaction) for their analysis. The visualization of polymorphisms identifiable with the markers is traditionally reached by electrophoresis technique, which it allows to separate the obtained DNA fragments according to their size and thanks to DNA negative charge. Molecular markers can be also classified according to the number of loci that can be analyzed in a single experiment. Multi-locus markers allow to analyze simultaneously several loci/alleles. Among these markers, there are RAPDs (Random Amplified Polymorphic DNA), RFLP (Restriction Fragment Length Polymorphism) and AFLP (Amplification Fragment Length Polymorphism) [35]. They are usually dominant, not allowing the distinction between the situation of homozygosity and that of heterozygosity of the analyzed locus. In contrast, codominant single-locus markers allow the analysis of a single locus, as in the case of markers based on gene sequences and SSR (Simple Sequence Repeats) [36]. Simple Sequence Repeats (SSRs) or microsatellites are tandemly repeat genetic loci of 1 to 6 base pairs, ubiquitous in prokaryotes' and eukaryotes' genomes [37]. SSR loci have mutation rates ranging from 10^{-7} to 10^{-3}, higher than the rate of point mutations that reach 10^{-8}–10^{-9} [37]. This rate seems to be affected by motif length, motif sequence, number of repeats, imperfection of SSR (e.g., imperfect repeats are more stable than pure repeats), and genomic location [38]. SSRs are extremely powerful and widely used genetic markers because of their co-dominant nature and high level of polymorphisms. Their expected stepwise evolution pattern can also be extremely useful to study how diversity is generated when populations originated from one or a few founder genotypes, as it has been done for the *Phytophthora ramorum* epidemic [39]. In addition, in some cases, it appears that some SSRs may be hypervariable, allowing to study the genetic structure of clonal (or mixed) populations [40]. Several approaches have been used to isolate SSR loci, including screening of SSR-enriched libraries with repeat-containing probes or RAPD-based protocols [41]. However, if a reference genome of the studied pathogen or a genome of a closely related species are available, screening and selection of SSR markers is straightforward, by using dedicated bioinformatic tools [38]. Without a reference genome or available sequences, Random Amplified Microsatellites (RAMS) technique has demonstrated to be useful for genotyping microbial genomes, and it has successfully applied to fungal plant pathogens [42].

The number of marker loci to be used in population genetic analysis should be tailored based on the available resources and the tested hypothesis. For an accurate assessment of population genetic structure, 6–20 marker loci distributed across different chromosomes are generally recommended, also taking into account a rarefaction analysis to assess their resolution potential in genotypic profiling [4]. Unlinked markers are preferred to provide a representative sample of variation in the whole genome and to allow analyses of linkage disequilibrium [4, 43]. Nucleotide sequences have more information on evolutionary history than categorical alleles (e.g., SSR), also providing insights on genetic similarity among haplotypes as well as on haplotype richness and evenness. Markers with high mutation rates such as SSRs may provide information into recent divergence, while markers with slower mutation rates (e.g., mitochondrial loci) are useful to infer on distant evolutionary history [43]. Markers are generally scored by using gel or capillary electrophoresis. Recently, high-resolution melting (HRM) analysis, as a rapid and sensitive polymorphism detection method based on quantitative PCR (qPCR) technologies, has been introduced as an alternative technique to capillary electrophoresis to investigate molecular marker polymorphisms [40, 44]. HRM analysis is well suited for genotype scanning and detection of single nucleotide polymorphisms (SNPs), exploiting the fact that PCR products with different sequences have distinct melting profiles [44].

In the case of GBS (e.g., RAD-seq), and when outputs of genotyping are composed by sequences, reads can be aligned to the reference genome of the pathogen species, if available, by using several bioinformatics tools such as Bowtie2 [45] or BWA [46]. Primary outputs of the read alignment process is a sortable SAM (Sequence Alignment/Map) file or its compressed binary version BAM. These files include information on barcodes, read quality, amplicon tags, and can be used as input for the software SAMtools [47], allowing to calculate read coverage on reference genome and useful for statistics and filtering purposes [47]. After mapping to reference genome, the variant calling process is needed to detect and annotate variants (e.g., SNPs, InDels) in the dataset. The most used pipeline to perform this step include the use of SAMtools and other popular variant callers such as GATK [48]. The produced output of the variant calling is generally composed by a Variant Call File (VCF) containing information about variants found at specific positions in a reference genome. However, when a reference genome is not available, other bioinformatics tools should be utilized to perform de novo assembly of the reads, in order to rebuild the sequences of candidate alleles. Two recent software that are able to deal with de novo assembly from RAD-seq data are RADassembler [49] and *PyRAD* [50].

2.3 Data Analysis

2.3.1 Genetic Diversity

The quantification of genetic diversity is one of the main analyses that should be performed on data obtained from scoring of molecular markers. As a generic rule, mutation, migration, and recombination increase genetic diversity, whereas natural selection and random genetic drift decrease diversity [17, 43]. Classical outputs of the scoring process encompass binary matrices (for dominant markers, such as RFLP), numerical matrices (for codominant markers based on different lengths, such as SSR), or specific file format resulting from the SNP calling (e.g., VCF files). Genetic diversity is constituted by the differences at locus level (gene/allele diversity) and by the variation of number of genetically distinct isolates/strains in populations (genotype or, in case of haploid organisms, haplotype diversity). The number of alleles per locus, allele frequencies per locus, and number of private alleles can be easily calculated from the above-mentioned input data. Introduced by Wright in the early 1950s, the fixation indices F_{is}, F_{it}, and F_{st} are generally used to evaluate allele diversity of the population. These indices offer an easy way to represent the structure of a population by evaluating the relationship between the number of homozygotes and the number of heterozygotes. The three indices can be seen as the subdivision into three components of the allelic variance of a given gene in the whole population (within the population, between subpopulations, and between individuals of a subpopulation). In particular, F_{st} is significant for assessing the level of heterozygosity occurring between the different subpopulations. In the case of populations of haploid organisms, such as fungi, it is more appropriate to define gene diversity rather than heterozygosity. Four indices, analogous to F_{st} ones, were formulated by Nei in 1973 [51]: H_E, or the expected gene diversity; H_S, the internal gene diversity of a subpopulation; H_T, the total gene diversity: and G_{ST}, or the gene diversity between the different subpopulations. The indices can be calculated according to the following formulas [45]:

$$H_E = 1 - \sum_{i=1}^{h} x_i^2 \quad H_S = 1 - \sum_{i=1}^{h} x_{ij}^2 \quad H_T = 1 - \sum_{i=1}^{h} \overline{x}_{ij}^2 \quad G_{ST} = \frac{H_T - \overline{H}_S}{H_T}$$

where h is the number of alleles, $x\,i$ is the frequency of the allele i at each locus, x_{ij} is the frequency of allele i in population j.

Other analogous indices based on F-statistics include R_{st}, useful when analyzing SSRs, that takes into account differences in the number of repeats between SSR alleles (allele size) [52].

The F-statistics are based upon comparison of gene/allele frequencies in populations. However, molecular data, in addition to information on the frequency of molecular markers, can be useful to quantify the mutational differences between genes. Analysis of Molecular Variance (AMOVA) is a statistical method developed by Excoffier and colleagues in 1992 [53] to estimate population differentiation from molecular data and testing hypotheses about such differentiation by analyzing differences between molecular sequences rather than simply evaluating deviation in gene frequencies. Compared to the traditional ANOVA, AMOVA variable is a haplotype, and since molecular data consist of Euclidean distances, data generally does not follow a normal distribution. A null distribution is therefore computed by resampling of the data [53]. Several molecular marker data (including RFLP, AFLP, RAMS, SSRs) and sequence data can be analyzed using this method [53]. The AMOVA can be assessed by using specific tools such as *GenAlEx* [54] that is a user-friendly excel-based software, or *Arlequin* [55], which has been specifically implemented to deal with SSR data. A large list of free software used for assessment of gene and genotpyic diversity are available, including some popular software such as POPGENE [56] and, more recently, PyPop [57], as well as several R packages (e.g., *poppr;* [58]).

Concerning gene flow, there is an indirect relation between this process and the index of subpopulations differentiation defined by the F-statistics, in a theoretical island model [59]. The following equation can be used to easily estimate gene flow:

$$Nm_{est} = \frac{(1/Fst - 1)}{4}$$

where Fst (or one of its analogs) represents fixation index, *N* is the effective population size, and *m* is the effective proportion of migrating individuals. If $Nm_{est} > 1$, gene flow is detected and effects of genetic drift are negated, and if $Nm_{est} > 4$, all subpopulations are included in one randomly mating large population [59]. It should be mentioned, however, that this estimation may be biased if requirements for the proposed island model (in which all populations are equally likely to give and receive migrants from all other populations, and no selection or mutations are present) are not met [60].

For nucleotide data, one of the most popular tests is Tajima's D-test [61]. It should be noted that the variance of Tajima's D depends on the demographic model [62]. Tajima's D test for recent selection allows to determine if a genomic region has evolved under a non-neutral model, thus showing footprint of introgression and selective sweeps. The R package *INTROGRESS* can be a powerful tool to detect introgressed blocks starting from different types of genotyping data [63]. To assess the role of selection on molecular data, the program KaKs calculator can be used to calculate divergence at synonymous/non-synonymous sites. The dN/dS ratio for each coding sequence is calculated with the following formulas:

$$d_N = -\frac{3}{4}\ln\left(1 - \frac{4p_N}{3}\right) \quad d_s = -\frac{3}{4}\ln\left(1 - \frac{4p_s}{3}\right) \quad dN/dS = \frac{d_N}{d_s}$$

where p_N and p_S were the proportions of non-synonymous and synonymous differences between the two analyzed sequences, respectively [64].

In case of hybridizing species, genome-wide four taxon Patterson's D statistics, that is, ABBA/BABA test, is a powerful tool to detect introgression events along entire genomes [65]. The test compares patterns of ancestral and derived alleles within four taxa, also including an outgroup. If gene flow is absent, equal numbers of sites for each pattern are expected. A positive/negative value of D is generally related to an excess of ABBA/BABA sites and might indicate introgression. For data obtained from GBS, the software *angsd* [66] uses the BAM files obtained from the alignment of genotype reads against the reference genome to perform the ABBA/BABA test.

Several different indices can be used to estimate genotypic diversity, mostly derived from ecological indices of species diversity. These measures include Stoddart and Taylor's index *G* and Shannon–Wiener index *H'*, both able to account for evenness and richness [43]. Evenness indices measure how uniformly genotypes are distributed within a population, while richness is a measure of the number of genotypes observed regardless of relative abundance [43]. Genotypic diversity is an important parameter for clonal or partially clonal populations, as observed for several plant pathogens. Clonality in the populations should be always evaluated and genotypic diversity should be calculated on clone corrected data. The *poppr* package in R is useful when dealing with data with high degree of clonality [58]. Usually, genotypic diversity increases with sample size, so it is useful to perform rarefaction analyses for genotypic diversity comparison from different size samples, and bootstrapping can be used to estimate confidence intervals for diversity estimates [17].

2.3.2 Genetic Distance and Structure

In order to infer relationships among genotypes/haplotypes, clustering methods including minimum spanning networks and bootstrapped dendrograms (trees) based on genetic distance are generally used. Different measures of genetic distance have been proposed, including Dice, Jaccard, and simple mismatch (Manhattan) distance [67]. Another well-known measure of genetic distance is Prevosti's distance [68], which was defined in the context of chromosome polymorphism. For dominant markers, the *restdist* program in Phylip is recommended as it assumes a nucleotide substitution model [69]. Pairwise genetic distance of populations is normally embedded in matrices, that can be used in subsequent statistical analysis. Non-metric multidimensional scaling (NMDS) and principal coordinate analysis (PCoA) are also used to summarize the overall variability among individuals, which includes both the divergence between groups and the variation occurring within groups. However, in order to test the relationships between different clusters focusing on between-group variability and neglecting within-group variation, Discriminant Analysis of Principal Components (DAPC) should be preferred over classical statistical methods. This is a multivariate method designed to precisely identify clusters of genetically related individuals, and it can be used for clonal or partially clonal populations [70]. DAPC can be carried out with the use of software like find.clusters function in *adegenet* R package [71].

Bayesian analysis, as implemented by the software STRUCTURE [72], is often used to investigate the population structure and to assign individuals into subpopulations. Successive K values (number of populations) are a priori tested to obtain the distinct clusters and to estimate the number of subpopulations. The online software STRUCTURE HARVESTER provides a way to assess and visualize likelihood values across multiple values of K and hundreds of iterations for detection of the number of genetic groups that best fit the data [73].

Haplotypes networks rely basically on two methods, the minimum spanning tree (MST) method and the method from Templeton, Crandall, and Sing 1992 (TCS) [74]. Starting from a matrix of pairwise distances among haplotypes, both methods allowed to obtain the shortest set of paths that link all observations, where the length of each link is represented by the pairwise distance. The TCS method, based on a model of evolution of the genetic characters measured on each individual, may have reticulations shaping alternative branchings, while MST does not allow reticulations. *PopART* (Population Analysis with Reticulate Trees) is a free population genetics software useful to infer and visualize haplotype network [75].

Correlation between genetic and spatial distance can give a clue to the introduction area of an invasive pathogen, on the way of spreading of a pathogen, the isolation by distance and on the potential dispersal of spores. The spatial autocorrelation analysis can be performed over a wide range of spatial scales, including local one, to identify sources of inoculum and way of spread [76]. Comparison and testing for correlation between the genetic and geographical distance matrices (e.g., by Mantel test) is useful when performing a genetic population analysis. The tool *SPAGeDi* (Spatial Pattern Analysis of Genetic Diversity) is able to infer relatedness or differentiation of populations by pairwise comparisons starting from marker data and spatial distance, and analyze how these values are related to spatial distances [77].

2.3.3 Linkage Disequilibrium and Estimation of Mode of Reproduction

The degree of association between alleles at different loci, or linkage disequilibrium (LD), is generated by the interaction between population history, mutations, and especially recombination, rate. Although high level of genotype diversity may be indicative that sexual reproduction may occur, LD measures are needed to assess the hypothesis of random mating. These tests require unlinked genetic markers, relatively large sample sizes (from hundreds to thousands of isolates), and an appropriate sampling method. A small sampling scale may lead to overrepresentation of clones, if dataset is not clone-corrected [4]. The LD can be assessed by using different software, such as POPGENE [56]. The observed linkage disequilibrium can be compared to expected distributions of linkage among loci by using the index of association (IA) [78]. Linkage disequilibrium analysis has been extensively used to infer the preferred mode of reproduction (sexual/asexual) of plant pathogens [40, 79]. Pairwise homoplasy index (*Phi*) test for global recombination embedded in *Splitstree* also assesses the probability that recombination occurs by comparing observed and permuted phi statistic (Φ), and LD can be evaluated through comparison as well [80]. The *cloncase* R package, for example, allows to estimate the expected clonal and potentially sexual reproduction using frequencies of multi-locus genotypes over time and seasons [81].

References

1. Hartl DL, Clark AG, Clark AG (1997) Principles of population genetics. Sinauer Associates, Sunderland
2. McDonald BA, Linde C (2002) Pathogen population genetics, evolutionary potential, and durable resistance. Annu Rev Phytopatol 40: 349–379
3. McDonald BA, Linde C (2002) The population genetics of plant pathogens and breeding strategies for durable resistance. Euphytica 124:163–180
4. McDonald BA (1997) The population genetics of fungi: tools and techniques. Phytopathology 87:448–453
5. Xu J (2006) Fundamentals of fungal molecular population genetic analyses. Curr Issues Mol Biol 8:75–90
6. Waples RS, Gaggiotti O (2006) INVITED REVIEW: what is a population? An empirical evaluation of some genetic methods for identifying the number of gene pools and their degree of connectivity. Mol Ecol 15: 1419–1439
7. Hill WG (1974) Estimation of linkage disequilibrium in randomly mating populations. Heredity 33:229–239
8. Crow JF (2010) Wright and Fisher on inbreeding and random drift. Genetics 184: 609–611
9. Loewe L, Charlesworth B, Bartolomé C et al (2006) Estimating selection on nonsynonymous mutations. Genetics 172:1079–1092
10. Dangl JL, Jones JD (2001) Plant pathogens and integrated defence responses to infection. Nature 411:826–833
11. Bergelson J, Kreitman M, Stahl EA et al (2001) Evolutionary dynamics of plant R-genes. Science 292:2281–2285
12. Jørgensen JH, Wolfe M (1994) Genetics of powdery mildew resistance in barley. Crit Rev Plant Sci 13:97–119
13. Thrall PH, Barrett LG, Dodds PN et al (2016) Epidemiological and evolutionary outcomes in gene-for-gene and matching allele models. Front Plant Sci 6:1084
14. Huang J, Si W, Deng Q et al (2014) Rapid evolution of avirulence genes in rice blast fungus Magnaporthe oryzae. BMC Genet 15: 1–10
15. Rep M, Kistler HC (2010) The genomic organization of plant pathogenicity in Fusarium species. Curr Opin Plant Biol 13:420–426
16. Milgroom MG, Peever TL (2003) Population biology of plant pathogens: the synthesis of plant disease epidemiology and population genetics. Plant Dis 87:608–617
17. Milgroom MG (2015) Population biology of plant pathogens: genetics, ecology, and evolution. APS Press, The American Phytopathological Society
18. Raffaele S, Kamoun S (2012) Genome evolution in filamentous plant pathogens: why bigger can be better. Nat Rev Microbiol 10: 417–430
19. Wallen RM, Perlin MH (2018) An overview of the function and maintenance of sexual reproduction in dikaryotic fungi. Front Microbiol 9: 503
20. Susi H, Burdon JJ, Thrall PH et al (2020) Genetic analysis reveals long-standing population differentiation and high diversity in the rust pathogen Melampsora lini. PLoS Pathog 16:e1008731
21. Zhan J (2016) Population genetics of plant pathogens. eLS 1-7
22. McDermott JM, McDonald BA (1993) Gene flow in plant pathosystems. Annu Rev Phytopathol 31:353–373
23. Stukenbrock EH (2016) Hybridization speeds up the emergence and evolution of a new pathogen species. Nat Genet 48:113–115
24. Feurtey A, Stevens DM, Stephan W et al (2019) Interspecific gene exchange introduces high genetic variability in crop pathogen. Genome Biol Evol 11:3095–3105
25. Currat M, Ruedi M, Petit RJ et al (2008) The hidden side of invasions: massive introgression by local genes. Evolution 62:1908–1920
26. Sillo F, Garbelotto M, Giordano L et al (2021) Genic introgression from an invasive exotic fungal forest pathogen increases the establishment potential of a sibling native pathogen. NeoBiota 65:109–136
27. Brasier CM (2001) Rapid evolution of introduced plant pathogens via interspecific hybridization: hybridization is leading to rapid evolution of Dutch elm disease and other fungal plant pathogens. Bioscience 51:123–133
28. Hurst LD (2002) The Ka/Ks ratio: diagnosing the form of sequence evolution. Trends Genet 18:486–486
29. Metzker ML (2010) Sequencing technologies—the next generation. Nat Rev Genet 11: 31–46

30. Grigoriev IV, Nikitin R, Haridas S et al (2014) MycoCosm portal: gearing up for 1000 fungal genomes. Nucleic Acids Res 42:699–704
31. Everhart S, Gambhir N, Stam R (2021) Population genomics of filamentous plant pathogens-a brief overview of research questions, approaches, and pitfalls. Phytopathology 111:12–22
32. Davey JW, Blaxter ML (2010) RADSeq: next-generation population genetics. Brief Funct Genomics 9:416–423
33. Hoban S, Gaggiotti O, ConGRESS Consortium et al (2013) Sample Planning Optimization Tool for conservation and population Genetics (SPOTG): a software for choosing the appropriate number of markers and samples. Methods Ecol Evol 4:299–303
34. Nadeem MA, Nawaz MA, Shahid MQ et al (2018) DNA molecular markers in plant breeding: current status and recent advancements in genomic selection and genome editing. Biotechnol Biotechnol Equip 32:261–285
35. Powell W, Morgante M, Andre C et al (1996) The comparison of RFLP, RAPD, AFLP and SSR (microsatellite) markers for germplasm analysis. Mol Breed 2:225–238
36. Schlötterer C (2004) The evolution of molecular markers-just a matter of fashion? Nat Rev Genet 5:63–69
37. Buschiazzo E, Gemmell NJ (2006) The rise, fall and renaissance of microsatellites in eukaryotic genomes. BioEssays 28:1040–1050
38. Gonthier P, Sillo F, Lagostina E et al (2015) Selection processes in simple sequence repeats suggest a correlation with their genomic location: insights from a fungal model system. BMC Genomics 16:1–12
39. Ivors K, Garbelotto M, Vries IDE et al (2006) Microsatellite markers identify three lineages of Phytophthora ramorum in US nurseries, yet single lineages in US forest and European nursery populations. Mol Ecol 15:1493–1505
40. Sillo F, Giordano L, Zampieri E et al (2017) HRM analysis provides insights on the reproduction mode and the population structure of Gnomoniopsis castaneae in Europe. Plant Pathol 66:293–303
41. Zane L, Bargelloni L, Patarnello T (2002) Strategies for microsatellite isolation: a review. Mol Ecol 11:1–16
42. Sillo F, Savino E, Giordano L et al (2016) Analysis of genotypic diversity provides a first glimpse on the patterns of spread of the wood decay fungus Perenniporla fraxinea in an urban park in northern Italy. J Plant Pathol 98: 617–624
43. Grünwald NJ, Everhart SE, Knaus BJ et al (2017) Best practices for population genetic analyses. Phytopathology 107:1000–1010
44. Luchi N, Pratesi N, Simi L et al (2011) High-resolution melting analysis: a new molecular approach for the early detection of *Diplodia pinea* in Austrian pine. Fungal Biol 115: 715–723
45. Langmead B, Salzberg SL (2012) Fast gapped-read alignment with Bowtie 2. Nat Methods 9: 357–359
46. Li H, Durbin R (2010) Fast and accurate long-read alignment with Burrows–Wheeler transform. Bioinformatics 26:589–595
47. Li H, Handsaker B, Wysoker A et al (2009) The sequence alignment/map format and SAMtools. Bioinformatics 25:2078–2079
48. Liu Q, Guo Y, Li J et al (2012) Steps to ensure accuracy in genotype and SNP calling from Illumina sequencing data. BMC Genomics 13:1–8
49. Li YL, Xue DX, Zhang BD et al (2018) An optimized approach for local de novo assembly of overlapping paired-end RAD reads from multiple individuals. R Soc Open Sci 5:171589
50. Eaton DA (2014) PyRAD: assembly of de novo RADseq loci for phylogenetic analyses. Bioinformatics 30:1844–1849
51. Nei M (1973) Analysis of gene diversity in subdivided populations. PNAS 70:3321–3323
52. Gaggiotti OE, Lange O, Rassmann K et al (1999) A comparison of two indirect methods for estimating average levels of gene flow using microsatellite data. Mol Ecol 8:1513–1520
53. Excoffier L, Smouse PE, Quattro JM (1992) Analysis of molecular variance inferred from metric distances among DNA haplotypes: application to human mitochondrial DNA restriction data. Genetics 131:479–491
54. Peakall ROD, Smouse PE (2006) GENALEX 6: genetic analysis in excel. Population genetic software for teaching and research. Mol Ecol Notes 6:288–295
55. Excoffier L, Laval G, Schneider S (2005) Arlequin (version 3.0): an integrated software package for population genetics data analysis. Evol Bioinformatics Online 1:47–50
56. Yeh FC, Yang RC, Boyle TB et al (1999) POPGENE version 1.32, the user-friendly shareware for population genetic analysis. In: Molecular biology and biotechnology centre. University of Alberta, Alberta
57. Lancaster AK, Single RM, Solberg OD et al (2007) PyPop update–a software pipeline for large scale multilocus population genomics. Tissue Antigens 69:192–197

58. Kamvar ZN, Tabima JF, Grünwald NJ (2014) Poppr: an R package for genetic analysis of populations with clonal, partially clonal, and/or sexual reproduction. PeerJ 2:e281
59. Slatkin M (1985) Gene flow in natural populations. Ann Rev Ecol Syst 16:393–430
60. Whitlock MC, Mccauley DE (1999) Indirect measures of gene flow and migration: FST≠ 1/(4Nm+ 1). Heredity 82:117–125
61. Tajima F (1989) Statistical method for testing the neutral mutation hypothesis by DNA polymorphism. Genetics 123:585–595
62. Nielsen R (2001) Statistical tests of selective neutrality in the age of genomics. Heredity 86:641–647
63. Gompert Z, Alex Buerkle C (2010) INTROGRESS: a software package for mapping components of isolation in hybrids. Mol Ecol Res 10:378–384
64. Nei M, Gojobori T (1986) Simple methods for estimating the numbers of synonymous and nonsynonymous nucleotide substitutions. Mol Biol Evol 3:418–426
65. Zheng Y, Janke A (2018) Gene flow analysis method, the D-statistic, is robust in a wide parameter space. BMC Bioinformatics 19:1–19
66. Korneliussen TS, Albrechtsen A, Nielsen R (2014) ANGSD: analysis of next generation sequencing data. BMC Bioinformatics 15:1–13
67. Kosman E, Leonard KJ (2005) Similarity coefficients for molecular markers in studies of genetic relationships between individuals for haploid, diploid, and polyploid species. Mol Ecol 14:415–424
68. Prevosti A, Ocana J, Alonso G (1975) Distances between populations of Drosophila subobscura, based on chromosome arrangement frequencies. Theor Appl Genet 45:231–241
69. Felsenstein J (2004) Inferring phylogenies. Sinauer associates, Sunderland
70. Jombart T, Devillard S, Balloux F (2010) Discriminant analysis of principal components: a new method for the analysis of genetically structured populations. BMC Genet 11:1–15
71. Jombart T (2008) Adegenet: a R package for the multivariate analysis of genetic markers. Bioinformatics 24:1403–1405
72. Pritchard JK, Stephens M, Donnelly P (2000) Inference of population structure using multilocus genotype data. Genetics 155:945–959
73. Earl DA (2012) STRUCTURE HARVESTER: a website and program for visualizing STRUCTURE output and implementing the Evanno method. Conserv Genet Res 4:359–361
74. Paradis E (2018) Analysis of haplotype networks: the randomized minimum spanning tree method. Methods Ecol Evol 9:1308–1317
75. Leigh JW, Bryant D (2015) popart: full-feature software for haplotype network construction. Methods Ecol Evol 6:1110–1116
76. Sillo F, Gianchino C, Giordano L et al (2018) Local epidemiology of the wood decay agent Laetiporus sulphureus in carob stands in Sicily. For Pathol 48:e12414
77. Hardy OJ, Vekemans X (2002) SPAGeDi: a versatile computer program to analyse spatial genetic structure at the individual or population levels. Mol Ecol Notes 2:618–620
78. Agapow PM, Burt A (2001) Indices of multilocus linkage disequilibrium. Mol Ecol Notes 1:101–102
79. Linde CC, Zala M, Ceccarelli S et al (2003) Further evidence for sexual reproduction in Rhynchosporium secalis based on distribution and frequency of mating-type alleles. Fungal Genet Biol 40:115–125
80. Huson DH (1998) SplitsTree: analyzing and visualizing evolutionary data. Bioinformatics 14:68–73
81. Ali S, Soubeyrand S, Gladieux P et al (2016) CLONCASE: estimation of sex frequency and effective population size by clonemate resampling in partially clonal organisms. Mol Ecol Res 16:845–861

Chapter 24

Microsatellite Genotyping in the Chestnut Blight Fungus *Cryphonectria parasitica*

Quirin Kupper and Simone Prospero

Abstract

This chapter describes the use of polymorphic microsatellite (simple sequence repeats, SSR) markers for genotyping isolates of *Cryphonectria parasitica*, the causal agent of chestnut blight. The SSR presented are particularly useful to characterize the genetic population structure of this invasive fungal pathogen, including invasion history (e.g., possible source population, introduced genotypes) and reproduction mode (sexual vs. asexual). Microsatellite markers can also be used to track fungal strains in laboratory and field experiments.

Key words Fungal pathogens, Biological invasion, Population genetics, Multiplex PCR

1 Introduction

The ascomycete *Cryphonectria parasitica* (Murr.) Barr, the causal agent of chestnut blight, is probably one of the best known invasive fungal pathogens in forests of Europe and North America [1]. This wound parasite infects stem and branches of susceptible chestnut (*Castanea*) species causing extensive bark lesions (so-called cankers, Fig. 1). To date, in many chestnut-growing regions of Europe, the disease is successfully controlled by a hyperparasitic mycovirus (*Cryphonectria hypovirus 1*, CHV1) that infects *C. parasitica* and reduces its virulence and sporulation [2]. Native to Eastern Asia, during the twentieth century *C. parasitica* has been spread around the globe most likely through infected chestnut plants. The invasion genetics of *C. parasitica* in Europe has been initially described using phenotypic markers (i.e., vegetative compatibility types; for a summary *see* [3]). In more recent years, European *C. parasitica* populations have been characterized using polymorphic microsatellite markers (e.g., [4–9]). These analyses revealed the invasion pathways of the pathogen, as well as main characteristics of the invasive populations, including their genetic diversity, the

Nicola Luchi (ed.), *Plant Pathology: Methods and Protocols*, Methods in Molecular Biology, vol. 2536,
https://doi.org/10.1007/978-1-0716-2517-0_24,

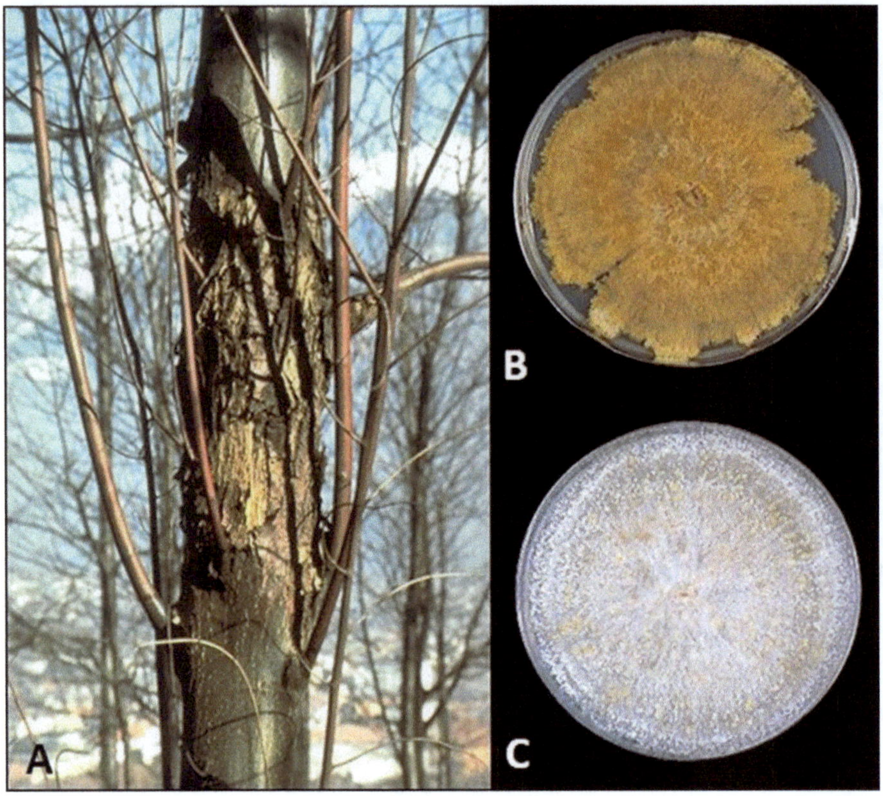

Fig. 1 (**a**) Bark lesion (canker) caused by a virulent strain of *Cryphonectria parasitica* on *Castanea sativa*; (**b**) Culture of a virulent (virus-free) strain of *C. parasitica* on PDA; (**c**) Culture of a hypovirulent (virus-infected) strain of *C. parasitica* on PDA (Photos: Phytopathology, WSL)

occurrence of sexual and/or asexual reproduction, and the genetic relatedness among founder genotypes. Microsatellite markers have also been used to track the spread of specific *C. parasitica* strains in laboratory and field experiments [10, 11].

Microsatellites, also known as simple sequence repeats (SSR), variable number tandem repeats (VNTR), and short tandem repeats (STR), are tandem repeats of 1–6 nucleotides found in the nuclear genomes of most taxa [12]. Thanks to their high level of polymorphism (multiallele loci), assumed neutrality, codominance, and abundance throughout genomes, microsatellites have proven to be valuable genetic markers [13, 14]. Moreover, microsatellites are easy to score and deliver robust and highly reproducible results. A microsatellite locus is typically surrounded by two flanking regions, whose sequences are generally species-specific. For this reason, microsatellites have to be developed de novo for each target species. The flanking regions allow to develop locus-specific forward and reverse primers (generally 18–25 bp) to amplify the microsatellites with PCR (Fig. 2). PCR amplification of a microsatellite locus usually results in relatively short (100–500 bp)

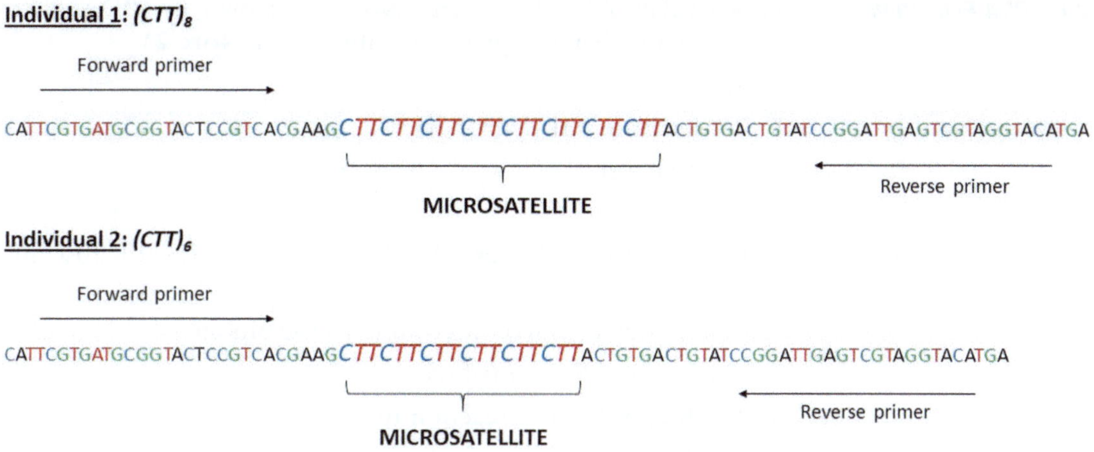

Fig. 2 Schematic representation of a microsatellite locus in a haploid organism (a diploid organism would have a complementary DNA sequence). The microsatellite, i.e., a tract of repetitive DNA in which certain DNA motifs (in this specific case, the trinucleotide motif *CTT*) are repeated typically 5–50 times, is surrounded by species-specific flanking regions. These regions are the target of forward and reverse primers. The length of the microsatellite (number of times that the motif is repeated) may vary among individuals

fragments. Microsatellite loci may be amplified both in single (each locus in a different reaction) and multiplex (several loci in the same reaction) PCRs and amplicons can be run and scored on agarose gels or, when using fluorescently labeled primer pairs, run in a capillary sequencer and scored with specific software.

Here, we provide a detailed protocol for genotyping *C. parasitica* at ten microsatellite loci that were specifically developed for this species (CPG3, CPG4, CPE1, CPE3, CPE4, CPE5, and CPE8, [15]; CPG6 and CPG14, [16]; I07-650, [17]). The ten loci are amplified in a total of three multiplex reactions (CpMix1 and CpMix2 including each three loci, and CpMix3 including four loci). Since in each locus-specific primer pair the forward primer is labeled with a fluorescent dye (either FAM-blue or HEX-green) at the 5′-end, amplicons are run on an ABI 3730 sequencer (Applied Biosystems, Carlsbad, CA), using an internal size standard (ROX-400). Allele sizes are then scored with the program GeneMapper (Applied Biosystems).

2 Material

Strictly separate between pre- and post-PCR applications to avoid contamination with previous PCR products. Avoid freeze-thaw cycles of DNA samples. Store DNA at 4 °C for short storage times (2–4 weeks) and at −20 °C only for long-term storage (1 month plus).

2.1 DNA Extraction

1. 5–12-day-old cultures (*see* **Note 1**) grown on PDA plates covered with a cellophane membrane (*see* **Note 2**).
2. Scalpel.
3. 70% ethanol (ethyl alcohol).
4. Freeze dryer.
5. Bead mill.
6. Pipettes for volumes: 1–10 μL; 5–50 μL; 20–200 μL; 100–1000 μL.
7. Pipette tips with filter according to the pipettes.
8. Vortex and Mini centrifuge.
9. "PCR clean" 2-mL reaction tubes.
10. "PCR clean" 1.5-mL reaction tubes.
11. Centrifuge for 2-mL reaction tubes, up to 20,000 × g.
12. Water bath, heating block or incubator, capable of heating up to 65 °C.
13. DNA extraction kits:
    (a) Column-based: DNeasy Plant Mini Kit (Qiagen, Hilden, Germany).
    (b) Magnetic bead-based: sbeadex™ Magnetic Bead Kit (LGC Biosearch Technologies, Hoddesdon, United Kingdom). An extraction kit based on magnetic beads requires an automated extraction instrument such as the KingFisher Flex System (Thermo Fisher Scientific, Waltham, Massachusetts, USA).

2.2 PCR

1. Specific primer sets for the ten SSR loci (Table 1).
2. Type-it Microsatellite PCR Kit (Qiagen, Hilden, Germany).
3. Water for molecular biology (DNase-free, RNase-free).
4. PCR plates.
5. Cap-strips for PCR plates or PCR sealing foil.
6. "PCR clean" 1.5-mL reaction tubes.
7. Pipettes for volumes: 1–10 μL; 5–50 μL; 20–200 μL.
8. Pipette tips with filter according to the pipettes.
9. Vortex and Mini centrifuge.
10. Plate centrifuge.
11. Thermo cycler with a 96-well block.

2.3 Fragment Analysis

1. Hi-Di™ Formamide (Applied Biosystems, Foster City, California, USA).
2. GeneScan™ 400HD ROX™ dye Size Standard (Applied Biosystems, Foster City, California, USA).

Table 1
Primer for multiplex 1, 2, and 3

Multiplex	Primer name[a]	Primer sequence 5′-3′	Fluorescent dye at 5′ end
CpMix1			
	CPE3-F	CAACTCGTCACCCACCTTG	FAM
	CPE3-R	ATAGGCATTCTCCGCTCTTG	
	CPE4-F	ACCTGAACCGACCTGGCTTTG	HEX
	CPE4-R	CCTTGTTGTGGTGGTGGAG	
	CPG6-F	ATCATCACGACGCAATGGTA	FAM
	CPG6-R	TCCGGGCATTCAGCAMAT	
CpMix2			
	CPE8-F	GCGAGCAAGCGTGATTCC	FAM
	CPE8-R	GCTCGTCCTGGTCCTCCT	
	CPG3-F	CGTAAGGCAGAGGCAGAGAC	HEX
	CPG3-R	TCCCTATGCCCAAGACACTC	
	I07-650-F	CATGCGAGAAATGCAGGAGTGTTG	FAM
	I07-650-R	GGGCTCCAGGATATCGAAGACATT	
CpMix3			
	CPG4-F	ATGCTCCGAAGGTTTGTCAC	HEX
	CPG4-R	CAACCGCAATCTGTTTCCTC	
	CPE1-F	GTCTCACCACACATCGCAAG	FAM
	CPE1-R	TCATCACGACAAAAGAAGACG	
	CPG14-F	TTCTGAAGGTGGTTGTGGTG	FAM
	CPG14-R	GGTCCGAACCATCAAAAGAC	
	CPE5-F	TGTCAACAACGGTCAACACC	HEX
	CPE5-R	CGGAGAGGAGAACTCTGAGAC	

[a] *F* forward primer, *R* reverse primer

3. Water for molecular biology (DNase-free, RNase-free).
4. Semiskirted PCR plate, compatible with the Applied Biosystems Genetic Analyzers.
5. Septa for semiskirted PCR plate, compatible with the Applied Biosystems Genetic Analyzers.
6. Pipettes for volumes: 1–10 μL.
7. Pipette tips according to the pipettes.

8. Genetic Analyzer, 3130 or 3500 series (Applied Biosystems, Foster City, California, USA).
9. GeneMapper™ Software 5 (Applied Biosystems, Foster City, California, USA).

3 Methods

3.1 DNA Extraction

For SSR genotyping, purified DNA of at least 1 ng/μL is required. We recommend to grow the axenic fungal cultures on PDA covered with a cellophane membrane (*see* **Note 1**), and to extract the DNA from the mycelium with a commercially available kit (*see* Subheading 2.1 DNA Extraction). Most silica column or magnet bead-based extraction methods produce satisfactory results.

1. Harvest the mycelium (piece of approx. 2 cm × 2 cm) from each plate (culture should be 5–12-day-old) (*see* **Note 2**) into a PCR clean 2 mL reaction tube using a sterile scalpel.
2. Freeze dry the harvested mycelium over night at 0.04 mbar.
3. Extract DNA according to the respective manufacturer's protocol for the used extraction method.

3.2 PCR

1. Dilute DNA 1:10 with molecular grade water.
2. Prepare three different PCR master mixes according to Tables 2, 3, and 4.
3. Prepare 9 μL master mix per reaction in a PCR plate.
4. Add 1 μL of diluted DNA for a total volume of 10 μL (*see* **Note 3**).
5. Vortex to mix and shortly spin down the plate using a plate centrifuge.
6. Run the PCR on a thermal cycler with the following parameters.
   (a) 5 min at 95 °C
   (b) 30 s at 95 °C
   (c) 1 min 30 s at 58 °C
   (d) 1 min 30 s at 72 °C
   (e) 30 min at 72 °C
   (f) 30 min at 60 °C
   Cycle steps (b) to (d) for 30 times.
7. Temporarily store the PCR reaction at 4 °C up to 1 week, for longer storage keep the reaction at −20 °C for up to 1 month.

Table 2
PCR reaction, CpMix1

CpMix1 (CPE3, CPE4, CPG6)		
Component	**μL per reaction**	**Final concentration**
2× Type-it Multiplex MM	5	1×
CPE3-F (10 μM)	0.1	0.1 μM
CPE3-R (10 μM)	0.1	0.1 μM
CPE4-F (10 μM)	0.1	0.1 μM
CPE4-R (10 μM)	0.1	0.1 μM
CPG6-F (10 μM)	0.1	0.1 μM
CPG6-R (10 μM)	0.1	0.1 μM
Water	3.4	–
Total	9	–

Table 3
PCR reaction, CpMix2

CpMix2 (CPE8, CPG3, I07-650)		
Component	**μL per reaction**	**Final concentration**
2× Type-it Multiplex MM	5	1×
CPE8-F (10 μM)	0.1	0.1 μM
CPE8-R (10 μM)	0.1	0.1 μM
CPG3-F (10 μM)	0.1	0.1 μM
CPG3-R (10 μM)	0.1	0.1 μM
I07-650-F (10 μM)	0.1	0.1 μM
I07-650-R (10 μM)	0.1	0.1 μM
Water	3.4	–
Total	9	–

3.3 Fragment Analysis

1. Dilute 10 μL of ROX400HD Size Standard in 1 mL of HiDi™ Formamide.
2. Mix well and short spin to remove residues on the cap of the tube.
3. Dilute the PCR product 1:2 with molecular grade water (*see* **Note 4**).
4. Transfer 10 μL of the ROX400-HiDi™ mixture per reaction on to the semiskirted PCR plate used for the Genetic Analyzer.

Table 4
PCR reaction, CpMix3

CpMix3 (CPG4, CPE1, CPG14, CPE5)		
Component	**μL per reaction**	**Final concentration**
2× Type-it Multiplex MM	5	1×
CPG4-F (10 μM)	0.1	0.1 μM
CPG4-R (10 μM)	0.1	0.1 μM
CPE1-F (10 μM)	0.1	0.1 μM
CPE1-R (10 μM)	0.1	0.1 μM
CPG14-F (10 μM)	0.1	0.1 μM
CPG14-R (10 μM)	0.1	0.1 μM
CPE5-F	0.1	0.1 μM
CPE5-R	0.1	0.1 μM
Water	3.4	–
Total	9	–

5. Add 1 μL of diluted PCR product per reaction.
6. Cover plate with septa and centrifuge with a short spin.
7. Analyze the plate on a Genetic Analyzer using the following settings:
   (a) Type: Fragment
   (b) Dye set: D
   (c) Injection time: 5 s
   (d) Run time: 19 min
   (e) Polymer: POP7

3.4 Data Analysis

1. Raw data are analyzed using GeneMapper™ Software 5. For each microsatellite locus, bin sets can be created according to expected fragment (allele) sizes (*see* Fig. 3).
2. Fragments at each SSR locus are visually checked for correct call and size assignment (*see* **Note 5**).
3. Once all samples have been analyzed, the Genotype Table from GeneMapper can be exported to an Excel file.
4. In Excel, fragments (alleles) at each locus can be concatenated using the function *Concat (text1; . . .)* and translated into multilocus genotypes (i.e., a unique combination of alleles across two or more loci).
5. Multilocus genotype data can be analyzed using specific population genetic software, e.g., R package poppr [18].

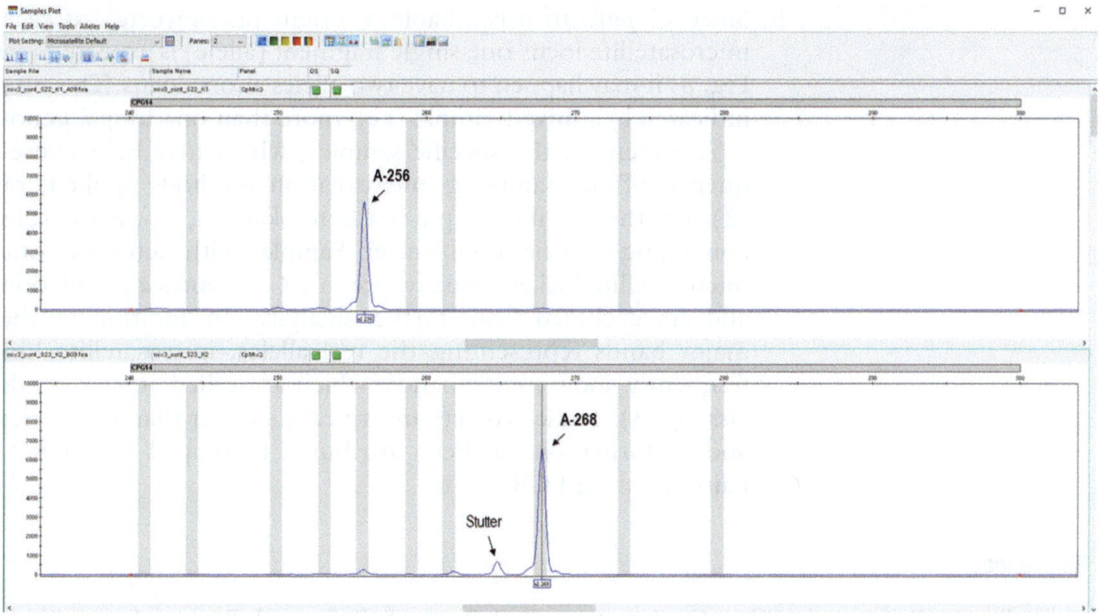

Fig. 3 Example of a polymorphic microsatellite locus (CPG14) in *Cryphonectria parasitica*. At this locus, the individual above has allele A-256, whereas the individual below has allele A-268. The horizontal gray bar indicates the range in which alleles are expected at this locus based on previous analyses. The vertical gray bars represent alleles already identified in previous analyses

4 Notes

1. For DNA extraction, *C. parasitica* cultures should be grown at 25 °C under light (if possible, a light intensity of 3330 lx and a 14 L:10 D photoperiod) for 5–12 days. Exposure to high light intensity promotes asexual sporulation [19, 20], increasing the amount of DNA.
2. Growing *C. parasitica* cultures on PDA (Potato Dextrose Agar, 39 g PDA per liter water) covered with a sterile cellophane membrane (autoclave cellophane membrane twice in water to remove potential toxic substances, discard the water after the first autoclaving), prevents agar residues in the DNA extract. Agar residues in the DNA extract can inhibit PCR reaction.
3. You may adjust the volumes of water and DNA to your needs.
4. The dilution of the PCR product might need to be adjusted depending on the Genetic Analyzer used. We noticed that due to a higher sensitivity on the 3500 series Genetic Analyzer (Applied Biosystems, Foster City, California, USA), a higher dilution of at least 1:5 is needed.

5. Since *C. parasitica* is a haploid fungus (ascomycete), at each microsatellite locus one single fragment (allele) is expected (*see* Fig. 3) it may happen to have two alleles at one locus. This may indicate (1) a mixed sample (i.e., more than one fungal genotype present in the specific sample), which may be a consequence of the sampling and isolation methods applied, or (2) that the specific sample is a heterokaryon, i.e., it has cells containing two or more nuclei. Samples with heterokaryotic microsatellite loci are usually rare in a *C. parasitica* population and are excluded from further analyses. In addition to the major bands representing the real alleles, microsatellite loci frequently show minor bands, which are called stutter bands (*see* Fig. 3). These artifacts are typically shorter than real alleles and its formation has been attributed to slipped-strand mispairing during PCR.

References

1. Rigling D, Prospero S (2018) *Cryphonectria parasitica*, the causal agent of chestnut blight: invasion history, population biology and disease control. Mol Plant Pathol 19:7–20
2. Rigling D, Robin C, Prospero S (2021) Mycovirus-mediated biological control. Encyclopaedia of virology, 4th edn. Elsevier
3. Robin C, Heiniger U (2001) Chestnut blight in Europe: diversity of *Cryphonectria parasitica*, hypovirulence and biocontrol. For Snow Landsc Res 76:361–367
4. Milgroom MG, Sotirovski K, Spica D et al (2008) Clonal population structure of the chestnut blight fungus in expanding ranges in southeastern Europe. Mol Ecol 17:4446–4458
5. Dutech C, Barres B, Bridier J et al (2012) The chestnut blight fungus world tour: successive introduction events from diverse origins in an invasive plant fungal pathogen. Mol Ecol 21: 3931–3946
6. Ježić M, Krstin L, Rigling D et al (2012) High diversity in populations of the introduced plant pathogen, *Cryphonectria parasitica*, due to encounters of genetically divergent genotypes. Mol Ecol 21:87–99
7. Prospero S, Rigling D (2012) Invasion genetics of the chestnut blight fungus *Cryphonectria parasitica* in Switzerland. Phytopathology 102:73–82
8. Prospero S, Lutz A, Tavadze B et al (2013) Discovery of a new gene pool and a high genetic diversity of the chestnut blight fungus *Cryphonectria parasitica* in Caucasian Georgia. Infect Genet Evol 20:131–139
9. Peters FS, Bußkamp J, Prospero S et al (2014) Genetic diversification of the chestnut blight fungus *Cryphonectria parasitica* and its associated hypovirus in Germany. Fun Biol 118: 193–210
10. Suzuki N, Cornejo C, Aulia A et al (2021) In-tree behavior of diverse viruses harbored in the chestnut blight fungus, *Cryphonectria parasitica*. J Virol 95:01962–01920
11. Meyer JB, Chalmandrier L, Fässler F et al (2019) Role of fresh dead wood in the epidemiology and the biological control of an invasive fungal tree pathogen. Plant Dis 103:430–438
12. Selkoe KA, Toonen RJ (2006) Microsatellites for ecologists: a practical guide to using and evaluating microsatellite markers. Ecol Lett 9: 615–629
13. Jarne P, Lagoda PJL (1996) Microsatellites, from molecules to populations and back. Trends Ecol Evol 11:424–429
14. Carneiro Vieira ML, Santini L, Lima Diniz A et al (2016) Microsatellite markers: what they mean and why they are so useful. Genet Mol Biol 39:312–328
15. Breuillin F, Dutech C, Robin C (2006) Genetic diversity of the chestnut blight fungus *Cryphonectria parasitica* in four French populations assessed by microsatellite markers. Mycol Res 110:288–296
16. Kubisiak TL, Dutech C, Milgroom MG (2007) Fifty-three polymorphic microsatellite loci in the chestnut blight fungus, *Cryphonectria parasitica*. Mol Ecol Notes 7:428–432

17. Davis JE, Kubisiak TL, Milgroom MG (2005) Polymorphic sequence-characterized codominant loci in the chestnut blight fungus, *Cryphonectria parasitica*. Mol Ecol Notes 5:195–197

18. Kamvar ZN, Tabima JF, Grünwald NJ (2014) Poppr: an R package for genetic analysis of populations with clonal, partially clonal, and/or sexual reproduction. PeerJ 2:e281

19. Hillman BI, Shapira R, Nuss DL (1990) Hypovirulence-associated suppression of host functions in *Cryphonectria parasitica* can be partially relieved by high light intensity. Phytopathology 80:950–956

20. Bryner SF, Rigling D (2011) Temperature-dependent genotype-by-genotype interaction between a pathogenic fungus and its hyperparasitic virus. Am Nat 177:65–74

Chapter 25

A Multiplex PCR Approach to Determine Vegetative Incompatibility Genotypes and Mating Type in *Cryphonectria parasitica*

Quirin Kupper and Carolina Cornejo

Abstract

This chapter presents a genotyping assay that uses DNA isolated from axenic cultures of *Cryphonectria parasitica*, which discriminates the six known diallelic *vic* loci and the two mating idiomorphs (*MAT* gene) based on (i) modified primer, labeled with a fluorescent dye, (ii) multiplex polymerase chain reaction (multiplex-PCR), and (iii) capillary electrophoresis system. Alternatively, we show that the same primer set is suitable for conventional PCR of each *vic* locus and *MAT* gene using nonmodified primer and agarose gel electrophoresis.

Key words Chestnut blight, Genetic diversity, Diagnostics, Biocontrol, Fluorochrome, Multiplex-PCR, Duplex-PCR

1 Introduction

The chestnut blight disease is being successfully controlled by infecting the virulent strain of *Cryphonectria parasitica* with the hyperparasitic mycovirus CHV1 (*Cryphonectria hypovirus 1*) in order to induce hypovirulence in the pathogen. *Cryphonectria parasitica* strains infected by CHV1 exhibit reduced virulence and sporulation capacity compared to isogenic noninfected strains [1, 2]. The artificial application of the hypovirus CHV1, however, requires that the vegetative compatibility (vc) types in the target *C. parasitica* population be determined beforehand. This is necessary because the hypovirus transmission is restricted between *C. parasitica* strains that have different vc types [3]. To date, six diallelic vegetative incompatibility (*vic*) loci have been identified, defining 64 different *vic* genotypes—also called EU vc types [4]. Conventional vc type analysis uses fungal cultures, isolated from infected material, which are subsequently cocultivated with tester strains of known vc type. But this technique is time

Nicola Luchi (ed.), *Plant Pathology: Methods and Protocols*, Methods in Molecular Biology, vol. 2536,
https://doi.org/10.1007/978-1-0716-2517-0_25,

consuming and not always reliable. Additionally, CHV1 is transmitted into asexual but not into sexual spores, and sexual reproduction may negatively impact the CHV1 transmission by increasing the vc type diversity through genetic recombination of polymorphic *vic* loci. For this reason, it is important to verify the mating type of field material previous to the application of CHV1. Testing the mating type is done by DNA extraction from isolates and conventional PCR. In order to reduce the workload and to optimize processes in the lab, the basic idea of this assay is to use DNA of *C. parasitica* isolates to analyze the *vic* loci and the *MAT* genes simultaneously and in high specimen number if needed (Fig. 1). To achieve this objective, we developed highly specific primer for each allele at the six known diallelic *vic* loci and both idiomorphs of the *MAT* genes of *C. parasitica* and labeled them with a fluorescent dye for genotyping by a capillary DNA analyzer.

Genotyping with fluorescent markers is a widely used method in molecular biology and it has a firm place in forensic, diagnostic, and scientific applications. This method is primarily used for the detection of SSR (Simple Sequence Repeats) or the so-called microsatellite loci for DNA profiling, paternity tests or population

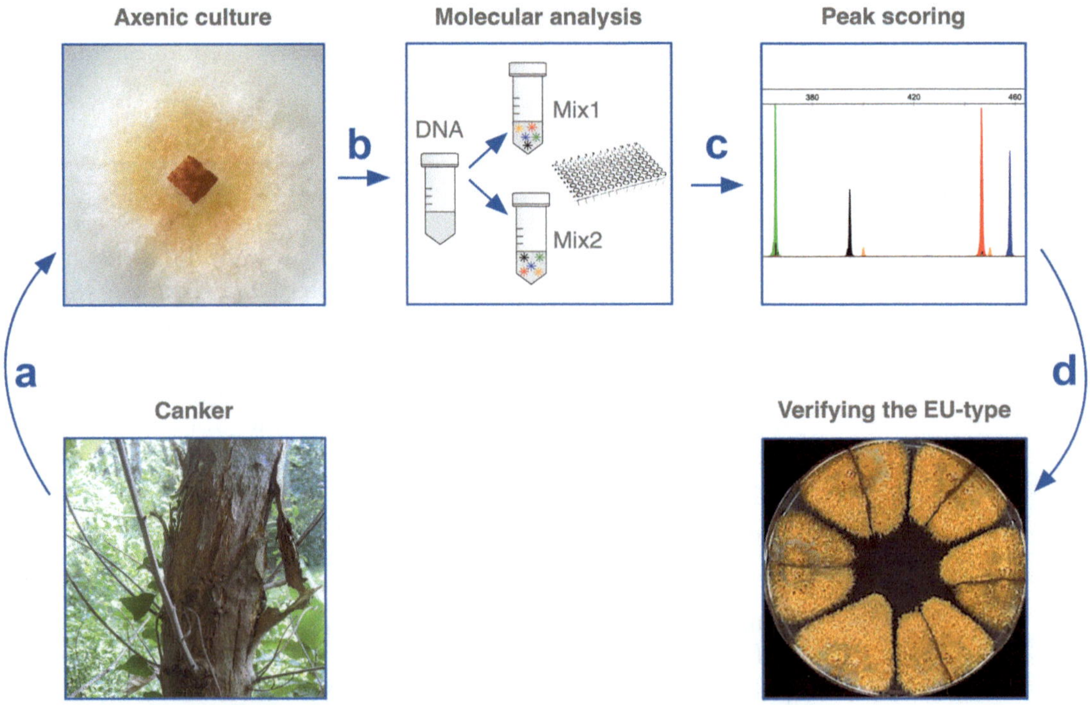

Fig. 1 Overview of the procedure to analyze the *vic* and *MAT* genes in *Cryphonectria parasitica* using the genotyping assay. (**a**) Isolation of *C. parasitica* from chestnut blight cankers. (**b**) DNA isolation from axenic culture, multiplex-PCRs, and detection of the fluorescent fragments in a genetic analyzer. (**c**) Automated peak scoring by a size-calling computer program. (**d**) A pairing test can be applied as a second chain of verification of the resulting EU vc type

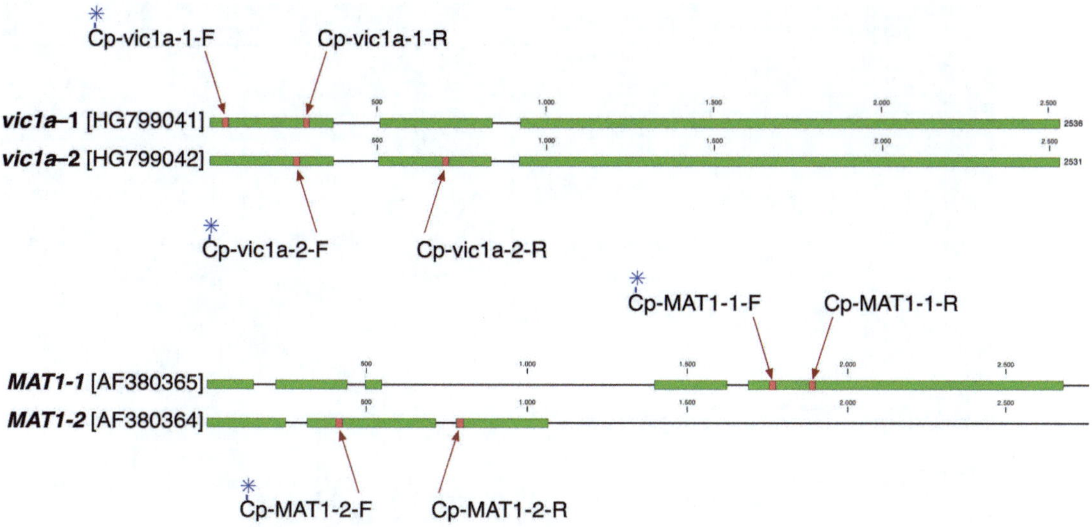

Fig. 2 Map of partial sequences of the *vic1a* locus and *MAT* gene of *Cryphonectria parasitica*, showing the process of primer design. Numbers in square brackets designate the GenBank accession of genes. Forward and reverse primers are pointed by red arrows. Blue asterisks highlight fluorescent modification at the 5′-end of forward primer. Black line = intron. Green bar = exon region. Red box = primer position

genetics studies. One primer of the primer pair is labeled with a fluorochrome (or fluorescent dye) that can re-emit light upon light excitation, convenient for automated DNA analyzer. The fluorochrome remains stable and safe during the PCR reaction and do not alter the amplicon movement in an electrophoretic capillary system. Instead of modifying primer, the use of a fluorescent tag or a universal sequence, such as M13 primer, is also a popular, suitable option [5, 6].

To obtain highly specific primer, we followed a strict development process (Fig. 2). First, we annotated exons and introns at each locus [for gene data, *see* 7–11]. Then we targeted, on the one hand, exon regions, which are more conserved than introns, to minimize allelic dropout, but, on the other hand, such areas within exons with the greatest sequence divergence between the two alleles at each locus.

Several drop-out effects are known, e.g., failure to amplify one of an individual's two alleles (allelic dropout), polymerase error rendering a "false" allele, or primer hybridization error due to sequence polymorphism within the DNA primer-binding sequence, which is more likely to happen in variable intron regions [12, 13]. Finally, the amplicon lengths had to differ sufficiently in order to distinguish fluorescence peaks from each other in electropherograms (Fig. 3), but move in a range between 100 and 500 bp.

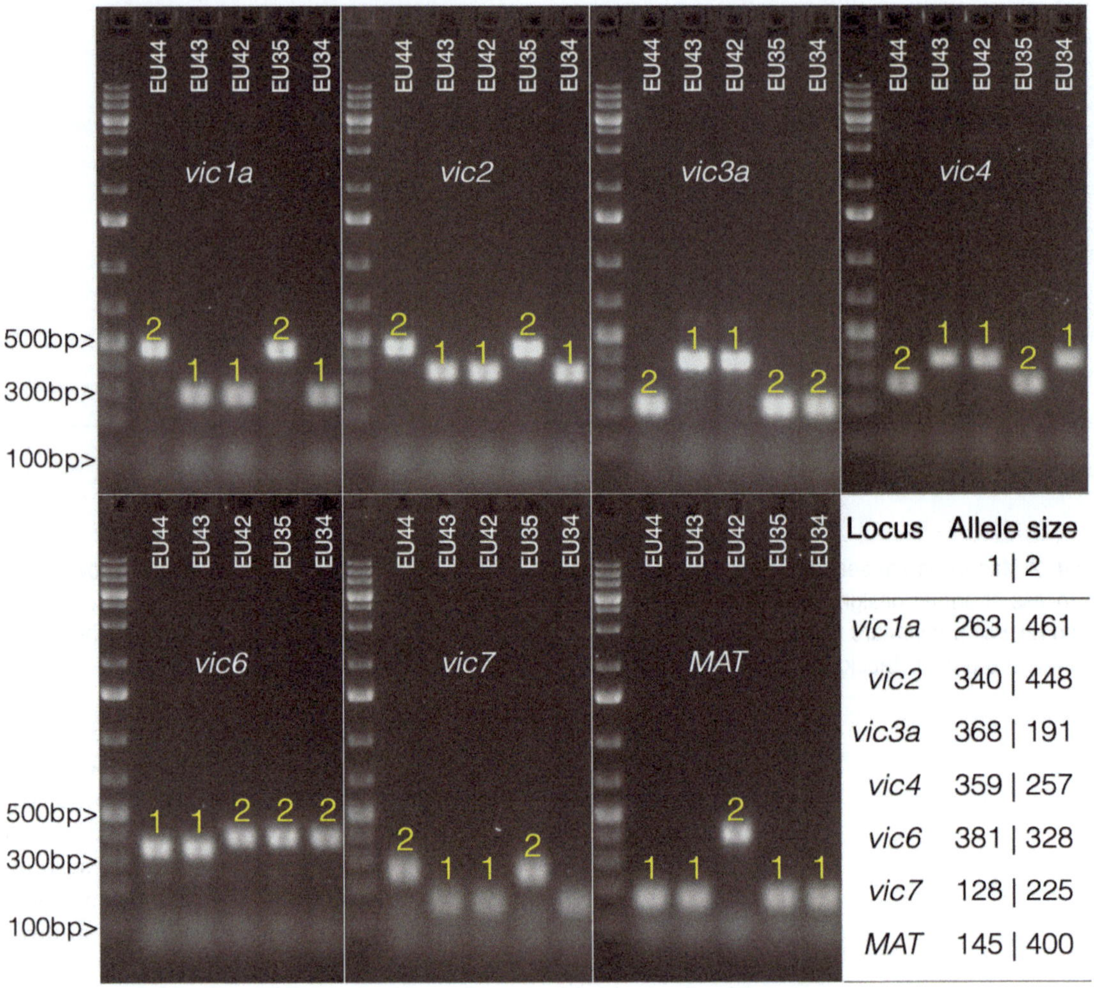

Fig. 3 About 1.5% agarose gel of a *vic*-loci and *MAT* genes duplex-PCR applied to DNA of *Cryphonectria parasitica*. The same primer set (Table 1) can be used for duplex- or monoplex-PCR, alternatively to the genotyping assay based on fluorescent primer

2 Materials

Strictly separate between pre- and post-PCR applications to avoid contamination with previous PCR products. Avoid freeze-thaw cycles of DNA samples and store DNA at 4 °C for short storage times (2–4 weeks) and at −20 °C only for long-term storage (1 month plus).

2.1 DNA Extraction

1. *Cryphonectria parasitica* cultures, grown on cellophane coated PDA plates (*see* **Note 1**), 5–12-days old (*see* **Note 2**).
2. Scalpel.
3. 70% Ethanol.

4. Freeze dryer (e.g., Martin Christ freeze dryer Alpha 1-2 LDplus, or any other standard freeze dryer).
5. Bead mill (e.g., TissueLyser II, Qiagen), or mortar and liquid nitrogen.
6. Pipettes for volumes: 1–10 μL; 5–50 μL; 20–200 μL; 100–1000 μL.
7. Pipette filter tips according to the pipettes.
8. Vortex and Mini centrifuge.
9. "PCR clean" 2-mL reaction tubes.
10. "PCR clean" 1.5-mL reaction tubes.
11. Centrifuge for 2-mL reaction tubes, up to 20,000 × g.
12. Water bath, heating block or incubator, capable of heating to 65 °C.
13. DNA purification kits, e.g.,
    (a) Column-based: Qiagen Plant Mini Kit (Qiagen, Hilden, Germany).
    (b) Magnetic bead-based: LGC sbeadex purification Kits (LGC Biosearch Technologies, Hoddesdon, United Kingdom). A purification kit based on magnetic beads also requires an automated extraction instrument such as the KingFisher Flex System (Thermo Fisher Scientific, Waltham, Massachusetts, USA).

2.2 PCR

1. Order primer set according to Table 1.
2. Prepare 10× primer mix for multiplex 1 according to Table 2 (*see* **Note 3**).
3. Prepare 10× primer mix for multiplex 2 according to Table 3.
4. Type-it Microsatellite PCR Kit, Cat. No. 206243/206246 (Qiagen, Hilden, Germany).
5. Water for molecular biology (DNase- and RNase-free).
6. PCR plates.
7. Cap-Strips for PCR plates or PCR sealing foil.
8. "PCR clean" 1.5-mL reaction tubes.
9. Pipettes for volumes: 1–10 μL; 5–50 μL; 20–200 μL.
10. Pipette filter tips according to the pipettes.
11. Vortex and mini centrifuge.
12. Plate centrifuge.
13. Thermo cycler with a 96-well block.

Table 1
Primer for multiplex 1 and 2

Gene	Allele	Primer name	Primer sequence 5′-3′	Fluorescent dye at 5′ end
Multiplex 1				
vic1a	1	Cp-vic1a-1	F: ATAGCAAAAGCCACAACCAC	FAM
			R: AGCCAACATATACACTCCG	–
	2	Cp-vic1a-2	F: TGCGACAGTCAAGGAAGATG	FAM
			R: ACGAGTGTCCTGCAAATTAA	–
vic2	1	Cp-vic2–1	F: TGCGACGGATAAGGTGACTA	PET
			R: TCGGTTGGCCTCGATATTG	–
	2	Cp-vic2–2	F: TATACGGCAAAAGGGAGGG	PET
			R: CTGGGAAGTCTTTGAGCAT	–
vic3a	1	Cp-vic3a-1	F: CTGCAGACAACAACCGGTC	VIC
			R: GGGCACGACAAATGGCACT	–
	2	Cp-vic3a-2	F: ACCGAAGACCACATGCGCAA	VIC
			R: AGGTTCCCCACCAGCGTCA	–
MAT1–1-2	1	Cp-MAT1–1	F: ACATCAGTCGTGGAAATGAAAA	NED
			R: ACATCACAAGTCGGCTCCCA	–
MAT1–2-1	2	Cp-MAT1–2	F: AGGAATGTGATCTGTGGTGTTCT	NED
			R: AATTTACTTTCACCGCGATTATG	–
Multiplex 2				
vic4	1	Cp-vic4–1	F: CTTTCTTCTCCAACCTGCT	PET
			R: TTCTTCTTCCGGCTTGCTT	–
	2	Cp-vic4–2	F: GCCGAGTCATCCAAAGCAG	PET
			R: GGGAGGCAGAATGGAAGAAA	–
vic6	1	Cp-vic6–1	F: CTTTTGCCTGCGAGGTCTTTCC	VIC
			R: TCAGCTCTTCTTCCTGTACGTG	–
	2	Cp-vic6–2	F: ATCCATGACCCTCTGCCAA	VIC
			R: CCAGCCACCGACTACTTTG	–
vic7	1	Cp-vic7–1	F: TTGTCGAAAGGAACGCGGCT	FAM
			R: TTCTTCCTCTACCATCATC TGTGG	–
	2	Cp-vic7–2	F: ACTTCAAACACGAATTCTGGC	FAM
			R: ACCATCTCCGTCTCGCATTCT	–

F forward primer, *R* reverse primer

Table 2
Primer mix 1. Total volume 10× primer mix 1: 500 μL

Primer	Initial concentration [μM]	Volume [μL]/500 μL-Mix	10× Primer concentration [μM]
Water	–	32.70	–
Cp-vic1a-1-F	10	31.50	0.630
Cp-vic1a-1-R	10	31.50	0.630
Cp-vic1a-2-F	10	31.50	0.630
Cp-vic1a-2-R	10	31.50	0.630
Cp-vic2-1-F	10	31.50	0.630
Cp-vic2-1-R	10	31.50	0.630
Cp-vic2-2-F	10	31.50	0.630
Cp-vic2-2-R	10	31.50	0.630
Cp-vic3a-1-F	10	15.75	0.315
Cp-vic3a-1-R	10	15.75	0.315
Cp-vic3a-2-F	10	23.65	0.473
Cp-vic3a-2-R	10	23.65	0.473
Cp-MAT1-1-F	10	31.50	0.630
Cp-MAT1-1-R	10	31.50	0.630
Cp-MAT1-2-F	10	36.75	0.735
Cp-MAT1-2-R	10	36.75	0.735

2.3 Fragment Length Analysis

1. Hi-Di™ Formamide, Cat. No. 4401457 (Applied Biosystems, Foster City, California, USA).
2. GeneScan™ 500 LIZ™ Dye Size Standard, Cat. No. 4322682 (Applied Biosystems, Foster City, California, USA).
3. Water for molecular biology (DNase- and RNase-free).
4. Semiskirted PCR plate, compatible with the Applied Biosystems Genetic Analyzers.
5. Septa for semiskirted PCR plate, compatible with the Applied Biosystems Genetic Analyzers.
6. Pipettes for volumes: 1–10 μL.
7. Pipette tips according to the pipettes.

Table 3
Primer mix 2. Total volume 10× primer mix 2: 500 μL

Primer	Initial concentration [μM]	Volume [μL]/500 μL-Mix	10× Primer concentration [μM]
Cp-vic4-1-F	10	51.00	1.020
Cp- vic4-1-R	10	51.00	1.020
Cp- vic4-2-F	10	48.00	0.960
Cp- vic4-2-R	10	48.00	0.960
Cp-vic6-1-F	10	33.00	0.660
Cp-vic6-1-R	10	33.00	0.660
Cp-vic6-2-F	10	30.00	0.600
Cp-vic6-2-R	10	30.00	0.600
Cp-vic7-1-F	10	40.00	0.800
Cp-vic7-1-R	10	40.00	0.800
Cp-vic7-2-F	10	48.00	0.960
Cp-vic7-2-R	10	48.00	0.960

8. Genetic Analyzer, 3130 or 3500 series (Applied Biosystems, Foster City, California, USA).
9. GeneMapper™ Software 5 (Applied Biosystems, Foster City, California, USA).

3 Methods

3.1 DNA Extraction

Purified DNA of at least 1 ng/μL is required. We recommend growing the axenic fungal cultures on PDA with cellophane membranes (*see* **Note 1**), and to extract the DNA from mycelium with a commercially available purification kit such as Qiagen Plant Mini Kit, or LGC sbeadex purification kits. Most silica column- or magnet bead-based purification methods produce satisfactory results.

1. Scrape 5–12-days old mycelium from the cellophane surface, using a sterilized scalpel, and introduce it into a "PCR clean" 2-mL reaction tube. An area of 2 cm by 2 cm mycelium is sufficient for DNA purification.
2. Freeze-dry harvested mycelium over night at 0.04 mbar and grind the dried mycelium using a bead mill. Alternatively, the mycelium can be ground using a mortar and liquid nitrogen.
3. Purify DNA according to the respective manufacturer documentation of the selected purification method.

3.2 PCR

1. Dilute DNA 1:5 with molecular grade water (*see* **Note 4**).
2. For each primer mix, prepare a master mix containing the following volumes per reaction (*see* **Note 5**):
   (a) 5 μL 2× Type-it Multiplex MM
   (b) 1 μL 10× primer mix 1, or primer mix 2
   (c) μL molecular grade water.
3. Distribute 8.75-μL master mix per reaction in a PCR plate.
4. Add 1.25 μL of diluted DNA for a total volume of 10 μL.
5. Use the Vortex to mix and shortly spin down the plate using a plate centrifuge.
6. Run the PCR on a thermal cycler with the following parameters:
   (a) 5 min at 95 °C
   (b) 30 s at 95 °C
   (c) 1 min 30s at 60 °C
   (d) 30 s at 72 °C
   (e) 45 min at 72 °C

   Cycle steps (b) to (d) for 25 times.
7. Temporarily store the PCR reaction at 4 °C up to 1 week, for longer storage keep the reaction at −20 °C for up to 1 month.

3.3 Fragment Length Analysis

1. Dilute 10 μL of LIZ500 Size Standard in 1 mL of HiDi™ Formamide.
2. Mix well and short spin to remove residues on the cap of the tube.
3. Dilute the PCR product 1:2 with molecular grade water (*see* **Note 6**).
4. Transfer 9 μL of the LIZ500-HiDi™ mixture per reaction into the semiskirted PCR plate used for the Genetic Analyzer.
5. Add 1 μL of diluted PCR product per reaction.
6. Cover plate with septa and centrifuge with a short spin.
7. Analyze the plate on a Genetic Analyzer using the following settings:
   (a) Type: Fragment
   (b) Dye set: G5
   (c) Injection time: 5 s
   (d) Run time: 20 min
   (e) Polymer: POP7.

3.4 Data Analysis

1. Analyze data using GeneMapper™ Software 5.
2. Bin sets can be created according to expected fragment sizes in Table 4.
3. Once analyzed and visually checked, the genotype table from GeneMapper can be exported to Excel.
4. In Excel, translate fragment sizes to alleles and allele code to EU vc types (*see* **Note 7**).

3.5 Nonfluorescent Duplex- or Monoplex-PCR

There are many reasons to use modified primer in a capillary Genetic Analyzer, i.e., they are applicable in multiplex-PCR as well as 96-well plate format, and resulting electropherograms can be scored automatically by a size-calling computer program. However, fluorescent dyes are expensive and out of reach for many laboratories. In order to overcome this financial burden, we suggest here to use nonfluorescent primer in a duplex-PCR for each *vic* locus and the *MAT* genes. While a conventional PCR approach reduces external costs, it is more time consuming. However, these circumstances are compensated for by precise, reliable results that still can be verified by conventional pairing tests as a second chain of evidence (Fig. 1).

1. For each *vic*-locus or *MAT* gene, prepare a master mix with the DNA polymerase of your choice, containing both primer pair per locus (duplex-PCR) or each primer pair separately (monoplex-PCR).
2. Run the PCR on a thermal cycler with the parameters required for the polymerase of your choice, but using 35 cycles and an annealing temperature of Ta = 60 °C or close to this temperature (*see* **Note 8**).
3. Load 5–10 μL of each PCR product on a 1–1.5% agarose gel.
4. Analyze the size length of duplex-PCRs according to the fragment length on Table 4 (*see* Fig. 3).

4 Notes

1. Grow *C. parasitica* cultures on potato dextrose agar (PDA; 39 g PDA per liter water), using a sterile cellophane membrane on the agar plate to prevent agar residues in the DNA extract. Before to use the cellophane, autoclave membranes twice in water and discard the water after the first autoclaving to remove potential contaminants. Agar residues in the DNA extract can inhibit PCR reaction.
2. To obtain higher amounts of DNA, *C. parasitica* cultures can be incubated at 25 °C under light for 5–12 days. Light exposure induces the formation of conidia spores, which increase the amount of DNA per area.

Table 4
Expected sizes of the PCR amplicons

Gene	Allele	Primer name	Fluorescent dye	Fragment length [bp]
Multiplex 1				
vic1a	1	Cp-vic1a-1	FAM	263
	2	Cp-vic1a-2	FAM	461
vic2	1	Cp-vic2–1	PET	340
	2	Cp-vic2–2	PET	448
vic3a	1	Cp-vic3a-1	VIC	368
	2	Cp-vic3a-2	VIC	191
MAT1–1-2	1	Cp-MAT1–1	NED	145
MAT1–2-1	2	Cp-MAT1–2	NED	400
Multiplex 2				
vic4	1	Cp-vic4–1	PET	359
	2	Cp-vic4–2	PET	257
vic6	1	Cp-vic6–1	VIC	381
	2	Cp-vic6–2	VIC	328
vic7	1	Cp-vic7–1	FAM	128
	2	Cp-vic7–2	FAM	225

3. The 10x primer premixes should be prepared beforehand and can be reused until fully consumed. Store at −20 °C.
4. Sensitivity tests showed that 20 pg/μL initial concentration of input DNA was just at the limit of detection [11].
5. You may adjust the volumes of water and DNA to your needs.
6. The dilution of the PCR product might need to be adjusted depending on the Genetic Analyzer used. We noticed that due to a higher sensitivity on the 3500 series Genetic Analyzer, a higher dilution of at least 1:5 is needed.
7. For routine lab work, it is highly recommended to use an Excel VBA Macro for this step. Upon request, the authors supply the code for macro that runs on Win10 machines.
8. The lower the annealing temperature, the higher the risk of primer dimer formation. Especially, a duplex-PCR with both *vic4* primer pairs is known to form primer dimer below Ta = 60 °C. Use instead monoplex-PCRs for *vic4*-1 and *vic4*-2 separately.

Acknowledgments

We appreciate the thoughtful comments of Daniel Rigling on an early version of the manuscript. This work was supported by the Swiss Federal Office for the Environment FOEN.

References

1. Peever TL, Liu Y-C, Cortesi P et al (2000) Variation in tolerance and virulence in the chestnut blight fungus-hypovirus interaction. Appl Environ Microbiol 66:4863–4869
2. Bryner S, Rigling D (2012) Hypovirus virulence and vegetative incompatibility in populations of the chestnut blight fungus. Phytopathology 102:1161–1167
3. Cortesi P, McCulloch CE, Song H et al (2001) Genetic control of horizontal virus transmission in the chestnut blight fungus, *Cryphonectria parasitica*. Genetics 159:107–118
4. Cortesi P, Milgroom MG (1998) Genetics of vegetative incompatibility in *Cryphonectria parasitica*. Appl Environ Microbiol 64: 2988–2994
5. Schuelke M (2000) An economic method for the fluorescent labeling of PCR fragments. Nat Biotechnol 18:233–234
6. Blacket MJ, Robin C, Good RT et al (2012) Universal primers for fluorescent labelling of PCR fragments—an efficient and cost-effective approach to genotyping by fluorescence. Mol Ecol Resour 12:456–463
7. Choi GH, Dawe AL, Churbanov A et al (2012) Molecular characterization of vegetative incompatibility genes that restrict hypovirus transmission in the chestnut blight fungus *Cryphonectria parasitica*. Genetics 190:113–127
8. Zhang D-X, Spiering MJ, Dawe AL et al (2014) Vegetative incompatibility loci with dedicated roles in allorecognition restrict mycovirus transmission in chestnut blight fungus. Genetics 197:701–714
9. Marra RE, Milgroom MG (2001) The mating system of the fungus *Cryphonectria parasitica*: selfing and self-incompatibility. Heredity 86: 134–143
10. Short DP, Double M, Nuss DL et al (2015) Multilocus PCR assays elucidate vegetative incompatibility gene profiles of *Cryphonectria parasitica* in the United States. Appl Environ Microbiol 81:5736–5742
11. Cornejo C, Šever B, Kupper Q et al (2019) A multiplexed genotyping assay to determine vegetative incompatibility and mating type in *Cryphonectria parasitica*. Eur J Plant Pathol 155:81–91
12. Blais J, Lavoie SB, Giroux S et al (2015) Risk of misdiagnosis due to allele dropout and false-positive PCR artifacts in molecular diagnostics: analysis of 30,769 genotypes. J Mol Diagn 17: 505–514
13. Miller CR, Joyce P, Waits LP (2002) Assessing allelic dropout and genotype reliability using Maximum Likelihood. Genetics 160:357–366

Part VII

Biocontrol

Chapter 26

A Detached Leaf Assay for Rapidly Screening Plant Pathogen-Biological Control Agents

Giovana Prado Fortuna Macan, Sammar Khalil, Pruthvi B. Kalyandurg, Nidhi Pareek, and Ramesh R. Vetukuri

Abstract

The detached leaf assay (DLA) is a nondestructive method for evaluating interactions between plants and disease-causing agents that allows quick characterization of potential pathogens' infectivity and plants' resistance to them. Here we show its utility for also assessing potential biological control agents (BCAs), by demonstrating its applicability for screening potential BCAs for the strawberry (*Fragaria* × *ananassa*) pathogen *Phytophtora cactorum*.

Key words Detached leaf assay, Screening, Biological control agents (BCA), Plant disease, Antagonist, *Phytophtora cactorum*

1 Introduction

The detached leaf assay (DLA) is an infection assay, in which excised leaves are inoculated by a pathogen suspension, and the development of disease symptoms is evaluated visually or by image analysis following incubation in appropriate conditions. It is widely applied in breeding programs and plant pathology studies as a simple method for assessing both pathogens' pathogenicity and plants' resistance to them [1–6]. It could have similar advantages for screening potential biological control agents (BCAs), but this possibility has not yet been exploited.

In recent years, microbial BCAs have received increasing attention due to their potential ability to control plant diseases sustainably. They can prevent pathogen infection and development through various mechanisms, such as: competition for space and nutrients; parasitism; production of antibiotics, enzymes, or toxins; and/or induction of host resistance [7–9]. Bacterial isolates may also prevent disease development by changing the behavior of pathogenic zoospores, including their motility, attachment, and

Nicola Luchi (ed.), *Plant Pathology: Methods and Protocols*, Methods in Molecular Biology, vol. 2536,
https://doi.org/10.1007/978-1-0716-2517-0_26,

cyst germination [10, 11]. Thus, BCAs offer possibilities to control plant pathogens in a highly targeted manner, without use of chemical agents.

It is also much less destructive than whole-plant inoculation, enabling both preservation of most of the sampled plants for further evaluation [12, 13] and rapid assessment of numerous isolates [14] (in the protocol presented here, disease symptoms can be assessed 4 days after inoculation). Moreover, it is inexpensive and less laborious than other screening approaches, so it can be easily performed more cheaply under controlled laboratory conditions, using equipment that is usually already available. Hence, this technique avoids the need for greenhouse facilities or field experiments, thereby reducing risks of pathogen dissemination and contamination [1, 2, 14].

The following sections present a reliable and straightforward DLA procedure for rapidly screening potential bacterial BCAs for controlling the strawberry (*Fragaria* × *ananassa*) pathogen *Phytophthora cactorum*.

2 Materials

2.1 Pathogen Inoculum

1. Oomycete pathogen: *Phytophthora cactorum* maintained on V8 agar plates (*see* **Note 1**).
2. V8 Juice Agar Medium: Prepared by adding 1.5 g $CaCO_3$ and 15 g agar to 100 mL of carrot juice, increasing the volume to 1 L with distilled water, stirring with a magnetic stirrer on low heat, and then autoclaving at 121 °C for 20 min.
3. V8 Juice Broth Medium: Prepared identically to V8 juice agar medium, but with no agar.
4. Laminar flow cabinet.
5. Petri dishes (90 × 15 mm and 90 × 25 mm).
6. Incubator.
7. Autoclaved strawberry cultivated soil extract solution.
8. Autoclaved distilled water.
9. Refrigerator (4 °C).
10. Hemocytometer (Hausser Scientific, Horsham, PA, USA).
11. Light microscope ((providing at least up to 400× magnification) Laborlux12 Leitz, Germany).

2.2 Bacterial Suspension

1. Tryptic Soy Broth (TSB): Medium prepared according to standard guidelines.
2. Falcon tubes (50 mL).
3. Rotary shaker (providing agitation at 220 rpm and 28 °C).

4. Benchtop centrifuge.
5. Autoclaved distilled water.
6. Spectrophotometer to measure optical density at 600 nm (OD_{600}).

2.3 Detached Leaf Assay (DLA)

1. Plastic containers.
2. Plastic or steel grids.
3. Moist paper.
4. Autoclaved distilled water.
5. Adjustable-volume micropipette.
6. Sterile pipette tips.
7. Parafilm.
8. Incubator.
9. Plant Material: Strawberry plants (*Fragaria* × *ananassa* cv. Sonata) grown in a greenhouse for 10 weeks in 9-L plastic pots containing peat (Weibull Horto, Sweden).
10. Pathogen inoculum and bacterial suspension (described above).
11. Digital camera or red-light imaging system (a Bio-Rad system is used in our laboratory).
12. Software: ImageJ.

3 Methods

3.1 Phytophthora cactorum *Inoculum/Zoospore Production*

1. Culture the *P. cactorum* strains on V8 juice agar media and incubate for 7 days at 21 °C (*see* **Note 2**).
2. To induce zoospore production, transfer three agar plugs (30 × 10 mm) from the periphery of the freshly growing mycelium to deep-form Petri dishes (90 × 25 mm) containing 20 mL of V8 juice broth media (*see* **Note 3**). Incubate the plates in the dark at room temperature for 3 days.
3. Wash the mycelium three times with autoclaved distilled water (Fig. 1) (*see* **Note 4**).
4. To promote sporangia development, add 20 mL of autoclaved soil extract solution to the deep-form Petri dishes (90 × 25 mm) (*see* **Notes 5** and **6**). Incubate the plates for 24 h under fluorescent light at 21 °C (*see* **Note 7**).
5. Remove the soil extract solution and cover the mycelium with autoclaved distilled water (20 mL). Incubate the plates for another 24 h under continuous fluorescent light at 21 °C (*see* **Note 8**) (Fig. 2).
6. To promote zoospore release, expose the plates to a cold shock in the refrigerator at 4 °C for 1 h and then incubate them at room temperature for another 1 h (*see* **Note 9**).

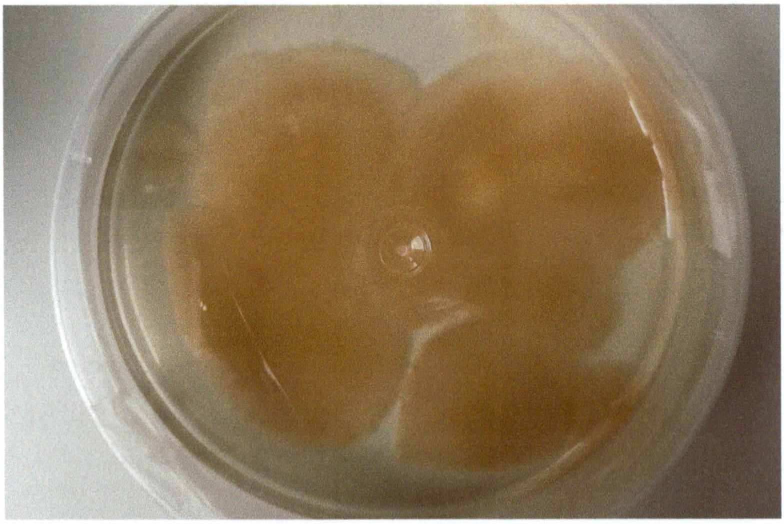

Fig. 1 *Phytophthora cactorum* mycelium after 3-days cultivation in V8 broth media and washing with autoclaved distilled water

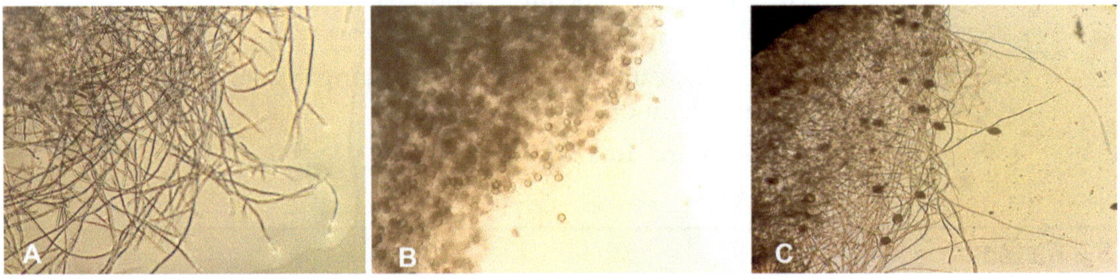

Fig. 2 *Phytophthora cactorum* on V8 juice agar plates viewed under a microscope (400×). (**a**) Vegetative hyphae. (**b**) Hyphae with oospores. (**c**) Hyphae with formed sporangia

7. To estimate the zoospore concentration, collect a sample in a microcentrifuge tube, vortex mix and count the encysted zoospores in a hemocytometer.
8. Adjust the zoospore suspension's density to 2×10^4 mL^{-1} or other desired concentration by adding autoclaved distilled water (*see* **Note 10**).

3.2 Bacterial Suspension

1. Grow each bacterial isolate selected for screening potential biocontrol agents overnight in culture tubes (50 mL) with 15-mL TSB medium at 28 °C on a rotary shaker set at 220 rpm (*see* **Note 11**).
2. Centrifuge the tubes for 10 min at 3000 rpm and resuspend the pelleted cells in sterile distilled water.
3. Measure the bacterial density (OD_{600}) and adjust it to 0.2 by adding autoclaved distilled water.

3.3 Detached Leaf Assay (DLA)

1. Collect young fully expanded leaves from 10-week-old strawberry plants (Fig. 3) (*see* **Note 12**).
2. Make small wounds on the abaxial epidermal surfaces of leaflets by gently scraping with a sterile pipette tip (*see* Fig. 4a, *arrow*) (*see* **Note 13**).
3. Prepare a mixture of the bacterial and zoospore suspensions with a final OD_{600} of 0.2 and density of 2×10^4 zoospores/mL, respectively.
4. Place 20 μL droplets of the mixture of bacterial and zoospore suspension on the injured surface of the leaflets on both sides of the midrib (Fig. 4b).
5. Prepare positive control consisting of leaflets inoculated solely with the *P. cactorum* zoospore suspension, and negative control inoculated only with sterile distilled water.
6. Include six leaflets per treatment and repeat the experiment three times.
7. Place the leaflets on grids in plastic boxes with moist paper. Seal the boxes with Parafilm and incubate them at 20 °C with a 16-h photoperiod (*see* **Note 14**).
8. Four days after inoculation, evaluate the severity of infection by capturing digital images using a digital camera (Fig. 5) or red-light imaging system (Bio-Rad) (Fig. 6) (*see* **Notes 15** and **16**).
9. Assess disease symptoms by image analysis using the ImageJ software.

3.4 Image Analysis

1. Import each image into ImageJ using menu commands "File" > "Open" (Fig. 7a and b).
2. Set the scale: "Straight Line" > Draw a line to a known distance (e.g., along a photographed scale) > "Analyze > Set Scale" > fill the "known distance", and select the unit (cm).
3. If necessary, crop the image focusing on the desired leaf: "Rectangular" > select the area > "image" > "crop".
4. Select a diseased area by calibrating the threshold: "Image" > "Adjust" > "Color Threshold" (Fig. 7c and d).
5. Make a binary image: "Process" > "Binary" > "Make Binary" (Fig. 7e).
6. Remove small fragments that are not of interest (not diseased areas): "Process" > "Noise" > "Remove Outliers" (Fig. 7f).
7. Measure the diseased area: "Analyze" > "Analyze particles" > select "Display results", "Summarize", and "in situ Show" > "OK" (Fig. 7g).

Fig. 3 Ten-week-old strawberry (*Fragaria* × *ananassa* cv. Sonata) plants

Fig. 4 (**a**) Small wounds (arrowed) on the abaxial epidermal surface of a strawberry (*Fragaria* × *ananassa* cv. Sonata) leaflet. (**b**) Droplets (20 μl) of the mixture of bacterial and zoospore suspension on either side of the leaf midrib

4 Notes

1. Cryogenic conditions (ideally in liquid nitrogen) are recommended for long-term storage of *Phytophthora* cultures [15]. In our laboratory, stock cultures are preserved in 20% (wt/vol) glycerol at −80 °C, while *P. cactorum* strains are maintained on V8 agar media amended with antibiotics (primaricin and rifampicin) and subcultured every 4 weeks.

Fig. 5 Strawberry leaflets placed on plastic grid in plastic container with moist paper (4 days after inoculation)

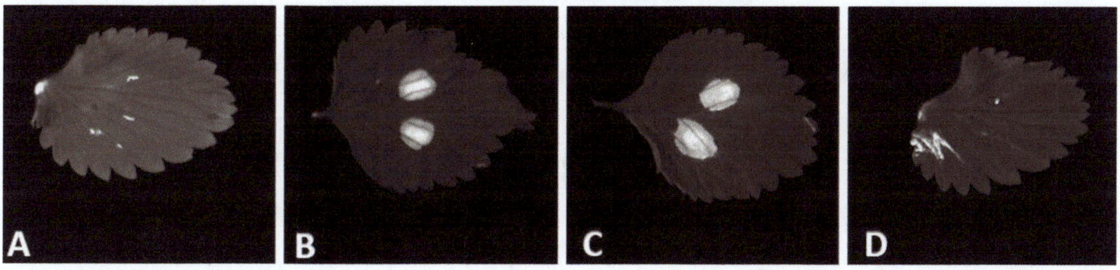

Fig. 6 Images of leaflets acquired with a near infrared imaging system (Bio-Rad), 4 days after inoculation. (**a**) Negative control (inoculated with autoclaved distilled water). (**b**) Positive control (leaflet inoculated with *P. cactorum* zoospore suspension). (**c**) BCA-test leaflets: inoculated with mixture of zoospore suspension and bacterial isolate (named B4), showing that this bacterial isolate was not able to prevent disease development. (**d**) BCA-test leaflets: inoculated with mixture of zoospore suspension and another bacterial isolate (named B8), showing that this isolate was an effective BCA, preventing the disease development

2. *Phytophthora cactorum* strains usually grow more slowly on V8 culture media than other *Phytophthora* species. For zoospore induction, freshly growing mycelium is needed, and 7-day-old cultures are most suitable.
3. Pea broth can also be used, but it provides lower sporangium yields than V8 broth.

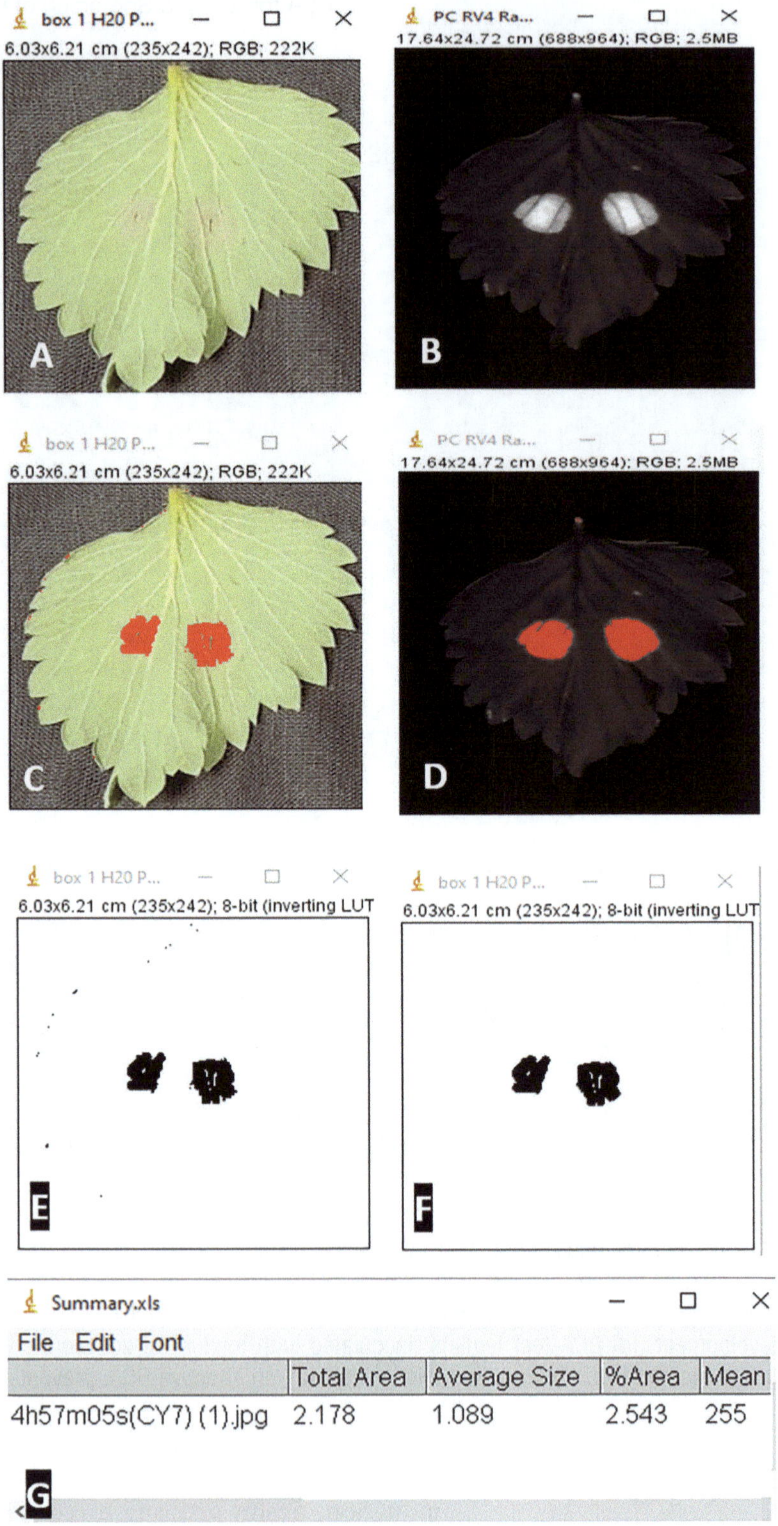

Fig. 7 Illustrative results of imaging analysis using ImageJ. (**a**) Visible light images (from camera). (**b**) Red-light images (from Bio-Rad system). (**c**) and (**d**) Infected areas selected by applying calibrated threshold. (**e**) Binary image. (**f**) Binary image after removing outliers. (**g**) Total selected infected areas

4. At this point, following cultivation on broth media, a mycelial culture devoid of oospores should be observable by microscopy (Fig. 2a).
5. *Phytophthora cactorum* is homothallic. A culture of a single isolate can produce sexual structures and abundant oospores [16]. Under the described conditions, we have observed production of large numbers of sexual spores even at margins of the freshly growing cultures (Fig. 2b). Thus, the production of the desired asexual spores can be challenging, and unsatisfactory results were obtained using various other tested methods and protocols for zoospore production.
6. The mycelia must be submerged in soil extract solution. The amount added can vary according to the dish size and depth. Generally, 20 mL should be enough.
7. Most *Phytophthora* species require light for sporangia production. Thus, incubation under continuous fluorescent light is usually essential. Make sure that all the plates are exposed to the light.
8. Sporangia start appearing within 2 days and their frequency peaks at day 3 (samples can be taken, and sporangia formation observed, under the microscope) (Fig. 2c).
9. Zoospores start to be released right after the incubation in the fridge and can be active for up to 6 h. However, to avoid loss of mobility, the zoospore suspension should be used within 2 h.
10. Zoospore motility is checked in the original and remaining suspension by microscopy right before as well as after plant inoculation.
11. The liquid culture can be obtained from a streak culture grown on a TSA plate agar plate at 25 °C for 2 days. Streak plates can be obtained from the frozen (−80 °C) stock cultures. To make a long-term stock of bacteria, a single colony is picked from a plate and cultured overnight in appropriate medium. Then, the tubes are centrifuged at 3000 rpm for 30 min. The supernatant is discharged, and 2 mL of freezing buffer is added. The culture is transferred to 2-mL Eppendorf tubes and kept at −80 °C.
12. Leaves should be collected and used right before suspension inoculation.
13. Surfaces of strawberry leaves have hydrophobic, protective cuticle. Thus, the leaves must be lightly wounded to remove this protective layer and allow drops of the mixture of bacterial and zoospore suspension to stay on their surface. The plastic boxes must also be moved carefully to preventing the drops rolling beyond the wounded surface.

14. A humid environment is important for preserving the leaves' turgidity during incubation and promotes disease development. Thus, the boxes must be tightly sealed.

15. Disease progression can also be followed by daily evaluations (capturing and analyzing images daily).

16. A near infrared imaging system can also be used for assessing disease symptoms. Moreover, according to Zahid et al. [17], this method allows the visualization and quantification of plant lesion and wounding.

References

1. Pettitt TR, Wainwright MF, Wakeham AJ, White JG (2011) A simple detached leaf assay provides rapid and inexpensive determination of pathogenicity of Pythium isolates to "all year round" (AYR) chrysanthemum roots. Plant Pathol 60:946–956
2. Kunkeaw S, Worapong J, Smith DR, Triwitayakorn K (2010) An in vitro detached leaf assay for pre-screening resistance to anthracnose disease in cassava (Manihot esculenta Crantz). Australas Plant Pathol 39:547–550
3. Browne RA, Cooke BM (2004) Development and evaluation of an in vitro detached leaf assay for pre-screening resistance to Fusarium head blight in wheat. Eur J Plant Pathol 110:91–102
4. Kumar K, Xi K, Turkington TK et al (2011) Evaluation of a detached leaf assay to measure fusarium head blight resistance components in barley. Can J Plant Pathol 33:364–374
5. Francis MI, Peña A, Graham JH (2010) Detached leaf inoculation of germplasm for rapid screening of resistance to citrus canker and citrus bacterial spot. Eur J Plant Pathol 127:571–578
6. Loyd AL, Benson DM, Ivors KL (2014) Phytophthora populations in nursery irrigation water in relationship to pathogenicity and infection frequency of Rhododendron and Pieris. Plant Dis 98:1213–1220
7. Guetsky R, Shtienberg D, Elad Y et al (2002) Improving biological control by combining biocontrol agents each with several mechanisms of disease suppression. Phytopathology 92:976–985
8. Narayanasamy P (2013) Mechanisms of action of bacterial biological control agents. In: Biological management of diseases of crops, vol 15. Springer, Dordrecht, pp 295–429
9. Köhl J, Kolnaar R, Ravensberg WJ (2019) Mode of action of microbial biological control agents against plant diseases: relevance beyond efficacy. Front Plant Sci 10:845
10. Zohara F, Akanda MAM, Paul NC et al (2016) Inhibitory effects of Pseudomonas spp. on plant pathogen Phytophthora capsici in vitro and in planta. Biocatal Agric Biotechnol 5: 69–77
11. Alcals LA (2008) Characterization and efficacy of bacterial strains for biological control of soilborne diseases caused by Phytophthora cactorum and Meloidogyne javanica on Rosaceous plants. Dissertation, Universitat de Girona
12. Miller-Butler MA, Smith BJ, Curry KJ, Blythe EK (2019) Evaluation of detached strawberry leaves for anthracnose disease severity using image analysis and visual ratings. HortScience 54:2111–2117
13. Miller-Butler MA, Smith BJ, Babiker EM et al (2018) Comparison of whole plant and detached leaf screening techniques for identifying anthracnose resistance in strawberry plants. Plant Dis 102:2112–2119
14. Maizatul-Suriza M, Seman IA, Madihah AZ, Hefni Rusli M (2017) Detached leaf assay for in vitro screening of potential biocontrol agents to control goosegrass weed (Eleusine indica). J Oil Palm Res 29:562–569
15. Ivors KL (ed) (2015) Laboratory protocols for phytophthora species. The American Phytopathological Society, California
16. Schmitthenner AF, Ravindra GB (1994) Useful methods for studying phytophthora in the laboratory. OARDC, Wooster
17. Zahid MA, Sandroni M, Vetukuri RR, Andreasson E (2021) A fast, nondestructive method for the detection of disease-related lesions and wounded leaves. BioTechniques 71:425–430

Chapter 27

Spray-Induced Gene Silencing to Study Gene Function in *Phytophthora*

Poorva Sundararajan, Pruthvi B. Kalyandurg, Qinsong Liu, Aakash Chawade, Stephen C. Whisson, and Ramesh R. Vetukuri

Abstract

RNA interference (RNAi) is a conserved cellular defense mechanism mediated by double-stranded RNA (dsRNA) that can regulate gene expression through targeted destruction of mRNAs (messenger RNAs). Recent studies have shown that spraying dsRNAs or small RNAs (sRNAs) that target essential genes of pathogens on plant surfaces can confer protection against pests and pathogens. Also called spray-induced gene silencing (SIGS), this strategy can be used for disease control and for transient gene silencing to study the function of genes in plant–pathogen interactions. Furthermore, as sRNAs can move locally, systemically, and cross-kingdom during plant–microbe interactions, SIGS allows quick detection and characterization of gene functions in pathogens and plants.

Key words Oomycetes, SIGS, Gene silencing, RNAi, Potato

1 Introduction

Spray-induced gene silencing (SIGS) is an RNAi-based strategy for plant trait improvement and disease control [1, 2] that does not require plant transformation. From a plant protection perspective, it involves inhibition of plant pathogens through exogenous application of dsRNAs or sRNAs targeting pathogen genes essential for survival and disease development. dsRNAs are sprayed directly on plant tissues and can either be directly taken up by the plant pathogen, triggering the pathogen RNAi machinery, and/or by the host plant RNAi machinery [1–4]. The pathogen RNAi machinery then targets the cognate mRNAs for destruction or inhibition. The use of SIGS as a tool for disease control and reverse genetics aimed at modulation of pathogen gene expression has been demonstrated with several plant pathogens, paving the way for developing SIGS as a transient silencing tool to study the function of genes [5].

Nicola Luchi (ed.), *Plant Pathology: Methods and Protocols*, Methods in Molecular Biology, vol. 2536,
https://doi.org/10.1007/978-1-0716-2517-0_27,

Here we demonstrate methods for using SIGS as a tool for gene knockdown studies, using the oomycete pathogen *Phytophthora infestans* and potato pathosystem. Oomycetes, also commonly called water moulds, superficially resemble fungi as they share features such as filamentous growth and reproduction via spores. However, they are more closely related to brown algae, diatoms, and protists [6]. *Phytophthora* species within the oomycetes are all plant pathogens that cause billions of dollars in losses to agriculture every year. The most notorious example is late blight disease on potato and tomato caused by *P. infestans*, made infamous by its role in the Irish potato famine in the mid-1800s. We recently applied SIGS for knockdown of genes in *P. infestans*. As a test case, we selected the guanine-nucleotide binding (G) protein β-subunit (*PiGPB1*; PITG_06376; XP_002998508). PiGPB1 is known to be involved in the signal transduction process and sporangial development during infection on potato [7] and could be used for host-induced gene silencing (HIGS) to control *P. infestans* infection. Through SIGS, it was demonstrated that targeting *PiGPB1* resulted in severe reduction in disease [8]. The dsRNAs for SIGS application can either be produced in vitro or in vivo. The most critical parameters to carefully consider for successful SIGS are the size and concentration of the dsRNA, and delivery methods which are discussed in detail in the method and notes sections.

2 Materials

2.1 Propagation of *P. infestans*

1. Rye grain (organic; produced without agrochemical application).
2. Distilled water.
3. Domestic blender.
4. Cheesecloth.
5. Strainer.
6. Sucrose.
7. Bacteriological agar (Saveen and Werner AB).
8. Autoclave.
9. Pimaricin (50-mg/mL stock suspension; Sigma-Aldrich).
10. Geneticin (100-mg/mL stock solution; Sigma-Aldrich).
11. Ampicillin (100-mg/mL stock solution; Sigma-Aldrich).
12. A scalpel with No. 11 or No. 21 blades.
13. Forceps – straight tip.
14. 20 °C incubator.
15. Sterile Milli-Q water.

16. 90-mm sterile Petri dishes.
17. Parafilm sealing film.

2.2 Plant Material

1. Potting compost.
2. 2.5-L plastic plant pots.
3. Potato tubers (e.g., susceptible cultivars such as Bintje or Désirée).
4. Climate control chambers.

2.3 dsRNA Synthesis

1. DNeasy Plant Mini Kit (Qiagen).
2. Phusion™ High-Fidelity DNA Polymerase (ThermoFisher Scientific).
3. 5× Phusion HF Buffer (ThermoFisher Scientific).
4. 10-mM dNTPs.
5. Forward and reverse primers.
   (a) In-Vitro Transcription:
   T7 PiGPB1 dsRNA.FOR: 5′-GTAATACGACTCACTATAGGGATGTTTATTTCGGGCTCGTGTGA-3′
   T7 PiGPB1 dsRNA.REV: 5′-GTAATACGACTCACTATAGGGTAGATATGCGCTCCGGAAGT-3′
   (b) Bacterial Cloning:
   PiGPB1.FOR: 5′- AAAAAAGCTTCTCTACGCTCCAGTTGGGTC-3′
   PiGPB1.REV: 5′- AAAAACTAGTGGTAGATATGCGCTCCGGAA-3′
   (c) qRT-PCR:
   PiGPB1 qPCR.FOR: 5′-TTCCGGAGCGCATATCTACC-3′
   PiGPB1 qPCR.REV: 5′-TCTTGAGTAGCGTGTCCCAG-3′
6. QIAquick® PCR Purification Kit (Qiagen).
7. MEGAscript RNAi Kit (Invitrogen).
8. L4440 plasmid (Addgene Plasmid #1654).
9. FastDigest restriction enzymes – HindIII & SpeI (ThermoFisher Scientific).
10. HT115 (DE3) *Echerichia coli* competent cells.
11. Tetracycline (100-mg/mL stock solution; Sigma-Aldrich).
12. LB (Luria-Bertani) medium (Duchefa Biochemie).
13. GeneJET Plasmid Miniprep Kit (ThermoFisher Scientific).
14. DreamTaq DNA Polymerase (ThermoFisher Scientific).

15. Isopropyl β-D-1-thiogalactopyranoside (IPTG) (100-mM stock solution; filter sterilized through a sterile 0.2-μm membrane filter).
16. Direct-zol™ RNA Miniprep Plus Kit (Zymo).
17. qScript cDNA SuperMix (Quantabio).
18. FastDigest restriction enzymes – HindIII & SpeI (ThermoFisher Scientific).
19. T4 DNA Ligase (ThermoFisher Scientific).
20. Gel DNA recovery kit (Zymo).
21. DyNAmo Flash SYBR Green qPCR Kit (ThermoFisher Scientific).
22. 6-well culture plates (e.g., Sigma-Aldrich cell culture plate).
23. PCR plate, 96-well, low profile, skirted (ThermoFisher Scientific).
24. 1.5-mL microcentrifuge tubes.
25. 2-mL microcentrifuge tubes.
26. 50-mL centrifuge tubes (e.g., Falcon tubes).
27. Mortar and pestle.
28. Plastic boxes for detached leaf assays (DLAs).
29. 1-mL Cuvettes.
30. Glass microscopic slides.
31. Coverslips.
32. Liquid nitrogen canister.
33. 50× TAE Buffer (Tris–acetate-EDTA) (Thermo Scientific) – dilute 20 mL of 50× stock in 980 mL of milli-Q water to make 1× TAE.
34. Nuclease-free water.
35. Standard agarose.
36. Gel Red (Biotium).
37. DNA Gel loading dye (6×) (Thermo Scientific).
38. RNA loading dye (2×) (Thermo Scientific).
39. 1-kb DNA ladder.
40. Trypan Blue (Sigma-Aldrich).
41. Lactic acid 85% (w/v) (Sigma-Aldrich).
42. Glycerol (Sigma-Aldrich).
43. Buffer saturated phenol (Invitrogen).

2.4 Equipment

1. Automizer.
2. Fume hood.

3. Orbital shaker.
4. Spectrophotometer.
5. Laminar flow cabinet.
6. 37 °C incubator and shaker.
7. Nano-drop spectrophotometer.
8. CFX96 qPCR machine (Bio-Rad).
9. Water bath with temperature control.
10. Heat block with temperature control.
11. Thermo Cycler for PCR (VWR).
12. Airbrush (Cocraft) and compressor (Biltema Mini Compressor MC.90).

3 Methods

3.1 Phytophthora infestans *– Culture*

1. Weigh 60 g of rye grain and wash thoroughly with distilled water. Soak the grains overnight at 21 °C in a beaker with water such that the grain is entirely submerged.
2. The next day, coarsely blend the soaked grain using a blender for 30 s, followed by heating at 50 °C for 3 h using a water bath.
3. Place a cheesecloth over a strainer and pass the heated mixture through to strain out the rye liquid. Make up the liquid with distilled H_2O to 800 mL and split equally into 400 mL each into two 1-L bottles.
4. Measure 10 g of sucrose and 7.5 g of agar separately for each bottle.
5. Add the weighed sucrose to the respective bottles and adjust pH to 7.0 using a pH meter. Next, make up the liquid to 500 mL with distilled H_2O in each bottle before adding the agar (*see* **Note 1**).
6. Mix well by shaking and autoclave at 121 °C for 30 min.
7. After autoclaving, allow the media to cool to around 55 °C before adding antibiotics. For 500 mL of rye agar, add to a final concentration 100-μg/mL ampicillin and 12-μg/mL pimaricin. Under sterile conditions, add antibiotics to rye agar and pour approximately 25 mL into individual 90-mm diameter Petri dishes. For growing *P. infestans* expressing green fluorescent protein (*Pi*-GFP), add 10-μg/mL geneticin (*see* **Notes 2–4**).
8. Cut 5-mm^2 plugs using a sterile blade from 2–3 week-old GFP-tagged *Phytophthora infestans* (*Pi*-GFP) agar-culture plates. Place the cut agar plugs with the side containing

mycelium facing down onto the rye agar plates, cooled down to room temperature.

9. Incubate the Petri dishes containing GFP-tagged *P. infestans* at 20 °C in darkness. Collect spores for further experiments after 2 weeks. Subculture, as explained above, every 2 weeks to maintain material.
10. Sporangia should be collected from (maximum age) two-week-old plates. Under sterile conditions, wash the plate with 2–5 mL of sterile water, scraping the surface and sides evenly and pass the solution through a 40-micron cell strainer into a 50-mL centrifuge tube.

3.2 Plant Material

1. Grow potatoes (e.g., cv. Bintje) from tubers in 2.5-L pots containing well-drained fertilized compost under greenhouse conditions of 21/19 °C day/night temperature, 16-h photoperiod, and 60% humidity.
2. Use four- to five-week-old plants to collect leaves for DLAs.

3.3 Phytophthora infestans DNA Extraction

1. Grow *P. infestans* in 6-well cell culture plate containing rye broth with 100-μg/mL ampicillin, 12-μg/mL pimaricin. Incubate at 20 °C without shaking for a week.
2. Collect the mycelia into a 2-mL microcentrifuge tube using a pair of forceps and weigh 50 mg. Freeze the sample in liquid nitrogen and grind immediately using a mortar and pestle. Extract *Phytophthora infestans* DNA using DNeasy Plant Mini Kit and elute in 100 μL of elution buffer.
3. Use a Nano-Drop spectrophotometer to determine the concentration of extracted DNA and store at −20 °C.

3.4 dsRNA Synthesis: In Vitro Transcription

1. Using a primer design tool for double-stranded RNA (e.g., SnapDragon – dsRNA Design [https://www.flyrnai.org/snapdragon]), design forward and reverse primer pairs amended with the T7 promoter sequence (GTAATACGACTCACTATAGGG) for *Phytophthora infestans* GPB1 (*PiGPB1*; PITG_06376; XP_002998508; positive control), and *P. infestans* test gene(s) with an amplicon size of approximately 200–400 bp.
2. Polymerase Chain Reaction can be carried out using Phusion™ High-Fidelity DNA Polymerase (*see* **Note 5**). Prepare a master mix containing 5× Phusion buffer, 10 mM dNTP, Phusion High-Fidelity DNA Polymerase, and distilled water. Add the respective forward and reverse primers individually and use 5–10 ng of *Phytophthora infestans* DNA obtained from previous steps as respective templates for GPB1 and other test gene (s) (*see* **Note 6**).

3. Use the following thermal conditions for PCR: Initial denaturation at 98 °C for 30 s; followed by denaturation (98 °C for 10 s); annealing (60 °C for 30 s); and extension (72 °C for 30 s) for a total of 30 cycles before proceeding to a final extension at 72 °C for 5 min.
4. Verify amplification of PCR products through gel electrophoresis (1% agarose in 1× TAE buffer) by running the samples alongside a 1-kb DNA ladder for reference. Incorporate a nucleic acid stain in the gel (e.g., GelRed), or stain after electrophoresis. Use a UV transilluminator to visualize the bands under ultraviolet light.
5. Before proceeding with dsRNA synthesis, purify the PCR product by QIAquick® PCR Purification Kit, and elute the purified PCR product in 20 μL of nuclease-free water. Determine the concentration using a Nano-Drop spectrophotometer.
6. Synthesize dsRNA for *PiGPB1* and other test gene(s) using 2 μL each of T7 enzyme mix, four ribonucleotides (ATP, CTP, GTP, UTP), and 10× T7 reaction buffer supplied with the MEGAscript RNAi Kit. Use 1 μg of respective purified PCR products as templates for dsRNA synthesis and make up the volume to a total of 20 μL using nuclease-free water. Treat the dsRNA with DNaseI and RNase provided in the kit to remove any DNA or ssRNA, before proceeding for purification. Elute the purified dsRNA in 100 μL of nuclease-free water (preheated to 95 °C).
7. As a positive control, synthesize dsRNA using the control template ($dsRNA^{Ct}$) provided in the kit.
8. Verify dsRNA synthesis by running a sample of the purified dsRNA with 6× loading dye or RNA dye in agarose gel (1% agarose in 1× TAE buffer) alongside a 1-kb ladder.
9. Measure the concentration of synthesized dsRNA using the Nano-Drop spectrophotometer before storing the dsRNA at −80 °C.

3.5 dsRNA Synthesis: Bacterial Cloning

3.5.1 PCR for Inserts

1. Using NCBI Primer-BLAST (https://www.ncbi.nlm.nih.gov/tools/primer-blast/), design forward and reverse primer pairs for *P. infestans GPB1* and other test gene(s) containing an additional restriction site at the 5′ end. In this case, for *PiGPB1,* the restriction recognition sites corresponding to HindIII (AAGCTT) and SpeI (ACTAGT) are added to the forward and reverse primers, respectively. The amplicon size should be approximately 200–400 bp (*see* **Note 7**).
2. As a negative control, use a nonspecific gene sequence from an unrelated species.

3. Carry out PCR using Phusion™ High-Fidelity DNA Polymerase, as explained in Subheading 3.4, steps **2–3**. Use 5–10 ng of *P. infestans* DNA obtained from previous steps as the template.
4. Verify amplification of PCR products through gel electrophoresis (1% agarose) by running the samples alongside a 1-kb ladder for reference.
5. Under UV light, swiftly excise the band corresponding to *PiGPB1* and other genes to be tested and transfer to a 1.5-mL microcentrifuge tube. Proceed with gel extraction and purification using the Zymoclean Gel DNA Recovery Kit. Weigh the excised gel and dissolve by incubating at 50 °C along with three volumes of ADB for each volume of excised agarose gel. Transfer to spin column for washing before elution in 10–15 μL of nuclease-free water. Store the purified PCR product at −20 °C until further use.

3.5.2 Plasmid Extraction

- Culture *E. coli* containing the L4440 plasmid in sterile LB broth containing 100-μg/mL ampicillin overnight at 37 °C in a shaker and extract plasmid using a GeneJET Plasmid Miniprep Kit.

3.5.3 Restriction Digestion

- Carry out restriction digestion of insert DNA (PCR products) and vector (L4440 plasmid) using FastDigest Restriction Digestion enzymes.
- Run the digested plasmid in 1% agarose in 1× TAE buffer. Swiftly excise the corresponding bands under UV light and transfer to a 1.5-mL microcentrifuge tube. Proceed with gel extraction and purification as explained in Subheading 3.5.1, step **5**.

3.5.4 Ligation

- Perform ligation of digested products using T4 DNA Ligase and a plasmid to insert ratio of 1:3 (calculate using http://nebiocalculator.neb.com/#!/). Incubate at 22 °C for 5 min using a heat block with temperature control.

3.5.5 Transformation of E. coli

1. Take HT115-DE3 competent cells from −80 °C storage and thaw on ice for 10–15 min.
2. Mix 5 μL of the ligated plasmid with 50 μL of HT115-DE3 competent cells in a sterile 1.5-mL microcentrifuge tube and incubate on ice for 30 min followed by heat shock at 42 °C for 30–60 s.
3. As an additional control, transform HT115-DE3 competent cells with empty L4440 plasmid.
4. Incubate on ice for 2 min.

5. Add 300 μL of fresh LB broth without any antibiotics to the transformed bacterial cells.
6. Incubate at 37 °C in a shaker for 1 h and plate on LB agar plates supplemented with tetracycline and ampicillin to a final concentration of 12 μg/mL and 100 μg/mL, respectively (*see* **Note 8**).
7. Incubate overnight at 37 °C and check for single colonies the next day.
8. Screen for positive colonies with the target gene insertion by colony PCR. Under sterile conditions, pick a single colony using a sterile inoculation loop and inoculate into 50 μL of sterile water. Use 2 μL of this as template for colony PCR and save the other 48 μL for setting up overnight cultures after verification. Proceed with PCR using DreamTaq DNA Polymerase kit. Electrophorese the amplified PCR product in 1% agarose gel alongside a 1-kb ladder and check if the bands correspond to the length of the designed target gene constructs.

3.5.6 Verification of Target Insertion

1. Set up overnight cultures of the transformed HT115-DE3 bacterial cells in LB broth containing 12-μg/mL tetracycline and 100-μg/mL ampicillin.
2. Extract plasmid the next day using the GeneJET Plasmid Miniprep Kit. Check the concentration of extracted plasmid using a Nano-Drop spectrophotometer.
3. Sequence the plasmid DNA using Sanger sequencing with primers designed in silico for dsRNA synthesis.
4. Using DNA sequence editing software (e.g., Snapgene or Benchling), prepare an in silico vector map including target insert. Align the sequences obtained from Sanger sequencing using Snapgene/Benchling to verify precise insertion.

3.5.7 dsRNA Synthesis

1. Set up cultures of the positive colonies in 3 mL of LB broth with ampicillin and tetracycline (final concentration of 100 μg/mL and 12 μg/mL, respectively). Incubate with shaking at 37 °C overnight.
2. The next day, fill a cuvette with 1 mL of the overnight culture and check the absorbance at 600-nm wavelength (optical density at 600 nm; OD_{600}) using a spectrophotometer.
3. Dilute the culture with fresh LB with antibiotics to an OD_{600} of 0.05 (make up to 20 mL) and incubate for approximately 3–4 h until an OD_{600} of 0.2 is obtained.
4. To induce dsRNA synthesis, add 2 μL of 100 mM IPTG to 20 mL of HT115 culture (final concentration of 10 mM) and incubate with shaking at 37 °C for 3–4 h.

5. Centrifuge the induced bacterial culture at 3500 rcf for 10 min at 20 °C. Decant the supernatant and use the pellet for RNA extraction using Direct-zol™ RNA Miniprep Plus Kit. Elute the RNA in 100 μL of elution buffer/nuclease-free water. Repeat elution in a new collection tube with 100 μL of fresh elution buffer/nuclease-free water.
6. Store the eluted RNA at −80 °C.
7. Dilute 2 μL of the eluted RNA in 8 μL of H_2O and 2 μL of 6× loading dye. Run in 1% agarose gel alongside a 1-kb ladder for reference. Visualize under UV light to verify if the bands correspond to the size of the target dsRNA sequence.

3.6 DLA

- Collect potato leaves, with six leaves for each treatment, from 4 to 5 weeks old potato plants (*see* **Note 9**). Place the collected leaves in the boxes.
- Prepare in vitro synthesized dsRNA for DLA to a final concentration of 20 ng μL^{-1} by diluting with distilled water accordingly. Spray the leaves using an automizer or an airbrush. Experimental controls are set by spraying $dsRNA^{Ct}$.
- For bacterial in vivo synthesized dsRNA, spray the leaves with 10 μg of total RNA per leaf using an airbrush. Experimental controls are set by spraying with RNA from HT115-DE3 carrying empty L4440, and L4440 with a cloned nonspecific gene target.
- Place the boxes with sprayed leaves in climate-controlled chambers with a temperature of 21 °C during the day and 19 °C at night, for 24 h.
- Collect *Pi*-GFP spores as explained in Subheading 3.1, step **10**; count the number of spores under a microscope with the help of a Fuchs Rosenthal chamber (0.200 mm depth) (*see* **Note 10**). Adjust the concentration of spores to 50,000 spores/mL by diluting with distilled water accordingly.

1. Twenty-four hours post spraying, infect the sprayed potato leaves by drop inoculation with 15 μL per leaf of freshly collected *Pi*-GFP spores (adjusted to 50,000 spores/mL). Carefully place them back in the climate chamber.
2. Five days after infection, observe the detached leaves for *P. infestans* infection lesions. Compare and quantify the lesion area of dsRNA-treated leaves to control leaves using software such as ImageJ (Fiji), to determine if the selected dsRNA target reduces *P. infestans* disease progression (Fig. 1).

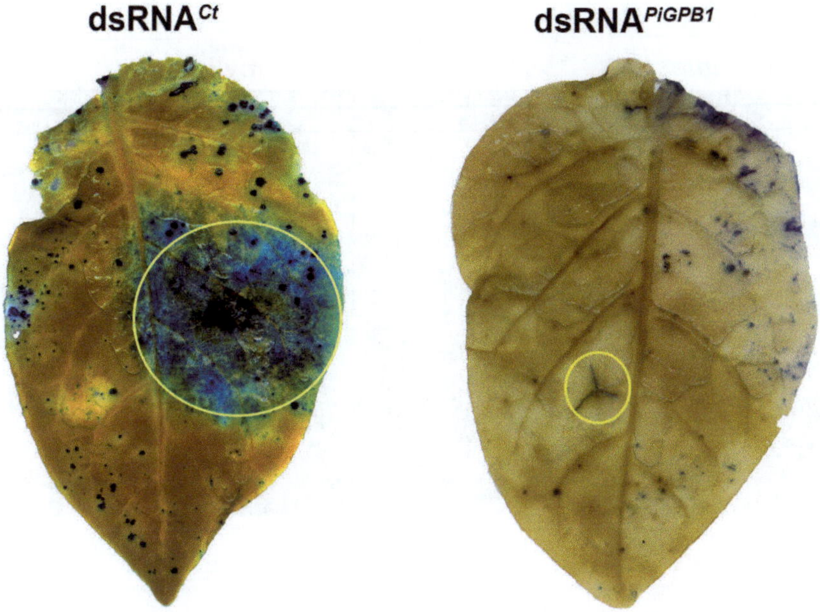

Fig. 1 Effect of control and test gene dsRNA sprays on *Phytophthora infestans* pathogenicity on potato leaves. Representative picture taken at 5-days postinoculation (dpi) of trypan blue stained potato leaves showing typical *P. infestans* infection on leaf sprayed with dsRNA^{Ct}, in comparison to decreased infection on leaf sprayed with dsRNA^{PiGPB1} (test). Yellow lines delineate the infected area on each leaf

3.6.1 Trypan Blue Staining

1. Prepare trypan blue staining solution containing 10 mL lactic acid, 10 mL buffer saturated phenol, 10 mL glycerol, 10 mL of distilled water, and 40 mg of trypan blue.
2. Five-days postinoculation, submerge infected potato leaves from each DLA treatment into individual 90-mm Petri dishes containing 20 mL of trypan blue stain solution for 30 min. Destain leaves with absolute ethanol in an orbital shaker overnight.
3. Transfer the destained leaves to 50% glycerol for 2 h to rehydrate. The rehydrated samples can be used for microscopy and long-term storage.

3.7 Confocal Microscopy

1. Preparation of samples for microscopy – Five-days postinoculation, take leaves from DLA to make 1 mm^2 cuts at the site of infection. Mount the cut leaf samples onto glass slides and add a drop of water before covering with a coverslip.
2. Using the confocal microscope (e.g., LSM880, Zeiss Microscopy GmbH, Germany), visualize disease lesions to confirm the effect of dsRNA^{PiGPB1} on *P. infestans* spore formation (or other test gene(s)) (Fig. 2). GFP and chlorophyll can be excited using lasers of 488 and 633 nm and detected at emission wavelengths of 499–552 and 647–721 nm, respectively.

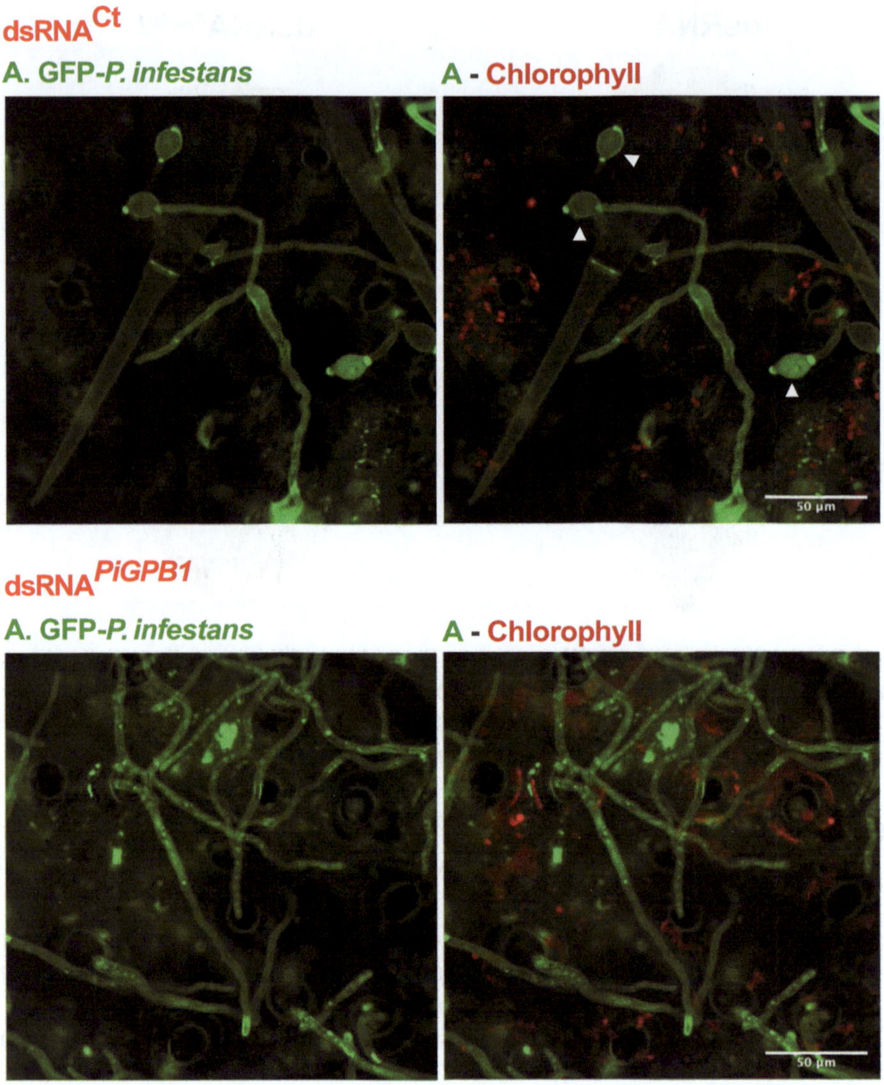

Fig. 2 Representative confocal microscopy images of *P. infestans* infection on potato leaves treated with dsRNA^{Ct} (control; **a**) or dsRNA^{PiGPB1} (**b**), taken at 5 dpi. Hyphal growth from the plant cells is visible in both control and treatment. Sporangia are formed in dsRNA^{Ct}-treated leaves (indicated as arrowheads) while there is no sporangia formation in the samples treated with dsRNA^{PiGPB1}, indicating possible gene knockdown of *GPB1* in *P. infestans*. Chloroplast autofluorescence is shown as red in these images; scale bars = 50 μm

3.8 RNA Extraction

1. Collect leaf discs (weighing 100 mg) from six infected detached leaf assay samples in individual 1.5-mL microcentrifuge tubes and freeze immediately using liquid nitrogen.
2. Grind the frozen samples into a fine powder in liquid nitrogen using mortar and pestle. Carry out RNA extraction using the Direct-zol™ RNA Miniprep Plus Kit. Once the ground samples are homogenized using TRIzol, centrifuge and transfer the

supernatant to a new 1.5-mL microcentrifuge tube (*see* **Note 11**). Elute RNA in 35 μL of nuclease-free water.

3. Measure the concentration of eluted RNA using a NanoDrop spectrophotometer before storing at −80 °C.

3.9 cDNA Synthesis

1. Generate cDNA from eluted RNA using qScript cDNA SuperMix. Combine 4 μL of the 5× cDNA SuperMix with respective RNA and make up to 20 μL using nuclease-free water. Use approximately 500 ng – 1 μg of extracted RNA as template for first-strand synthesis.
2. Incubate as follows: 25 °C for 5 min; 42 °C for 30 min; 85 °C for 5 min.
3. Store the cDNA at −20 °C.

3.10 qRT-PCR

- Design forward and reverse primers for qPCR using the Primer-BLAST tool from NCBI (https://www.ncbi.nlm.nih.gov/tools/primer-blast/) (*see* **Notes 12** and **13**).
- Set up a 20 μL reaction using DyNAmo flash SYBR Green qPCR kit. The components are as follows.

Component	Volume
2× Master mix	10 μL
50× ROX dye	0.12 μL
Forward primer (0.5-μM final concentration)	1 μL
Reverse primer (0.5-μM final concentration)	1 μL
Template (cDNA) (20–25 ng)	variable
Water	Make up to 20 μL

- Use 20 ng of respective cDNA as template. Set up qPCR reactions targeting both reference and target genes individually; with four biological replicates each of cDNA synthesized from DLA samples (*see* **Note 14**). The thermal conditions are as follows.

Step	Temperature	Time	No of cycles
Initial denaturation	95 °C	5 min	1
Denaturation	95 °C	10 s	40
Annealing/extension	60 °C	30 s	
Final extension and melt curve	65 °C to 95 °C in increments of 0.5 °C		

3.10.1 qRT-PCR Analysis

- Analyze the qPCR results using the $2^{-\Delta\Delta C}$ method [9, 10].
  - Obtain C_t values from the Bio-Rad CFX machine.
  - Calculate the average C_t value for each biological replicate.
  - Calculate $\Delta C_t = C_t$ (target) – C_t (reference) for the same biological replicate.
  - Calculate average ΔC_t of the control samples.
  - Calculate $\Delta\Delta C_t$.

    $\Delta\Delta C_t$ (control) = ΔC_t (each biological replicate, control sample) – Average ΔC_t of control sample.

    $\Delta\Delta C_t$ (test) = ΔC_t (each biological replicate, test sample) – Average ΔC_t of control sample.
  - Calculate $2^{-\Delta\Delta Ct}$.
  - To obtain fold change values, calculate geometric mean of $2^{-\Delta\Delta Ct}$ values of control and test samples individually.
  - Additionally, calculate standard deviation or standard error of the mean for the respective $2^{-\Delta\Delta Ct}$ values.

4 Notes

1. Exclude adding agar before autoclaving if preparing rye broth instead of rye agar.
2. When excited by light in the blue to ultraviolet range, the green fluorescent protein emits a bright green fluorescence, hence deriving its name.
3. Green fluorescent protein is expressed under the control of *Ham34* promoter [11].
4. Add antibiotics to the same final concentration if preparing rye broth.
5. For synthesis of dsRNA, several PCR reactions may be needed to generate enough amplified product for in vitro transcription.
6. GPB1 in *Phytophthora infestans* does not contain any introns and so genomic DNA can be used for PCR. For other *P. infestans* genes, check for presence of introns in FungiDB [https://fungidb.org/fungidb/app].
7. An additional 4 bp of random sequence needs to be added to the 5′ end of the primer to ensure correct restriction digestion after PCR.
8. Evenly spread transformed HT115-DE3 bacteria onto the prepared LB agar plates using a sterile L-shaped spreader.
9. Take airtight boxes and line the bottom with water-dampened tissue paper, overlaid with mesh. This maintains a microclimate

in the box, mimicking greenhouse conditions and preventing the leaves from drying and wilting quickly.

10. Load 10 μL of spore sample into each of the two chambers and cover using a coverslip. Each chamber has 16 squares of 1 mm length, represented by three grid lines and each of these 16 squares are further divided into 16 smaller squares. Count the number of sporangia in all 16 1-mm squares. Avoid including the spores on the triple-lined boundaries. Take an average of these numbers and use the formula below to determine the number of sporangia present in 1 mL of spore solution.

    $$\text{Spores/mL} = (\text{Average per square} \times 16)/0.0032$$

11. This eliminates the chance of any particulate debris clogging the column and reducing the yield of extracted RNA.
12. Primers for qRT-PCR need to be outside the region used for generating the dsRNA.
13. Include primers for reference genes in *P. infestans* and *S. tuberosum* while designing qRT-PCR.
14. Use three technical replicates per biological replicate.

Acknowledgments

This work has been supported by Nova Nordisk Fonden (0074727), FORMAS (2019-01316), The Swedish Research Council (2019-04270), Carl Tryggers Stiftelse för Vetenskaplig Forskning (CTS 20:464), The Crafoord foundation (20200818), Partnerskap Alnarp (1317/Trg,VO/2020, 1353/VO/2021), and Alnarp stipendiekommitténs. SCW acknowledges financial support from the Scottish Government Rural and Environment Science and Analytical Services Division (RESAS). The authors have no conflicts of interest to declare.

References

1. Koch A, Biedenkopf D, Furch A et al (2016) An RNAi-based control of Fusarium graminearum infections through spraying of long dsRNAs involves a plant passage and is controlled by the fungal silencing machinery. PLoS Pathog 12:e1005901
2. Wang M, Jin H (2017) Spray-induced gene silencing: a powerful innovative strategy for crop protection. Trends Microbiol 25(1):4–6
3. Sang H, Kim JI (2020) Advanced strategies to control plant pathogenic fungi by host-induced gene silencing (HIGS) and spray-induced gene silencing (SIGS). Plant Biotechnol Rep 14:1–8
4. Cagliari D, Dias NP, Galdeano DM et al (2019) Management of pest insects and plant diseases by non-transformative RNAi. Front Plant Sci 10:1319
5. Gebremichael DE, Haile ZM, Negrini F et al (2021) RNA interference strategies for future management of plant pathogenic fungi: prospects and challenges. Plan Theory 10(4):650
6. Assinder SJ (2004) Molecular genetics and genomics of phytophthora. In: Arora DK, Khachatourians GG (eds) Applied mycology and biotechnology, fungal genomics, vol 4. Elsevier, pp 137–160

7. Latijnhouwers M, Govers F (2003) A Phytophthora infestans G-protein β subunit is involved in sporangium formation. Eukaryot Cell 2:971–977
8. Kalyandurg PB, Sundararajan P, Dubey M et al (2021) Spray-induced gene silencing as a potential tool to control potato late blight disease. Phytopathology 111(12):2168–2175. https://doi.org/10.1094/PHYTO-02-21-0054-SC
9. Livak KJ, Schmittgen TD (2001) Analysis of relative gene expression data using real-time quantitative PCR and the $2^{-\Delta\Delta C_T}$ method. Methods 25(4):402–408
10. Vandesompele J, De Preter K, Pattyn F et al (2002) Accurate normalization of real-time quantitative RT-PCR data by geometric averaging of multiple internal control genes. Genome Biol 3(7):RESEARCH0034
11. Avrova AO, Boevink PC, Young V et al (2008) A novel Phytophthora infestans haustorium-specific membrane protein is required for infection of potato. Cell Microbiol 10:2271–2284

Chapter 28

Plant Sampling for Production of Essential Oil and Evaluation of Its Antimicrobial Activity In Vitro

Felicia Menicucci, Eleonora Palagano, Aida Raio, Gabriele Cencetti, Nicola Luchi, Andrea Ienco, and Marco Michelozzi

Abstract

Essential oils (EOs) and oleoresins are complex mixtures mainly made up of terpenes, synthesized by a wide variety of plants. Individual terpenes may show broad-spectrum activity against different plant pathogens, and their combination into EO and oleoresin mixtures enhances plant chemical defense. The interest in EOs has significantly increased due to the trend of using natural products as herbicides, insecticidal and antimicrobial agents. In addition, the use of plant mixtures is an emerging approach to face the problem of antimicrobial resistance in agriculture. This chapter reports guidelines about plant sample collection for the production of EOs and provides protocols to test their activity as antimicrobial agents against bacteria and fungi. It also describes a solvent-free method for the inclusion of EOs into β-cyclodextrins. This type of formulate is prepared to turn liquid EOs into easily manageable water-soluble powders, and to control the release of volatile compounds, aiming to increase EOs' applications in agriculture.

Key words Essential oils, Conifers, Oleoresins, Aromatic plants, Terpenes, Antimicrobial activity, Phytopathogens, Chemotype

1 Introduction

Essential oils (EOs) and oleoresins are plant mixtures mainly made up of terpenes. They are produced in different plant tissues as xylem, cortex, and foliar tissue, and occur in various organs such as flowers, leaves, branches, stems, and roots [1].

Terpenes represent the largest class of natural products, including more than 80,000 compounds [2].

These compounds exhibit enormous structural and functional diversity, despite their common biosynthetic origin deriving from the two 5-carbon (C_5) units, namely, isopentenyl pyrophosphate (IPP) and its isomer dimethylallyl diphosphate (DMAPP). The fusion of multiple basic C_5 units results in C_{10} monoterpenes, C_{15} sesquiterpenes, C_{20} diterpenes, C_{30} triterpenes, C_{40} tetraterpenes,

Nicola Luchi (ed.), *Plant Pathology: Methods and Protocols*, Methods in Molecular Biology, vol. 2536,
https://doi.org/10.1007/978-1-0716-2517-0_28,

and polyterpenes, containing more than 40 carbons. The volatile terpenes (C_{10} monoterpenes and C_{15} sesquiterpenes) characterize the EO composition, whereas conifer oleoresins are constituted by non-volatile C_{20} diterpenes, in addition to mono- and sesquiterpenes [3] (Fig. 1).

Some plant terpenes are primary metabolites, such as the growth regulator abscisic acid and gibberellins, and those found within the carbon skeleton of chlorophyll, steroids, and carotenoids [4]. However, the vast majority of terpenes, including mono-, sesqui-, and diterpenes, are considered "secondary" metabolites. Several of these secondary terpenes were demonstrated to be toxic, repellent, or attractive, and thus indispensable for the adaptive responses of the plant to environmental challenges and for the survival of the species [4, 5].

Although individual terpenes may show broad-spectrum activity against different phytopathogens, their combination into terpene mixtures displays empowered antimicrobial and insecticidal properties, due to an additive/synergistic effect [6].

EOs are synthesized by a wide range of plant species belonging to all taxonomic groups, whereas the oleoresin is typical of conifer plants, such as the *Pinus* species [7]. In physiological conditions, conifers produce and store large amounts of constitutive (or primary) oleoresin. Additionally, induced (or secondary) oleoresin occurs when the plant reacts against pests and diseases, producing new and different terpene mixtures. Constitutive and induced oleoresins represent effective chemical defenses that conifers have evolved over hundreds of millions of years to deter herbivores and pathogens [8].

In general, plants may induce qualitative and quantitative variations in the composition of an oleoresin or EO (intra-plant, intra- and inter-provenances, and inter-species) originating different terpene profiles, or chemotypes, to enhance their protection. This strategy hampers herbivores to evolve resistance toward a specific terpene mixture [9].

Concerning EOs, the interest for their antimicrobial properties has considerably increased due to the trend of using natural products as food preservatives and as insecticidal, antimicrobial, and cytotoxic agents [10–12]. These properties are related to the lipophilic nature of the hydrocarbon skeleton and the functional groups of the EO components. Indeed, the hydrophobic characteristics of these bioactive compounds facilitate their passage through the biological membranes and, hence, their activity within the cell [13]. The complex composition of EOs makes it difficult to explain their specific mechanism of action, but, on the other hand, it is a positive factor that may limit the development of microorganism resistance, typically exerted by synthetic drugs [14, 15].

Therefore, there is an urgent need to develop suitable antimicrobial alternatives to be used alone or in combination with the

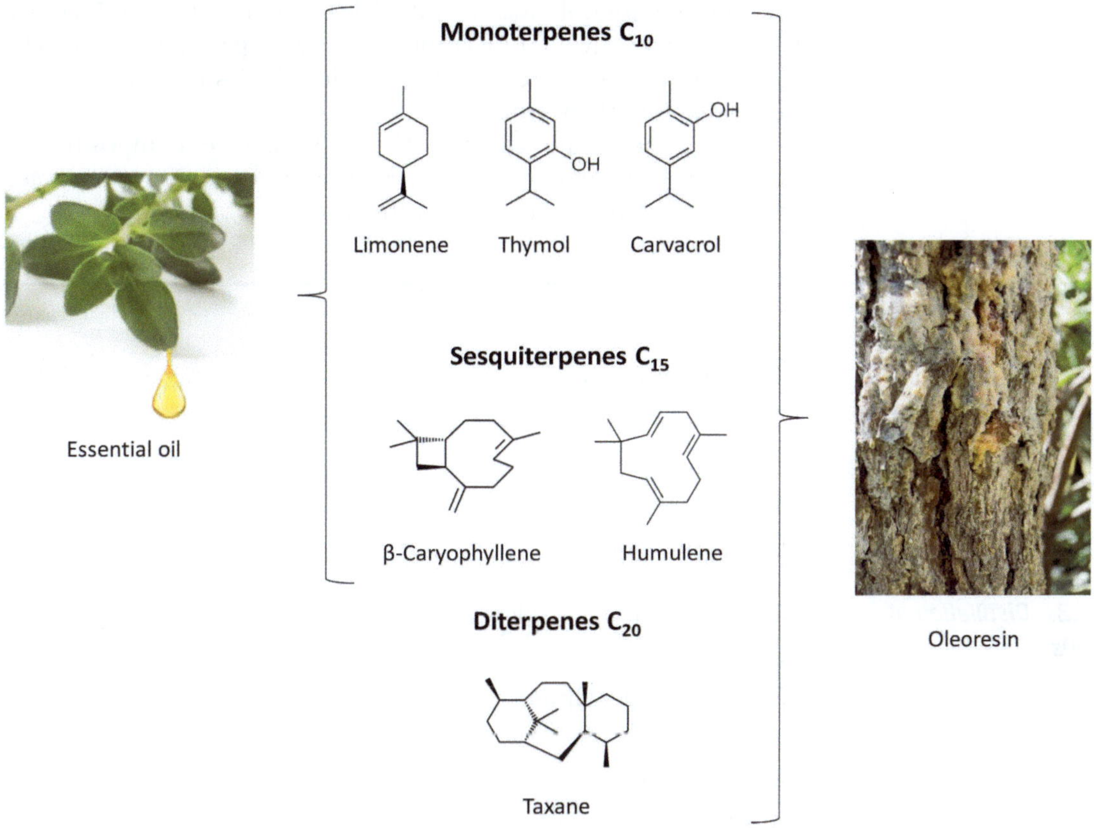

Fig. 1 Examples of mono-, sesqui-, and diterpene compounds. EO mixtures are composed of volatile mono- and sesquiterpenes, whereas oleoresins consist of volatile mono- and sesquiterpenes plus non-volatile diterpenes

classical ones, improving drugs' efficacy against a wide variety of plant pathogens [16]. In this context, plant EOs could represent a valuable natural resource to be exploited [17]. Another possible application of EOs concerns their use in plant treatment as non-synthetic pesticides and herbicides, as the massive use of the traditional ones implies serious consequences on both the environment and human health [18–20].

On the other hand, the application of EOs poses some issues, specifically related to their high volatility and scarce aqueous solubility. The high volatility leads to a rapid loss of EOs' efficacy; moreover the hydrophobicity of their constituents strongly limits EOs' use in agriculture, as most of the conventional chemicals applied are watery products [21, 22].

In this perspective, the use of EO-based formulations has been recently experimented in agriculture, trying to overcome the aforementioned drawbacks [23–25]. The aim of this work is to provide practical guidelines for sampling and collecting plant tissues to produce EOs. Information about how to use these natural products for microbiological assays and the technique to produce the β-cyclodextrin-thyme EO complex are also provided [26]. This solid formulation is prepared to turn liquid EOs into easy-to-handle, water-soluble powders.

Throughout the chapter, the EO mixture composition is defined as terpene constituents, expressed as relative percentages of the total terpenes composing the mixture.

2 Materials

2.1 Plant Sample Collection

1. Disposable gloves.
2. Ice pack.
3. Plastic bags.
4. Scissors.
5. Thermal bag.

2.2 Distillation of EOs

1. Deionized or distilled water.
2. Distillator.
3. Graduated cylinder.
4. Glass vials.
5. Heat source.
6. Plant tissues.

2.3 In Vitro Test: Determination of the Antimicrobial Activity of EOs

The procedure requires strict adherence to aseptic techniques: sterilize all materials and use a laminar flow hood to prevent any contamination. Optimum temperature growth and time of incubation for plant pathogenic fungi and bacteria may change in relation to the microbial species.

1. Antimicrobial compound as reference for activity comparison (*see* **Note 1**).
2. Antimicrobial compound (e.g., thyme EO).
3. Nutrient Glucose Agar medium (NAG): in 1 L of deionized water contained in a conical flask, dissolve 25 g of NB powder, 18 g of agar and 2.5 g of glucose.
4. Potato Dextrose Agar medium (PDA): in 1 L of deionized water contained in a conical flask, dissolve 39 g of PDA powder.
5. Saline solution (0.8% NaCl): in 100 mL of deionized water contained in a conical flask, dissolve 0.8 g of NaCl salt.
6. Deionized or distilled water.
7. Fresh fungal culture.
8. Fresh bacterial culture.

9. Cork-borer (0.5 mm diameter).
10. Dissecting needle.
11. Cuvettes.
12. Pipettes.
13. Tips for pipettes.
14. Sterile loops.
15. Sterile Petri dishes (90 × 15 mm).
16. Laboratory glassware.
17. Parafilm.
18. Magnets.
19. Magnetic stirrer.
20. Ruler.
21. Incubator.
22. Spectrophotometer.
23. Autoclave.

2.4 In Vitro Test: Determination of the Minimum Inhibitory Concentration (MIC)

The procedure requires strict adherence to aseptic techniques: sterilize all materials and use a laminar flow hood to prevent any contamination.

1. Antimicrobial compound as reference for activity comparison (*see* **Note 1**).
2. Antimicrobial compound (e.g., thyme EO).
3. Fresh fungal culture.
4. Fresh bacterial culture.
5. Nutrient Glucose Agar medium (NAG): in 1 L of deionized water contained in a conical flask, dissolve 25 g of NB powder, 18 g of agar and 2.5 g of glucose.
6. Nutrient Glucose Broth (NGB): in 1 L of deionized water contained in a conical flask, dissolve 25 g of NB powder and 2.5 g of glucose.
7. Potato Dextrose Agar medium (PDA): in 1 L of deionized water contained in a conical flask, dissolve 39 g of PDA powder.
8. Potato Dextrose Broth medium (PDB): in 1 L of deionized water contained in a conical flask, dissolve 24 g of PDB powder.
9. Methanol.
10. Deionized or distilled water.
11. Resazurin.
12. Cork-borer (0.5 mm diameter).
13. Dissecting needle.

14. Eppendorf tubes.
15. Glass coverslips.
16. Hemocytometer.
17. Parafilm.
18. Pipettes.
19. Tips for pipettes.
20. Screw cap tube.
21. Sterile 96-well microplates.
22. Sterile Petri dishes (90 × 15 mm).
23. Sterile spreader loop.
24. 1.5 mL Eppendorf tubes.
25. Loops.
26. Multi-well microplate reader spectrophotometer.
27. Screw cap tube.
28. Optical microscope.
29. Autoclave.
30. Spectrophotometer.
31. Incubator.
32. Vortex.

2.5 Production of β-Cyclodextrin-EO Complex

1. Grinding steel jars with push-fit lids (volume: 10 mL).
2. H_2O/EtOH buffer: mixture contains ultrapure water and pure ethanol in the ratio 2:1.
3. Laboratory mixer mill.
4. Liquid nitrogen.
5. Pestle.
6. Pipettes.
7. Polystyrene box with lid.
8. Steel balls (2 mm diameter).
9. Thyme EO.
10. Tips for pipettes.
11. β-cyclodextrin.

3 Methods

3.1 Sampling Method of Plant Tissues for EO Production

It is possible to sample wild plants in their native habitat, if harvesting is not difficult and too expensive. Alternatively, samples can be collected easily from plants growing in experimental fields, such as plantations of provenances, varieties, and clones (*see* **Note 2**).

1. Cut plant samples with scissors or garden pruners and store into plastic bags.
2. Promptly place the collected samples in a thermal ice bag to bring them to the laboratory.
3. Store samples at 4 °C if plant material is processed by 24 h from the collection. For prolonged preservation, store the material at −20 °C.

3.1.1 Factors to Consider in Sample Collection: Plant Age

Variations in constitutive terpene profiles were detected between adult trees and juvenile seedlings. Age effects seem to be irrelevant after the juvenile stage, therefore sampling should be restricted to adult plants or to same age plants [27].

3.1.2 Factors to Consider in Sample Collection: Tissue Age and Tissue Source

Constitutive terpene composition depends on the age of plant tissue (Fig. 2): for instance, these differences were detected studying young and mature foliar and cortical samples of Mediterranean pines of the group "*halepensis*" *Pinus halepensis* Mill., *Pinus brutia* Ten., and *Pinus eldarica* Medw. (Fig. 3).

In addition, different types of tissues show variations in constitutive terpene profiles, like those observed among mature 1-year foliar and cortical samples of Mediterranean pines of group "*halepensis*" (Fig.4) (*see* **Note 3**).

3.1.3 Factors to Consider in Sample Collection: Season

Absolute amounts (expressed, e.g., as mg of terpene g^{-1} dry weight tissue) of constitutive terpenes can be affected by environmental abiotic factors, while relative contents (e.g., each single monoterpene expressed as percentage of the total monoterpenes) are under strong genetic control and little influenced by abiotic stressors [3, 27–29]. Therefore, constitutive terpene profiles in mature tissues and plants have been used as biochemical markers in chemosystematics to characterize species, provenances, hybrids, and clones [30–32].

For example, the relative contents of constitutive β-pinene, β-myrcene, and limonene in mature foliar and cortical tissues of *P. halepensis*, *P. brutia*, and *P. eldarica*, showed little variation between sampling times during the course of seasons throughout the 3 years of the study.

However, the variations in the percentage of monoterpenes between different sampling times within a pine species were much lower than those related to the percentage of the same monoterpenes among the different pine species of the "*halepensis*" group (Fig. 5).

3.2 Steam Distillation of EOs

1. Place the water in the boiler until the established level.
2. Place the plant evenly on the grill and close the lid of the distiller.

Fig. 2 *Taxus baccata* L. with young (light green) and mature needles (dark green). Taxol ($C_{47}H_{51}NO_{14}$) is a diterpenoid with strong antitumor activity, synthesized by *T. baccata* and other species of the *Taxus* genus [28]

3. Fasten the steel collector and the glass burette.
4. Turn on the heat source and, when the temperature reaches 50 °C, activate the cooling of the system.
5. Keep the temperature at 100 °C for about 1 h, to allow the hydrosol (or hydrolate) and oil collection into the burette (*see* **Note 4**).
6. Turn off the heat source, when the increase of the oil level into the burette is irrelevant.
7. Collect into a glass vial the extracted hydrosol from the bottom of the graduated cylinder (*see* **Note 5**).
8. After hydrosol collection, recover the EO as described at **step 7** (Fig. 6).

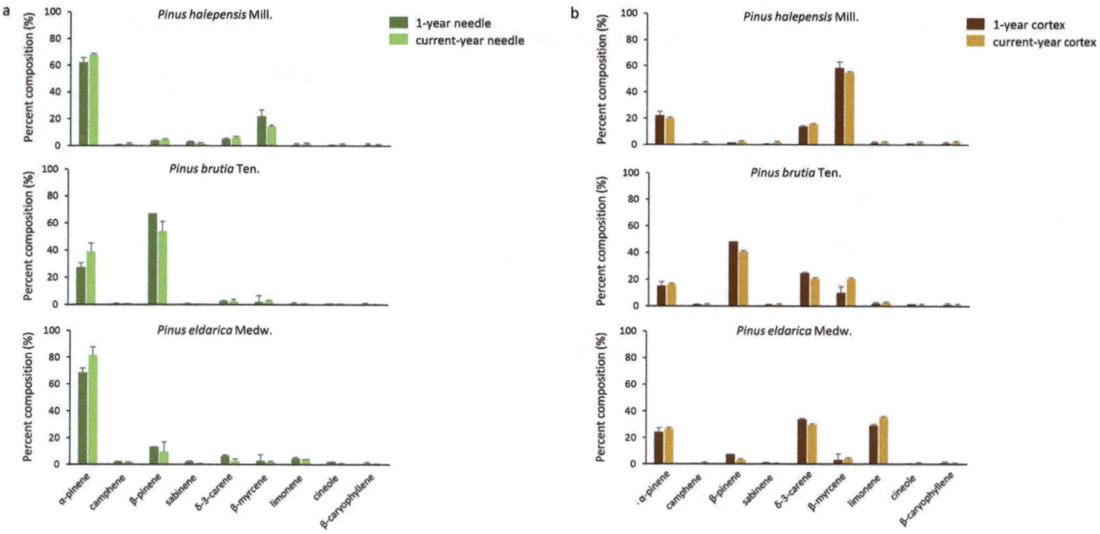

Fig. 3 Variations in the relative content of eight monoterpenes (α-pinene, camphene, β-pinene, sabinene, δ-3-carene, myrcene, limonene, and cineole) and one sesquiterpene (β-caryophyllene) between mature (1-year tissue) and young (current-year tissue) needle (**a**) and cortex (**b**) samples of *Pinus halepensis* Mill., *Pinus brutia* Ten., and *Pinus eldarica* Medw

3.3 In Vitro Test: Determination of the Antimicrobial Activity of EOs

3.3.1 Protocol for Fungi

1. Prepare the required amount of PDA, stir and then autoclave at 121 °C for 15 min.
2. Pour the medium into sterile Petri dishes (about 15 mL/dish) and let solidify the medium at room temperature under the laminar hood.
3. Leave the plates without the lid under the laminar hood until the medium surface is dry.
4. After medium solidification, cut two wells of 0.5 cm diameter at the opposite sides of the Petri dishes, using a sterile cork-borer.
5. Fill in the wells with the amount of EO to be tested (e.g., 20 μL per well).
6. Fill in the wells with the reference antimicrobial compound using the same concentration as the EO's one.
7. Fill in the wells with sterile distilled water (positive control dishes).
8. Cut a 0.5 cm disk from the edge of the fungal colony using a sterile cork-borer (*see* **Note 6**). Transfer the disk in the middle of the Petri dish between the two wells using a dissecting needle. Prepare at least three replicates (i.e., three Petri dishes for the EO, three Petri dishes for the reference antimicrobial compound, and three Petri dishes as positive control).
9. Seal the Petri dishes with three layers of Parafilm, incubate at the optimum growth temperature for each fungal species and monitor the fungal growth daily.

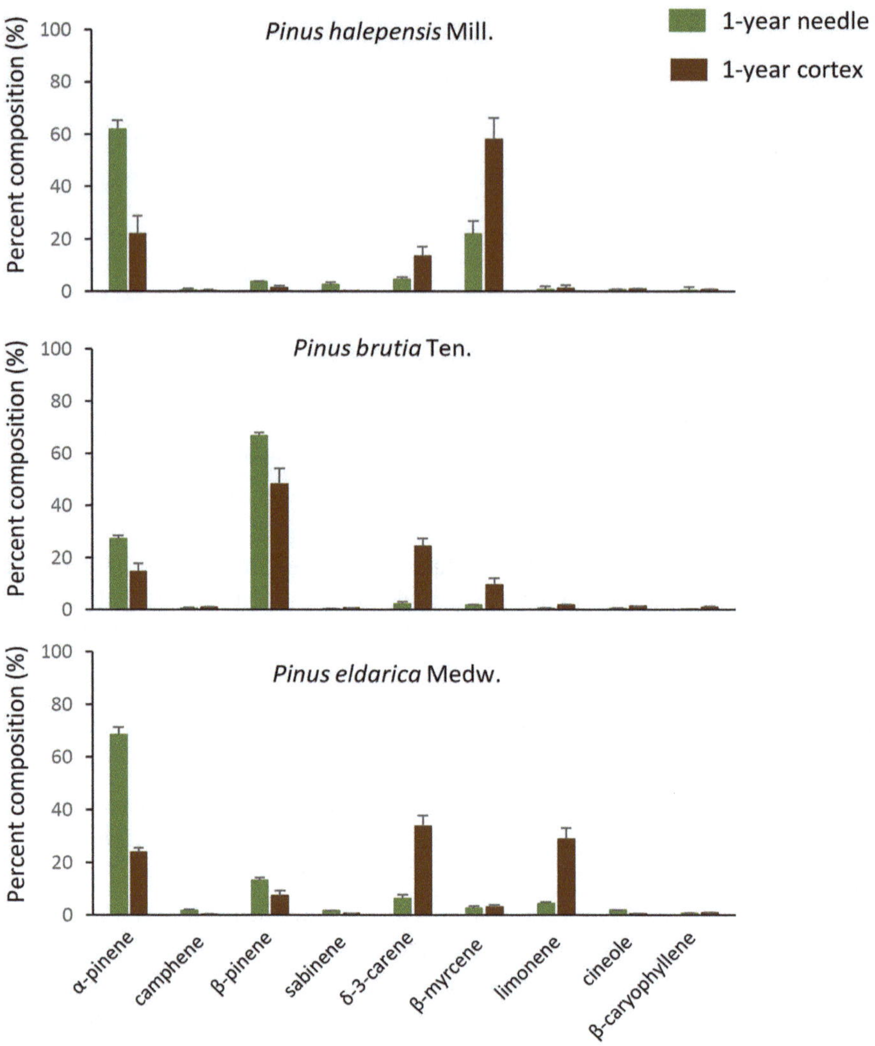

Fig. 4 Differences in the relative content of eight monoterpenes (α-pinene, camphene, β-pinene, sabinene, δ-3-carene, myrcene, limonene, and cineole) and one sesquiterpene (β-caryophyllene) between mature (1-year tissue) needle and mature (1-year tissue) cortex samples of *Pinus halepensis* Mill., *Pinus brutia* Ten., and *Pinus eldarica* Medw

10. Using a ruler, measure two diameters for each fungal colony and calculate the average diameter obtained from the three replicates.
11. Compare the collected data (average diameter of colonies grown in plates with EO, average diameter of colonies grown in plates with the reference antimicrobial compound and average diameter of colonies of positive control dishes) to evaluate the inhibiting activity of the tested EO against the fungal species selected for the experiment.

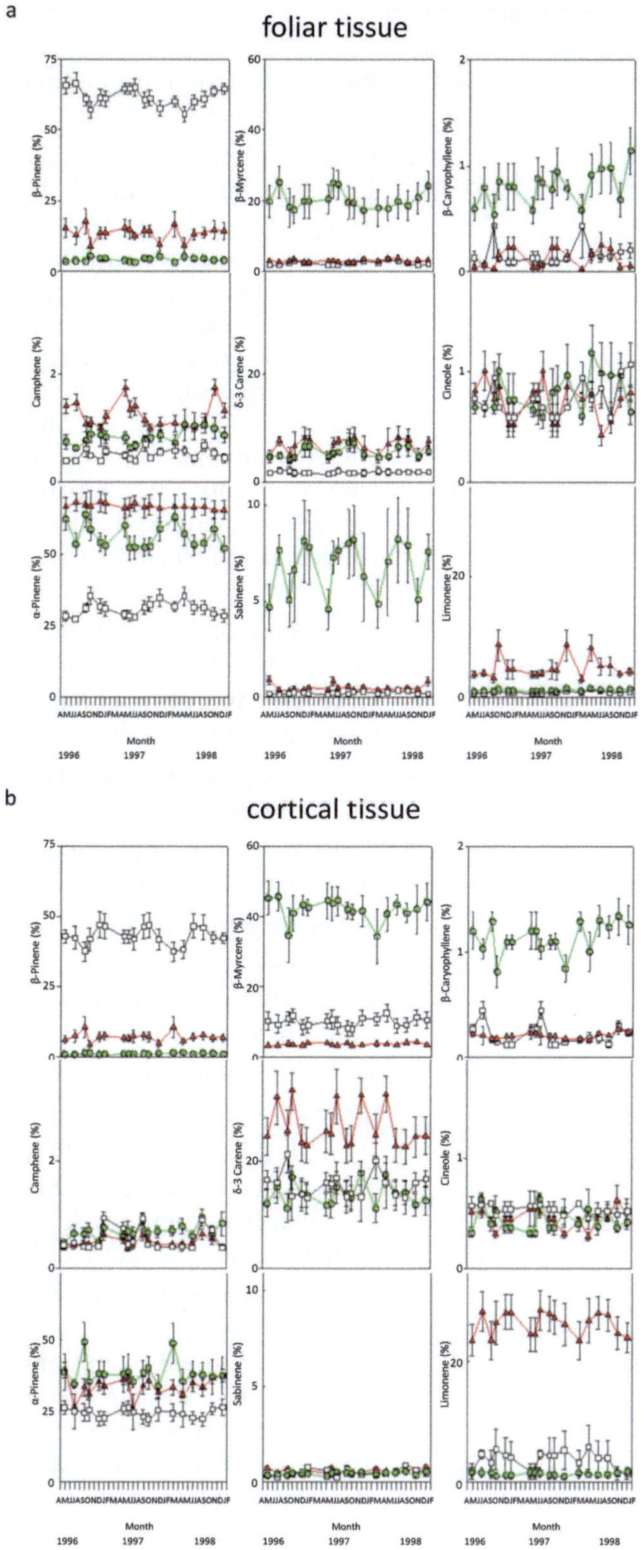

Fig. 5 Variations in the relative content of eight monoterpenes (α-pinene, camphene, β-pinene, sabinene, δ-3-carene, myrcene, limonene, and cineole) and one sesquiterpene (β-caryophyllene) of mature foliar (**a**) and cortical (**b**) tissue samples collected during three consecutive years (1996–1998) from adult trees of *Pinus halepensis* Mill., *Pinus brutia* Ten., and *Pinus eldarica* Medw. (Adapted from Michelozzi et al. [33])

3.3.2 Protocol for Bacteria

1. Prepare the required amount of Nutrient Glucose Agar (NGA), stir and then autoclave at 121 °C for 15 min.
2. Prepare the bacterial suspension as follow: before starting the antimicrobial test, refresh the bacterial strain from the long storage culture by streaking a loop of the suspension on the surface of NGA medium and incubate at the optimum growth temperature. Prepare an appropriate amount of saline solution (0.8% NaCl) and autoclave at 121 °C for 15 min. After 48 h incubation collect a little amount of the fresh bacterial culture from the agar surface using a sterile loop and prepare a suspension in saline solution. Measure the optical density of the bacterial suspension by a spectrophotometer, setting the wavelength at 530 nm. Prepare the appropriate amount of the bacterial suspension needed for the experiment and adjust the concentration to an optical density of 0.1, which corresponds to about 1×10^8 cfu/mL.
3. Put 1.5 mL of the bacterial suspension at the bottom of a sterile Petri dish, add 15 mL of NGA cooled at about 55 °C and gently mix to obtain a homogeneous distribution of the bacterial suspension into the medium.
4. Leave the plate without the lid under the laminar hood until the medium surface dries.
5. After medium solidification, cut two wells of 0.5 cm diameter at the opposite sides of the Petri dish, using a sterile cork-borer.
6. Fill in the wells with the amount of EO to be tested (e.g., 20 μL per well).
7. Fill in the wells with the reference antimicrobial compound using the same concentration as the EO's one.
8. Fill in the wells with sterile distilled water (positive control dishes).
9. Prepare at least three replicates (i.e., three Petri dishes to test the efficacy of the EO, three Petri dishes for the reference antimicrobial compound, and three Petri dishes as positive control). Seal the Petri dishes with three layers of parafilm and incubate at the optimum temperature of each bacterial species.
10. Active compounds determine the formation of a clear zone surrounding the wells. Using a ruler, measure the size of the inhibition halo between the well and the margin of the area of active bacterial growth. Make at least two measurements for each well, and calculate the average value obtained from the three replicates.
11. Compare the collected data to evaluate the inhibiting activity of the tested EO against the bacterium selected for the experiment (Fig. 7).

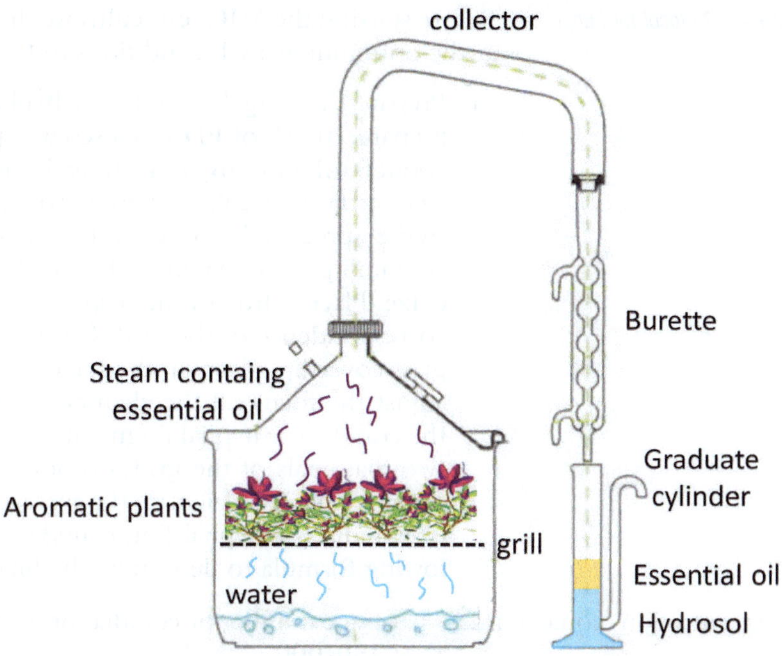

Fig. 6 Schematic setup for the distillation of EOs

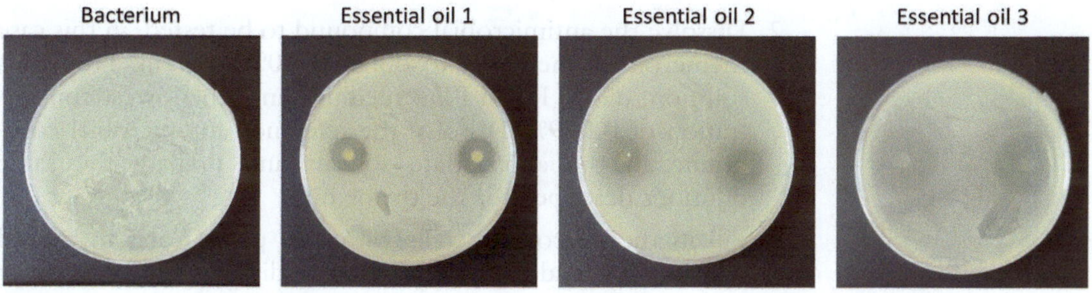

Fig. 7 Antibacterial test: inhibition halos (clear zones) determined by three different EO treatments

3.4 In Vitro Test: Determination of the Minimum Inhibitory Concentration (MIC)

MIC is the lowest concentration of an antibacterial agent expressed in mg/L (μg/mL) which, under strictly controlled in vitro conditions, completely prevents visible growth of the tested strain [34]. This value depends on the chemical characteristics of the antimicrobial compound and on the microbial species/strain, therefore it is a useful parameter. Here, we report a commonly used protocol to determine the MIC value, namely the 96-well microplate dilution method, which is suitable for both bacteria and fungi. The protocol is the same for fungi and bacteria, except for **step 1**, that concerns the suspension preparation, and **step 5**, which describes how to evaluate the growth of the microorganism.

3.4.1 Protocol for Fungi

Before starting the MIC test, cultivate the fungus on PDA medium at the optimum growth conditions to have a fresh culture.

1. Prepare the fungal (conidial or hyphal) suspension as follows: prepare 10 mL of PDB in a screw cap tube and sterilize. Pour about 5 mL of PDB on the fungal colony and scrape it from the agar surface to collect conidia (or hyphal fragments) using a sterile spreader loop. Transfer the suspension into the tube containing the remaining PDB medium and vortex to homogenize. Place a drop of the suspension into a hemocytometer, in correspondence of the grid. Cover the hemocytometer with a glass coverslip, place on the stage of an optical microscope and adjust the focus till the chamber grid becomes visible. Count the conidia (or hyphal fragments) within the squares along the two diagonals of the grid; do not consider those outside the squares (Fig. 8). Calculate the average value, that is the number of conidia (or hyphal fragments)/squares, and apply the following formula to determine the final concentration:

$$\text{Concentration (conidia/mL)} = \text{average number of conidia (or hyphal fragments)} \textit{per} \text{ square} \times 250{,}000$$

   Adjust the concentration of the suspension to about 1×10^6 conidia/mL.

2. Dissolve the antimicrobial compound to be tested, in this case thyme oil, in methanol to obtain a 10% solution. From this solution, dilute 1:1 in PDB medium until the lowest concentration of 0.039%. Dissolve the reference antimicrobial compound in the appropriate solvent and prepare the same dilutions described for the thyme oil.
3. Fill in the microplate wells with 100 μL of each thyme oil dilution and add 100 μL of the conidial suspension. Repeat the same procedure for the reference antimicrobial compound. For positive control, fill in the microplate wells with 100 μL of PDB medium plus 100 μL of the conidial suspension. For negative control, fill in the microplate wells with 200 μL of PDB medium. Test each dilution at least in triplicate (Fig. 9).
4. Seal the 96-well microplate with a double layer of parafilm and incubate at the optimum growth conditions. Check the plates daily.
5. Determine fungal growth by both visual reading and spectrophotometric readings and compare the data. For visual reading, turbid suspensions in the wells (in comparison with the negative control wells) indicate not-inhibiting concentrations of the tested compound, as the fungus has grown into these wells. Transparent wells correspond to inhibiting concentrations, able to prevent the fungus growth. MIC value can be visually determined as the lowest concentration in correspondence of

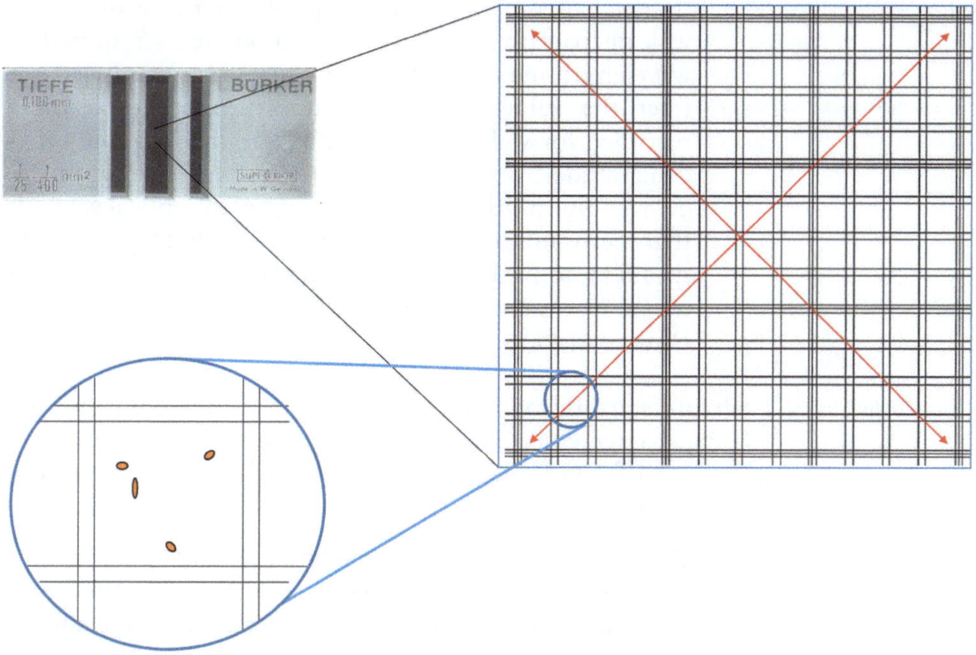

Fig. 8 Hemocytometer and scheme of the counting grid. The counting grid consists of 144 squares; counts of conidia must be performed along the two main diagonals (indicated by the red arrows). An average of four conidia per square (represented as orange small dots) corresponds to a final concentration of 1×10^6 conidia/mL

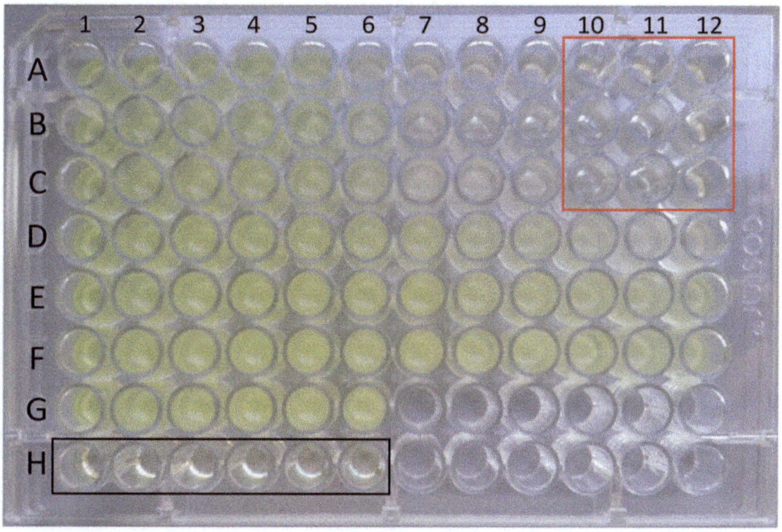

Fig. 9 MIC test performed into a 96-well microplate. The antimicrobial effect of different EOs tested is clearly noticeable in the red box (lines A, B, and C; columns 10, 11, and 12). Here the conidial suspension results transparent as the negative control wells filled up with Potato Dextrose Broth only (line H; columns 1–6, black box)

transparent wells. For spectrophotometric readings, a multi-well microplate reader spectrophotometer is required and the wavelength must be set at 450 nm. After data collection, subtract the value of the medium turbidity (i.e., the values recorded for wells filled up with PDB medium only) from the other values: optical values close to zero correspond to concentrations inhibiting the fungus growth, optical values greater than zero indicate that the fungus is able to grow and correspond to non-inhibiting concentrations. The MIC corresponds to the lowest concentration showing optical reading values close to zero.

3.4.2 Protocol for Bacteria

Before starting the MIC test, refresh the bacterial strain from the long storage culture by streaking a loop of the suspension on the surface of NGA medium and incubate at the optimum growth conditions.

1. Prepare the bacterial suspension as follows: prepare 10 mL of Nutrient Glucose Broth (NGB) in a screw cap tube and sterilize. Take a loop of the fresh bacterial culture using a sterile loop, suspend it in the NGB medium and vortex to homogenize the suspension. Measure the optical density of the bacterial suspension by a spectrophotometer, setting the wavelength at 530 nm. Adjust the concentration to an optical density of 0.1, which corresponds to about 1×10^8 cfu/mL.
2. Dissolve the antimicrobial compound to be tested, in this case thyme oil, in methanol to obtain a 10% solution. From this solution, dilute 1:1 in NGB medium until the lowest concentration of 0.039%. Dissolve the reference antimicrobial compound in the appropriate solvent and prepare the same dilutions described for the thyme oil.
3. Fill in the microplate wells with 100 μL of each thyme oil dilution and add 100 μL of the bacterial suspension. Repeat the same procedure for the reference antimicrobial compound. For positive control, fill in the microplate wells with 100 μL of NGB medium plus 100 μL of the bacterial suspension. For negative control, fill in the microplate wells with 200 μL of NGB medium. Test each dilution at least in triplicate.
4. Seal the 96-well microplate with a double layer of parafilm and incubate at 27 °C for 24 h.
5. Determine bacterial growth by visual reading of the wells before adding resazurin: turbid suspensions (in comparison with the negative control wells) indicate bacterial growth and then not-inhibiting concentrations of the antimicrobial compound. Transparent wells correspond to inhibiting concentrations of the tested compound. MIC value can be visually determined as

the lowest concentration in correspondence of transparent wells.

6. Add 30 μL of a resazurin solution (0.015%) to each well and incubate the microplate for 3 h at 27 °C. Record the color of the suspension observed for each well: blue color indicates inactive/killed bacterial cells, red-pink color indicates metabolically active cells. Therefore, MIC can be visually determined as the lowest concentration in correspondence of blue wells.
7. Compare the data collected by visual reading with those collected after the addition of resazurin dye in order to determine the MIC of the tested antimicrobial agent.

3.5 ß-Cyclodextrin-EO Complex

1. Insert one steel ball in the grinding steel jars and put in a polystyrene box. Add liquid nitrogen, close the box and leave to cool down for 20 min (*see* **Note 7**).
2. Add the ß-cyclodextrin powder and the EO in the ratio 1:1 (*see* **Note 8**) with an appropriate volume of H_2O/EtOH buffer. For example, for 1 g of ß-cyclodextrin powder add 139 μL of thyme EO and 500 μL of H_2O/EtOH buffer (*see* **Note 9**).
3. Place the grinding jars filled with the material to be milled and the ball between the rubber supports of the clamping device and clamp firmly. Set the vibrational intensity at 70 amplitude (about 20 Hz) and start the machine with the hood closed for 20 min. The movement of the grinding jar combined with the movement of the ball results in the intensive mixing of the sample.
4. Recover the ß-cyclodextrin-EO complex and put it on an aluminum-foil and into the stove at 35–40 °C for 2 h. After that, leave the aluminum-foil with the complex at room temperature overnight.
5. The day after, grind the complex with a pestle until a fine powder is obtained. Put the ß-cyclodextrin-EO complex in a glass vial, seal with parafilm and conserve at 4 °C.

4 Notes

1. The EO's activity should be compared with that of an antimicrobial compound chosen among those usually applied to control the target microbial species.
2. Collection of tissue samples with morphological lesions and/or visible symptoms of infection should be avoided. On the contrary, the damaged tissues should be collected to investigate the variation of the terpene mixtures in response to physical wounding, pest attacks, and diseases.

3. Sample collection should be restricted to the same elevation and horizontal position on stem and in crown of the plant.
4. The hydrosol, or hydrolate, is the water-based residue of EO distillation. For a long time, the hydrosol was considered a waste product, but recently, it was successfully experimented as an antimicrobial agent [35, 36].
5. It is recommended to use inert materials, namely glass vials and laboratory glassware; plastic materials should be avoided.
6. For fungi, cultivate the strains on PDA medium for 1–7 days (depending on the species) at the optimum growth conditions, to have fresh cultures for the antimicrobial test.
7. Fill in the polystyrene box with liquid nitrogen till ¼ of the jars' height. This step is necessary to prevent the jar overheating during the mechanical shaking.
8. The amount of EO to be added should be calculated considering the molecular weight of the most representative constituent. For example, in the case of thyme oil, thymol is the most abundant compound.
9. The sample quantity should not be less than 25% of the milling cup volume.

References

1. Dhifi W, Bellili S, Jazi S et al (2016) Essential oils' chemical characterization and investigation of some biological activities: a critical review. Medicines (Basel) 3:25
2. Zhou F, Pichersky E (2020) More is better: the diversity of terpene metabolism in plants. Curr Opin Plant Biol 55:1–10
3. Langenheim JH (1994) Higher plant terpenoids: a phytocentric overview of their ecological roles. J Chem Ecol 20(6):1223–1280
4. Croteau R, Kutchan TM, Lewis NG (2000) Secondary metabolites – Chap 24. In: Biochemistry & molecular biology of plants. American Society of Plant Physiologists, pp 1250–1318
5. Hartmann T (1996) Diversity and variability of plant secondary metabolism: a mechanistic view. Entomol Exp Appl 80:177–188
6. Gershenzon J, Dudareva N (2007) The function of terpene natural products in the natural world. Nat Chem Biol 3:408–414
7. Franz C, Novak J (2010) Sources of essential oils. In: Baser KHC, Buchbauer G (eds) Handbook of essential oils: science, technology, and applications. CRC Press/Taylor & Francis Group, Boca Raton, pp 39–82
8. Phillips MA, Croteau RB (1999) Resin-based defenses in conifers. Trends Plant Sci 4: 184–190
9. Cates RG (1996) The role of mixtures and variation in the production of terpenoids in conifer-insect-pathogen interactions. In: Romeo JT, Saunders JA, Barbosa P (eds) Phytochemical diversity and redundancy in ecological interactions. Recent advances in phytochemistry, vol 30. Springer, Boston. https://doi.org/10.1007/978-1-4899-1754-6_7
10. Mehdizadeh L, Moghaddam M (2018) Essential oils: biological activity and therapeutic potential. In: Therapeutic, probiotic, and unconventional foods. Academic Press, pp 167–179
11. Dagli N, Dagli R, Mahmoud RS et al (2015) Essential oils, their therapeutic properties, and implication in dentistry: a review. J Int Soc Prev Community Dent 5(5):335–340. https://doi.org/10.4103/2231-0762.165933
12. Edris AE (2007) Pharmaceutical and therapeutic potentials of essential oils and their individual volatile constituents: a review. Phytother Res 21(4):308–323

13. Knobloch K, Pauli A, Iberl B et al (1989) Antibacterial and antifungal properties of essential oil components. J Essent Oil Res 1(3): 119–128
14. Ghavam M, Manca ML, Manconi M et al (2020) Chemical composition and antimicrobial activity of essential oils obtained from leaves and flowers of Salvia hydrangea DC. ex Benth. Sci Rep 10(1):1–10
15. Rossiter SE, Fletcher MH, Wuest WM (2017) Natural products as platforms to overcome antibiotic resistance. Chem Rev 117(19): 12415–12474
16. Salem N, Kefi S, Tabben O, Ayed A, Jallouli S, Feres N, Hammami M, Khammassi S, Hrigua I, Nefisi S, Sghaier A, Limam S, Elkahoui S (2018) Variation in chemical composition of Eucalyptus globulus essential oil under phenological stages and evidence synergism with antimicrobial standards. Ind Crop Prod 124:115–125
17. Omonijo FA, Ni L, Gong J et al (2018) Essential oils as alternatives to antibiotics in swine production. Anim Nutr 4:126–136
18. Isman MB (2000) Plant essential oils for pest and disease management. Crop Prot 19: 603–608
19. Kotan R, Dadasoĝlu F, Karagoz K et al (2013) Antibacterial activity of the essential oil and extracts of Satureja hortensis against plant pathogenic bacteria and their potential use as seed disinfectants. Sci Hortic 153:34–41
20. De Mastro G, El Mahdi J, Ruta C (2021) Bioherbicidal potential of the essential oils from Mediterranean lamiaceae for weed control in organic farming. Plants (Basel) 10:818
21. Mena P, Galindo A, Collado-González J et al (2013) Sustained deficit irrigation affects the colour and phytochemical characteristics of pomegranate juice. J Sci Food Agric 93: 1922–1927
22. Menicucci F, Michelozzi M, Raio A et al (2021) Thymol-loaded lipid nanovectors from the marine microalga Nannochloropsis sp. as potential antibacterial agents. Biocatal Agric Biotechnol 32:101962
23. Sarwar M, Salman M (2015) Toxicity of oils formulation as a new useful tool in crop protection for insect pests control. Int J Chem Biomol Sci 1:297–302
24. Sharma A, Dubey S, Iqbal N (2020) Microemulsion formulation of botanical oils as an efficient tool to provide sustainable agricultural Pest management. In: Nano-and microencapsulation-techniques and applications. IntechOpen. https://doi.org/10.5772/intechopen.91788
25. Tanovic B, Gasic S, Hrustic J et al (2013) Development of a thyme essential oil formulation and its effect on *Monilinia fructigena*. Pestic Phytomed (Belgrade) 28(4):273–280. https://doi.org/10.2298/PIF1304273T
26. Wadhwa G, Kumar S, Chhabra L et al (2017) Essential oil–cyclodextrin complexes: an updated review. J Incl Phenom Macrocycl Chem 89:39–58
27. Squillace AE (1976) Analyses of monoterpenes of conifers by gas-liquid chromatography. In: Miksche JP (ed) Modern methods in forest genetics. Proceedings in life sciences. Springer, Berlin, Heidelberg
28. Akbar S (2020) *Taxus baccata* L. (Taxaceae). In: Handbook of 200 medicinal plants. Springer, Cham, pp 1753–1761
29. Hanover JW (1992) Applications of terpene analysis in forest genetics. New For 6(1–4): 159–178
30. Baradat P, Marpeau A, Walter J (1991) Terpene markers. In: Muller-Starck G, Ziehe M (eds) Genetic variation in European populations of forest trees. Sauerlander's Verlag, Frankfurt am Main, pp 40–66
31. Barbero F, Maffei M (2016) Biodiversity and chemotaxonomic significance of specialized metabolites. In: Plant specialized metabolism. CRC Press, pp 35–76
32. Nikolić JS, Zlatković BK, Jovanović S et al (2021) Needle volatiles as chemophenetic markers in differentiation of natural populations of *Abies alba*, *A. x borisii-regis*, and *A. cephalonica*. Phytochemistry 183:112612
33. Michelozzi M, Tognetti R, Maggino F et al (2008) Seasonal variations in monoterpene profiles and ecophysiological traits in Mediterranean pine species of group "halepensis". iForest Biogeosci For 1(1):65
34. EUCAST Definitive Document (1998) Methods for the determination of susceptibility of bacteria to antimicrobial agents. Terminology. Clin Microbiol Infect 4:291–296
35. Acheampong A, Borquaye LS, Acquaah SO et al (2015) Antimicrobial activities of some leaves and fruit peels hydrosols. Int J Chem Biomol Sci 1:158–162
36. Di Vito M, Bellardi MG, Mondello F et al (2019) Monarda citriodora hydrolate vs essential oil comparison in several anti-microbial applications. Ind Crop Prod 128:206–212

Index

A

B

C

D

E

F

Nicola Luchi (ed.), *Plant Pathology: Methods and Protocols*, Methods in Molecular Biology, vol. 2536,
https://doi.org/10.1007/978-1-0716-2517-0,

Q

R

S

T

V

W

X

MIX
Papier aus verantwortungsvollen Quellen
Paper from responsible sources
FSC® C105338

If you have any concerns about our products,
you can contact us on
ProductSafety@springernature.com

In case Publisher is established outside the EU,
the EU authorized representative is:
Springer Nature Customer Service Center GmbH
Europaplatz 3, 69115 Heidelberg, Germany

Printed by Libri Plureos GmbH
in Hamburg, Germany